Unobserved Components and Time Series Econometrics

Unobserved Components and Time Series Econometrics

Edited by

Siem Jan Koopman

and Neil Shephard

OXFORD
UNIVERSITY PRESS

OXFORD
UNIVERSITY PRESS

Great Clarendon Street, Oxford, OX2 6DP,
United Kingdom

Oxford University Press is a department of the University of Oxford.
It furthers the University's objective of excellence in research, scholarship,
and education by publishing worldwide. Oxford is a registered trade mark of
Oxford University Press in the UK and in certain other countries

First Edition published in 2015

Published in the United States of America by Oxford University Press
198 Madison Avenue, New York, NY 10016, United States of America

British Library Cataloguing in Publication Data
Data available

Library of Congress Control Number: 2015941107

ISBN 978–0–19–968366–6

CONTENTS

LIST OF FIGURES

LIST OF TABLES

1 Introduction

Siem Jan Koopman and Neil Shephard

1.1 An overview of the volume

This volume was written as a tribute to Professor Andrew Harvey. It brings together papers written for a conference held in Andrew's honor at Oxford University in June 2012. At least one of the authors on each paper is a long-term friend, coauthor, or former student of Andrew's.

Andrew Harvey has been an active researcher for four decades, writing on many aspects of time series modeling with a particular focus on economic and, more recently, financial applications. As well as generating path-breaking research articles, he has been unusually influential in writing research monographs and textbooks, and building and documenting software. A distinctive voice, he is perhaps the most influential scholar in the area of time series modeling using unobserved components in economics.

The structure of this book is as follows.

At our invitation, Andrew has kindly written a unique personal account of his research. This is our Chapter 2. It gives a glimpse at how he thinks about research problems.

In Chapter 3, Stella and Stock develop a parsimonious bivariate model for time series of inflation and unemployment. The dynamic model captures persistent variation in trend inflation and in the non-accelerating inflation rate of unemployment. The model includes five unobserved components which are associated with trend and stochastic volatility. The reduced form of the model can be viewed as a time-varying vector autoregression model for changes in the rates of inflation and unemployment. The parameters and latent variables of the model can be divided into three blocks and can be conveniently estimated using Bayesian methods and, in particular, using a Gibbs sampling procedure. The empirical evidence is provided for a quarterly data set from 1960Q1 to 2011Q3. It is found that the implied backwards-looking Phillips curve has a time-varying slope that is steeper in the 1970s than in the 1990s. Pseudo out-of-sample forecasting experiments indicate improvements upon univariate benchmarks. Since 2008, the implied Phillips curve has become steeper and the non-accelerating inflation rate of unemployment has increased. The pseudo out-of-sample forecasts are accurate; this parsimonious model improves upon the forecasts of delivered by a set of well-regarded univariate models.

Next, De Rossi investigates the omission of treating the errors-in-variables problem in the measurement of the "true" net asset value of an exchange traded fund. This omission causes a strong upward bias in the estimation of a fund's tracking error which is defined as the standard deviation of its active returns. The error-in-variable problem is typically due to management fees, tax liabilities, asset illiquidity, asynchronous prices, and exchange rate fluctuations. The issue has crucial practical implications for investors who need to select an exchange traded fund based on its tracking error. An unobserved time series model is formulated that can account for this omission by introducing an unobserved components model with stochastic volatility. A key implication of the presented solution is that the exchange traded funds can potentially track the relevant index well over long investment horizons.

Diebold and Yilmaz develop a framework for the measurement of real output connectedness for a set of six developed countries, between 1962 and 2010. They base their approach using concepts and technology from modern network theory. The details of their approach are presented clearly. In their empirical study it is shown that global connectedness is sizable and time-varying over the business cycle. The nature of the time variation is also studied, relative to the ongoing discussion about the changing nature of the global business cycle. For example, they find that connectedness corresponding to transmissions to others from the United States and Japan is disproportionately important.

Ansley and de Jong in Chapter 6 argue that the use of the well-known Butterworth filter for smoothing the climatic time series has undesirable properties. This method produces smoothed estimates of the signal component with high standard errors and possibly with considerable bias. Then they continue to advocate the use of standard time series models to avoid such undesirable problems. They propose a state space model that can be treated by standard maximum likelihood methods. They show that it fits well with a long history of temperature data. Their proposed modeling framework is able to treat questions such as whether there has been global warming in the last ten to twenty years. Even after accounting for parameter uncertainties, their answer is in the affirmative. However, based on their methods, they also provide empirical evidence that warming has not been accelerating in the last ten to twenty years.

Mesters and Koopman study the forecasting of the yearly outcome of the Boat Race between Cambridge and Oxford. The relative performance of different dynamic models are studied in the context of forty years of forecasting the Boat Race. Each model is defined by a binary density conditional on a latent signal that is specified as a dynamic stochastic process with fixed predictors. The out-of-sample predictive ability of the models is compared between each by using a variety of loss functions and predictive ability tests. The model with its latent signal specified as an autoregressive process cannot be outperformed by the other specifications. This model has been successful in correctly forecasting thirty-one out of forty outcomes of the Boat Race.

In Chapter 8, Fiorentini and Sentana consider static factor models with semi-parametrically specified elliptical distributions. In an in-depth exposition they derive algebraic expressions for the score test of serial correlation in both the levels and squares of the common factors and also of the idiosyncratic factors. The model is semi-parametric but the analysis is likelihood-based, while the computations partly rely on simulations. Gaussian tests against non-normality are also considered. A Monte Carlo study investigates the level of reliability and the power of tests in finite samples. An empirical illustration is presented for monthly U.S. stock returns. They find clear evidence of first order serial correlation in the common factors, including its volatility, while this evidence is weaker for the idiosyncratic factors.

Lemke and Godsill continue with the development of a general framework for inference in the presence of α-stable processes which are represented as conditionally Gaussian distributions. The specification relies on (exact) series representations of the stable laws and the corresponding stochastic integrations in terms of infinite summations of random Poisson process arrival times. The analysis can therefore be governed effectively by Bayesian inference based on auxiliary variables, particle filtering, and Markov chain Monte Carlo methods. The authors claim that their treatment is remarkably simple, because of their introduction of a novel approximation of the series residual terms based on exact moment calculations. The framework is applicable for variables in continuous-time and in discrete-time. They present the details for fitting discrete-time autoregressive processes with α-stable noise and for sequential state estimation in continuous-time autoregressive processes driven by Lévy processes.

Chapter 10 introduces a new class of martingale unobserved component models. Shephard generalizes the familiar linear Gaussian unobserved component models or structural time series models to a wider class where, for example, the disturbances can be subject to stochastic volatility. For the basic illustrations of this model, it leads to forecast functions that have a time-varying (or local) rate of discounting the past observations. It is shown in an elegant way that this model-based treatment can be based on an auxiliary particle filter for which many, say 1000, competing Kalman filters are applied in parallel. The new methodology is illustrated for a time series of U.S. inflation with the purpose to extract the underlying rate of inflation and to forecast.

Poncela and Ruiz establish the relationship between the mean square error of the factor estimates produced by the Kalman filter and the number of variables included in the dynamic factor model. For the case of knowing the model parameters, it is shown that filter uncertainty only decreases marginally when the cross-sectional dimension is beyond a relatively small number. Furthermore, the authors present a consistency proof for the filtered estimates of, possibly, non-stationary factors in a dynamic factor model with serially uncorrelated idiosyncratic disturbances with weak cross-sectional dependence. Finally, it is argued and evidence is provided that the contribution of the parameter estimation uncertainty can represent a considerable percentage of the uncertainty in the estimation of the factors.

Busetti, in Chapter 12, treats the end of sample problem when testing for parameter instabilities. Various modifications of existing tests are considered. For example, Wald-type tests of a one-time shift and locally most powerful (LMP) tests against the hypothesis of a random walk parameter are examined. Then he introduces new statistics that have particularly high power when changes occur at the end of the sample. These tests are functionals of the Wald and LMP statistics such that either the set of possible change points is restricted to the last part of the sample or the occurrence of change points is given increasing weight throughout the sample. Asymptotic critical values of the tests are provided and their properties are evaluated in finite samples. Busetti continues with considering in detail cases of a known and an unknown change point at the end of the sample. He presents detailed results for how the different tests behave, mostly in terms of power, under these different situations. The empirical illustration considers two macroeconomic variables for which the tests are used to detect structural changes at the time of the "Great Recession."

Helske and Nyblom investigate the so-called plug-in prediction intervals for autoregressive processes with Gaussian disturbances. It is well known that the coverage probabilities for these prediction intervals fall below the nominal ones. By means of simulation experiments, they show that ordinary linear regression theory can help to yield one-step prediction intervals that have much more precise coverage probabilities. These intervals can be viewed as posterior predictive intervals in a Bayesian paradigm with uniform priors for both autoregressive coefficients and for the log-variance of the disturbances. The authors generalize their initial results further for the construction of multi-step prediction intervals which are obtained by simulation from the posterior density or from an importance density. The new methods are illustrated for the task of forecasting gross domestic product growth in the United Kingdom and Spain.

Ma and Nelson explore, in Chapter 14, models that can be represented as $y = \gamma \cdot g(\beta, x) + \varepsilon$ and focus on the standard Wald test for $\hat{\beta}$. In earlier work it is found that the Wald test has systematically wrong sizes in finite samples when the identifying parameter γ is small relative to its estimation error. Here, the authors take the step of considering the Lagrange multiplier test based on linearization of $g(.)$ for this model class. This LM test may be interpreted as an approximation to an exact test for a ratio of regression coefficients. It is impressive to find that their proposed LM test has nearly the correct size in many different models including nonlinear regression, ARMA, GARCH, and unobserved components models, while in all these cases the Wald test performs poorly.

The final chapter concerns generalized linear spectral models. Proietti and Luati study parametric spectrum estimators for a generalized linear model for exponential random variables with, power link. They point out that Whittle estimation of the coefficients is unfeasible but that estimation based on an ARMA representation for the power transformation of the spectrum can be easily carried out by maximizing the Whittle likelihood. In a standard fashion, information criteria can be used for the selection of a spectral model. In their first illustration they deal with the Box-Jenkins *Series A* and estimate the inverse autocovariance function. In their second illustration they study the quarterly growth rate of U.S. gross domestic product for the extraction of the business cycle.

Finally, we would like to thank various colleagues for refereeing the papers in this volume. The conference on which this volume is based was supported by the Bank of England, the Oxford-Man Institute, and the Journal of Applied Econometrics. We are very grateful for this.

1.2 **Andrew Harvey's main contributions**

In this section, we list Andrew Harvey's main written contributions to the fields of econometrics and statistics. They include all of the papers which have most influenced us, together with a list provided by Andrew. We have split the section into pieces to allow the main messages from the work to be more easily accessed. The list includes our favorite time series book of all time, *Time Series Models*, which was the introductory time series book we studied as students of Andrew's at the London School of Economics in the 1980s.

1.2.1 BOOKS

- *The Econometric Analysis of Time Series*, 1981, Deddington: Philip Allan. Second edition, 1990. German translation, Oldenbourg Verlag, 1994. Chinese translation, Wu-Nan, 1998.
- *Time Series Models*, 1981, Deddington: Philip Allan. Japanese translation, Tokyo University Press, 1985. Second edition, 1993. German translation, Oldenbourg Verlag, 1995.
- *Forecasting, Structural Time Series Models and the Kalman Filter*, 1989, Cambridge: Cambridge University Press.
- *Dynamic Models for Volatility and Heavy Tails*, 2013, Econometric Society Monograph. Cambridge: Cambridge University Press.

1.2.2 EDITED VOLUMES

- *Time Series*, 1994, Cheltenham: Edward Elgar.

- *State Space and Unobserved Component Time Series Models*, (with S. J. Koopman and N. Shephard), 2004, Cambridge: Cambridge University Press.
- *Readings in Unobserved Components Models*, (with T. Proietti), 2005, Oxford: Oxford University Press.

1.2.3 SOFTWARE AND SURVEYS

- Stochastic volatility, E. Ghysels and E. Renault, Statistical Methods in Finance, in *Handbook of Statistics, vol. 14*, G. S. Maddala, Rao, C. R, and H. D. Vinod (Eds.), pp. 119–91, Amsterdam: North Holland.
- *STAMP Structural Time Series Analyser, Modeller and Predictor*, (S. J. Koopman, J. A. Doornik, and N. Shephard), 2000:2010, London: Timberlake Consultants.
- Forecasting with unobserved components time series models, in *Handbook of Economic Forecasting*, G. Elliot, C. W. J. Granger, and A. Timmermann (Eds.) (2006), pp. 327–412, Amsterdam: North Holland.

1.2.4 HIGHLIGHTS ON LINEAR UNOBSERVED COMPONENT MODELS

- Forecasting economic time series with structural and Box Jenkins models, P. H. J. Todd (1983), *Journal of Business and Economic Statistics*, 1, 299–315. Reprinted in *Modelling Seasonality*, S. Hylleberg (Ed.), 1992, pp. 341–58, Oxford: Oxford University Press.
- A unified view of statistical forecasting procedures, *Journal of Forecasting*, 1984, 3, 245–75.
- Trends and cycles in macroeconomic time series, *Journal of Business and Economic Statistics*, 1985, 3, 216–27.
- Time series models for count or qualitative observations, C. Fernandes, *Journal of Business and Economic Statistics*, 1989, 7, 409–22.
- Seemingly unrelated time series equations and a test for homogeneity, F. J. Fernandez, *Journal of Business and Economic Statistics*, 1990, 8, 71–82.
- On the probability of estimating a deterministic component in the local level model, N. Shephard, *Journal of Time Series Analysis*, 1990, 11, 339–47.
- Diagnostic checking of unobserved components time series models, S. J. Koopman, *Journal of Business and Economic Statistics*, 1992, 10, 377–89.
- Detrending, stylized facts and the business cycle, A. Jaeger, *Journal of Applied Econometrics*, 1993, 8, 231–47.

- Seasonality in dynamic regression models, A. Scott, *Economic Journal*, 1994, 104, 1324–45.
- Intervention analysis with control groups, *International Statistical Review*, 1996, 64, 313–28.
- Trends, cycles and autoregressions, *Economic Journal*, 1997, 107, 192–201.
- The modelling and seasonal adjustment of weekly observations, S. J. Koopman and M. Riani, *Journal of Business and Economic Statistics*, 1997, 15, 354–68.
- Tests of common stochastic trends, J. Nyblom, *Econometric Theory*, 2000, 16, 176–99.
- Signal extraction and the formulation of unobserved components models, S. J. Koopman, *Econometrics Journal*, 2000, 3, 84–107.
- Testing for the presence of a random walk in series with structural breaks, F. Busetti, *Journal of Time Series Analysis*, 2001, 22, 127–50.
- Computing observation weights for signal extraction and filtering, S. J. Koopman, *Journal of Economic Dynamics and Control*, 2003, 27, 1317–33.
- General model-based filters for extracting trends and cycles in economic time series, T. Trimbur, *Review of Economics and Statistics*, 2003, 85, 244–55.
- Seasonality tests, F. Busetti, *Journal of Business and Economic Statistics*, 2003, 21, 420–36.
- Cyclical components in economic time series: a Bayesian Approach, T. Trimbur and H. K. van Dijk, *Journal of Econometrics*, 2007, 140, 618–49.
- Testing for Trend (with F. Busetti), *Econometric Theory*, 2008, 24, 72–87.
- Computing the mean square error of unobserved components extracted by misspecified time series models (with D. Delle Monache), *Journal of Economic Dynamics and Control*, 2009, 33, 283–95.
- Specification and misspecification of unobserved components models, D. Delle Monache, in *Economic Time Series: Modeling and Seasonality*, W. R. Bell, S. H. Holan, and T. S. McElroy (Eds.), 2012, pp. 83–108, London: Chapman and Hall/CRC.

1.2.5 HIGHLIGHTS ON TIME SERIES METHODS

- A simple test for serial correlation in regression analysis, G. D. A. Phillips, *Journal of the American Statistical Association*, 1974, 69, 935–9.
- A comparison of the power of some tests for heteroscedasticity in the general linear model, G. D. A. Phillips, *Journal of Econometrics*, 1974, 307–16.

- Maximum likelihood estimation of regression models with autoregressive-moving average disturbances, G. D. A. Phillips, *Biometrika*, 1979, 66, 49–68.
- Testing for serial correlation in simultaneous equation systems, G. D. A. Phillips, *Econometrica*, 1980, 747–59.
- The Kalman filter and its applications in econometrics and time series analysis, *Methods of Operations Research*, 1981, 44, 3–18.
- Estimating missing observations in economic time series, R. Pierse, *Journal of the American Statistical Association*, 1984, 79, 125–31.
- A unified view of statistical forecasting procedures, A. Harvey, *Journal of Forecasting*, 1984, 3, 245–83.
- Continuous time autoregressive models with common stochastic trends, J. Stock, *Journal of Economic Dynamics and Control*, 1988, 12, 365–84.
- Quantiles, expectiles and splines, G. Rossi, *Journal of Econometrics*, 2009, 152, 179–85.
- Tracking a changing copula, *Journal of Empirical Finance*, 2010, 17, 485–500.
- When is a copula constant? A test for changing relationships, F. Busetti, *Journal of Financial Econometrics*, 2011, 9, 106–131.
- Kernel density estimation for time series models, Oryshchenko, *International Journal of Forecasting*, 2012, 28, 3–14.

1.2.6 HIGHLIGHTS ON STOCHASTIC VOLATILITY

- Estimating regression models with multiplicative heteroscedasticity, A. Harvey, *Econometrica*, 1976, 461–5.
- Multivariate stochastic variance models, E. Ruiz and N. Shephard, *Review of Economic Studies*, 1994, 61, 247–64. Reprinted in (i) *ARCH Models*, R. F. Engle (Ed.), 1995, Oxford: Oxford University Press; (ii) *Recent Developments in Time Series*, S. Leybourne and T. Mills (Eds.), 2003, Cheltenham: Edward Elgar.
- Estimation of an asymmetric stochastic volatility model for asset returns, N. Shephard, *Journal of Business and Economic Statistics*, 1996, 14, 429–34.
- Testing for a slowly changing level with special reference to stochastic volatility, M. Streibel, *Journal of Econometrics*, 1998, 87, 167–89.
- Long memory in stochastic volatility, in *Forecasting volatility in Financial Markets*, J. Knight and S. Satchell (Eds.), 1998, 307–20, Oxford: Butterworth-Heineman.
- Unobserved component time series models with ARCH disturbances, E. Ruiz and E. Sentana, *Journal of Econometrics*, 1992, 52, 129–57.

1.2.7 HIGHLIGHTS ON APPLICATIONS

- Forecasting hourly electricity demand using time varying splines, S. J. Koopman, *Journal of the American Statistical Association*, 1993, 88, 1228–36.
- The effects of seat belt legislation on British road casualties, J. Durbin, *Journal of the Royal Statistical Society, Series A*, 1986, 149, 187–227.
- Estimating the underlying change in unemployment in the UK, with discussion, C. H. Chung, *Journal of the Royal Statistical Society, Series A*, 2000, 163, 303–39.
- Convergence in the trends and cycles of euro zone income, V. M. Carvalho, *Journal of Applied Econometrics*, 2005, 20, 275–89.

2 The development of a time series methodology: from recursive residuals to dynamic conditional score models

Andrew Harvey

I arrived at the University of Kent in late 1971 as a new lecturer. I had not done a PhD because I had gone straight from a Masters at the London School of Economics (LSE) to work for two years in the Kenyan Central Bureau of Statistics. I was supposed to be studying statistics at the LSE but in reality I was spending most of my time trying to topple capitalism.[1] I failed in this endeavor, but at least passed the Masters—just. While in Kenya I spent two years planning, implementing, and analyzing household budget surveys. Nairobi was a great experience but it meant that when I arrived in Canterbury I had little idea of what academic life entailed or how I should set about doing research. Fortunately, Garry Phillips was in the department as a senior lecturer and after my first year there he told me that econometrics was an exciting new field and would I like to join him on a research project. Garry had noticed that the one-step ahead predictions in a static time series regression were serially independent. He was very excited about this result and somewhat disappointed to find a little later that Gauss had discovered it in the early nineteenth century. However, Gauss had not used what we called recursive residuals as the basis of a test for serial correlation and he certainly had not computed its power over a range of regression models. This is what Garry and I did and the paper was published in the *Journal of the American Statistical Association* (JASA) in 1973. The late Jim Durbin was also working in recursive residuals and he gave a paper on this topic at a meeting of the Royal Statistical Society at around this time. Peter Young, who had an engineering background, pointed out that control engineers knew all about this kind of thing because the least squares

[1] The year 1968–9 was one in which the LSE was occupied by the students in the Michaelmas term. The authorities responded by installing iron gates at various key points in the Houghton Street complex so that parts of the building could be isolated in the event of another occupation. This move backfired because the students tore down the gates one Friday evening in Lent term. As a result, the school was closed for three weeks.

recursions were just a special case of the Kalman filter. The state space and Kalman filtering literature in engineering employed very different notation and conventions from those in econometrics and statistics and so the ideas there were not readily accessible. However, it was clear that Peter was right and reading the Kalman filtering literature introduced me to a fascinating, albeit strange, new world.

Although the least squares recursions were obviously a special case of the Kalman filter in which the state vector was fixed, there was one issue that puzzled me. The least squares recursions started off after an initial estimate had been constructed from the first k sets of observations, where k is the number of explanatory variables in the regression. On the other hand, most of the engineering literature took the initial conditions as given. Such an assumption was not unreasonable. After all if NASA wanted to track a rocket going to the moon, they knew the starting point. In other cases observations might be coming in online and assumptions about the initial conditions were relatively unimportant once a large numbers of observations had been processed. Unfortunately, long time series were rarely available in economics. Furthermore, economic variables are often nonstationary and in such cases a Kalman filter cannot be initialized with a known unconditional distribution. After a while I realized that the solution was to initialize the Kalman filter with a large covariance matrix, the so-called "diffuse prior." This device seemed to work in practice, but care had to be taken to ensure that entering large numbers at the beginning did not lead to excessive rounding errors and numerical instability. In the following years exact algorithms were developed by econometricians and statisticians, the principal agents being Robert Kohn, Craig Ansley, Piet de Jong, and Siem Jan Koopman.

Garry Phillips and I used the diffuse prior in an article, published in the 1979 issue of *Biometrika*, on the exact maximum likelihood estimator of the parameters in a static regression model with autoregressive-moving average (ARMA) disturbances. I think it was Johannes Ledolter who had pointed out to me that ARMA models could be put in state space form. Garry and I realized that the state space form enabled the exact likelihood function to be computed for a stationary ARMA process. The 1979 article combined this insight with the work we had been doing on recursive residuals.

After a while it dawned on me that the flexibility of the state space form opened up the possibility of developing time series models that were different from the Box-Jenkins autoregressive integrated moving average (ARIMA) models that then dominated the subject. What many people did not realize was that the Box-Jenkins models were, in fact, rooted in earlier models that provided a statistical rationale for exponential smoothing methods. For example, a first-order moving average in first difference is equivalent to a random walk plus noise. Such a model is perhaps the simplest example of what I later called structural time series models. However, at the time, the Kalman filter had barely been developed and, in any case, its implementation presented serious computational problems. By constructing "reduced form" models that depended on a single disturbance term, Box and Jenkins were able to develop a computationally feasible time series modeling methodology. The downside of this methodology was that it ended up putting too much emphasis on selection of models from within the whole ARIMA class and many of these models were difficult to interpret. By this time I had moved to the LSE statistics department, where I had many fruitful conversations on this topic with Jim Durbin. Jim had never been convinced by the Box-Jenkins approach and since he was aware of how the subject had developed following the papers of the early 1960s, he was sure that the structural approach was the way to go. In 1986, Jim and I used structural time series models to analyze the effects of the seat belt law. The results were set out in a report for the Department for Transport and the methodology was presented to a meeting of the Royal Statistical Society; see Harvey and Durbin (1986). Further discussion on the technical and methodological issues involved in the comparison between ARIMA and structural time series models can be found in my 1989 book and the books by Durbin and Koopman (2001, 2012). The way in which structural time series models provide a unified framework for understanding and developing forecasting models was first set out in Harvey (1983), a paper inspired by what I heard at the Forecasting conference in Istanbul in 1982. That paper showed that methods such as exponential smoothing and Holt-Winters that were popular with practitioners were simply filters arising from applying state space methods to structural time series models. Seeing forecasting methods in this way enables them to be put on a firm statistical basis and opens the way to generalization in many directions. Unfortunately this message has still not got through in many quarters and some textbook writers persist in compartmentalizing procedures which are just special cases of a well-formulated statistical model.

During the 1980s and early 1990s I worked on structural time series models with a number of PhD students. Many of the findings were incorporated in the structural time series analyzer, modeller, and predictor (STAMP) package, which was begun by Simon Peters and continued by Siem Jan Koopman. In the 1990s the teaching staff in the LSE Statistics department included Neil Shephard, Siem Jan Koopman, Piet de Jong, and myself. This was a great period, but by the end of the 1990s we had all left: Neil to Oxford, Siem Jan to the Netherlands, Piet to Australia, and me to Cambridge.

Although Jim Durbin had officially retired in the late 1980s, he continued to work with Siem Jan and they brought out their book on state space models in 2001. The first part of the book gave a thorough treatment of linear Gaussian models, while the second part dealt with new material on the estimation of nonlinear models by simulation based techniques. Siem Jan and I continued to work together off and on. We produced a number of joint papers and updated the STAMP package, which was now part of OxMetrics, at regular intervals.

However, we both had our own research agendas, albeit within the overall framework of unobserved components models. Over the last few years, we have been pursuing similar ideas, but coming at them from somewhat different directions. These ideas concern the development of a class of observation driven models that may be regarded as approximations to nonlinear unobserved components models, but which when analyzed as models in their own right turn out to have very attractive properties. These dynamic conditional score models will, I believe, have a significant impact on the way in which nonlinear time series models are formulated. They deal with the problems posed by heavy-tailed distributions when modeling location and they provide a comprehensive theory for changing scale, and hence the volatility models used in financial econometrics. Finally, they offer exciting possibilities for capturing time-varying relationships in both macroeconomics and finance.

3 A state-dependent model for inflation forecasting

Andrea Stella and James H. Stock

3.1 Introduction

Recent research on inflation forecasting in the postwar United States suggests empirical support for four conclusions. First, it is difficult to improve over univariate time series models, at least on average. This point was made most dramatically by Atkeson and Ohanian (2001), who showed that a random walk model for the annual rate of inflation beat multivariate models using measures of economic activity as predictors (Phillips curve models). Second, when multivariate models do improve upon univariate models, they tend to do so episodically. For example, Stock and Watson (2009, 2010) and Dotsey et al. (2011) find that Phillips curve models improve upon univariate models during periods of slack, but typically not otherwise. Ball and Mazumder (2011) suggest that the time variation in the Phillips Curve slope occurs at a decadal, not business cycle, frequency, although they only consider contemporaneous relations, not forecasting specifications. Third, it is not clear whether the apparent episodic predictability is better modeled as a stationary nonlinear model, as suggested by Bos and Shephard (2006), Fuhrer and Olivei (2010), and Stock and Watson (2010), or as a result of time variation (or, relatedly, state dependence with time-varying unobserved states) as in Cogley and Sargent (2002, 2005), Levin and Piger (2004), Stock and Watson (2007), Cogley et al. (2010), Ball and Mazumder (2011), and D'Agostino et al. (2013), or both as in Granger and Jeon (2011). Fourth, predictors other than activity measures appear to be generally unreliable and unpromising (for a survey, see Stock and Watson (2009)). One reason for this murky state of affairs is that there is limited variation in the data with which to identify the nature of the time variation and/or nonlinearity, and that models with many parameters tend to be overwhelmed by estimation uncertainty and thus produce poor forecasts.

This paper takes up the challenge of developing a tightly parameterized model that is capable of capturing the time-variation in the inflation-activity predictive relation. The specific model proposed here builds on two streams of literature. The first stream is papers that model the trend component of inflation, which can be thought of as target inflation, as a latent random walk; recent examples of papers that take this approach include Stock and Watson (2007), Cogley and Sbordone (2008), Cogley et al. (2010), Sbordone et al. (2010), and D'Agostino et al. (2013). Specifically, we follow Stock and Watson (2007) by adopting a stochastic volatility model that allows the variance of trend inflation to change over time, in effect allowing the degree of anchoring of inflation expectations to vary over time. The second literature stream is work on estimating the natural rate of unemployment, in which the natural rate (typically interpreted as the non-accelerating inflation rate of unemployment, the NAIRU) is modeled as a latent random walk trend in the rate of unemployment. This latent time-varying NAIRU is modeled as a random walk in Staiger et al. (1997a) and Gordon (1998), and we adopt that specification here with the extension that the innovation variance to the NAIRU also can evolve over time as a stochastic volatility process. The only paper we are aware of that merges both a time-varying trend in inflation and a time-varying NAIRU is Harvey (2011). Relative to Harvey (2011), we extend the model to include time-varying volatility, so that the projection coefficients (the Phillips curve in terms of the observables) vary over time in a parsimonious specification.

In the model of this paper, the deviations of the inflation and unemployment rates from their trends—that is, the inflation and unemployment gaps—are linked through a common cyclical component, and the inflation and unemployment rates are both potentially measured with error. The variances of the disturbances to the latent components are modeled as stochastic volatility processes. Given the innovation variances of the latent components, the model implies a vector autoregressive (VAR) representation of the changes of the rates of inflation and unemployment, the inflation equation of which is a forecasting Phillips curve. Because the innovation variances evolve over time, the implied VAR parameters also evolve, which implies a backwards-looking Phillips curve forecasting relation with time-varying coefficients. Our base model has three parameters to be estimated, plus three tuning parameters which are weakly identified and are therefore fixed.

Other than Harvey (2011), the models most closely related to ours are those in Cogley et al. (2010), D'Agostino et al. (2013), and in Granger and Jeon (2011). Cogley et al. (2010) specify a trivariate VAR of the rates of inflation, unemployment, and interest, with twenty one time-varying VAR coefficients and twenty four stochastic volatility processes; D'Agostino et al. (2013) use the same model, but with only three stochastic volatility processes. With an eye toward forecasting, we focus on a much tighter parameterization, with two time-varying parameters (the trends), and, in our base model, three stochastic volatility processes. Granger and Jeon (2011) consider time-varying parameter models without stochastic volatility for the inflation equation (the Phillips curve,) but (like Cogley et al. (2010)) their specifications fix the lag length at two whereas the VAR representation implied by our model is infinite-order with time-varying coefficients at all lags, thereby allowing for time-varying dependence.

Because our model nests and extends existing work, some of our results connect with the existing literature. Consistent with Stock and Watson (2007, 2010), we find considerable time variation in the innovation to the permanent component of inflation, consistent with drifting expectations in the 1970s and anchored expectations in the 2000s. Consistent with Staiger et al. (1997a) and Gordon (1998), we find low frequency variation in the NAIRU, with the NAIRU highest in the 1970s and early 1980s, and lowest in the 1990s. The novel aspects of our results are the implied time-varying projections and simulated forecasting performance. The base model estimates considerable time variation in the slope of the predictive Phillips curve, which is measured as the sum of the coefficients on unemployment in the linear projection of inflation on current and past changes of unemployment, after subtracting estimated target inflation and the NAIRU. The empirical time variation is at a lower frequency than the cyclical variation suggested in Stock and Watson (2009, 2010) and Dotsey et al. (2011), and the pattern is also different from the decadal swings of Ball and Mazumder (2011): according to the estimates in Section 3.4, the Phillips curve slope is moderate in the 1960s, steeper in the 1970s and early 1980s, and flat for the great moderation period. In the most recent data, the predictive Phillips curve is estimated to be as steep in 2010 as it was in the 1970s, and the NAIRU is rising.

We also report the results of a pseudo out-of-sample forecasting comparison of the performance of this model, relative to the Atkeson and Ohanian (2001) random walk forecast and the Stock and Watson (2007) univariate unobserved components-stochastic volatility model. The base model generally outperforms these two univariate benchmarks, particularly at short forecasting horizons. This finding is robust to various modifications of the base model. The model is estimated using a Bayesian Gibbs sampling procedure; since the posterior distribution of the model is rather complex and hard to sample from, we draw the parameters and latent variables sequentially from tractable conditional posterior distributions.

The rest of this paper is organized as follows. The model and identification are discussed in Section 3.2. Section 3.3 discusses estimation by Bayesian Markov Chain Monte Carlo methods, and Section 3.4 presents the results. Section 3.5 concludes.

3.2 The model

We model unemployment and inflation symmetrically as the sum of a random walk trend, a common cyclical component, and serially uncorrelated measurement error, so that the two observed series are represented in terms of five unobserved components. In the most general model the variances of the innovations of the unobserved components all follow stochastic volatility processes, with latent variance processes that evolve over time. This multivariate unobserved components stochastic volatility (UC-SV) model is:

$$\text{Inflation} \begin{cases} \pi_t = \tau_t + \lambda\delta_t + \eta_t, & \eta_t = \sigma_{\eta,t}\xi_{\eta,t} \\ \tau_t = \tau_{t-1} + \epsilon_t, & \epsilon_t = \sigma_{\epsilon,t}\xi_{\epsilon,t} \\ ln(\sigma^2_{\eta,t}) = ln(\sigma^2_{\eta,t-1}) + \nu_{\eta,t} & \\ ln(\sigma^2_{\epsilon,t}) = ln(\sigma^2_{\epsilon,t-1}) + \nu_{\epsilon,t} & \end{cases} \tag{3.1}$$

$$\text{Unemployment} \begin{cases} u_t = \tilde{\tau}_t + \delta_t + \tilde{\eta}_t, & \tilde{\eta}_t = \sigma_{\tilde{\eta},t}\xi_{\tilde{\eta},t} \\ \tilde{\tau}_t = \tilde{\tau}_{t-1} + \tilde{\epsilon}_t, & \tilde{\epsilon}_t = \sigma_{\tilde{\epsilon},t}\xi_{\tilde{\epsilon},t} \\ ln(\sigma^2_{\tilde{\eta},t}) = ln(\sigma^2_{\tilde{\eta},t-1}) + \nu_{\tilde{\eta},t} & \\ ln(\sigma^2_{\tilde{\epsilon},t}) = ln(\sigma^2_{\tilde{\epsilon},t-1}) + \nu_{\tilde{\epsilon},t} & \end{cases} \tag{3.2}$$

$$\text{Common cyclical component} \begin{cases} \delta_t = \alpha_1\delta_{t-1} + \alpha_2\delta_{t-2} + \zeta_t, & \zeta_t = \sigma_{\zeta,t}\xi_{\zeta,t} \\ ln(\sigma^2_{\zeta,t}) = ln(\sigma^2_{\zeta,t-1}) + \nu_{\zeta,t} & \end{cases} \tag{3.3}$$

where $\xi_t = \{\xi_{\eta,t}, \xi_{\epsilon,t}, \xi_{\tilde{\eta},t}, \xi_{\tilde{\epsilon},t}, \xi_{\zeta,t}\}$ is i.i.d. N(0,I) and $\nu_t = \{\nu_{\eta,t}, \nu_{\epsilon,t}, \nu_{\tilde{\eta},t}, \nu_{\tilde{\epsilon},t}, \nu_{\zeta,t}\}$ is i.i.d. $N(0, \gamma I)$ and ξ_t and ν_t are independent.

With $\lambda = 0$, the inflation block of the model is the same as in Stock and Watson (2007); with $\lambda \neq 0$, the inflation block can be interpreted as a Phillips curve. The unemployment block parallels the inflation block.

The multivariate UC-SV model can be interpreted in various ways. The trend in inflation τ_t can be thought of as representing inflation expectations (which makes (1) akin to a new Keynesian Phillips curve (NKPC); this is the interpretation given in Stock and Watson (2007, 2010) or as a central bank's inflation target (e.g., as in Cogley and Sbordone (2008) and Sbordone et al. (2010)). The trend in the unemployment rate, $\tilde{\tau}_t$, provides a time-varying NAIRU, in the sense that δ_t is the unemployment gap that potentially affects inflation. The random walk specification for the NAIRU is the same specification as in Staiger et al. (1997a) and Gordon (1998), extended to allow for stochastic volatility. The assumption that the innovations to inflation expectations and to the NAIRU are independent reflects the distinct institutional and economic processes that underpin a central bank's long-term inflation target and credibility on the one hand, and, on the other hand, the changes in productivity and labor market composition that are generally considered to drive the NAIRU. The independence of the serially uncorrelated disturbances ϵ_t and $\tilde{\epsilon}_t$ is motivated by viewing these as measurement error arising from the independent surveys from which the series are constructed. The common cyclical component δ_t links the inflation and unemployment gaps over business cycle frequencies, and λ plays the same role in this model as does the coefficient on the activity variable in the NKPC, when the NKPC is specified in terms of an activity gap or the unemployment rate.

From a statistical perspective, the model provides a parsimonious parameterization of time variation in the joint process for inflation and unemployment, with four parameters $(\lambda, \alpha_1, \alpha_2, \gamma)$. Given fixed values of the innovation variances, that is, at a given point in time, the unobserved components model implies specific joint autocovariances of $(\Delta\pi_t, \Delta u_t)$, which in turn implies a VAR representation of $(\Delta\pi_t, \Delta u_t)$. The parameters of the VAR representation depend on the innovation variances because the VAR coefficients essentially arise from a multivariate signal extraction problem, which in turn depends on the various signal-to-noise ratios. See Koopman and Harvey (2003a) for a discussion of the relevant issues.

Assuming $\alpha_1, \alpha_2, \gamma$ are nonzero, the model is identified. The likelihood, however, turns out to be quite flat in some of the parameters. We therefore imposed the following three restrictions:

(i) $\gamma = 0.2$;
(ii) $\sigma^2_{\tilde{\epsilon}} = 0.01$; and
(iii) $\sigma^2_{\tilde{\eta}} = 0.2$.

Restriction (i) is taken from Stock and Watson (2007), extended here to the multivariate system. Restriction (ii) eliminates stochastic volatility in the NAIRU. This restriction is consistent with the previous work cited above on unobserved components models of the NAIRU in which there was no evidence of time-varying low frequency variances. Restriction (iii) is imposed to resolve an empirical problem of separately estimating the measurement error and the cyclical component of the unemployment rate. Section 3.4 reports sensitivity analysis for various alternative treatments of $\sigma^2_{\tilde{\eta}}$, including time-varying estimation. With these three restrictions imposed, the model has three estimated parameters, α_1, α_2, and λ.

3.3 Estimation strategy

The model is estimated using a Gibbs sampling procedure. The parameters and latent variables are divided in three blocks. In the first block, we draw the latent variables $\tilde{\tau}$, δ, and τ conditioning on the inflation and unemployment series, the parameters, α_1, α_2, and λ, and the stochastic volatilities. We first determine the joint posterior distribution of inflation, unemployment, and the three latent variables conditional on the parameters and the stochastic volatilities; we then sample from the conditional distribution of the latent variables. Next, we condition on the latent variables and the stochastic volatilities and we sample from the posterior of the parameters. We assume normal conjugate priors for the parameters, with zero mean and variance equal to one hundred. Also, α_1 and α_2 are constrained to be in the triangle $-1 \leq \alpha_2 \leq 1 - |\alpha_1|$ so that the cyclical process δ_t is stationary. Finally, we draw the stochastic volatilities — à la Kim et al. (1998), conditioning on the latent variables and the parameters. Ghysels et al. (1996) and Shephard (2005a) provide literature reviews of the econometrics of stochastic volatilities. Bos and Shephard (2006) discuss some of the issues involved in estimating unobserved component models with stochastic volatilities.

3.4 **Data description and empirical results**

The data set is quarterly from 1960Q1 to 2011Q3. Inflation is measured by the gross domestic product (GDP) deflator and the unemployment rate is the quarterly average of the monthly total civilian unemployment rate. The data are plotted in Figure 3.1.

3.4.1 IN-SAMPLE RESULTS ON HISTORICAL TIME VARIATION

Figures 3.2 and 3.3 plot the estimated time paths of the volatilities that are not fixed under the base model; these and the following estimated time paths are posterior means and 95% intervals from the nonlinear, non-Gaussian smoother and are conditional on the full data set. The variance of the permanent innovation to inflation has an estimated time path that is similar to the one in Stock and Watson (2007), with the variance of the permanent component large in the 1970s and early 1980s and smaller in the 1960s, 1990s, and early 2000s; these latter periods correspond to periods of relatively well-anchored inflation expectations. The variance of the transitory innovation in inflation shows more variability than in Stock and Watson (2007). The variance of the cyclical component follows a recognizable time path: it was large before the mid-1980s, it diminished during the Great Moderation, and it increased substantially since 2007 to pre-Great Moderation levels. It must be noted that the confidence intervals around the stochastic volatilities are wide, which means that there is considerable uncertainty on the conclusions regarding the time-varying patterns of volatility.

Estimates of the trend and common cyclical components are plotted in Figure 3.4. As in Stock and Watson (2007) and Cogley et al. (2010), trend inflation closely tracks the low-frequency swings in inflation. The estimated NAIRU (Figure 3.5a), fluctuates through a range of just over one percentage point. The NAIRU is estimated to be greatest during the 1970s, to be lowest during the early 2000s, and to be rising recently. Figure 3.5b shows the estimated NAIRU when we relax restriction (iii). Estimating σ_{η}^2 does not change

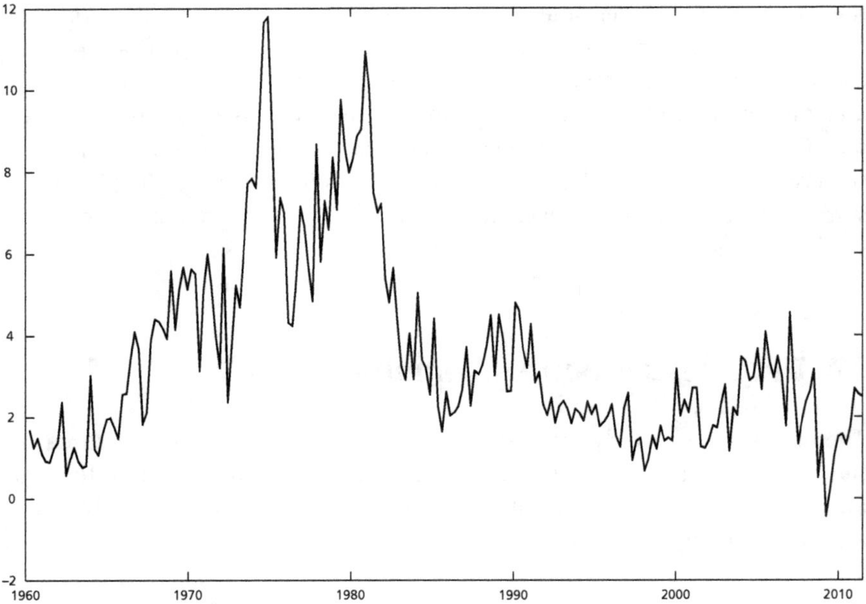

(a) Rate of inflation (%, annual rate)

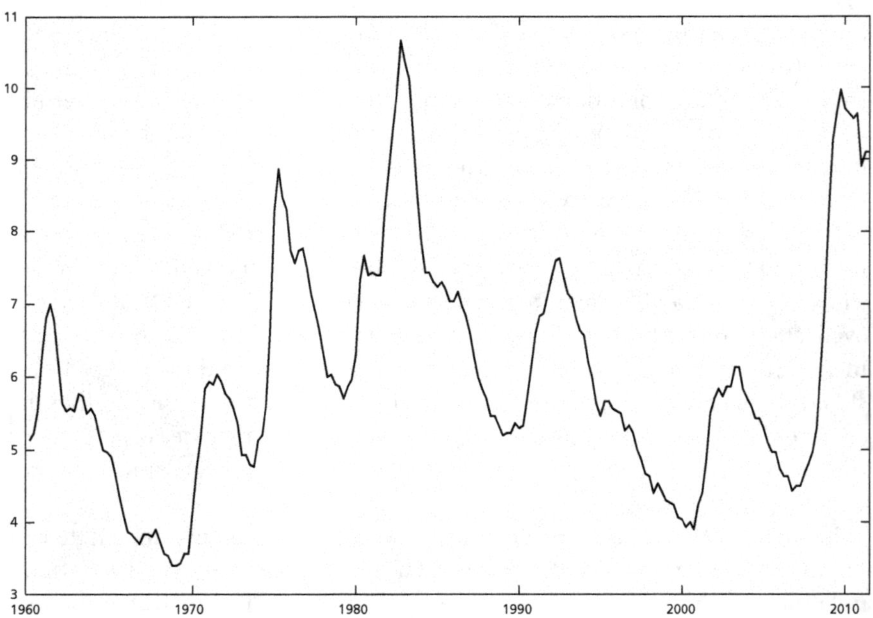

(b) Rate of unemployment (%, annual rate)

Figure 3.1. U.S. GDP deflator inflation and unemployment rates, quarterly, 1960Q1 to 2011Q3

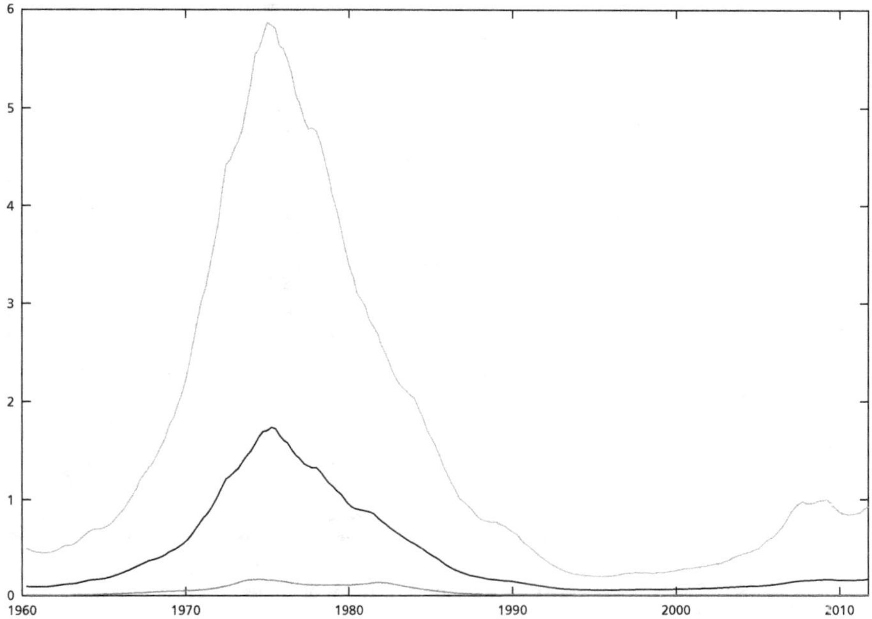

(a) Variance of permanent innovation to inflation

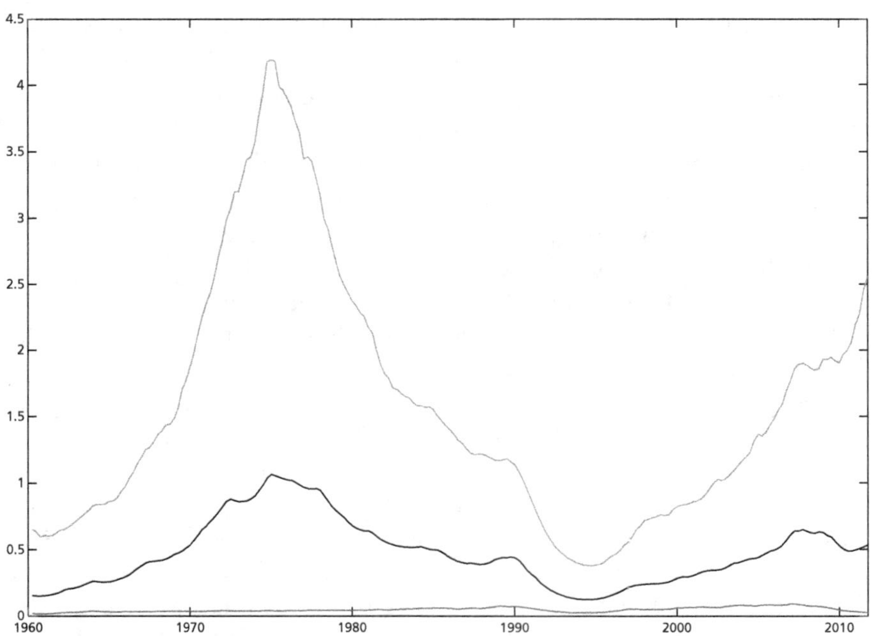

(b) Variance of transitory innovation to inflation

Figure 3.2. Posterior mean and 95% interval of time-varying variances of innovations

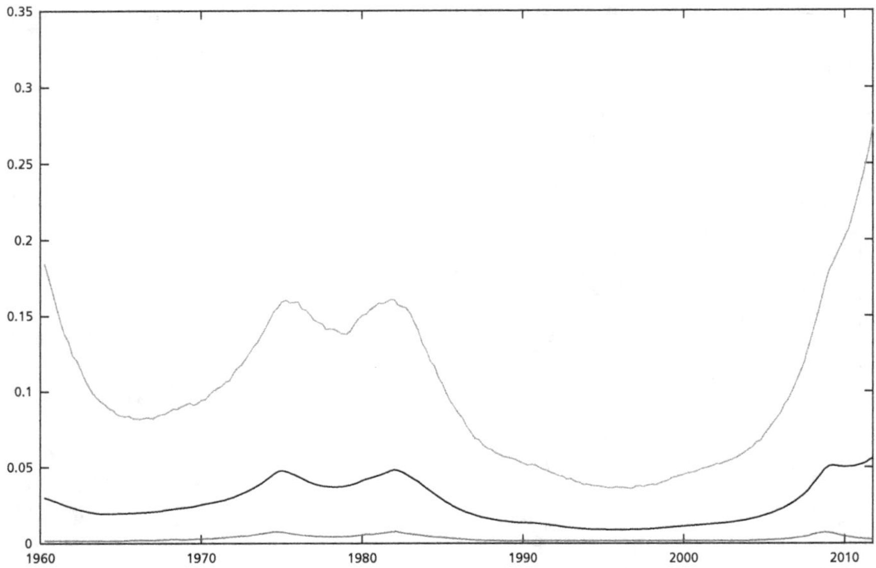

Variance of innovation to cyclical component

Figure 3.3. Posterior mean and 95% interval of time-varying variances of innovations

the low frequency properties of the estimated NAIRU, but it adds considerable high frequency noise to it. When we allow $\sigma^2_{\tilde{\eta}}$ to be unrestricted, the estimation attributes all the high frequency noise in the unemployment rate to $\tilde{\tau}$, rather than to the measurement error $\tilde{\eta}$. We think that this result is due to weak identification and we therefore fix $\sigma^2_{\tilde{\eta}}$ to 0.2. The forecasting performance of the model is only marginally affected by this restriction, as shown later in the paper.

Figure 3.6 presents the posterior distributions of the parameters. Most notably, the posterior for λ has essentially all its mass less than zero, with a posterior mean of –0.37 and a 95% Bayesian credible set of (–0.58, –0.17). The posteriors for all three parameters are much tighter than the priors, with different means.

One aim of this model is to allow for the possibility of time-varying VAR coefficients, in particular for a time-varying slope of the Phillips curve. Figure 3.7 plots the slope of the Phillips curve, as measured by the time-varying sum of the coefficients on lagged unemployment in the implied VAR for detrended inflation and unemployment, $(\pi_t - \tau_t, u_t - \tilde{\tau}_t)$. The Phillips curve slope κ is steepest (the implied sum-of-coefficients is most negative) in the 1970s, is the flattest in the 1990s, and has increased in steepness since 2007. It must be noted that there is considerable uncertainty around the estimate of κ, as shown by the 95% confidence intervals.

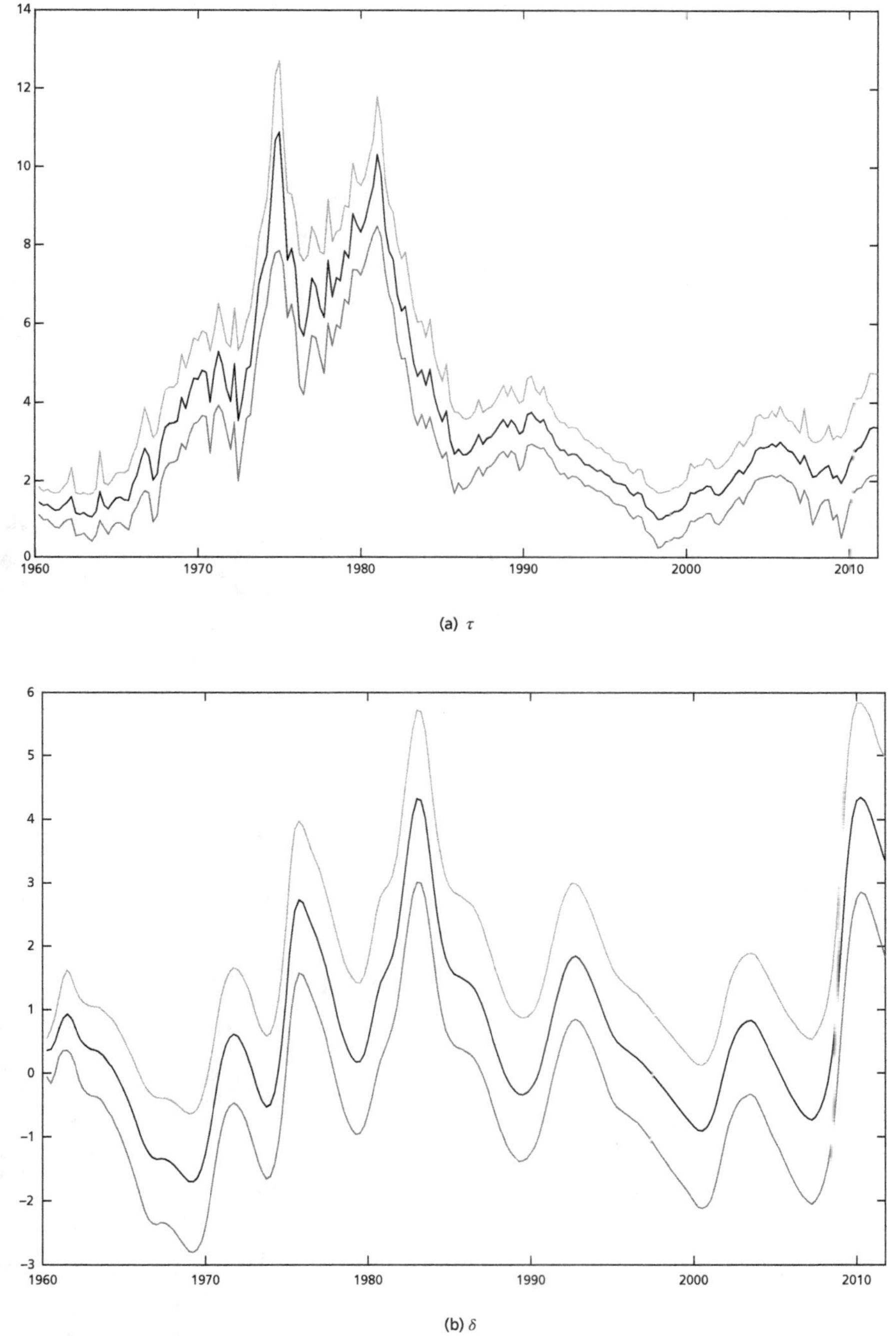

Figure 3.4. Posterior mean and 95% interval of unobserved components

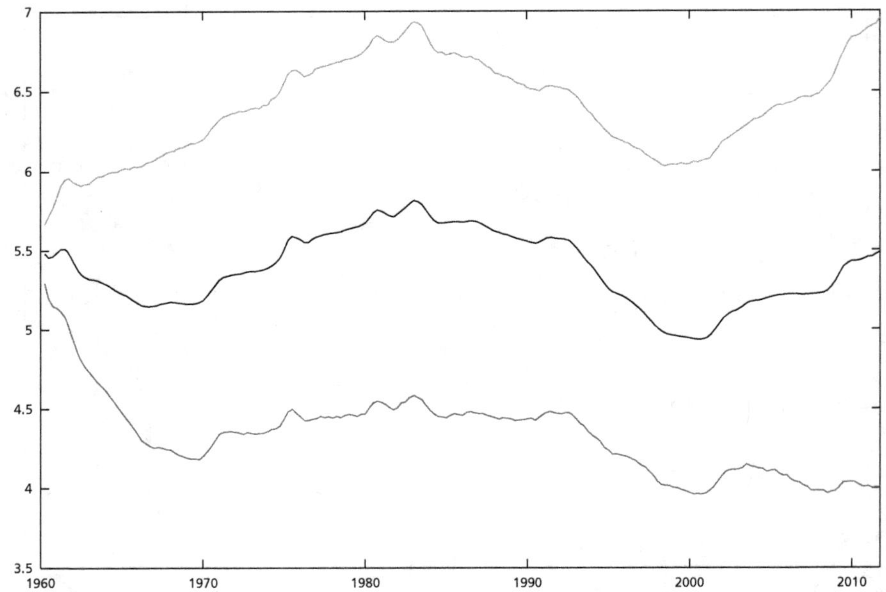

(a) $\bar{\tau}$

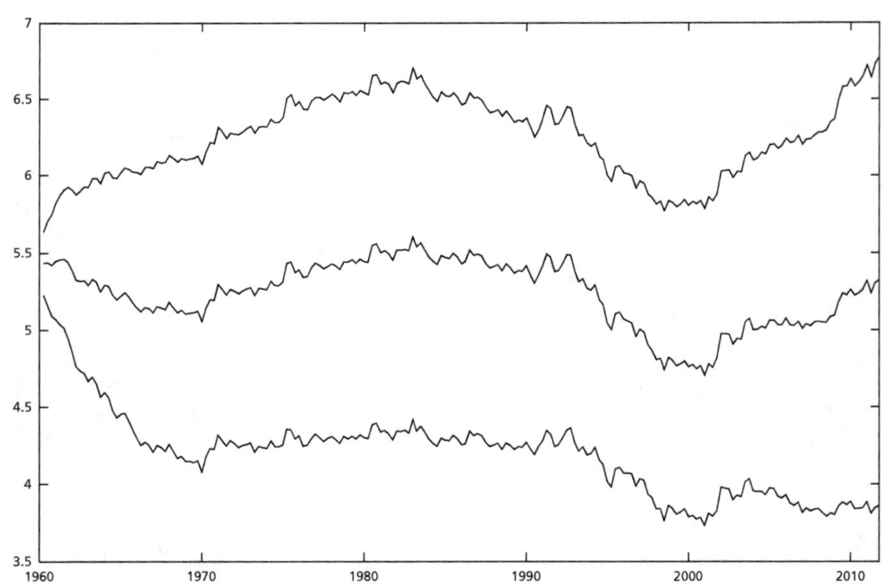

(b) $\bar{\tau}$ with ^{2}

Figure 3.5. Posterior mean and 95% interval of unobserved components

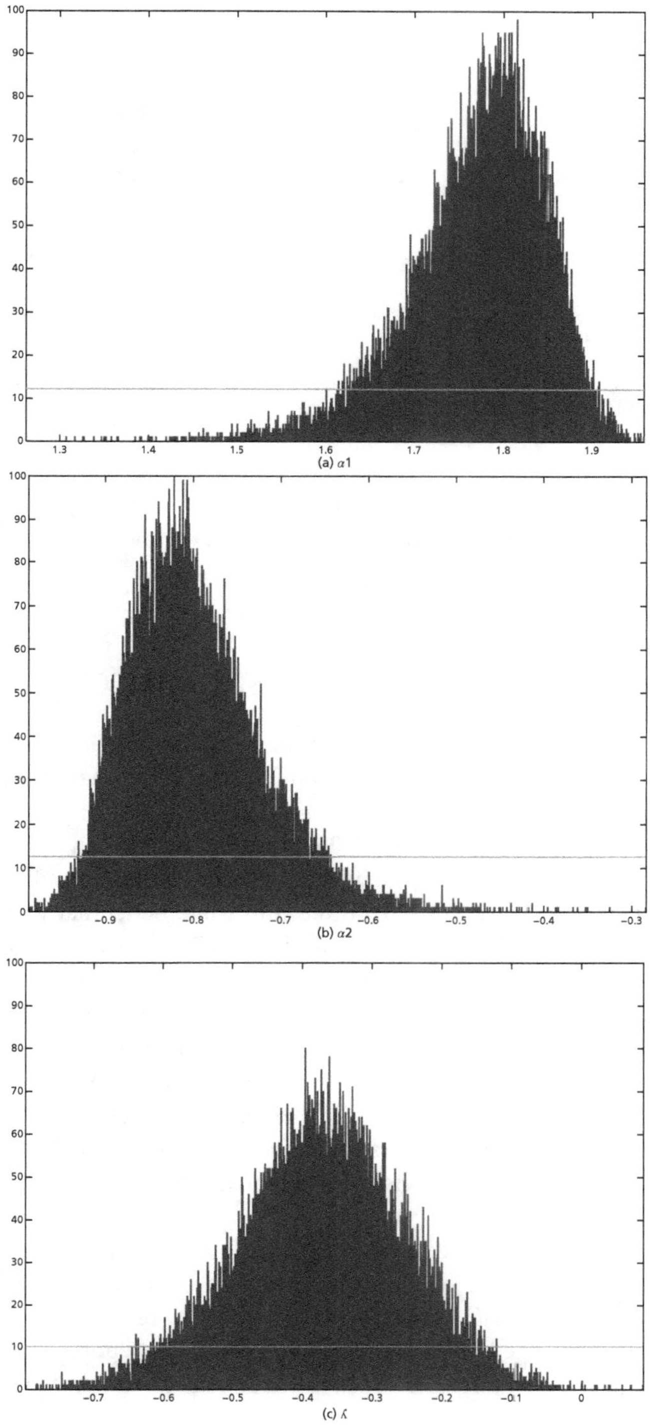

Figure 3.6. Full-sample posterior and diffuse prior distributions of parameters

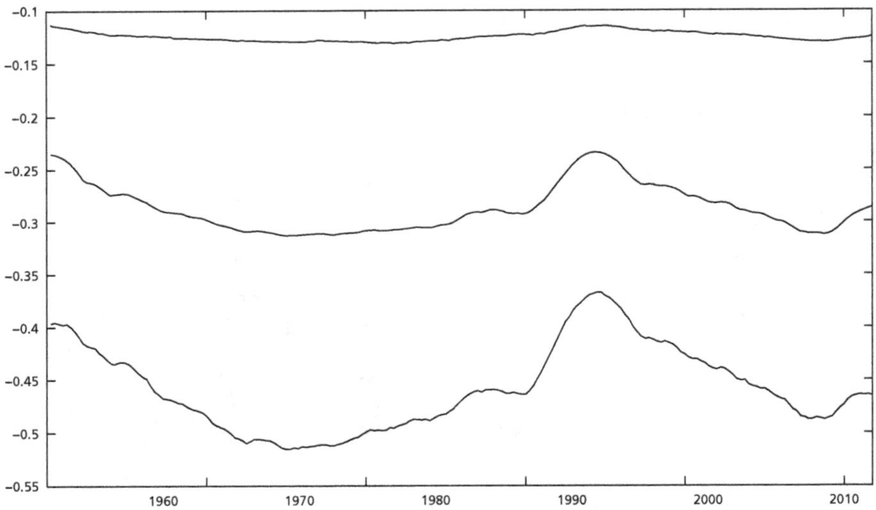

Figure 3.7. Slope of the Phillips curve κ for $(\pi_t - \tau)$ and 95% interval

3.4.2 PSEUDO OUT-OF-SAMPLE FORECAST EVALUATION

This section reports the results of a pseudo out-of-sample inflation forecasting exercise. Specifically, the model parameters were estimated using Gibbs sampling, with a quarterly expanding estimation sample starting with 1960Q1–1979Q4. The model was simulated 25,000 times after a burn-in period of 5,000 draws. The analysis focuses on the base model of Section 3.2 and one variant described below. We also consider three univariate benchmark models, the Stock and Watson (2007) univariate UC-SV model and two versions of the Atkeson and Ohanian (2001) quarterly random walk model. The first AO model (AO(i)) forecasts h-quarter ahead inflation using lagged four-quarter inflation; the second AO model (AO(ii)) forecasts h-quarter ahead inflation using lagged h-quarter inflation. For horizons greater than one quarter, the object being forecast is the multi-period percentage rate of inflation at annual rates, for example at four-quarters the object of interest is $100ln(P_{T+4}/P_T)$. This pseudo out-of-sample exercise uses the most recent data with no real-time data adjustments.

The results are summarized in Table 3.1 to Table 3.4. The first row of each block reports the pseudo out-of-sample root mean squared forecast error (RMSFE) of the univariate UC-SV model, and the remaining rows report RMSFEs relative to the UC-SV benchmark. Results are reported for one, two, four, and eight-quarter ahead forecasts, and for various subsamples.

Table 3.1 Pseudo out-of-sample forecasting performance: January 1980 to April 1989

	January 1980 to April 1989			
Model	h =1	h = 2	h = 4	h = 8
SW 2007	1.1375	1.1144	1.1813	1.5171
Relative RMSFE				
SW 2007	1.0000	1.0000	1.0000	**1.0000**
AO (i)	1.0652	1.0880	1.1172	1.0939
AO (ii)	1.0292	0.9994	1.1172	1.2821
No lags - $\sigma_{\eta}^2 = 0.2$	**0.9893**	0.9904	0.9997	1.0330
No lags - σ_{η}^2 estimated	0.9894	0.9939	1.0079	1.0600
2 lags - $\sigma_{\eta}^2 = 0.2$	0.9902	**0.9861**	**0.9885**	1.0502
2 lags - σ_{η}^2 estimated	0.9897	1.0004	1.0141	1.0737

Notes: Entries are RMSFEs, relative to the Stock and Watson (2007) univariate UC-SV. Bold entries are the smallest relative RMSFE for the indicated series/period/horizon.

Table 3.2 Pseudo out-of-sample forecasting performance: January 1990 to April 1999

	January 1990 to April 1999			
	h =1	h = 2	h = 4	h = 8
SW 2007	0.5818	0.5019	0.5083	0 6252
Relative RMSFE				
SW 2007	1.0000	1.0000	1.0000	1.0000
AO (i)	0.9948	0.9972	1.0470	1.0278
AO (ii)	1.0988	1.0401	1.0470	1.1325
No lags - $\sigma_{\eta}^2 = 0.2$	0.9555	0.9186	0.9027	0.9603
No lags - σ_{η}^2 estimated	0.9547	0.9205	0.9022	0.9701
2 lags - $\sigma_{\eta}^2 = 0.2$	0.9652	**0.9048**	**0.8491**	**0.9160**
2 Lags - σ_{η}^2 estimated	**0.9543**	0.9149	0.8864	0.9633

Notes: See the notes in Table 3.1

Over the full sample, the base model provides modest improvements over the UC-SV and AO models, particularly at the one and two-quarter horizons. Table 3.1 also reports results for three variants of the base model. The first variant provides a different treatment of σ_{η}^2 (see (iii) in Section 3.2), allowing σ_{η}^2 to be estimated. This change generally makes little difference to the pseudo out-of-sample forecast RMSFEs, a result that is consistent with the difficulty we found in identifying this parameter using the full data set; a notable exception is represented by the 2000s, when the model with unrestricted σ_{η}^2 does better. The second and third variants allow for a more flexible specification for how δ_t enters the inflation equation (1), in particular, the term $\lambda\delta_t$ in (1) is replaced by the distributed lag, $\lambda_1\delta_t + \lambda_2\delta_{t-1} + \lambda_3\delta_{t-2}$; σ_{η}^2 is fixed to 0.2 in the second

Table 3.3 Pseudo out-of-sample forecasting performance: January 2000 to March 2011

	January 2000 to March 2011			
	h =1	h = 2	h = 4	h = 8
SW 2007	0.8884	0.7906	0.7637	0.8175
Relative RMSFE				
SW 2007	1.0000	1.0000	1.0000	1.0000
AO (i)	1.0104	1.0114	1.0133	1.0021
AO (ii)	1.1190	1.0230	1.0133	1.0760
No lags - $\sigma_\eta^2 = 0.2$	1.0021	1.0060	1.0068	0.9832
No lags - σ_η^2 estimated	0.9691	0.9501	0.9384	**0.9420**
2 lags - $\sigma_\eta^2 = 0.2$	1.0402	1.0733	1.1109	1.0720
2 lags - σ_η^2 estimated	**0.9508**	**0.9260**	**0.9207**	0.9448

Notes: see the notes in Table 3.1

Table 3.4 Pseudo out-of-sample forecasting performance: Whole sample

	Whole sample			
	h =1	h = 2	h = 4	h = 8
SW 2007	0.8980	0.8384	0.8612	1.0602
Relative RMSFE				
SW 2007	1.0000	1.0000	1.0000	**1.0000**
AO (i)	1.0367	1.0539	1.0817	1.0689
AO (ii)	1.0714	1.0117	1.0817	1.2276
No lags - $\sigma_\eta^2 = 0.2$	0.9895	0.9875	0.9911	1.0151
No lags - σ_η^2 estimated	0.9776	0.9719	0.9777	1.0277
2 lags - $\sigma_\eta^2 = 0.2$	1.0051	1.0061	1.0093	1.0398
2 lags - σ_η^2 estimated	**0.9713**	**0.9675**	**0.9754**	1.0373

Notes: see the notes in Table 3.1

variant and estimated in the third. The model with a distributed lag does slightly better, above all in the 1990s.

3.5 **Conclusions**

We consider these results to be encouraging. The multivariate UC-SV model captures several of the features found in the inflation forecasting literature, most notably a time-varying state-dependent Phillips curve. The time-varying latent variables have natural interpretations as inflation expectations (or target inflation), the NAIRU, and a common cyclical component. The pseudo out-of-sample forecasting exercise suggests that this tightly parameterized model can improve upon the forecasts of what we understand to be the best univariate models.

ACKNOWLEDGMENTS

The views in this paper are solely the responsibility of the authors and should not be interpreted as reflecting the views of the board of governors of the federal reserve system or of any other person associated with the federal reserve system. We would like to thank Eduardo Morales, Giulio Tiozzo, and Mark Watson for helpful comments.

4 Measuring the tracking error of exchange traded funds

Giuliano De Rossi

4.1 Introduction

The tracking error of a mutual fund is defined as the standard deviation of the difference between the fund's return and the return to the benchmark it tracks.[1] The aim of this paper is to address an empirical question: How should the tracking error of an exchange traded fund (ETF) be estimated? Given the recent growth in the ETF market—and considering the importance of tracking error as a selection criterion—this is a crucial issue faced by a large number of asset managers on a daily basis. Our contribution is to propose a novel solution based on an unobserved components approach.

ETFs are a variety of listed fund, typically tracking a chosen benchmark. Two main features make them attractive to investors seeking exposure to a broad asset class. First, the fact that they are traded on an exchange provides liquidity and transparency. Second, ETFs are typically characterized by low management fees, usually deducted when dividends are distributed. A number of academic studies have been devoted to this type of funds, e.g., Gastineau (2001), Poterba and Shoven (2002).[2]

When selecting an ETF, portfolio managers need to assess how accurately the available funds are able to track the benchmark index over time. To this end, tracking error measures the dispersion of a portfolio's returns relative to the returns of the manager's benchmark. Several approaches to the estimation of tracking error, surveyed, among others, by Pope and Yadav (1994) and Shin and Soydemir (2010), have been advocated in the literature. As for the data, empirical studies have used either ETF price returns or the return to the net asset value of the fund in the estimation. The net asset value is considered a better proxy of the value of the fund as it is typically less noisy than the market price.

However, the net asset value is not accurately represented due to distortions introduced by agency costs, tax liabilities, asset illiquidity, and asynchronous prices. This, as we shall argue below, results in a severe errors-in-variables

[1] Somewhat confusingly, the return difference between fund and benchmark is also sometimes referred to as tracking error.

[2] See Engle and Sarkar (2006) for a concise description of the mechanism that allows certain institutional investors to redeem or create shares in the fund; a distinctive feature of ETFs.

problem (particularly for global ETFs) which is often ignored by practitioners and in published academic research.

The problem is typically solved in the empirical literature by using low frequency observations. For example, in an empirical paper on emerging markets ETFs, Blitz and Huij (2012) use quarterly or even annual data to mitigate the effect of the distortion. Unfortunately, such a solution comes at a high cost because it discards a large amount of information which is potentially useful, particularly if the goal of the analysis is to test for changes in the tracking error over time.

This paper is related to the analysis of Engle and Sarkar (2006), who model the *true* net asset value of an ETF as a latent variable and use a Kalman filter to estimate its value. Their focus, however, is on premia and discounts rather than on tracking error. In addition, their state space model does not allow for heteroscedasticity while we build stochastic volatility into our approach.

An alternative approach to the estimation of ETF premia or discounts is considered by Delcoure and Zhong (2007). They build on the work of Goetzmann et al. (2001), who used an instrumental variable estimator to recover the unobserved net asset value of international mutual funds.

Finally, the return decomposition proposed in this paper is reminiscent of the approach of Moise and Russell (2012), who use high frequency data to decompose the intraday price returns of the Standard & Poor's Depositary Receipts ETF into *fundamental return* and *price deviation* components.

We show that our method is successful in dealing with the errors-in-variables problem and that it can be used to track changes in the tracking error of a fund over time. It also results in a useful decomposition of tracking error into permanent and transitory deviations. Our findings suggest that the relative importance of each component varies significantly across different ETFs. As a consequence, tracking error can be viewed as a function of the investment horizon: Short-term investors should try to minimize temporary return discrepancies while long-term investors should focus predominantly on the magnitude of permanent errors.

The remainder of the paper is organized as follows. Section 4.2 defines tracking error and highlights the errors-in-variables problem. Our proposed solution is illustrated in Section 4.3. Section 4.4 presents the empirical results of an analysis of the tracking error of two emerging markets ETFs. Section 4.5 concludes.

4.2 **ETF tracking error in the existing literature**

Define $p_{I,t}$, the log of the value of the index at time t, and $p_{F,t}$, the log of the *true*, unobservable, net asset value (NAV) of the fund. The observed one-period active return is just the difference $\tilde{r}_{F,t} - r_{I,t}$, where $\tilde{r}_{F,t} = \Delta\tilde{p}_{F,t}$, $r_{I,t} = \Delta p_{I,t}$ and $\tilde{p}_{F,t}$ is the observed NAV (or price) of the fund. The aim of this section is to estimate the tracking error (TE), defined as the volatility of the active return. The variable of interest to us is the *cumulated active return of the fund*, defined as

$$\lambda_t \equiv p_{F,t} - p_{I,t}. \tag{4.1}$$

A common estimator of TE, adopted, among others, by Vardharaj et al. (2004), is the standard deviation of the active return:

$$TE_0 = \sqrt{\frac{1}{T-1}\sum_{t=1}^{T}\left(\tilde{r}_{F,t} - r_{I,t}\right)^2} \tag{4.2}$$

where T is the number of observations.

Shin and Soydemir (2010) use three alternative measures:

$$\begin{aligned} TE_1 &= \frac{1}{T}\sum_{t=1}^{T}\left|\tilde{r}_{F,t} - r_{I,t}\right| \\ TE_2 &= \widehat{s} \\ TE_3 &= \sqrt{\frac{1}{T-1}\sum_{t=1}^{T}\left(RD_t - \overline{RD}\right)^2} \end{aligned} \tag{4.3}$$

where $\widehat{s}$ is the standard error in the regression $\tilde{r}_{F,t} = a + br_{I,t} + u_t$ and $RD_t = \left|\tilde{r}_{F,t} - r_{I,t}\right|$. The advantage of TE_1 is that it is robust to outliers. TE_2 is similar to TE_0 but it can deal with funds whose beta to the benchmark index is persistently different from one. TE_3, the standard deviation of the absolute differences in return between the fund and the index, combines the advantages of TE_0 and TE_1.

Vardharaj et al. (2004) estimate the TE of a sample of U.S. domestic equity mutual funds between 1992 and 2002. According to their analysis, the *typical* levels of TE, measured as the annualized standard deviation TE_0, are below 2% for an enhanced index fund and between 5 and 10% for an actively managed fund. An index fund should theoretically have a TE of zero. They also note that TEs tend to increase when the volatility of the benchmark increases.

The performance of an ETF in tracking the index is affected by a number of factors. Some have a *permanent* effect on the price, as is the case for management fees and administrative expenses, which are typically deducted directly from the dividends. Another example is replication error: some funds hold a subset of the constituents of the index, selected through an optimization procedure. Whenever one of the stocks excluded from the selection is affected by an idiosyncratic shock, the ETF experiences a permanent return deviation from the benchmark. Rebalancing costs, due to corporate actions and the periodic rebalances of the benchmark index, also have permanent effects.

Other factors tend to cause a *temporary* discrepancy between benchmark return and fund return. For example, the ETF can trade at a premium or at a discount because of supply and demand factors. Another notable example is the effect of asynchronous pricing of the stocks and exchange rates, described in detail by Delcoure and Zhong (2007). Both effects are particularly severe for global ETFs.[3]

Finally, dividends gathered from the stocks held in the fund may not be reinvested immediately (the typical assumption underlying most index construction methodologies), but at regular time intervals. As a result of all the aforementioned distortions, the TE of an ETF is not zero in practice.

Consider two of the most liquid ETFs tracking the Morgan Stanley Capital International (MSCI) Emerging Markets Index, a broad basket of global stocks: Vanguard and iShares. A well known analytics platform used in the financial industry[4] estimates their TE to be, respectively, 24.3% and 15.7% when price data is used. Such levels of TE are so grossly exaggerated that the funds would appear to be aggressively active rather than passive trackers. NAV data yields TEs of 17.1% and 6.25%, which would still suggest that Vanguard displays an extreme level of TE.

Figure 4.1 compares the total return index to each fund with the corresponding benchmark index. Graphically, both ETFs appear to do a good job at tracking the MSCI index.

The estimates displayed in Table 4.1 were obtained by evaluating the TE formulas (4.2) and (4.3) from daily NAV data for the period October 6, 2009 to October 6, 2010. Again, the TE of the Vanguard fund appears to have been importantly overestimated.

Active returns are measured with error due to the use of asynchronous stock prices and asynchronous exchange rates. De Rossi et al. (2010) show that by simply fitting an MA model to the data it is possible to obtain a more reliable TE estimate. Such a simple time series model, however, cannot account for the time variation in TE that stems from the dynamics of market volatility. We set out to address this problem in the Section 4.3 by building a nonlinear state space model which incorporates stochastic volatility.

3 Engle and Sarkar (2006) document larger and more persistent premia on international ETFs.

4 Bloomberg, estimates obtained on June 28, 2012.

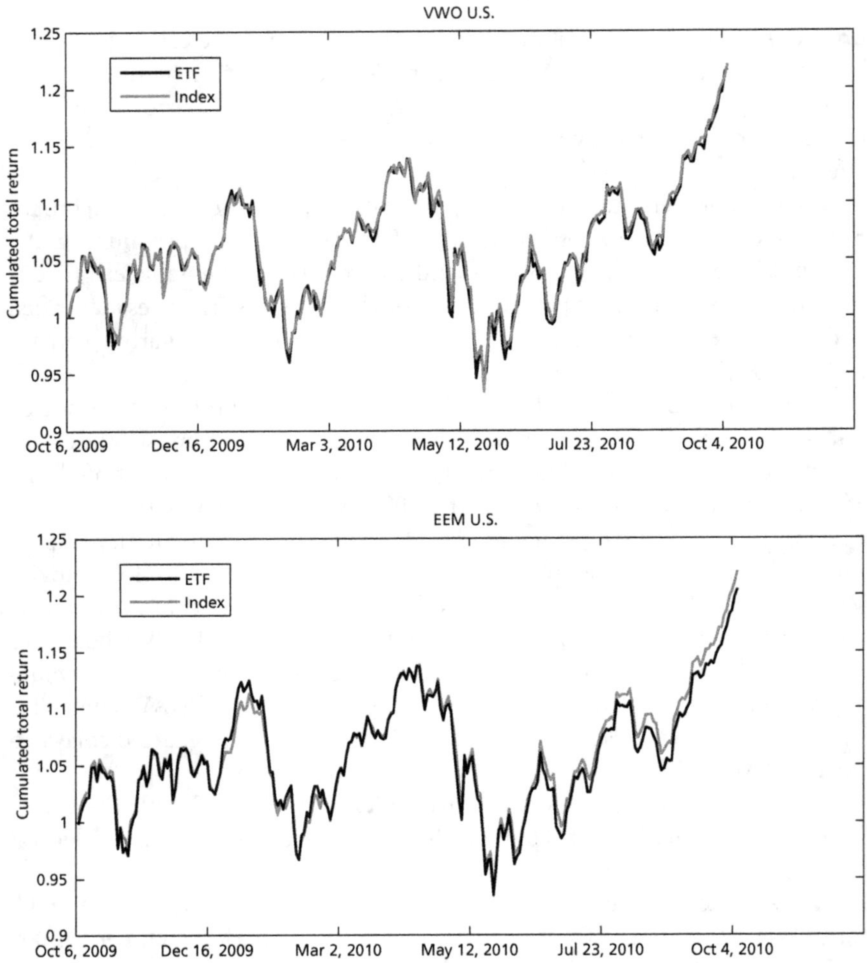

Figure 4.1. Total return to two ETFs tracking the MSCI Emerging Markets Index (Bloomberg ticker NDUEEGF): Vanguard (VWO U.S., top panel) and iShares (EEM U.S., bottom panel). Daily NAV data between October 6, 2009 and October 6, 2010

Source: Bloomberg, UBS estimates.

Table 4.1 Estimated tracking errors (percent) of the Vanguard (VWO) and iShares (EEM) funds tracking MSCI emerging markets using the traditional measures (4.2) and (4.3). Daily NAV data between October 6, 2009 and October 6, 2010. The figures are scaled by $\sqrt{260}$

	VWO	EEM
TE_0	12.81	6.23
TE_1	9.63	4.78
TE_2	12.81	6.21
TE_3	8.43	3.99

Source: Bloomberg, UBS estimates.

4.3 **An unobserved components model**

4.3.1 BUILDING TOWARD OUR MODEL

We propose to describe the dynamics of TE by means of a simple unobserved components model with stochastic volatility, similar to the one adopted by Stock and Watson (2007) to analyze U.S. inflation. Stochastic volatility, unlike other modeling approaches such as generalized autoregressive conditional heteroskedasticity (GARCH), lends itself naturally to estimation in a state space framework and therefore it is a convenient device to introduce time varying volatility in this context.

We assume that

$$\begin{bmatrix} p_{I,t} \\ p_{F,t} \end{bmatrix} = \begin{bmatrix} \mu_I \\ \mu_I + \mu_F \end{bmatrix} + \begin{bmatrix} p_{I,t-1} \\ p_{F,t-1} \end{bmatrix} + \begin{bmatrix} \eta_{I,t} \\ \eta_{I,t} + \eta_{F,t} \end{bmatrix} \tag{4.4}$$

where μ_I, μ_F are constants, while η_I and η_F are mutually independent random variables with zero mean and variances σ_I^2 and σ_F^2. If the NAV were observable then we would be able to measure directly the difference between fund return and benchmark return at any time t.

Equation (4.4) describes the evolution of two state variables that are modeled as random walks plus drift with correlated innovations. The assumption that the ETF NAV and the index are not cointegrated is a natural one in this context[5] because fees, administrative costs, and replication error will drive the two series apart in the long term. As a result, we expect μ_F to be different from zero and in particular $\mu_F < 0$.

While we do observe the index p_I, we can only obtain a noisy realization $\tilde{p}_F$ of the NAV (or price) p_F. The measurement equation is then:

$$\begin{bmatrix} p_{I,t} \\ \tilde{p}_{F,t} \end{bmatrix} = \begin{bmatrix} p_{I,t} \\ p_{F,t} \end{bmatrix} + \begin{bmatrix} 0 \\ \sigma_t \epsilon_t \end{bmatrix} \tag{4.5}$$

where ϵ_t is standard normal. Expressions (4.4) and (4.5) imply that the one-period active return at time t can be written as

$$\tilde{r}_{F,t} - r_{I,t} = \Delta \tilde{p}_{F,t} - \Delta p_{I,t} = \mu_F + \eta_{F,t} + \sigma_t \epsilon_t - \sigma_{t-1} \epsilon_{t-1}.$$

The quantity $\mu_F + \eta_F$ can be thought of as the *permanent* component of the return deviation between fund and index, while $\sigma_t \epsilon_t - \sigma_{t-1} \epsilon_{t-1}$ represents the transitory component.

The variable σ_t introduces stochastic volatility in the measurement error that affects the NAV of the ETF. The state variable h_t drives the evolution of σ_t according to the equations:

[5] In contrast, cointegration is a natural assumption when analyzing the relation between NAV and price, as argued by Engle and Sarkar (2006). The premium represents the error correction term in their cointegrated system.

$$\sigma_t = \exp(h_t/2)$$
$$h_t = \omega + \phi h_{t-1} + \eta_{h,t} \tag{4.6}$$

where ω, ϕ are constant and $\eta_{h,t}$ is normally distributed with mean zero and variance q. The innovations $\eta_{h,t}$, $\eta_{I,t}$, $\eta_{F,t}$ and the noise term ϵ_t are assumed to be mutually independent.

The observed cumulated active return is defined as $\tilde{\lambda}_t \equiv \tilde{p}_{F,t} - p_{I,t}$. Over an interval of τ periods we obtain, conditional on $h_1, \ldots, h_t$,

$$\mathrm{E}\left(\tilde{\lambda}_t - \tilde{\lambda}_{t-\tau}\right) = \tau\mu_F$$
$$\mathrm{Var}\left(\tilde{\lambda}_t - \tilde{\lambda}_{t-\tau}\right) = \tau\sigma_F^2 + \exp(h_{t-\tau}) + \exp(h_t). \tag{4.7}$$

We regard the term $\exp(h_{t-\tau}) + \exp(h_t)$ as the transitory component of the variance (4.7) as it stems from the measurement errors at the beginning and at the end of the interval, while the permanent component consists of the term $\tau\sigma_F^2$. The square root of (4.7) is the population value of TE_0 in our model. Our TE forecast is obtained by plugging the parameter estimates into the expression.

From equations (4.1), (4.4), (4.5), and (4.6) we obtain a nonlinear state space model:

$$\tilde{\lambda}_t = \lambda_t + \exp(h_t/2)\,\epsilon_t$$
$$\lambda_t = \mu_F + \lambda_{t-1} + \eta_{F,t} \tag{4.8}$$
$$h_t = \omega + \phi h_{t-1} + \eta_{h,t}$$

where $\epsilon_t \sim N(0,1)$, $\eta_{F,t} \sim N\left(0, \sigma_F^2\right)$ and $\eta_{h,t} \sim N\left(0, q\right)$. We need to estimate the parameter vector $\psi = \{\mu_F, \sigma_F^2, \omega, \phi, q\}$.

4.3.2 INFERENCE

Given the nonlinear and non-Gaussian nature of the state space model (4.8) we resort to a Markov Chain Monte Carlo (MCMC) approach in order to obtain the posterior distribution of ψ. Our goal is to sample from a Markov chain whose invariant distribution is the required posterior. Gilks et al. (1996) and Robert and Casella (2005) contain detailed accounts of the approach.

The proposed methodology is a slight modification of Kim et al. (1998) and Chib et al. (2002).[6]

[6] One of the stochastic volatility models considered in those papers can be obtained by setting $\lambda_t = 0$ $\forall t$ in (4.8). We augment their Gibbs sampler in order to obtain draws for μ_F, σ_F^2, and λ.

The object of interest is the augmented posterior

$$\pi(\psi, h, \lambda) \propto \pi(\psi) \prod_{t=1}^{T} f\left(\tilde{\lambda}_t | h_t, \lambda_t, \psi\right) f(\lambda_t | \lambda_{t-1}, \psi) f(h_t | h_{t-1}, \psi).$$

We obtain draws from the posterior in two blocks: $\{\mu_F, \sigma_F^2, \lambda\}$ and $\{\omega, \phi, q, h\}$. The advantage of this blocking design is that the target density of each state variable can be approximated by building a linear Gaussian state space model and running the simulation smoother of de Jong and Shephard (1995).

In order to draw the first block of variables, $\psi^{(1)} = \{\mu_F, \sigma_F^2\}$, we use the Metropolis-Hastings algorithm. First we run the Kalman filter on the linear system

$$\tilde{\lambda}_t = \lambda_t + \exp(h_t/2)\,\epsilon_t$$

$$\lambda_t = \mu_F + \lambda_{t-1} + \eta_{F,t} \tag{4.9}$$

conditional on h, treating λ_t as the state variable. We then find the value m of $\{\mu_F, \sigma_F^2\}$ that maximizes the (log) density $g(\tilde{\lambda}_t)$ in the linear Gaussian state space (4.9). The matrix V is obtained by inverting the Hessian of the same function at the optimum.

The proposal draw $\psi^{(1)\prime}$ is then sampled from $N(m, V)$ and accepted with probability

$$\min\left\{\frac{g\left(\psi^{(1)\prime}\right)\pi\left(\psi^{(1)\prime}\right) f_N\left(\psi^{(1)}; m, V\right)}{g\left(\psi^{(1)}\right)\pi\left(\psi^{(1)}\right) f_N\left(\psi^{(1)\prime}; m, V\right)}, 1\right\}.$$

Next, we draw the state λ by running the simulation smoother of de Jong and Shephard (1995) in (4.9).

A similar procedure is used to draw $\{\omega, \phi, q, h\}$. Here we use the linearization proposed e.g., in Harvey et al. (1994)

$$y_t = \log\left(\left(\tilde{\lambda}_t - \lambda_t\right)^2\right) = h_t + z_t \tag{4.10}$$

$$h_t = \omega + \phi h_{t-1} + \eta_{h,t}$$

where $z_t \equiv \log(\epsilon^2)$.

The non-Gaussian density of z is approximated by a mixture of seven normal distributions as detailed in Chib et al. (2002).[7] This introduces a new state variable $s_t \in \{1, \ldots, 7\}$. Conditional on s_t, z_t is assumed to be Gaussian with mean and variance depending on s_t.

[7] In a more recent paper, Omori et al. (2007) adopt a ten-point mixture instead.

We then draw the state h by running the simulation smoother in (4.10). The MCMC algorithm can be summarized as follows:

1. Initialize $\psi^{(1)} = \{\mu_F, \sigma_F^2\}, \lambda, \psi^{(2)} = \{\omega, \phi, q\}, h$
2. Sample $\psi^{(1)}$ and λ from $\psi^{(1)}, \lambda|\tilde{\lambda}, h$
3. Sample s from $s|y, h, \psi^{(2)}$
4. Sample $\psi^{(2)}$ and h from $\psi^{(2)}, h|\tilde{\lambda}, s$
5. Go back to 2.

4.3.3 CHOICE OF PRIORS

A normal prior is used for ω. In particular, the prior mean of ω implies an equilibrium value for σ_t of 50 basis points when $\phi = 0.9$.

The parameters σ_F^2 and q have lognormal priors. The prior mean of σ_F^2 implies an annualized contribution of 1.5% to the TE. The prior mean of q is chosen so as to set the standard deviation of the stationary distribution of h to four when $\phi = 0.9$. As a result, if h_t were one standard deviation away from its equilibrium value then σ_t would equal 3.69%.

Finally, along the lines of Kim et al. (1998), we imposed a rescaled beta prior with support (0, 1) on ϕ. We have chosen the hyperparameters so as to concentrate the probability mass between 0.75 and 1.

4.4 **Empirical analysis**

4.4.1 DATA

The data consists of daily NAVs and dividends, in addition to the total return index series, for two funds tracking the MSCI Emerging Markets Index: Vanguard and iShares. We obtained a sample covering the complete history of both funds until May 25, 2012 from Bloomberg.[8] This choice of funds is particularly interesting because of the different replication strategies adopted by the two fund providers. While Vanguard has always used complete replication, iShares has moved from partial replication to complete replication between 2010 and 2011.[9] Indeed, both funds currently hold *more stocks* than the index as they select several equity lines for the same company, e.g., when American depository receipts are available.

[8] Data on the Vanguard fund starts on March 11, 2005, while the data on iShares MSCI EM is available from April 11, 2003.

[9] Data on the number of stocks held in the fund can be obtained from the company website.

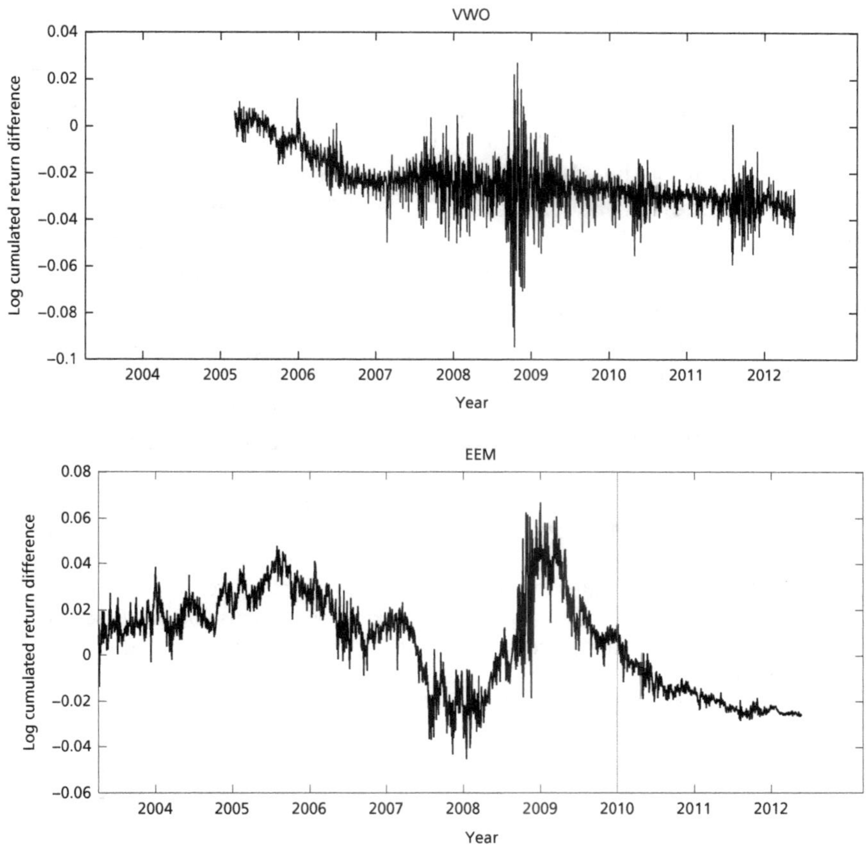

Figure 4.2. Cumulative outperformance ($\tilde{\lambda}_t$ in the notation used in the text) of the Vanguard (VWO) and iShares (EEM) emerging markets ETFs

Source: Bloomberg, UBS estimates.

One of the goals of our analysis is to assess whether the change in replication strategy has improved the ability of the iShares fund to track the benchmark.

Figure 4.2 plots the observed cumulated active return of each fund, $\tilde{\lambda}_t$. An increase in $\tilde{\lambda}_t$ indicates a period of outperformance for the ETF. In practice, because of the presence of fees and administrative costs, we expect the ETF to underperform steadily over time. The visual impression is that each series is characterized by a strong stochastic trend and a considerable amount of noise, particularly in the case of Vanguard. The noise component displays some clear volatility clusters, notably at the end of 2008 when market volatility spiked.

4.4.2 ESTIMATION RESULTS

All computations were carried out in R although the functions that perform the Kalman filter and return the log likelihood of a linear state space model were written in C and linked to the main code.

In implementing the MCMC sampler we discarded the first 1000 draws and analysed the next 8000. Tables 4.2 and 4.3 report the quantiles of the distributions of the MCMC draws for each parameter. The estimated posteriors are markedly different from the priors, e.g., the mass for the autoregressive parameter ϕ is concentrated in a much narrower range.

The estimation results suggest that, compared to Vanguard, the TE volatility of iShares is more persistent and its innovations have a higher dispersion (q is centered around 0.1, compared to 0.05 in the case of Vanguard). The posterior distribution of the iShares parameter ω is centered around a higher value.

Table 4.2 Quantiles of the posterior distribution of the model parameters, Vanguard MSCI emerging markets ETF. The model is specified in Equation (4.8). μ_F is the drift term in the transition equation for the state λ. σ_F, the standard deviation of the innovations, is expressed as an annualized standard deviation assuming 260 trading days a year. ω, ϕ, and q are the constant term, autoregressive coefficient and variance of the disturbances in the transition equation for the state h. $\overline{\sigma} = \exp(0.5\omega/(1-\phi))$ is the equilibrium value σ_t reverts to in the long term. $q/(1-\phi^2)$ is the variance of the unconditional distribution of h_t

Parameter	10%	25%	50%	75%	90%
$\mu_F \times 10^5$	−3.220	−2.648	−2.101	−1.507	−0.994
$\sigma_F \times 100$	0.498	0.538	0.590	0.647	0.698
ω	−0.268	−0.228	−0.187	−0.148	−0.117
ϕ	0.974	0.978	0.982	0.986	0.989
$q \times 100$	3.557	4.090	4.785	5.564	6.393
$\overline{\sigma} \times 100$	0.436	0.480	0.529	0.583	0.638
$q/(1-\phi^2)$	1.025	1.164	1.361	1.628	1.993

Source: UBS estimates.

Table 4.3 Quantiles of the posterior distribution of the model parameters, iShares MSCI emerging markets ETF. Variables defined in Table 4.2

Parameter	10%	25%	50%	75%	90%
$\mu_F \times 10^5$	−3.481	−2.344	−1.108	−0.055	0.966
$\sigma_F \times 100$	1.764	1.821	1.890	1.955	2.018
ω	−0.170	−0.143	−0.112	−0.086	−0.060
ϕ	0.986	0.988	0.991	0.993	0.995
$q \times 100$	7.139	8.183	9.470	10.817	12.052
$\overline{\sigma} \times 100$	0.147	0.189	0.233	0.292	0.362
$q/(1-\phi^2)$	3.585	4.173	5.105	6.703	8.747

Source: UBS estimates.

The drift term μ_F is estimated to be negative, as expected. Considering the median of the posterior, our estimate is -0.21 basis points for the Vanguard ETF, which corresponds to an annual drift of 55 basis points. The dispersion of the posterior suggests that the parameter is estimated with much less precision in the case of iShares.

In order to be able to compare the estimated dynamics of the two stochastic processes, λ, it is better to consider transformations of the estimated parameters. Hence in Tables 4.2 and 4.3 we display the quantiles of the estimated posteriors of two quantities of interest: the quantity $\overline{\sigma} = \exp(0.5\omega/(1-\phi))$ and the long-term dispersion of h, $q/(1-\phi^2)$.

The former can be thought of as an equilibrium value: if σ_t exceeds that level then mean reversion in h will pull it back down. The latter is just the unconditional variance of the state h.

The comparison leads to two main conclusions. First, the *permanent* component of TE is on average larger for iShares. This follows on from the fact that the distribution of σ_F for iShares concentrates its mass on a range of values that are higher compared to Vanguard.

By inspecting the posterior distribution of $\overline{\sigma}$ we can draw the second conclusion, i.e., that, in the long term, the Vanguard fund tends to display a higher level of *transitory* TE.

Moreover, the posterior distributions of the parameters suggest that the unconditional second moment of h appears to be higher for iShares. We interpret this result as implying that, regarding the temporary component of TE,

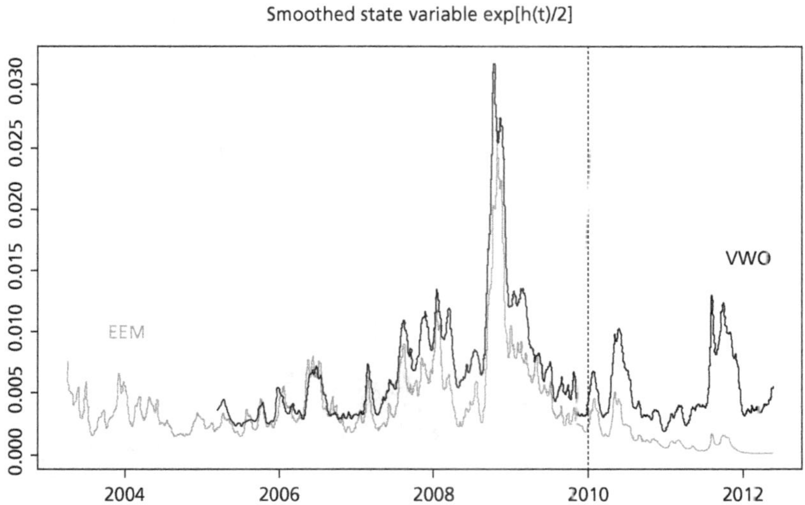

Figure 4.3. Smoothed state variable $\exp(h_t/2)$. The plot shows the median of the distribution of the MCMC draws

Source: UBS estimates

the iShares fund displays a lower long-term volatility on average but the uncertainty surrounding its future volatility appears to be greater.

Finally, we turn our attention to the posterior distributions of the state h_t. Figure 4.3 plots the median of the distributions of the MCMC draws for the quantity $\exp\left(\widehat{h}_t/2\right)$, which can be interpreted as an estimate of σ_t, the standard deviation of measurement error. Both series appear to show considerable

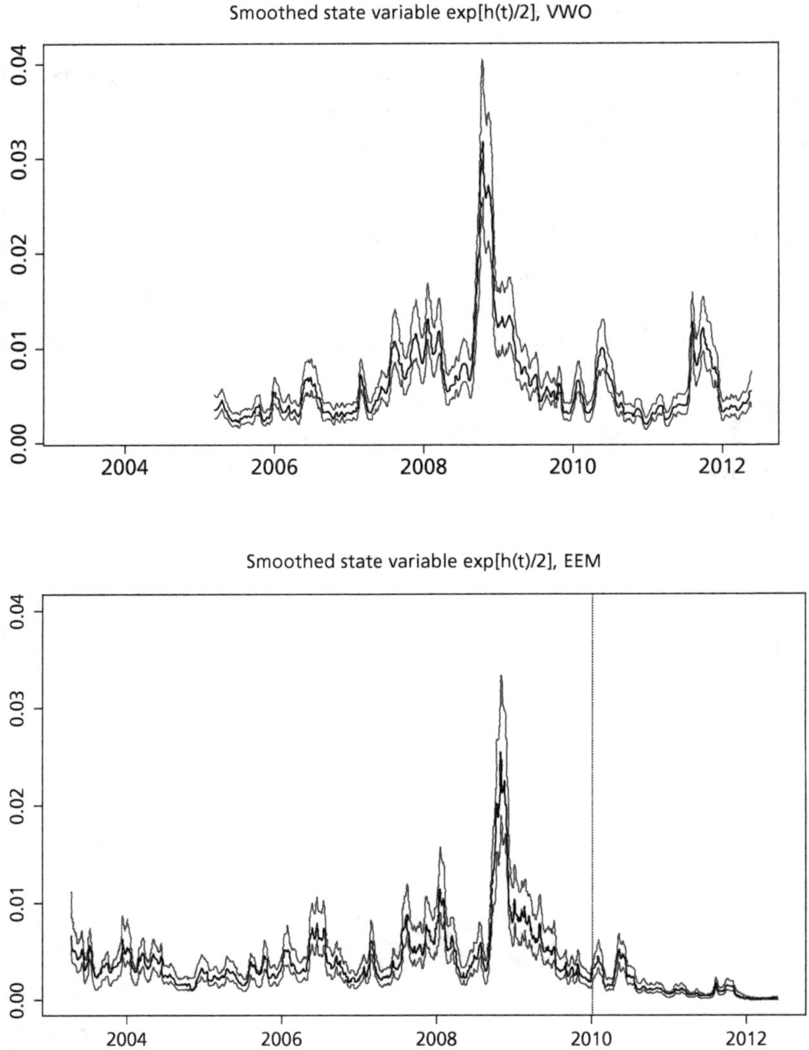

Figure 4.4. Smoothed state variable $\exp(h_t/2)$. The chart shows the median and 10% and 90% quantiles of the distribution of the MCMC draws

Source: UBS estimates.

variation over time and to be driven by market volatility: the overall level starts to rise at the beginning of the credit crisis in mid 2007 and reaches its peak in the last quarter of 2008 when Lehman Brothers collapsed. The chart also reveals a noticeable common pattern in the evolution of h_t, which suggests that the factors affecting temporary TE are largely common to the two funds. In general, the estimated value of h tends to be higher for the Vanguard fund.

According to our estimates, during the period in which the iShares fund increased the number of stocks held in its portfolio, the two series behaved differently. Since the beginning of 2010 the estimated temporary TE has trended gently downwards for iShares, while the corresponding series for Vanguard has displayed two marked peaks during the summers of 2010 and 2011. Further analysis would be needed in order to draw a definite conclusion but it is interesting to note that the difference in temporary TE levels that appeared between the two funds starting in 2010 is consistent with a reduction in replication error which may stem from the change in replication style.

Figure 4.4 depicts the same posterior medians together with 10% and 90% quantiles of the approximate distribution. The posteriors appear to be relatively tight, except around the historical peaks of volatility. The results can be interpreted as evidence of significant time variation in the TE of both funds.

4.5 Conclusions

Estimating the TE of international ETFs is problematic because of a severe errors-in-variables issue that results in a strong upward bias. The existing empirical literature (and a widely used commercial analytics platform) either fail to acknowledge the problem or, at best, avoid it by using low frequency data.

We develop an unobserved components model that treats the NAV of an ETF as a latent variable so as to overcome the errors-in-variables problem. Moreover, we assume that the measurement error on the fund's NAV is characterized by stochastic volatility. This feature of the model generates persistence in TE, a phenomenon which is clearly present in the actual data. The intuition behind TE clustering hinges on the notion that, as Vardharaj et al. (2004) argue, TE varies with market volatility. Hence the well documented persistence in stock price volatility should in principle generate persistence in TE.

Our econometric approach yields an estimate of the level of TE persistence. Even though we do not pursue this idea further, it is worth noting that this feature is potentially useful for forecasting purposes as we can exploit the mean reversion effect that characterizes the dynamics of TE, captured by the unobserved stochastic volatility factor.

We apply our proposed methodology to two ETFs tracking the MSCI emerging markets equity index. The results seem to provide realistic values of the TE for passive tracker funds. In addition, we document significant time variation in TE for both funds and a strong common pattern. It is worth noting that, for the two funds considered, the historical peaks in TE occur during periods of heightened market volatility. A possible interpretation of this result is that when prices and exchange rates are highly volatile the distortion caused by asynchronous pricing is exacerbated.

One of the advantages of our approach is that it extracts the information content from daily data efficiently and, as a consequence, it can be used to track changes in the TE of a fund over relatively short periods of time. While we do not carry out a formal test, we do observe that the estimated TE of one of the ETFs decreased steadily when the fund changed its replication strategy so as to reduce the impact of replication error.

In our model the effect of the temporary component of TE (which represents factors such as asynchronous pricing and discounts or premia) is independent of the length of the holding period.[10] In contrast, the effect of the permanent component of TE (which accounts e.g., for optimization error and rebalance costs) compounds over time. Our contribution is to derive a formula that, given the estimated parameter values, expresses TE as a function of the investment horizon.

A potential extension of our work would entail introducing stochastic volatility in the innovations (η_F in our notation) of the latent variable that represents the cumulative active return. Particularly when the benchmark is not fully replicated, permanent return deviations between fund and benchmark may be heteroscedastic, for example, because replication errors may tend to have stronger effects when market volatility is high.

[10] Conditional on a realization of the volatility process.

5 Measuring the dynamics of global business cycle connectedness

Francis X. Diebold and Kamil Yilmaz

5.1 Introduction

The empirical literature on business cycle measurement has been particularly vibrant since the early 1980s, with increased emphasis on international aspects in the 1990s, producing, for example, evidence of commonality in the macroeconomic fluctuations of industrial and developing countries. Indeed that finding is widespread, ranging from classical analyses of pairwise gross domestic product (GDP) correlations (e.g., Backus et al. (1995), Baxter (1995), inter alia) to more recent analyses of dynamic latent factor models (e.g., Kose et al. (2003), Canova et al. (2007), Kose et al. (2008), inter alia).

As the evidence of commonality in cross-country business cycle fluctuations accumulated, the literature started to focus on the effect of globalization on international business cycles. Kose et al. (2003), for example, find that the impact of the world factor on the cross-country correlation of macroeconomic aggregates (output, consumptions, and investment) increased from the 1990s onward.

Of course the evidence is not uniformly one-sided. Doyle and Faust (2005) find no evidence of increased output growth-rate correlations for the G-7 countries. Stock and Watson (2005) argue that that the comovement of macroeconomic aggregates actually *declined* during the globalization era of 1984–2002, but rather than linking their results directly to the globalization process, they conclude that their results are likely due to the diminished importance of common shocks among the G-7 countries. Eickmeier (2007) emphasizes that the impact of globalization on international propagation of macroeconomic shocks is unclear and needs to be studied further.

Here, we contribute by using a new connectedness measurement technology that is fundamentally grounded in modern network theory (Diebold and Yilmaz (2014)) to measure real output connectedness for a set of six developed countries, 1962–2011. We show that global connectedness is sizable and time-varying over the business cycle, and we study the nature of the time variation vis-à-vis the changing nature of the global business cycle. We also show that connectedness corresponding to transmissions to others from the United States and Japan is disproportionately important.

In particular, we define and track connectedness of monthly industrial production for the G-6 countries (G-7 less Canada). Our approach is different from earlier studies of international business cycles in that, rather than finding a common world factor or indicator that measures international business cycles, our approach helps one identify how output shocks in one country affect output in other countries, with a lag. Thus we can study directional connectedness, both pairwise and in the aggregate. Time variation in connectedness is potentially of great interest as the intensity of business cycle connectedness is likely to vary over time. We show that business cycle connectedness among G-6 countries is important, that connectedness is indeed time-varying, and that the United States and Japan are the major transmitters of business cycle shocks to other countries.

In addition, our analysis differs from most earlier work in terms of the data used. We use industrial production indexes at monthly frequency rather than the quarterly data from the national income accounts. There are two reasons for this choice. First, the use of monthly data allows us to capture the connectedness of business cycle shocks much faster, as seen in the latest economic crisis. Second, the use of monthly data allows us to have more observations in calculating the connectedness index for each rolling sample window.

The structure of this paper is very simple. We introduce our framework for connectedness measurement in Section 5.2, discussing both population and sample (estimation) issues. Then we apply it to global business cycles in Section 5.3, emphasizing dynamic aspects. We conclude in Section 5.4.

5.2 Measuring connectedness

Here we introduce basic aspects of connectedness measurement as relevant for our subsequent multi-country business-cycle analysis. The connectedness measurement framework was originally developed in Diebold and Yilmaz (2009, 2012, 2014).

5.2.1 THE CONNECTEDNESS TABLE

The variance decomposition indicates the amount of information each variable contributes to the other variables in the autoregression. It determines how much of the forecast error variance of each of the variables can be explained by exogenous shocks to the other variables. Our approach to connectedness is

Table 5.1 Connectedness table schematic

	x_1	x_2	$\dots$	x_N	From others
x_1	d_{11}	d_{12}	$\cdots$	d_{1N}	$\Sigma_{j=1}^{N} d_{1j}, j \neq 1$
x_2	d_{21}	d_{22}	$\cdots$	d_{2N}	$\Sigma_{j=1}^{N} d_{2j}, j \neq 2$
$\vdots$	$\vdots$	$\vdots$	$\ddots$	$\vdots$	$\vdots$
x_N	d_{N1}	d_{N2}	$\cdots$	d_{NN}	$\Sigma_{j=1}^{N} d_{Nj}, j \neq N$
To others	$\Sigma_{i=1}^{N} d_{i1}$ $i \neq 1$	$\Sigma_{i=1}^{N} d_{i2}$ $i \neq 2$	$\cdots$	$\Sigma_{i=1}^{N} d_{iN}$ $i \neq N$	$\frac{1}{N} \Sigma_{i,j=1}^{N} d_{ij}$ $i \neq j$

based on assessing shares of forecast error variation in various locations due to shocks arising *elsewhere*. This is intimately related to the familiar econometric notion of a variance decomposition: the H-step forecast error variance share d_{ij} is just the fraction of i's H-step forecast error variance due to shocks in variable j. The full set of variance decompositions produces the core of what we call the "connectedness table." All of our connectedness measures—from simple pairwise to system-wide—flow from the connectedness table.

The simple Table 5.1, the connectedness table, proves central to understanding the various connectedness measures and their relationships. Its main upper-left $N \times N$ block contains the variance decompositions. For future reference we call that upper-left block a "variance decomposition matrix," and we denote it by $D = [d_{ij}]$. The connectedness table simply augments D with a rightmost column containing row sums, a bottom row containing column sums, and a bottom-right element containing the grand average, in all cases for $i \neq j$.

To understand and interpret the information conveyed by the connectedness table, it is helpful to cut through the notational clutter via a simple example, as in the example connectedness table, Table 5.2 with $N = 4$. The twelve off-diagonal entries in the upper-left 4×4 D matrix are the twelve pieces of the four forecast-error variance decompositions, d_{ij}. From a connectedness perspective, they measure "pairwise directional connectedness." The 3,2 entry of 14, for example, means that shocks to x_2 are responsible for 14% of the H-step-ahead forecast error variance in x_3. We write $C_{3 \leftarrow 2} = 14$. In general the pairwise directional connectedness from j to i is $C_{i \leftarrow j} = d_{ij}$. Note that in general $C_{i \leftarrow j} \neq C_{j \leftarrow i}$. Hence there are $N^2 - N$ separate pairwise directional connectedness measures. They are analogous to bilateral imports and exports for each of a set of N countries. Sometimes we are interested in *net* pairwise directional connectedness, in a fashion analogous to a bilateral trade balance. For example, for x_2 and x_3 we have $C_{23} = C_{3 \leftarrow 2} - C_{2 \leftarrow 3} = 14 - 1 = 13$. In general we have $C_{ij} = C_{j \leftarrow i} - C_{i \leftarrow j}$. There are $\frac{N^2 - N}{2}$ net pairwise directional connectedness measures.

Table 5.2 Connectedness table example

	x_1	x_2	x_3	x_4	From others
x_1	96	1	2	1	4
x_2	28	67	1	3	32
x_3	14	14	70	1	29
x_4	18	11	5	65	34
To others	60	26	8	5	25

In our subsequent analysis, however, we will be interested in more aggregative connectedness measures. The eight off-diagonal row and column sums, labeled "from" and "to" in the connectedness table, are the eight *total directional connectedness* ("directional connectedness" for short) measures. The value of 32 in the second entry of the rightmost column, for example, means that x_2 receives 32% of its variation from others (x_1, x_3, and x_4). We write $C_{2\leftarrow\bullet} = 28 + 1 + 3 = 32$. In general, the total directional connectedness from others to i is

$$C_{i\leftarrow\bullet} = \sum_{\substack{j=1 \\ j\neq i}}^{N} d_{ij},$$

and the total directional connectedness from j to others is

$$C_{\bullet\leftarrow j} = \sum_{\substack{i=1 \\ i\neq j}}^{N} d_{ij}.$$

Hence there are $2N$ total directional connectedness measures, N "to others" and N "from others," analogous to "total exports" and "total imports" for each of a set of N countries. Just as with pairwise directional connectedness, sometimes we are interested in net total directional effects. For x_2, for example, we have $C_2 = C_{\bullet\leftarrow 2} - C_{2\leftarrow\bullet} = 26 - 32 = -6$. In general, net total directional connectedness is $C_i = C_{\bullet\leftarrow i} - C_{i\leftarrow\bullet}$. There are N net total directional connectedness measures, analogous to the total trade balances of each of a set of N countries.

Finally, we consider the most aggregative connectedness measure, the grand total of the off-diagonal entries in D (equivalently, the sum of the "from" column or "to" row), which measures "total (system-wide) connectedness". We typically express this total cross-variable variance contribution, given in the lower right cell of the connectedness table, as a percentage of total variation. Hence total connectedness in our example is $C = \frac{99}{400} \times 100 = \frac{99}{4} = 24.8$. Note that total variation is just one hundred times N, the number of variables in the system, because each of the N rows sums to one hundred. Conversion to percent eliminates the one hundred so that ultimately total connectedness is simply average total directional connectedness (whether "from" or "to"). In general we have

$$C = \frac{1}{N} \sum_{\substack{i,j=1 \\ i \neq j}}^{N} d_{ij}.$$

There is just a single total connectedness measure, as total connectedness distills a system into a single number analogous to total world exports or total world imports (the two are of course identical).

The connectedness table makes clear how one can begin with the most disaggregated (e.g. microeconomic, firm-level pairwise directional) connectedness measures and aggregate them in various ways to obtain macroeconomic economy-wide total directional and total connectedness. Different agents may be disproportionately interested in one or another of the measures. For example, firm i may be maximally interested in total directional connectedness from others to i, $C_{i \leftarrow \bullet}$, whereas regulators might be more concerned with total directional connectedness from i to others, $C_{\bullet \leftarrow i}$, or in total connectedness C.

5.2.2 IDENTIFYING SHOCKS

As already emphasized, our approach is based on variance decompositions. An H-step forecast error variance decomposition, d_{ij}, answers an interesting and important question: what fraction of H-step forecast error variance of variable i is due to shocks in *another* variable j? This must, of course, be done within the context of a model. If shocks in reduced-form vector autoregressions were orthogonal, variance decomposition calculations would be trivial. That is, variance decompositions are easily calculated in orthogonal VARs, because orthogonality ensures that the variance of a weighted sum is simply an appropriately-weighted sum of variances.

The much more realistic case involves correlated VAR shocks. Consider therefore a data-generating process with correlated shocks, with moving-average representation $x_t = \Theta(L)\varepsilon_t$, $E(\varepsilon_t \varepsilon_t') = \Sigma$. The variance decomposition calculations are more involved, because we first need to isolate the independent shocks that underlie the observed system. One way or another, we must transform the shocks to orthogonality to calculate variance decompositions. The orthogonalization can be handled in several ways, and we now sketch two.

Cholesky factorized variance decompositions

This time-honored orthogonalization method traces at least to Sims (1980). The correlated-shocks model above is mathematically identical to the orthogonal-shocks model, $x_t = A(L)u_t$, $E(u_t u_t') = I$, $A(L) = \Theta(L)Q$, $u_t = Q^{-1}\varepsilon_t$, where the lower triangular matrix Q is the Cholesky factor of Σ; that is, $QQ' = \Sigma$. Hence a simple Cholesky factor transformation orthogonalizes the system.

Variance decompositions based on Cholesky factorization may be sensitive to ordering, as has been well-appreciated at least since Sims (1980). Interestingly, we often find that total connectedness is robust to ordering (that is, the range of total connectedness estimates across orderings is often quite small), whereas directional connectedness often appears more sensitive to ordering. Hence so-called "generalized variance decompositions," which we now introduce, may be more useful when studying directional connectedness.

Generalized variance decompositions

The generalized variance decomposition (GVD) framework of Koop et al. (1996) and Pesaran and Shin (1998) produces variance decompositions invariant to ordering. The GVD approach does not require orthogonalized shocks; rather, it allows and accounts for correlated shocks using the historically observed error distribution, under a normality assumption.

The GVD matrix $\Delta = [\delta_{ij}]$ has entries

$$\delta_{ij} = \frac{\sigma_{jj}^{-1} \sum_{h=0}^{H-1} (e_i' A_h \Sigma e_j)^2}{\sum_{h=0}^{H-1} (e_i' A_h \Sigma A_h' e_i)},$$

where σ_{jj} is the standard deviation of ε_j, e_j is a selection vector with j^{th} element unity and zeros elsewhere, A_h is the coefficient matrix multiplying the h-lagged shock vector in the infinite moving-average representation of the non-orthogonalized VAR, and Σ is the covariance matrix of the shock vector in the non-orthogonalized VAR.

Because shocks are not necessarily orthogonal in the GVD environment, sums of forecast error variance contributions (that is, row sums in GVD matrices) are not necessarily unity ($\sum_{j=1}^{N} \delta_{ij} \neq 1$). Hence we base our generalized connectedness indexes not on Δ, but rather on $\tilde{\Delta} = [\tilde{\delta}_{ij}]$, where $\tilde{\delta}_{ij} = \frac{\delta_{ij}}{\sum_{j=1}^{N} \delta_{ij}}$. Note that, by construction, $\sum_{j=1}^{N} \tilde{\delta}_{ij} = 1$ and $\sum_{i,j=1}^{N} \tilde{\delta}_{ij} = N$. Armed with $\tilde{\Delta}$, we can immediately calculate generalized connectedness measures $\tilde{C}$, $\tilde{C}_{\bullet \leftarrow j}$, $\tilde{C}_{i \leftarrow \bullet}$, $\tilde{C}_i$, $\tilde{C}_{i \leftarrow j}$, $\tilde{C}_{j \leftarrow i}$, and $\tilde{C}_{ij}$.

In essence the GVD does not impose orthogonality of shocks. Therefore, in the GVD framework all variables in a system are subject to shocks simultaneously. This in return amounts to obtaining impulse responses and variance decompositions for each variable treating each variable as the leading variable in the VAR. GVDs moreover require normality, and hence may be more useful for assessing connectedness of log volatilities, which are well-approximated as Gaussian, than for returns, which are not.

5.2.3 SAMPLE CONNECTEDNESS

Clearly C depends on x, H, and $A(L)$, so we have written $C(x, H, A(L))$. In reality A is unknown and must be approximated (e.g., using a finite-ordered vector autoregression). Recognizing the centrality of the approximating model adopted, we write $C(x, H, A(L), M(\theta))$, where θ is a finite-dimensional parameter.

However, everything we have written thus far is in *population*, whereas in reality we must use an approximating model estimated using data $1:T$, so we write $\widehat{C}_t(x, H, A_t(L), \widehat{M}_{1:T}(\hat{\theta}_t))$. To economize on notation we henceforth drop $A(L)$, because it is determined by nature rather than a choice made by the econometrician, relying on the reader to remember its relevance and writing $\widehat{C}_t(x, H, \widehat{M}_{1:T}(\hat{\theta}_t))$.

Hence we now consider estimation of connectedness. In addition, we want to allow for time-variation of connectedness, effectively allowing the connectedness table and all of its elements to vary over time, so we write $C_t(x, H, A_t(L), M(\theta_t))$.

Estimation

Choice of x—the object of interest to be studied—has important implications for the appropriate approximating model; for example, x may (or may not) be strongly serially correlated, conditionally heteroskedastic, or highly disaggregated. Here we study real output growth across countries, for which serial correlation, in particular, is surely important.

Connectedness measurements are defined only with respect to a reference universe (specific set of x's). In general they will not—and should not—be robust to choice of reference universe. Hence, given a decision as to the type of x to be examined, a second important issue is precisely which (and hence how many) x's to use. For example, in cross-country analyses we may want to use sufficiently many countries to ensure that we have good global coverage. Whether this requires a small or large number of x's depends on the distribution of activity across countries.

The predictive horizon, H, is also important, and one must take a stand on it as well. The horizon is important particularly because it is related to issues of dynamic connectedness (in the fashion of contagion) as opposed to purely contemporaneous connectedness. To take a simple pairwise example, shocks to j may impact the forecast error variance of i only with a lag, so that $C_{i\leftarrow j}$ may be tiny for small H but nevertheless large for larger H. Intuitively, as the horizon lengthens there may be more chance for things to become connected. Such dynamic phenomena, and the rich patterns that are possible, are closely related to aspects of multi-step Granger causality, as treated, for example, in Dufour and Renault (1998), Dufour and Taamouti (2010), and the references therein.

In a sense, varying H lets us break down connectedness into "long-run," "short-run," etc. More precisely, as H lengthens we obtain a corresponding sequence of conditional prediction error variance decompositions for which the conditioning information is becoming progressively less valuable. In the limit as $H \to \infty$, we obtain an *un*conditional variance decomposition.

The remaining issue—choice of empirical approximating model $\widehat{M}_{1:T}(\hat{\theta}_t)$—is the obvious elephant in the room. For real activity, conditional mean dynamics in $M(\theta)$ will surely be important and will need to be modeled. We will use the obvious workhorse model, a vector autoregression, $VAR(p)$. We emphasize, however, that almost any model could be used, including much more deeply structural models, so long as underlying orthogonal shocks can be identified.

Time-variation

Connectedness is just a transformation of system coefficients. Hence if the coefficients are time-varying, so too will be connectedness. Tracking ("nowcasting") real-time connectedness movement is of central interest. One can capture parameter variation by using a rolling estimation window; we write $\hat{C}(x, H, \widehat{M}(\theta; w))$, where w denotes window width. We will take this approach, estimating the model repeatedly, at each time using only the most recent w observations.

There are of course both benefits and costs of a rolling-window approach. Benefits include both tremendous simplicity and coherence with a wide variety of possible data-generating processes (DGPs) involving time-varying parameters. Costs include that, given a particular DGP, rolling is generally suboptimal, and it requires selection of w. Too large a w produces "oversmoothing," and too small a w produces "undersmoothing," in a manner precisely analogous to bandwidth choice in density estimation.

5.3 Global business cycle connectedness

To study business cycle connectedness we use monthly seasonally adjusted industrial production (IP) for the G-7 countries less Canada (G-6), from January 1958 to December 2011, as shown in Figure 5.1. We exclude Canada because its IP growth is very highly correlated with that of the United States.

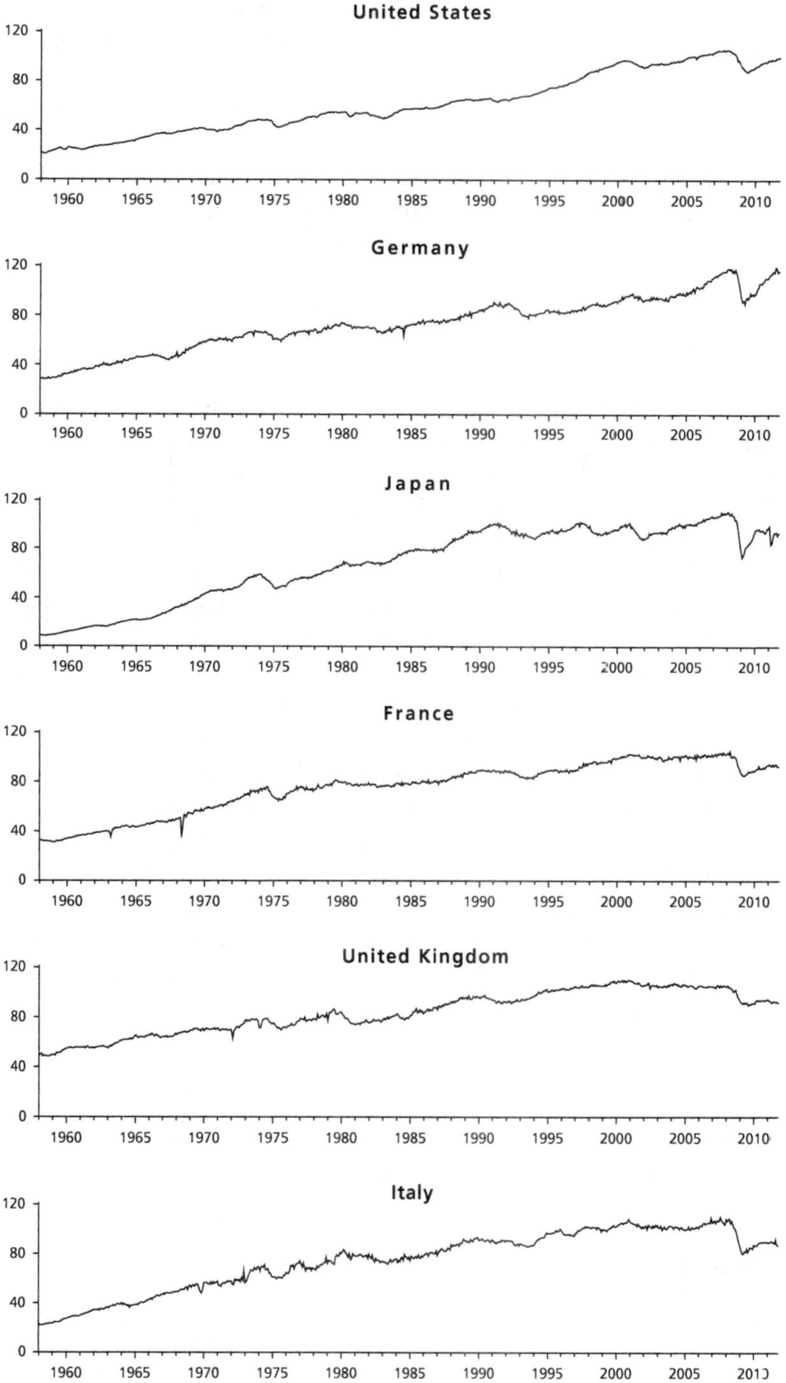

Figure 5.1. G-6 industrial production, January 1958 to December 2011

Indeed year-on-year industrial production growth rates for the U.S. and Canada have a correlation of 0.87, which is much higher than the correlations for other G-6 country pairs.

We begin in Section 5.3.1 by assessing the cointegration status of the IP data, which has implications for the structure of our subsequent analysis in Sections 5.3.2, 5.3.3, and 5.3.4. In Section 5.3.2 we examine static connectedness, in Section 5.3.3 we examine dynamic total connectedness, and in Section 5.3.4 we examine dynamic directional connectedness.

5.3.1 COINTEGRATING RELATIONSHIPS

Before proceeding to examine business cycle connectedness, we must ascertain the likely integration and cointegration status of our G-6 (log) IP data. We first tested for unit roots in each log IP series, using augmented Dickey-Fuller (ADF) tests with augmentation lag length selected using Akaiki information criterion (AIC), and allowing for linear trend under the alternative. To conserve space we do not show the unit root test results. There was no evidence against the unit root in any log IP series, and overwhelming evidence against the unit root in every differenced log IP series (allowing for nonzero mean under the alternative).

We now assess cointegration status. We show the results of Johansen's maximum eigenvalue and trace tests in Table 5.3. The hypothesis of 0 cointegrating relationships is clearly rejected, the hypothesis of at most 1 conintegrating relationship is not rejected at the 5% level (but it is rejected by the trace test at the 10% level), and there is no evidence against the hypothesis of at most R conintegrating relationships for $R > 1$. All told, then, it appears that there is only one cointegrating relationship among the G-6 IP series. Hence we proceed to adopt a vector error-correction specification for our approximating model, although we note that it is not far from a simple VAR in first differences (i.e., 0 cointegrating relationships).

Table 5.3 Tests of the number of cointegrating relationships

CI rank	Max eval	(P-value)	Trace	(P-value)
0	48.9	(0.004)	115.6	(0.0011)
At most 1	29.1	(0.168)	66.7	(0.0865)
At most 2	17.4	(0.550)	37.6	(0.3191)
At most 3	14.7	(0.310)	20.3	(0.4054)
At most 4	5.4	(0.687)	5.6	(0.7473)
At most 5	0.12	(0.723)	0.12	(0.7230)

Note: We assume linear trends in industrial production series and intercepts in any cointegrating relationships.

5.3.2 STATIC CONNECTEDNESS

In the empirical analysis of business cycle connectedness we first estimate the vector error correction (VEC) model for the full sample and report the connectedness index and the directional connectedness in Table 5.4 along with the underlying generalized variance decomposition. The connectedness index for the full sample period is 29.1%, indicating that less than one-third of the total variance of the forecast errors for the G-6 countries is explained by the connectedness of shocks across countries, whereas the remaining 70.9% is explained by idiosyncratic shocks.

It is important at this stage to note that the connectedness index for the whole sample is sensitive to the inclusion of new observations in the sample. The connectedness index for the period from January 1958 to December 2008 is only 27%. When the sample is extended to May 2009, the connectedness index for the full sample jumps to 69%. Finally, the inclusion of observations from June 2009 to December 2011 lowers the index to 29.1%.

In terms of the directional connectedness to others (measured by $\tilde{C}^H_{\bullet\leftarrow i}$) throughout the full sample, Japan is the country that contributed the most to other countries' forecast error variance (58.1 points, which is close to 10% of the total forecast error variance to be explained), followed by France (35.4 points). According to the full sample directional connectedness measures, the U.S., Germany, and the U.K. contributed at similar rates (29.2, 23.9, and 17.1 points, respectively), followed by Italy (10.7 points).

Table 5.4 Static connectedness, G-6 industrial production, January 1958 to December 2008

	U.S.	Germany	Japan	France	U.K.	Italy	From others
U.S.	89.2	0.8	3.5	3.3	2.2	1.1	10.8
Germany	5.6	55.7	23.9	8.4	6.0	0.5	44.3
Japan	8.1	5.6	77.4	6.0	1.9	1.0	22.6
France	3.5	10.1	11.0	64.2	4.2	7.0	35.8
U.K.	7.7	3.3	4.2	3.8	80.0	1.0	20.0
Italy	5.9	2.0	14.3	13.2	4.0	60.6	39.4
To others	30.9	21.7	56.8	34.7	18.3	10.5	
Net = To − from	20.0	−22.5	34.1	−1.1	−1.7	−28.8	Index = 28.8%

Note: Each cell in the upper-left 6 × 6 matrix reports the relative (percent) contribution of the "column" country to the variance of the forecast error for the "row" country. The "directional from others" column reports the total forecast error variance shares of the row countries attributable to shocks from other countries. The "directional to others" row reports the total contributions of each column country to all other countries' forecast-error variance. Each cell in the "net directional connectedness" (to − from) row reports the difference between the corresponding cells in the "directional to others" row and the ones in the "directional from others" column. The total connectedness index in the lower right is the average of the elements of the "directional from others" column (or equivalently, the "directional to others" row), multiplied by 100%.

In terms of the directional connectedness received from others (measured by $\tilde{C}^H_{i\leftarrow\bullet}$), the U.S. appears to be the country that received the lowest percentage of shocks from other countries (10.7 points, equivalent to just 1.8% of the total forecast error variance to be explained), followed by the U.K. (22.4 points) and Japan (22.6 points). Germany received the highest percentage (43.5 points) of shocks from other countries, followed by Italy (39.4 points) and France (35.7 points).

Finally, we calculate the difference between the column-wise sum (the "contribution from others") and the row-wise sum ("contribution to others") to obtain the "net directional connectedness" given by $\tilde{C}^H_i$. Japan (35.4 points) and the U.S. (18.4 points) are net transmitters of industrial production shocks to other countries, while the U.K. (−5.3 points) and France (−0.3 points) received very low percentage of business cycle shocks in net terms. Italy (−28.7 points) and Germany (−19.6 points), on the other hand, are definitely the leading net recipients of business cycle shocks over the full sample.

5.3.3 DYNAMIC TOTAL CONNECTEDNESS

The connectedness table for the full sample provides important clues as to how the connectedness index is calculated and interpreted. However, we want to focus more on the dynamics of business cycle connectedness over time. The fact that the inclusion of new observations in the sample leads to significant jumps in the connectedness index definitely highlights the need to study the dynamics of connectedness over time.

As the VEC is the correct model for the full sample, the dynamic analysis of connectedness also relies on the variance decomposition from the VEC model estimated over rolling five-year windows. Here is how the connectedness plot is obtained: we estimate the VEC model for the first five-year sub-sample window (April 1958–April 1963) and obtain the value for the generalized variance decomposition-based connectedness index (from now on, the connectedness index). Moving the sub-sample window one month ahead, we estimate the VEC model again and calculate the connectedness index for the new sub-sample, and so on. Graphing the connectedness index values for all sub-sample windows leads to the connectedness plot.

So far we have discussed the connectedness plot based on the underlying VEC(1) model, estimated over five-year rolling windows with a twelve-month forecast horizon. Next we want to discuss the appropriateness of our assumptions and the robustness of our results to these assumptions. Let us start with the underlying VEC(1) model. Earlier we reported unit root and cointegration test results for the full sample from January 1958 to December 2011. While the test results indicated that the correct underlying model is the VEC(1) for the full sample, this does not necessarily imply that VEC(1) is the correct model for each five-year rolling window. For that reason, we repeated the unit root and cointegration tests for all five-year rolling windows considered. The ADF test fails to reject the presence of unit roots in the log IP series for an overwhelming majority of the windows considered for all countries. In the case of first-differenced log IP series, the ADF test rejects the presence of unit roots for almost all countries. The only exception is Japan. In the case of Japan, for a non-negligible number of rolling windows the ADF test fails to reject the presence of unit roots in the first-differenced log IP series. While this is a cause for concern, we do not test for unit roots in further differenced IP series. Instead, we proceed with the Johansen cointegration test over the rolling windows. In the case of the U.K., the ADF test rejects the unit root in the first-differenced log IP series so strongly for all windows that the p-value is very close to zero.

Similarly, both the trace and maximum test statistics reject the null of no cointegration relationship among the six log IP series at the 5% level for an overwhelming majority of rolling windows considered. This means that the test prefers a VEC model to a VAR in first differences. In contrast, the trace statistic, in particular, fails to reject the null of at most one cointegration equation linking all six log IP series. Therefore, the Johansen cointegration tests indicate that there is either one or two cointegration equations among the six IP series. Based on these results, we expand the connectedness index analysis to rolling windows based on the VEC(1) model. Later on, we will show the differences in the connectedness indexes for different models.

The dynamic connectedness index based on the VEC(1) model is plotted in Figure 5.2. We also calculate an alternative connectedness index based on the Cholesky variance decomposition. Even though we do not plot it here, we can report that the two indexes move in tandem, with the difference between the two indexes seldom exceeding ten percentage points. Therefore, it would be sufficient to focus on the generalized VD-based connectedness index for the rest of the paper.

Turning to Figure 5.2, the first thing one observes about the connectedness plot is the absence of a long-run trend. The connectedness plot clearly shows that while there are periods during which shocks to industrial production are substantially transmitted to others, there are yet other periods during which the connectedness of business cycle shocks was much less important. Actually, during or after all U.S. recessions (indicated by the shaded bars in Figure 5.2), the connectedness index recorded significant upward movements. The only exception is the 1969–70 recession, during which the index moved down. In addition, the index went up in late 1993, and after a brief correction in late 1994, it went up again in 1995. While there was no U.S. recession during this period, France, Germany, Italy, and Japan experienced recessions ending in late 1993 or early 1994 (see the Economic Cycles Research Institute's website <http://www.businesscycle.com/resources/cycles/>). As a result, the upward

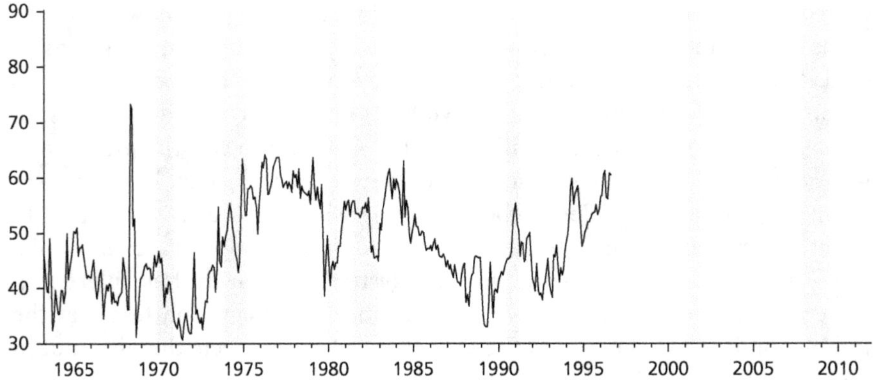

Figure 5.2. Dynamic total connectedness, G-6 industrial production, January 1958 to December 2011

Note: We use a 5-year rolling window. The shaded areas indicate U.S. recessions. See Section 5.3.3 for details.

movement in the connectedness index is most likely due to the connectedness originating from these countries.

Second, while the connectedness index fluctuates over time, it is possible to differentiate between several trends. First, during the 1973–75 recession the connectedness index increases by almost 20–25 percentage points and fluctuates around 50% after the 1981–82 recession. Starting in 1984, the connectedness index declines all the way to 33%. This result is consistent with the findings of McConnell and Perez-Quiros (2000) and Blanchard and Simon (2001) that the volatility of the U.S. GDP declined after 1984 (the great moderation). As the volatility of GDP declines, the connectedness index declines to pre-1973 levels.

Third, after the great moderation of the late 1980s, the behavior of the connectedness index reflects the influence of globalization. From 1989 onward, the band within which the connectedness index fluctuates starts to move upward with the current wave of globalization that started in earnest in the early 1990s. As the sample windows are rolled to include 1996, the index reaches 60%, but declines to 40% as the data for the late 1990s and 2000 are included. The index starts to increase again toward the end of the mild recession of 2000–01, reaching 60% by the end of 2002. However, as the other G-6 countries followed the quickly recovering U.S. economy to a major expansion, the connectedness index reached 65% in 2004Q2. The index then declines to 60% again as the window is rolled to include 2004H2, and then gradually moves down, reaching its bottom around 40% from 2006Q4 until 2008Q1.

During the era of globalization, from the late 1980s to 2007, the connectedness index followed three distinct cycles. Each cycle lasted longer and had a larger bandwidth than the previous one. During the first cycle, which lasted from 1989 to the end of 1992, the index fluctuated between 33% and 53%, while in the second cycle, which lasted from 1993 to 1999, the index fluctuated between 37% and 60%. Finally, during the third cycle from 2001 to 2007, the index fluctuated between 44% and 65%.

This result is consistent with Kose et al.'s (2003) finding that with the globalization process, business cycles have become more synchronized. It basically indicates that the comovement of industrial production fluctuations has tended to be more significant since the late 1980s. In other words, when there is a shock to industrial production in one or more countries in the G-6 group, its tendency to be transmitted across other countries increases as one move from 1989 toward 2007. This result can also be interpreted as consistent with Doyle and Faust's (2005) conclusion that the correlation coefficients among the industrial production series have not increased much since the late 1980s. The output fluctuations tend to move together during periods of high connectedness indexes, compared to the periods with low connectedness indexes. When one analyzes the period since the late 1980s as a whole, one may not obtain high correlation coefficients. Actually, for the period from 1989 to 2007 the connectedness index is only 36%.

Next, we focus on the behavior of the connectedness index since June 2008 (see Figure 5.3). We want to focus on its most recent behavior, not only because it provides us with more clues about the business cycle connectedness since the beginning of the sub-prime crisis in the U.S., but also because, in 2008 and 2009, the index recorded the biggest jump in its history. The index increased sharply from 41% in May 2008 to 53% in July, to 68% in September, and then to 80% in December. With the inclusion of January 2009 through October 2009 in

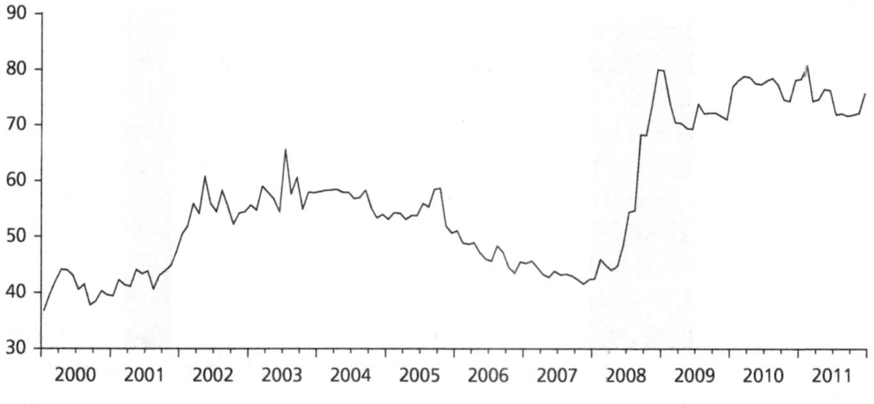

Figure 5.3. Dynamic total connectedness, G-6 industrial production, January 2000 to December 2011

Note: We use a 5-year rolling window. The shaded areas indicate U.S. recessions. See Section 5.3.3 for details.

the analysis, the index declined slightly to 71%. As the economic recovery went underway in the G-6 countries in 2009H2, the index moved slightly upward again, reaching 75% by December 2009.

The behavior of the index during the Great Recession of 2007–09 is in stark contrast to its behavior in previous recessions. It increased thirty seven points from April to December 2008. The jump in the index during the Great Recession is an indication of how the G-6 countries were pulling each other down. To give an example, during the recession following the first oil price hikes, in a matter of three and a half years from 1972 to 1976, the connectedness index recorded a relatively smaller increase, from a low of thirty two in August 1972 to a high of sixty four in April 1974.

So far we have only discussed the total connectedness index. However, as we argued in the introduction, the analysis of directional connectedness provides us with quite interesting results to discuss in some detail. However, before going ahead with the analysis of directional connectedness plots, we want to make sure that the results we obtained from the total connectedness analysis are not due to some special characteristics of the VAR framework we use. For that reason, we now report the robustness of the total connectedness index with respect to the model choice, the window width, the forecasting horizon, and the ordering of variables.

Let us start the robustness analysis with the choice of the underlying model. Along with the single cointegration equation (VEC1) model, we calculated the total connectedness index under a VAR in first differences (DVAR), a VEC2, and a VEC5 model. We present the dynamic connectedness plots in Figure 5.4. The dynamic connectedness plot obtained from the underlying DVAR model differs substantially from the one obtained from the VEC1 model. Given the fact that the null hypothesis of no cointegration equation was very strongly rejected by Johansen's cointegration test using both trace and maximum eigenvalue statistics, it is not very surprising that the dynamic connectedness index obtained from the DVAR model is quite different from the one obtained with the VEC1 model.

Cointegration test results, in general, preferred the VEC1 model to the VEC2 and VEC5 models, but there were many instances, especially in the case of maximum eigenvalue statistics, where the null of at most one cointegration equation was rejected in favor of two or more cointegration equations. Given that the Johansen test results did not indicate an overwhelming preference in favor of the VEC1 model, we decided to compare the behavior of the total connectedness index from the underlying VEC1 model with the ones obtained from the VEC2 and VEC5 models. As can be seen, there is a level difference between the connectedness index obtained from the VEC1 model and the ones obtained from the VEC2 and VEC5 models. As the level difference is not changing substantially over time, there is little difference between the time

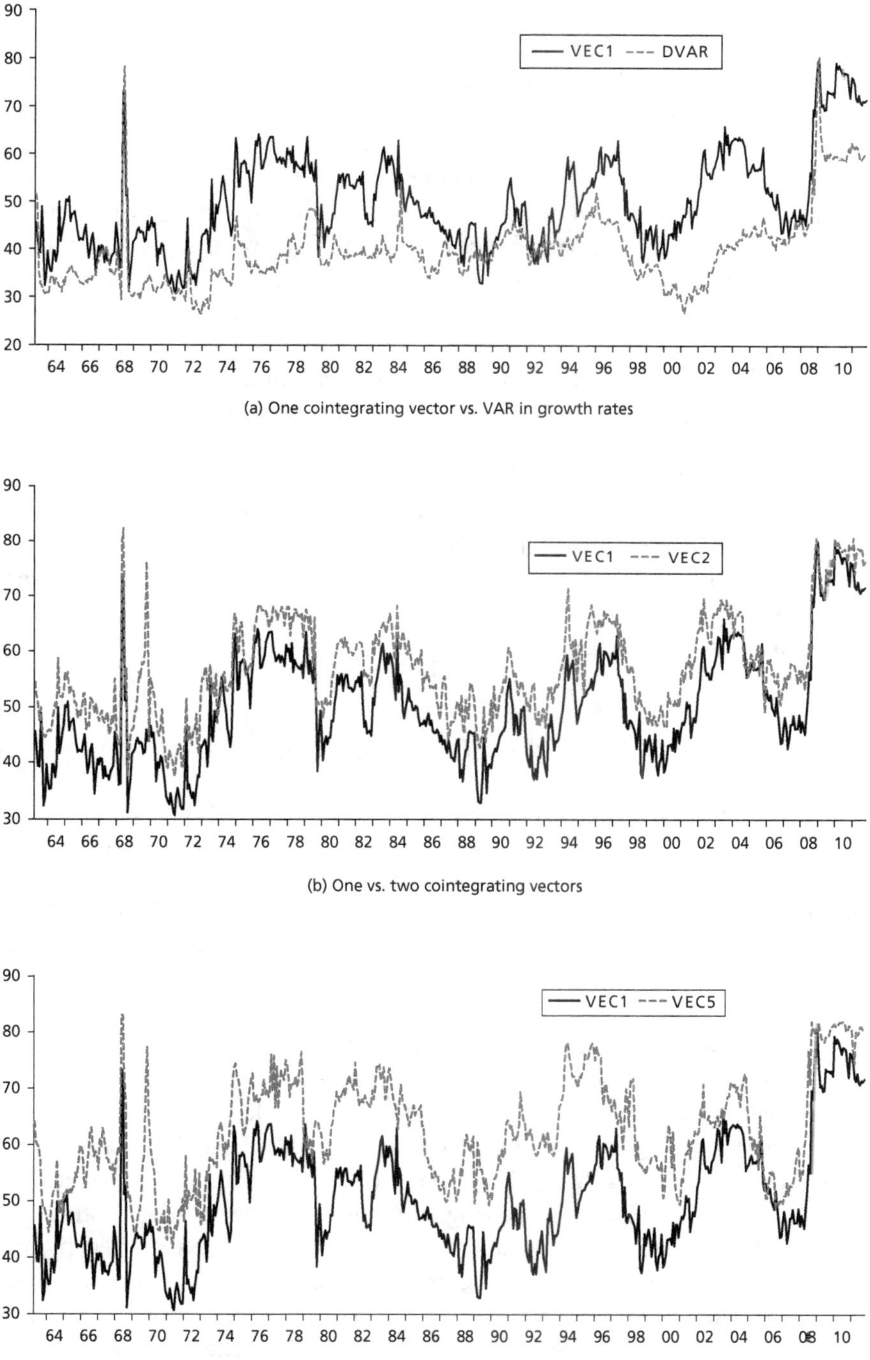

Figure 5.4. Robustness of estimated dynamic total connectedness to VECM specification

Note: In each sub-panel, a solid line denotes our VECM with one cointegrating vector, and a dashed line denotes an alternative model.

behavior of the VEC1-based connectedness index and the ones obtained from the VEC2 and VEC5 models. Based on these results, we decided to use the VEC1 as the main underlying model.

Next, we present the robustness checks with respect to the window width and the forecast horizon in Figure 5.5. In this robustness exercise we consider four, six, and seven years as the alternatives to our benchmark window width of five years. In the case of the forecast horizon, we consider six, and eighteen months in addition to our benchmark forecast horizon of twelve months. In total we plot the dynamic connectedness index in twelve sub-graphs. In addition to the connectedness index, which is based on generalized variance decomposition, we plot the tenth and ninetieth percentile values of the Cholesky-based connectedness index out of one hundred random orderings.

Irrespective of the forecast horizon and window width considered, the generalized and Cholesky variance decomposition-based connectedness indexes follow very similar patterns. This comparison assures us that the use of the generalized variance decomposition-based connectedness index leads to quite sensible results. Figure 5.5 also assures us that the result we obtained for the benchmark values of the window width and the forecast horizon carries over when we use other values for these two important parameters of our connectedness index methodology.

5.3.4 DYNAMIC DIRECTIONAL CONNECTEDNESS

Following the analysis of the total connectedness index, we can now focus on the directional connectedness of business cycles across countries. Directional connectedness indexes are critical in understanding the respective roles of each of the G-6 countries in spreading shocks to their local industrial outputs to other countries. In Figure 5.6 we present all three indexes of connectedness: "connectedness to others," "connectedness from others," and "net connectedness to others," which we will discuss in some detail.

Throughout the 1970s, Japan was the most important source country of net connectedness (Figure 5.6), followed by France and Germany. During the second half of the 1970s, the gross connectedness from Japan to others reached as high as 180%, whereas the connectedness received by Japan from others was only around 40–50%, leading the net connectedness from Japan to reach as high as 150%. While Germany had high directional connectedness to others in the late 1960s, early 1970s, and late 1970s, its net connectedness was negative in the mid-1970s, immediately after the first oil price shock. France, on the other hand, had significant directional connectedness to others after the first oil price hikes in 1973–74 as well as the second half of the 1970s. The U.S., on the other hand, had negative directional connectedness to others, and, therefore, was a net recipient of business cycle shocks over most of the 1970s.

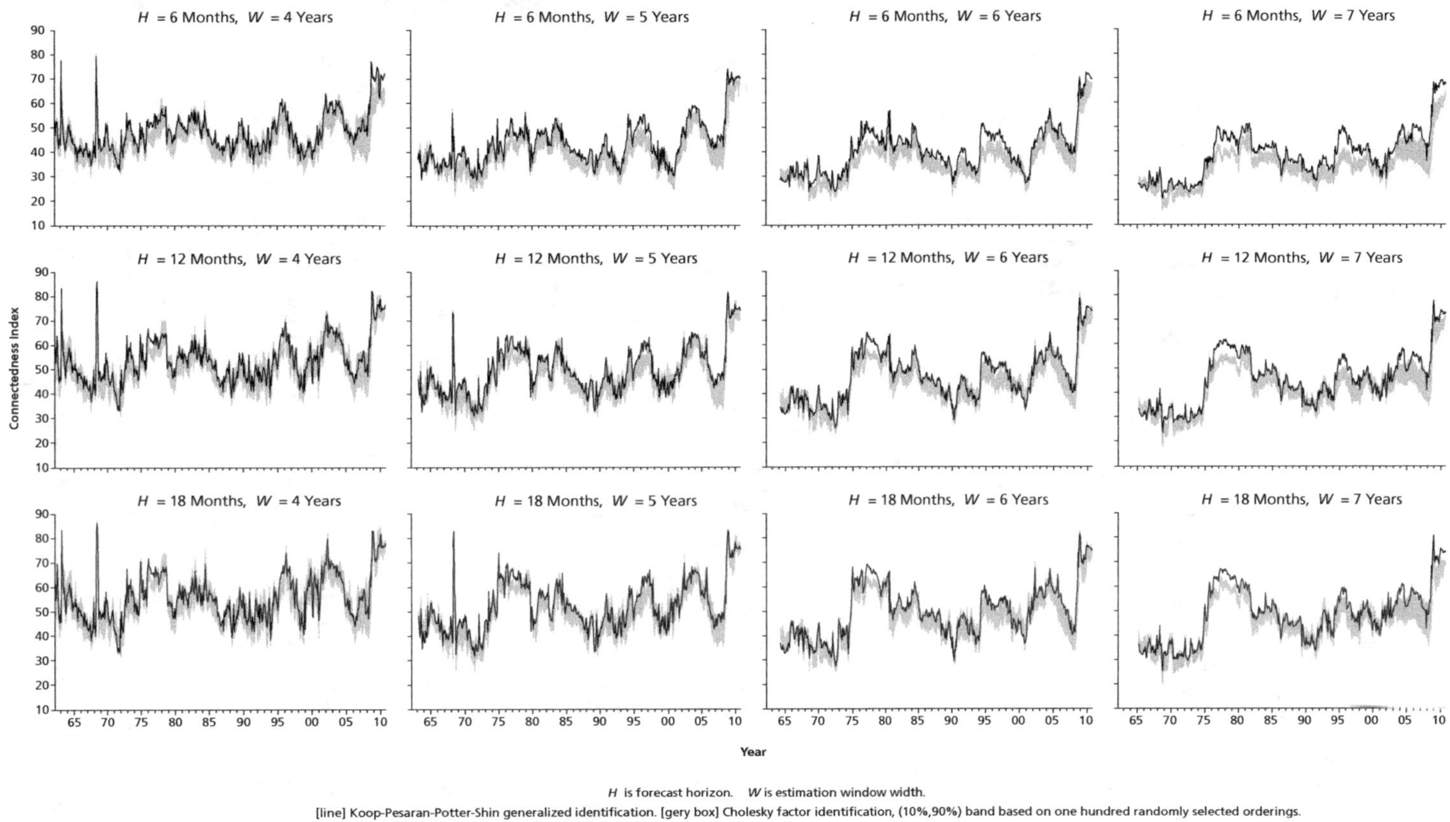

Figure 5.5. Robustness of estimated dynamic total connectedness to window width, forecast horizon, and VECM identification

Note: We explore window widths of four, five, six, and seven years (across columns), and horizons of six, twelve, and eighteen months (across rows). Each sub-figure contains a solid line based on generalized identification, and a 90% region based on one hundred randomly selected Cholesky factor identifications. See Section 5.3.3 for details.

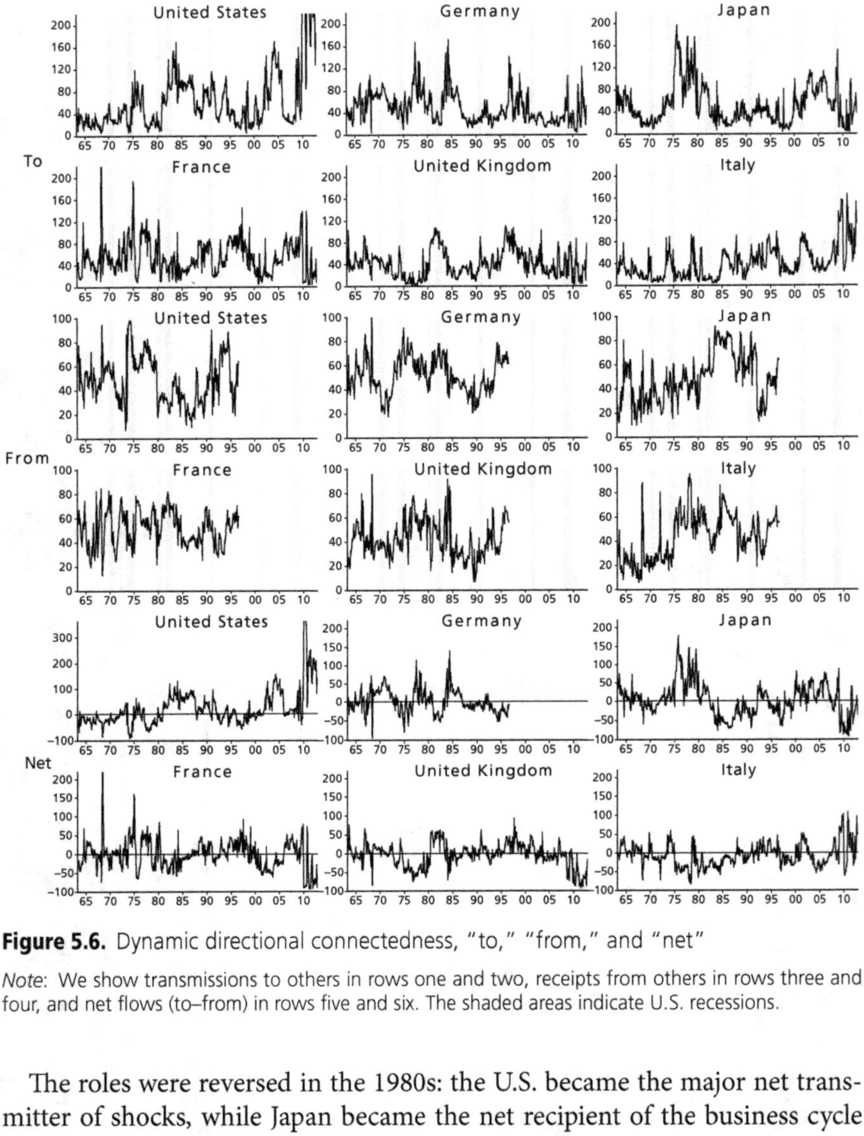

Figure 5.6. Dynamic directional connectedness, "to," "from," and "net"

Note: We show transmissions to others in rows one and two, receipts from others in rows three and four, and net flows (to–from) in rows five and six. The shaded areas indicate U.S. recessions.

The roles were reversed in the 1980s: the U.S. became the major net transmitter of shocks, while Japan became the net recipient of the business cycle shocks. Following the 1981–82 U.S. recession, gross connectedness transmitted by the U.S. to others jumped above 160%, and net connectedness from the U.S. fluctuated between 50 and 120%. Japan's net connectedness, on the other hand, declined to as low as –80% after the 1982 recession and stayed at low levels until the end of 1987. Germany and the U.K. also had positive net directional connectedness to others after the 1981–82 recession, but their roles were rather secondary compared to that of the U.S.

Throughout the 1990s, Japan's net directional connectedness was positive, but it was rather low. This fact is consistent with the decade-long recession

Japan suffered while the other G-6 countries continued to attain higher growth rates. Neither the U.S. nor Germany was one of the countries that had net directional connectedness in the 1990s. Rather, France and the U.K. had sizable net directional connectedness in the 1990s, even though their net connectedness was not as significant and as persistent as the ones the U.S., Japan, and Germany attained in the 1970s and 1980s. The role these countries played during the 1990s is closely related to the aftermath of the ERM crisis of 1992 and the ensuing slowdown in these economies.

Moving forward in time, the U.S. and Japan returned to their locomotive roles in the 2000s. In particular, with a high net connectedness to others, the U.S. was a net transmitter of business cycle shocks after the 2001 recession. In response to the slowdown in the economy in early 2001, the Federal Reserve lowered the fed funds rate from 5.5% in January 2001 all the way down to 2% in November 2001. This aggressive policy stance was effective in stimulating domestic demand. As a result, industrial production grew at a monthly rate of between 0.5 and 1.0% in 2002H1. The 2001 recession lasted for only eight months, from March to November 2001. With this rapid turnaround, the U.S. started to generate substantial connectedness in 2002H1, with the net connectedness from the U.S. reaching 100%. After a brief lull in 2003, the net connectedness from the U.S. increased again to surpass 130%. Being the driver of worldwide demand, the U.S. had an impact on other countries until the end of 2006, as the net connectedness of the U.S. declined to almost zero. Japan also had a positive net connectedness in the first seven years of the 2000s: its net connectedness fluctuated around 50%. Germany's net directional connectedness was negative throughout the 2000s and during the global recession of 2008–09. France, Italy, and the U.K. were also net recipients of business cycle shocks before the global recession.

Lately, with a net connectedness measure lower than −50% since 2007, Japan has become a net recipient rather than a net transmitter of business cycle connectedness. In the meantime, the net connectedness from the U.S. has gradually increased with the intensification of the sub-prime crisis since mid-2007. As emphasized above, from April to December 2008, the total connectedness index jumped up substantially to reach 80%. The U.S. was the most important contributor to the increase in business cycle connectedness, with a net connectedness contribution of more than 150%. The gross directional connectedness from the U.S. jumped close to 180% following the collapse of Lehman Brothers in September 2008.

While the U.S. was the major net transmitter of shocks to others, France and Italy had also become important net transmitters during the global recession of 2008–09. Actually, the net connectedness of both countries stayed high after late 2008, fluctuating between 50 and 100%. According to Figure 5.6, since the global financial crisis, the net connectedness of Germany, the U.K., and Japan declined rapidly, dropping all the way to −100% in early 2010

5.3.5 USING COUNTRY REAL-ACTIVITY FACTORS

In order to check the robustness of our results obtained from the industrial production indexes, in this short section we use a set of alternative measures of the behavior of each of the G-7 economies over the business cycle. Recently, using data on major macroeconomic variables at monthly and quarterly frequency Aruoba et al. (2011) estimated dynamic factor models for the G-7 countries and derived country factors for each of the countries using thirty-seven monthly indicators. Aruoba et al. (2011) showed that the country factors captured the main macroeconomic developments over a period of forty years and their behavior over time were fairly consistent with the business cycle narrative for each of the countries.

Applying the connectedness methodology to monthly country factors obtained by Aruoba et al. (2011), we calculated the total connectedness index for different window lengths. In Figure 5.7 we present the total connectedness index obtained from a VAR of the country factors over a sample window of five years through seven years, along with total connectedness indexes obtained from the VEC1 model of monthly industrial production indexes.

When we use five-year and six-year rolling windows to calculate the connectedness index, the behavior of the country-factor and industrial-production-based connectedness indexes is quite similar. In the case of seven-year window there is a level difference between the two indexes. The industrial-production-based index tends to be higher than the country-factor-based index. Despite that level difference, however, the two indexes behave quite similarly over time. Based on Figure 5.7, we can conclude that the connectedness of business cycles across the G-6 countries are well captured by the use of industrial production data.

5.3.6 INTERNATIONAL TRADE AND DIRECTIONAL CONNECTEDNESS

Germany has been the biggest economy and the manufacturing powerhouse of Europe. It is therefore not easy to reconcile some of the above dynamic directional connectedness results with the image of Germany as the engine of growth in the E.U. Now, let us discuss the logic behind this result in some detail. Trade flows play a key role in the transmission of shocks across countries. When there is a shock to domestic demand in country i, this shock is transmitted to other countries through the trade channel. As the aggregate demand in country i takes the hit, the demand for imports is affected as well. As a result, the domestic shock is likely to be transmitted to other countries that are major exporters to that country.

As can be seen in Table 5.5, from 1999 to 2008, Germany's average trade surplus in manufacturing vis-à-vis the other five countries was equivalent to

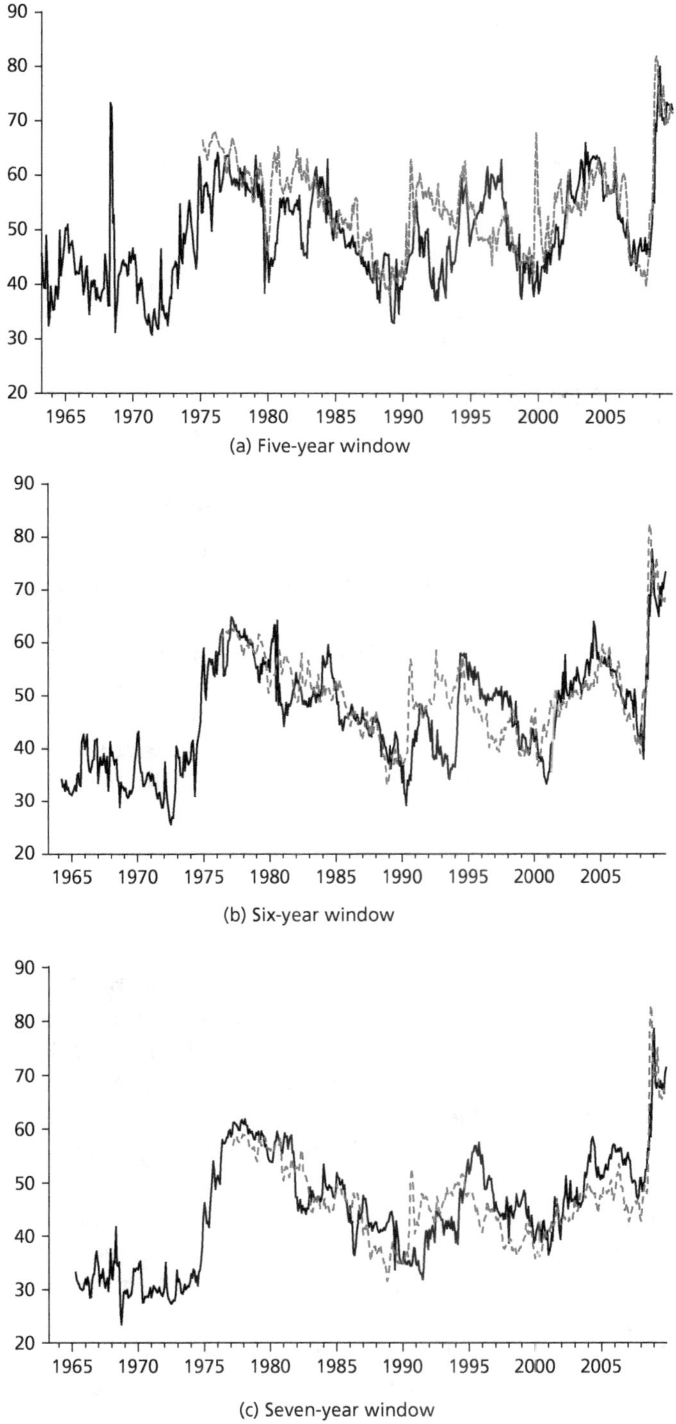

(a) Five-year window

(b) Six-year window

(c) Seven-year window

Figure 5.7. Total connectedness as assessed using country factors vs. industrial production

Note: We show the country-factor version as a dashed line and the industrial production version as a solid line. See Section 5.3.5 for details.

Table 5.5 Bilateral manufacturing trade balance relative to local manufacturing production, 1999–2008 average

	U.S.	Germany	Japan	France	U.K.	Italy
U.S.	–	2.4	2.7	1.4	1.6	0.6
Germany	–0.6	–	0.1	–2.4	–5.0	–1.7
Japan	–1.2	–0.2	–	–0.1	–1.6	–0.1
France	–0.1	1.6	0.02	–	0.1	0.8
U.K.	–0.1	1.5	0.3	0.7	–	0.9
Italy	–0.2	1.1	0.02	–0.2	–1.2	–
Total	–2.0	6.5	3.1	–0.5	–6.1	0.5

Note: Each cell shows the manufacturing trade balance of the column country with the row country, divided by the industrial production of the column country. For example, Germany's manufacturing trade surplus with the U.S. is 2.4% of German industrial production, the corresponding U.S. manufacturing trade deficit with Germany is 0.6% of U.S. industrial production.
Source: Authors' calculations based on OECD data.

6.5% of its industrial output. Over the same period the U. K., the U. S., and France ran manufacturing trade deficits, while Japan and Italy ran manufacturing trade surpluses vis-à-vis the other G-6 countries. Germany happens to be the most important exporter of manufacturing goods to France, the U.K., and Italy, and it ranks only second or third among exporters to the U.S. and Japan. As a result, when there is a negative shock to industrial production in one or more of the G-6 countries, this shock is likely to be transmitted, first and foremost, to Germany and then to the other countries. From this perspective, it is logical for Germany to have a higher connectedness from others compared to its connectedness to others.

In order to better understand the possible link between the trade balance and the business cycle connectedness, we undertake a linear regression analysis. In the regressions, the dependent variable is logarithmic net directional connectedness ($Log(C_{\bullet\leftarrow i}/C_{i\leftarrow\bullet})$). On the right-hand side we include only the logarithmic trade balance ($Log(Exports/Imports)$), with a lag of twelve months. Regression results for the full sample (April 1963 to February 2010) as well as for the sub-sample that covers the period from January 1990 to the end of the sample are presented in Table 5.6.

The results for the full sample (April 1963 to February 2010) are not encouraging. Only the U.S. has a statistically and economically meaningful coefficient estimate with the expected negative sign. The estimated coefficient implies that when the ratio of one-year lagged U.S. exports to U.S. imports increases by 1%, the ratio of the connectedness from the U.S. to the connectedness received by the U.S. will decline by 1.76%. The estimated coefficient for the U.K. (6.43) is also statistically significant, and its positive sign implies that a decrease in the export–import ratio will lead to a decrease in the connectedness from the U.K. relative to the connectedness received by the U.K.

With the globalization process underway, trade flows became more and more important in the 1990s. Therefore, it makes sense to focus on the recent

Table 5.6 Directional connectedness and the trade balance

	April 1963–February 2010		January 1990–February 2010	
	Coefficient	R^2	Coefficient	R^2
France	2.45	0.70	−4.95	0.80
	(2.91)		(4.36)	
Germany	−3.85	0.70	−6.31**	0.61
	(2.63)		(2.4)	
Italy	0.59	0.68	−4.10**	0.56
	(1.42)		(1.17)	
Japan	−1.16	0.75	−6.98**	0.74
	(1.34)		(2.46)	
U.K.	6.43*	0.74	6.46+	0.71
	(2.61)		(3.45)	
U.S.	−1.76**	0.76	−2.59**	0.63
	(0.49)		(0.91)	

Note: We regress logarithmic net directional connectedness ($Log(C_{\bullet\leftarrow i}/C_{i\leftarrow\bullet})$) on the logarithmic trade balance (*Log(Exports / Imports)*), lagged twelve months. Standard errors are presented in parentheses. **, * and + indicate statistical significance at one, five, and ten percent levels, respectively.

decades. For that reason, we restrict the sample to the post–1990 period. Five out of six estimated coefficients for the post–1990 period had negative signs, as expected. For four of these countries (France is the exception), the negative coefficient estimates are statistically different from zero at the 1% significance level. The estimated elasticity for the U.K. is still positive, but it is different from zero at the 10% level of significance only. The U.K. runs chronic deficits in merchandise trade, which is financed by chronic trade surpluses in services. Being a deficit country in merchandise trade, the U.K. is likely to be a net transmitter of shocks to other countries. However, given the large size of its trade surplus in services (service exports revenue reached $249 billion in 2009 compared to $117 billion in merchandise exports), the U.K. is likely to be a net receiver of shocks in services sectors from other countries. As a result, when we regress the log of the ratio of connectedness transmitted and received by the U.K. on the log of the export–import ratio for goods, the coefficient turns out to be positive but statistically insignificant.

5.4 **Concluding remarks**

We have applied the connectedness methodology to the study of business cycles in several leading economies. In doing so, we have made several contributions to the literature on international business cycles.

First, the connectedness methodology is different from, and complements, the empirical approaches typically used. While the factor model approach aims to obtain a world business cycle measure, the connectedness framework distinguishes between idiosyncratic shocks to industrial production and the connectedness of industrial production shocks across countries. Furthermore, the connectedness index is based on a multivariate VECM, which is better able to capture the increased comovement of business fluctuations in more than two countries compared to analyses based on pairwise correlations.

Second, the analysis sheds new light on the nature of business cycles, clearly showing that the cross-country comovement of business fluctuations is not constant over time, and that it does follow an upward trend. Rather, business cycle connectedness fluctuates substantially over time. However, the band within which the connectedness index fluctuates has increased since 1984. This result is consistent with the findings of both Kose et al. (2003) and Doyle and Faust (2005): when shocks in individual countries are small, they cannot be expected to be transmitted to other countries regardless of the integration among countries, but when the shocks are large enough they are transmitted, and the cross-country correlation of macroeconomic aggregates increases.

Third, we use directional connectedness measures to identify each country as a gross and/or net transmitter of business cycle shocks to other countries as well as a gross/net recipient of business cycle shocks from other countries over different time periods. The directional connectedness measures show that the U.S. (1980s and 2000s) and Japan (1970s and 2000s) are the major net transmitters of shocks to other countries, while Germany is the major net receiver of shocks in the 2000s.

Fourth, with an unprecedented jump between May and December 2008, the business cycle connectedness index captures well the global nature of the current recession. The quick and large increase in cross-country connectedness suggests that recovery from "the great global recession" may require coordinated policy actions among the major industrial and emerging market economies.

Fifth, we showed that the sign of a country's net business cycle connectedness is closely related to the country's trade balance. Those countries that run trade surpluses tend to be net recipients of shocks, whereas countries that run trade deficits are likely to be transmitters of shocks.

ACKNOWLEDGMENTS

For helpful guidance we thank the editors, Siem Jan Koopman and Neil Shephard, and two anonymous referees. For support we thank the Real Time Data Research Center at the Federal Reserve Bank of Philadelphia. Yilmaz thanks the Turkish Scientific and Technological Research Council (TUBITAK) for Grant No. 111K500. The usual disclaimer applies.

6 Inferring and predicting global temperature trends

Craig Ansley and Piet de Jong

6.1 Introduction

The article is motivated by the desire to improve on smoothing methods proposed and used in the climatic time series literature. Smoothing methods are used to infer and predict global temperature trends. Current methods include "minimum roughness smoothing" (MRS) and variants proposed in Mann (2004) and Jones and Mann (2004). MRS is based on Butterworth filters with parameters chosen on the basis of frequency domain properties. The climatic time series literature proposes MRS or variants be applied with specific choices of parameters to series padded at either end to overcome boundary problems. This article shows that these procedures have poor statistical properties. Further this article displays, discusses, and implements more defensible methods of smoothing including joint smoothing of multiple time series based on Butterworth filters.

Gómez (2001) and Harvey and Trimbur (2003) have displayed classes of time series models for which the Butterworth filter delivers optimal (minimum mean square error) estimates. Using this class of models, it is possible to obtain optimal trend and slope estimates efficiently using the Kalman filter and related smoothing algorithms, even at the ends of the series. The time series approach has the additional advantage of providing an established framework for the identification of an appropriate member of the class and estimation of unknown parameters. We compare the results so obtained with MRS smoothing. This article further displays, discusses, and fits variants of Butterworth filter models including where a number of climatic time series are analyzed jointly for improved common trend and slope estimation and prediction.

Contributions to the analysis of climatic data have appeared in the statistics and econometrics literature including Hendry (2010) who notes that global temperatures are rising on a high variance stochastic trend. Based on just four temperature series, our joint model in Section 6.6 indicates a rising trend, but the associated standard error is relatively high. This is consistent with Hendry's observations. He also discusses the problem of shifting distributions, which to a certain extent is addressed through the use of non-stationary stochastic trends in our models.

Magnus et al. (2011, 2013) note that global warming is the net effect of two components: the (global) greenhouse effect and the (local) solar radiation effect. They construct a simple statistical model encompassing these effects, and apply it to data from a large number (over 1300) of recording stations worldwide for the period 1959–2002. They find that both components are substantial, but the net effect is an increase in global temperatures. The model is able to handle data from many sources simultaneously, but it accommodates only very simple time series effects. It is possible that time series models such as those discussed in this paper could improve the results. In particular, there are missing data problems, which can be handled simply within a Kalman filter time series framework.

Hendry and Pretis (2013) point out major flaws in a particular statistical analysis of climate data in the earth sciences literature, and warn of a range of problems that can arise from the use of inappropriate statistical methods.

The further sections of this paper are structured as follows. Section 6.2 introduces climatic time series used to illustrate procedures and outline problems of the existing approaches. Section 6.3 describes the methods based on the Butterworth filter proposed by Mann (2004), Jones and Mann (2004), and Mann (2008), including the use of "padding." Smoothed series based on the proposed padding methods are shown to have undesirable statistical properties. Section 6.4 displays the models for which Butterworth filtering is optimal and discusses and fits the models to climatic time series and compares results to MRS smoothing. Variants of Butterworth models are developed, discussed, and estimated in Section 6.5. Finally, Section 6.6 introduces a joint Butterworth model where a common trend is fitted to major temperature series simultaneously. This joint model is used to fit and forecast global temperature out to the beginning of the next century.

6.2 **Data background**

The analysis in this paper is illustrated with four climatic time series, displayed in Figure 6.1. The left panel displays annual Northern Hemisphere (NH) temperature from 1856 through to 2003. Temperatures are in anomaly form[1] i.e.,

[1] Plotting anomalies rather than temperatures exaggerates the appearance of trends. In this paper "temperature" and "anomaly" are used interchangeably as they differ from one another only by a constant. The appropriate value of this constant changes according to whether we measure temperature in Celsius, Fahrenheit, or as a departure from a historical average. The chosen constant has no bearing on the results presented in this article.

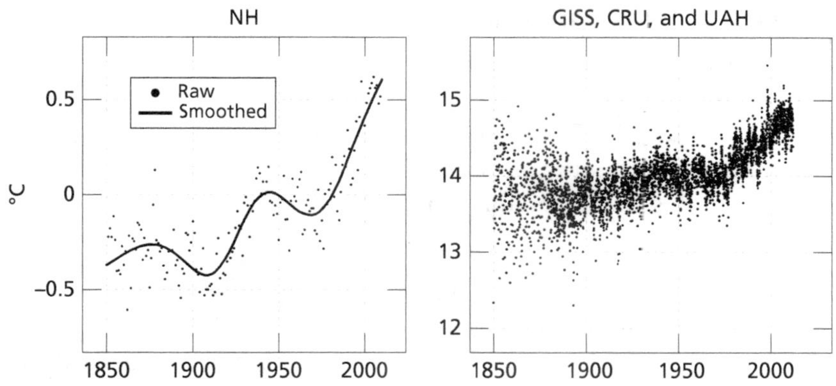

Figure 6.1. Climatic time series

Note: Left hand panel displays yearly Northern Hemisphere temperature anomalies as well as a smoothed version. The right panel displays monthly global time series with different starting points.

deviation of the actual temperature from a historical average over a specified period of time. Also displayed in this panel is temperature smoothed using a procedure advocated in Mann (2004) which is further discussed and criticized below.

The right panel of Figure 6.1 displays three monthly global time series starting at different years but all ending in January 2012:

- **GISS**—Goddard Institute for Space Studies, Hansen (2010). This dataset is combined land-surface air and sea-surface water temperature anomalies in 0.01°C. These are converted to actual temperatures by dividing by one hundred and adding fourteen, the average for the period 1951–80. The time series starts January 1880 for a total of 1585 months.
- **CRU**—Climate Research Unit University of East Anglia, Jones (2010). Monthly temperature index from January 1850 for a total of 1945 months. Data is scaled in °C to match GISS as discussed below.
- **UAH**—University of Alabama in Huntsville satellite temperature dataset, UAH (2010). Monthly satellite global temperature data starting December 1978 for a total of 398 months. The data is also scaled to °C to match GISS.

The three monthly temperature time series are derived from series recorded with different bases. To make the levels directly comparable the CRU are UAH series are rebased to that of GISS by adding estimated offsets. The common base is arbitrary and plays no role in the analysis. The base is chosen to match that of GISS as discussed in Section 6.6. The plot of monthly temperatures suggests decreasing volatility.

6.3 Smoothing approaches based on the Butterworth filter

Mann (2004) advocates Butterworth filters for smoothing climatic time series. Butterworth filters smooth time series using a moving average with weights w_j where the weights have generating function

$$B(z) \equiv \sum_{j=-\infty}^{\infty} w_j z^j = \frac{1}{1 + r^2\{(1-z)(1-z^{-1})\}^m} \,. \tag{6.1}$$

Here $m \geq 0$ and $r \geq 0$ determine the weights w_j in the implied two-sided filter, and z is a complex variable. The smoothed value at any t is $B(z)y_t$ where now z is interpreted as the lag operator. Since the filter is "two-sided," the smoothed value $B(z)y_t$ at time t requires time series values either side of t.

6.3.1 PROPERTIES OF THE BUTTERWORTH FILTER

Note: $B(1) = 1 = \sum_{j=-\infty}^{\infty} w_j$ and the filter is symmetric $w_{-j} = w_j$. If $r = 0$ then $B(z) = 1$. The function $B(e^{i\lambda})$, as a function of λ, is called the frequency response function and for $j = 0, \pm 1, \ldots$

$$w_j = \frac{1}{2\pi} \int_{-\pi}^{\pi} B(e^{i\lambda}) e^{-ij\lambda} d\lambda \approx \frac{1}{N} \sum_{\ell=0}^{N-1} B(e^{i2\pi\ell/N}) e^{-ij2\pi\ell/N} \,. \tag{6.2}$$

The expression (6.2) on the right is evaluated simultaneously, for all relevant j, with the discrete Fourier transform. In computations N is chosen suitably large given m and r.

Figure 6.2 shows, for different m and r, the frequency response function in the left panel and the implied weights w_j in the right panel. In each case it is convenient to write $r = \{2\sin(\lambda_r/2)\}^{-m}$ where λ_r is called the "cutoff" frequency or halfwidth. With this m and r, $B(e^{i\lambda_r}) = B(e^{i0})/2 = 1/2$. Hence $\lambda_r = 2\sin^{-1}(r^m/2)$ is where the frequency response function passes from being closer to one to closer to zero. Further, for given λ_r, as m increases, that is as the "boxiness" of the frequency response function increases, r approaches infinity.

The parameter r controls the "bandwidth" of the filter—i.e., the frequencies passed by the filter—while m controls the "boxiness" of the frequency response function of the filter. The frequency response function increasingly concentrates near $\lambda = 0$ as m increases. Choosing a smoothing filter based on the shape of $B(e^{i\lambda})$ runs into three issues:

- Model selection: what class of filters, and what member of the class? White noise or red noise? Jones and Mann (2004).

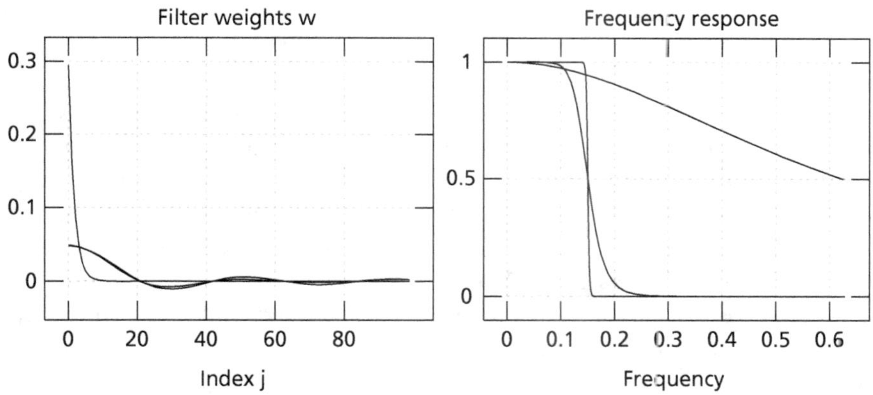

Figure 6.2. Filters w_j and frequency response functions $B(e^{i\lambda})$ of Butterworth filters

Note: The w_j with a spike at the origin corresponds to $m = 1$ and $r = 1.625$ and period $2\pi/\lambda_r = 2$ and frequency response function which declines gradually. The other two plots in each panel are for $m = 2$ and $m = 50$, both with halfwidths $\lambda_r = 0.15$. The w_j for $m = 2$ and $m = 50$ are barely discernible

- Parameter estimation (loss–pass bandwidth).
- Smoothing near the end of the series. This is a problem at either end of the series, particularly at the latest end of the series where smoothed values, indicative of the trend, are required for the latest possible times. In the climate context primary interest is on the trend in global temperature, say, over the last decade.

6.3.2 SMOOTHING AT THE END OF THE SERIES

Given m and r in (6.1), an important issue is how to deal with smoothing near the ends of the series: the so-called end point problem. Mann (2004) suggests padding the series so as to extend the series at either end. The padded values are linearly transforms of existing values. For example the "minimum roughness strategy" (MRS) extends the recent end of the series with artificial values $\tilde{y}_{n+k} = 2y_n - y_{n-k}$, $k = 1, 2, \ldots$. Geometrically this corresponds to first mirroring the series about $t = n$ and then reflecting the mirrored series about $y = y_n$. Mann (2004, 2008) also suggest other variants such as "minimum slope smoothing" or combinations of methods. In this paper we focus on analyzing MRS: the other proposals can be easily shown to have similar shortcomings and seen to have no advantages over well established time series approaches.

With MRS the future k–step change at $t = n$ is assumed equal to the most recent k–step change:

$$\tilde{y}_{n+k} - y_n = y_n - y_{n-k}, \qquad k = 1, 2, \ldots. \tag{6.3}$$

Hence the change at the start of the padded portion is the same as the change at the end of the observed series. Thus slope information is replicated into the future.

Mann (2004) suggests the padded series provides a reliable estimate of slope at the end of the series. Soon et al. (2004) have shown that MRS can produce poor results at the end of the series. Mann (2008) proposes a longer padding, as well as using a weighted average of the three padding methods. As shown below, padding is inefficient. Further sections show, in terms of established time series practice, padding is superfluous.

Applying MRS, i.e., (6.1) with z interpreted as the lag operator, and using (6.3), we obtain a final smoothed value

$$\hat{\mu}_n = \sum_{j=0}^{\infty} w_j y_{n-j} + \sum_{j=1}^{\infty} w_j \tilde{y}_{n+j} = \sum_{j=0}^{\infty} w_j y_{n-j} + \sum_{j=1}^{\infty} w_j (2y_n - y_{n-j}) = y_n \, .$$

A similar calculation shows the smoothed value at $t = n - 1$, one step back from the latest time, is

$$\hat{\mu}_{n-1} = (1 - w_0 - w_1) y_n + \sum_{j=0}^{\infty} (w_j - w_{j+2}) y_{n-j-1} ,$$

indicating substantial weight on the latest value and insignificant weight on the previous values, especially if the weights decline slowly as occurs if m is large. The smoothed estimate of the final slope is

$$\hat{\mu}_n - \hat{\mu}_{n-1} = (w_0 + w_1) y_n - \sum_{j=0}^{\infty} (w_j - w_{j+2}) y_{n-j-1}$$

$$= \sum_{j=0}^{\infty} (w_j + w_{j+1})(y_{n-j} - y_{n-j-1}) \, ,$$

a weighted average of observed changes. With any reasonable choice of Butterworth weights (6.1), the more recent changes are weighed more heavily with the rate of decline determined by m and r.

6.3.3 STATISTICAL PROPERTIES OF MRS

The above expressions for the final level and slope estimates permit insight into the bias and estimation error properties of MRS. Consider a simple signal plus error model for the time series $y_t = \mu_t + \varepsilon_t$ where μ_t is a smoothly varying trend and ε_t is mean zero measurement noise with variance σ^2 say, independent of t. Then using the above notation, the estimate of the final level is $\hat{\mu}_n = y_n = \mu_n + \varepsilon_n$ which is unbiased but has variance σ^2. Further, the estimated final slope is equal to

$$\hat{\mu}_n - \hat{\mu}_{n-1} = \sum_{j=0}^{\infty}(w_j + w_{j+1})(\mu_{n-j} - \mu_{n-j-1})$$

$$+ (w_0 + w_1)\varepsilon_n - \sum_{j=1}^{\infty}(w_{j-1} - w_{j+1})\varepsilon_{n-j}\,. \qquad (6.4)$$

To minimize bias, $w_0 + w_1$ must be made large and all other $w_j + w_{j+1}$ small. This motivates a small m. If m is large the bias is large unless the μ_t are constant or at least constant for the more recent past where the $w_j + w_{j-1}$ are small. Hence large m results in significant bias if the slope varies over time.

With constant variance and uncorrelated measurement error, $\varepsilon_t \sim (0, \sigma^2)$, the variance of the slope estimator is from (6.4)

$$\sigma^2\left\{(w_0 + w_1)^2 + c\right\}, \qquad c \equiv \sum_{j=1}^{\infty}(w_{j-1} - w_{j+1})^2\,. \qquad (6.5)$$

The variance is dominated by $(w_0 + w_1)^2$ especially if m is large, for then the differences $w_{j-1} - w_{j+1}$ are negligible. However if m is large then the bias is large. Varying m trades off between bias and variance with large m increasing bias but reducing variance and vice versa.

6.3.4 APPLICATION OF MRS TO NH

To illustrate the bias and variance, assume σ is estimated as the difference between the raw and MRS smoothed versions NH as displayed in Figure 6.1, yielding $\hat{\sigma} = 0.114$. Arguably this is an underestimate of σ, since both ends of the series are, as shown above, "under smoothed." With this estimate of σ then the standard deviation attached to the final smoothed value is 0.114, suggesting a 95% normal based confidence interval of

$$\hat{\mu}_n \pm 2\hat{\sigma} = 0.61 \pm 2 \times 0.114 = (0.38, 0.83).$$

This compares to the NH mean of 0.115. Thus the estimated current level is considerably higher than the observed historical past, but the width of the confidence interval does not inspire confidence as to the actual level. Further, using the Butterworth filter with m and r as used by Mann (2004) and the MRS smoothing leads to a final slope estimate of 0.02 and associated 95% confidence interval $(-0.18, 0.22)$ and hence the current slope is not significantly positive. While MRS smoothing procedure may well provide unbiased estimates of the

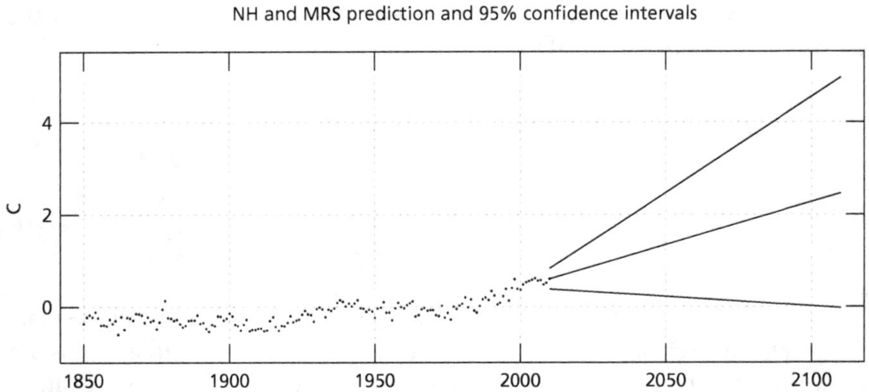

Figure 6.3. Predicted future NH temperature and 95% prediction interval based on BW$(5, r)$ with $\lambda_r = 0.15$ and MRS

final level and slope, the standard errors attached thereto make the estimates of little value.

It is natural to use the latest level and slope estimates for forecasting. If $k > 0$ is the lead time, the forecast of temperature at time $n + k$ is $\hat{\mu}_n + k(\hat{\mu}_n - \hat{\mu}_{n-1})$ is as displayed for different k in Figure 6.3. Also displayed are the 95% confidence bands. These are the forecast plus or minus twice the standard error derived as follows. The covariance between $\hat{\mu}_n$ and $\hat{\mu}_n - \hat{\mu}_{n-1}$ is $k(w_0 + w_1)\sigma^2$ and hence the variance of the forecast is

$$\sigma^2 \left[1 + k^2\{(w_0 + w_1)^2 + c\} + 2k(w_0 + w_1)\right]$$
$$= \sigma^2 \left[\{1 + k(w_0 + w_1)\}^2 + k^2 c\right],$$

where $c \geq 0$ is given in (6.5). Supposing $c = 0$ yields an approximate but over–tight 95% confidence interval for the k–step ahead prediction of

$$\hat{\mu}_n + k(\hat{\mu}_n - \hat{\mu}_{n-1}) \pm 2\hat{\sigma}\{1 + k(w_0 + w_1)\},$$

which is linear in k. The number of standard errors the level at $t = n + k$ is away from the current estimate $\hat{\mu}_n$ is hence

$$\frac{\hat{\mu}_n - \hat{\mu}_{n-1}}{\hat{\sigma}\left\{\frac{1}{k} + (w_0 + w_1)\right\}} < \frac{\hat{\mu}_n - \hat{\mu}_{n-1}}{\hat{\sigma}(w_0 + w_1)} = \frac{0.02}{0.114 \times 0.097} = 1.81.$$

Thus the MRS analysis of the NH series provides only weak evidence that future expected temperature will be higher than the current level. In particular, a test of the hypothesis that average temperature in one hundred years' time will be equal to the latest value versus it will be higher has a p–value greater than 0.04, computed using normal assumptions.

The ad hoc nature and imprecision of MRS and associated filtering motivates the use of more refined smoothing methods. Previous authors have shown Butterworth filtering is optimal for a certain class of time series models. These models suggest standard statistical methods for choosing m and r, together with smoothing procedures that correctly handle the critical end point issue. These procedures are described and applied in the following sections.

6.4 Butterworth filter and minimum variance trend estimation

Gómez (2001) and Harvey and Trimbur (2003) have shown that the Butterworth filter $B(z)$ in (6.1) is optimal for a certain class of time series models. Suppose that μ_t and y_t are generated according to the model

$$y_t = \beta + \mu_t + r\varepsilon_t\,, \qquad (1-z)^m \mu_t = \eta_t\,, \qquad \varepsilon_t,\, \eta_t \sim (0, \sigma^2)\,. \qquad (6.6)$$

Then $B(z)y_t$ is the minimum mean square error (MMSE) estimate of the signal μ_t. Here, ε_t and η_t are serially and contemporaneously uncorrelated, both with mean zero and constant variance σ^2. Model (6.6), apart from the location β and scale σ, depends on the two parameters m and r and is denoted $y_t \sim \mathrm{BW}(m, r)$. With this model y_t is, apart from observational noise ε_t, a signal μ_t whose m-th difference is white noise. If $\sigma = 0$ then $\mathrm{BW}(m, r)$ produces an order $m - 1$ polynomial corrupted by noise. Further, if $\sigma = r = 0$ then y_t is generated by a polynomial of order $m - 1$.

The smoothed value $B(z)y_t$ requires data either side of time t. If only limited data are available the MMSE estimate is still defined but the computation of $B(z)y_t$ must be modified as discussed below.

The MMSE prediction is a polynomial of order $m - 1$. If $m = 1$ the smoothed series is a flat line, with level equal to the latest inferred level. With $m = 2$ the prediction is a straight line with level and slope equal to the latest inferred level and slope, and so on. Thus the $\mathrm{BW}(1, r)$ model, in the absence of a forcing function, is inconsistent with future global warming: global warming requires $m > 1$.

In the $\mathrm{BW}(m, r)$ model (6.1), r is the noise to signal ratio: the ratio of the standard deviation of measurement noise to that of η_t, the noise driving the signal μ_t. For r near 1, $\ln r$ is the approximate percentage the standard deviation of $r\varepsilon_t$ is above or below that of η_t. For large m, small signal shocks η_t have large impacts on the signal and, in turn, the measurement y_t. Hence if m is large we would expect r to be large. In practice $\ln(r)/m$ is usually moderate; roughly independent of m for feasible model choices.

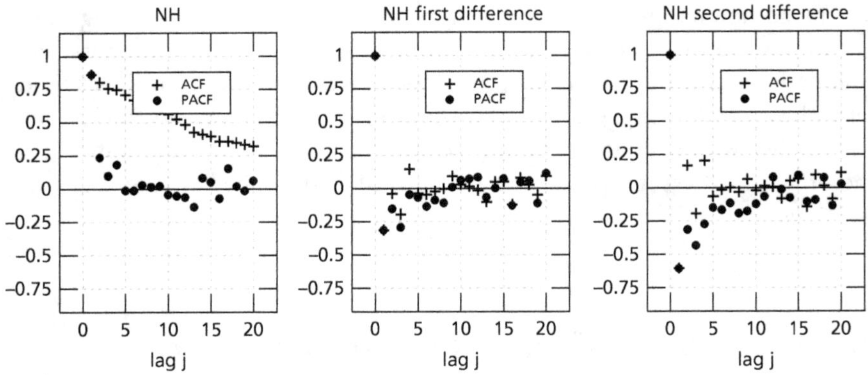

Figure 6.4. Estimated auto (ACF) and partial autocorrelations (PACF) functions of NH and its first two differences

6.4.1 THE TIME SERIES APPROACH VERSUS MRS

An advantage of the time series model approach is that there is an established time series framework for the identification of an appropriate m and the estimation of r. To illustrate on the NH series, Figure 6.4 shows the auto– and partial autocorrelation functions of the series and its first two differences. These suggest first differencing ($m = 1$) may be appropriate but $m = 2$ is likely "overdifferencing."

Given m, the noise to signal ratio r is estimated using maximum likelihood where the likelihood can be evaluated with the Kalman filter. It is assumed that the disturbances ε_t and η_t are normal. The advantage of using the Kalman filter is that together with the fixed interval smoothing algorithm it delivers predictions and smoothed estimates, as well as associated confidence intervals, for each t, properly allowing for end effects. There is no requirement, as there is with the Butterworth filter, for time series values to be available on either side of any given time point. Further, the algorithms apply to more generalized models including those applicable where more than one series is observed.

6.4.2 ESTIMATED BUTTERWORTH MODELS FOR NH

Table 6.1 presents results of likelihood calculations for different BW(m, r) models. These calculations assume the disturbance terms ε_t and η_t are normally distributed.

Of the models listed in Table 6.1, the BW$(1, r)$ has the highest likelihood. Further, the Mann r value derived from the halfwidth $\lambda_r = 0.15$ is significantly

Table 6.1 NH estimates and likelihoods for different BW(m, r) models

m	r	loglik	$\frac{\ln r}{m}$	$\frac{\ln \sigma}{m}$	$r\sigma$	λ_r	$\frac{2\pi}{\lambda_r}$
5	13230	479.9	1.90	−2.34	0.111	0.15	41.9
5	42582	485.0	2.11	−2.55	0.115	0.12	52.0
2	25.32	502.1	1.63	−2.72	0.112	0.20	31.9
1	1.625	513.9	0.49	−2.88	0.091	0.63	10.0

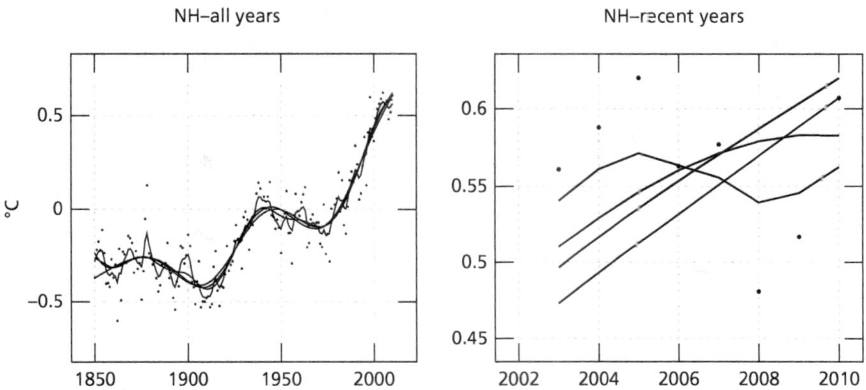

Figure 6.5. NH temperature and smoothed values

Note: The left panel displays the entire series and smooths. The right panel focuses on the latest years. Dots in each paned are actual temperatures while the four lines connect smoothed values derived from MRS smoothing and models in Table 6.1. The very smooth almost straight line passing through the final observation point corresponds to MRS smooth. The three remaining smooths have proper end effect adjustment and correspond to final three BW(m, r) models of Table 6.1. The most erratic line corresponds to $m = 1$ and gives the lowest final smoothed value.

different from the maximum likelihood value[2] with a difference in the log-likelihoods of 5.1.

The panels of Figure 6.5 display the MMSE smoothed with $r = 9.49$ corresponding to $m = 5$ and $\lambda_r = 0.15$. Note that the smoothed series does not pass through the end points as it would be forced to using the MRS strategy. Instead the smooth is calculated optimally from the available data. The smooth in the left hand panel is that with the maximum likelihood estimate (mle) $\hat{r}$. The smooth is more irregular but more believable partly on account of the correct end effect adjustments and partly on account of the use of the mle of r.

[2] The $m = 5$, σ values reported in Table 6.1 are close to but smaller than the standard deviation of the difference between the measured and smoothed temperature values discussed in Section 6.3. The σ estimate suggested in Section 6.3 is exact only if the smoothed values are exact.

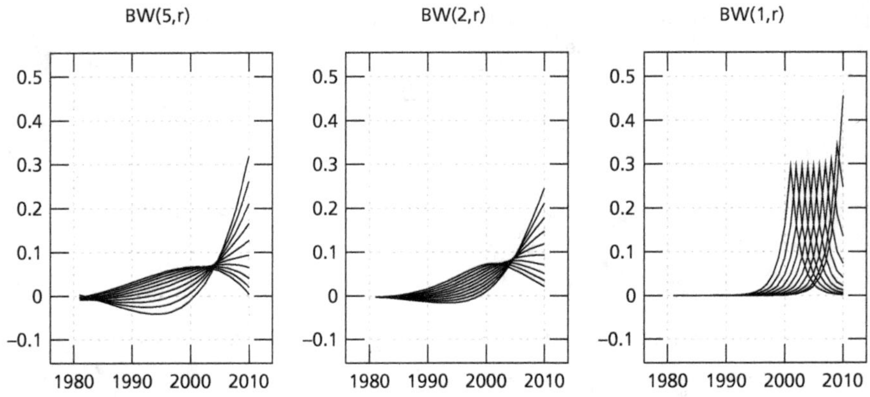

Figure 6.6. MMSE smoothing weights near the end of the NH series

Note: Different plots in each panel displays MMSE weights for $j = 0, \pm 1, \dots$ on each y_{t-j} for each μ_t as t approaches the end point n. For each model, r is the mle value as per Table 6.1.

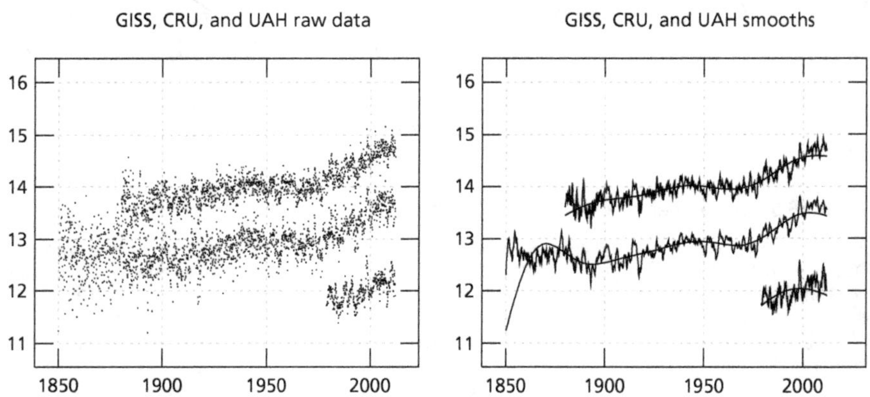

Figure 6.7. Left panel displays CRU, GISS, and UAH series with arbitrary origins (CRU is longest series, UAH shortest). Right panel displays smooths based on the mle BW$(1, r)$ model (erratic lines) and MRS with cutoff frequency of $\lambda_r = 0.15/12$ (smooth lines)

The implicit smoothing weights for the different BW(m, r) models in Table 6.1 are displayed in Figure 6.6; these are computed as in Koopman and Harvey (2003b). Near the end of the series, weights are adjusted implicitly and zero weight is given to unobserved future temperatures. In the case of MRS an entirely different approach is used as explained in Section 6.3.

To conclude this section similar methods are applied to the three monthly temperature series displayed in the right panel of Figure 6.1. The left panel of Figure 6.7 redisplays the three time series. The series are given arbitrary origins for visual clarity. The right panel displays (the more erratic lines) the exact MMSE estimates, including time points near the ends, calculated with the

BW(1,r) model, where in each case r is estimated using maximum likelihood and the smooths are exact MMSE estimates for all time points including those near the ends. The smoother lines in the right panel of Figure 6.7 are the BW(1,r) smooths using MRS where r corresponds to a cycle of forty years, as recommended in Mann (2004). These latter smoothed values clearly do not provide an accurate appraisal of trends.

6.5 **Variants and fits of the Butterworth model**

There are two models closely related to the BW$(2, r)$ model.

6.5.1 RED MEASUREMENT NOISE

The BW(m, r) model is generalized by replacing ε_t in (6.1) with $a(\alpha_t + r\varepsilon_t)$ where α_t is a stationary autoregressive or red noise component $\alpha_{t+1} = \rho\alpha_t + \nu_t$ with $|\rho| < 1$. Similar to the other disturbances in (6.1), ν_t is also assumed mean zero, variance σ^2, and independent of the other noise terms. The model is denoted BW$_a(m, r)$. Busetti and Harvey (2007) suggest a model for global temperature data that is an integrated random walk with "red noise" replacing ε_t. This model is in a wider class permitting deterministic or stochastic trend, and includes Butterworth smoothing models of degree one or two. Model selection and parameter estimation within the general class suggested by Busetti and Harvey (2007) are standard time series procedures.

The relative importance of the red noise α_t compared to ε_t is determined by r while a controls the importance of the total measurement noise. The red noise α_t aims to capture non-persistent serial dependence and has no direct effect on long-term forecasting. However red noise may make for more accurate modeling and less biased parameter estimates and forecasts.

The case $a \to 0$ but $ar > 0$ corresponds to BW(m, ar). As $\sigma \to 0$ then y_t approaches a straight line trend observed with autocorrelated error. In general the BW$_a(m, r)$ models produce MMSE forecasts which are polynomial of order $m - 1$ in the lead time. The difference from the BW(m, r) model forecast function is that for small lead times forecasts deviate from a polynomial by the transient influence of the red noise.

6.5.2 CUBIC SPLINE

MMSE smoothing with a slightly modified BW$(2, r)$ model is equivalent to fitting a cubic spline. A cubic spline is made up of cubic polynomial pieces smoothly joined points, called knots. Suppose the knots are at $t = 1, 2 \ldots, n$ and the joins are such that the first two derivatives exist. Then the MMSE smooth with the following model is a cubic spline, (Wecker and Ansley (1983); Brown and de Jong (2001)):

$$y_t = \beta + \mu_t + r\varepsilon_t \quad \mu_{t+1} = \mu_t + \delta_t + \frac{\eta_t}{3 - \sqrt{3}} \quad \delta_{t+1} = \delta_t + \eta_t \,. \qquad (6.7)$$

Model (6.7) differs from the BW$(2, r)$ model in (6.6) only in that the noise term η_t enters both the level and slope, the former with a known coefficient. Similar to the BW(m, r) model, apart from the location β and scale σ, there is one unknown parameter, r, controlling the degree of smoothness and interpreted as a "roughness penalty."

6.5.3 FITS OF BUTTERWORTH MODELS TO TEMPERATURE DATA

Table 6.2 displays the fits of different models to the four temperature series in Figure 6.1. Reasonable fits in each case appear to be the BW$_a(1, r)$ model. The cubic spline model does not appear to be an improvement over the BW(1,r) model as judged by the log-likelihood. With the BW$_a(1, r)$, the forecast function from the latest level is eventually horizontal when plotted against the lead time. Thus BW$_a(1, r)$ models are inconsistent with a future global warming trend, and suggests the consideration of BW$_a(m, r)$ models with $m \geq 2$.

Table 6.2 Fits of models to temperature time series

	loglik	$\ln a$	$\ln \frac{1+\rho}{1-\rho}$	$\ln r$	$\ln \sigma$
NH ($n = 161$)					
BW(1,r)	514.3			0.49	−2.88
Cubic spline	502.1			3.24	−5.43
BW$_a(1, r)$	518.6	0.64	−0.65	−0.53	−3.09
BW$_a(2, r)$	519.2	3.52	−0.44	−1.84	−5.74
GISS ($n = 1585$)					
BW(1,r)	4039.2			0.63	−2.67
Cubic spline	3800.4			3.07	−4.92
BW$_a(1, r)$	4138.7	2.15	1.08	0.05	−4.42
BW$_a(2, r)$	4143.8	7.53	2.05	0.06	−9.80
CRU ($n = 1945$)					
BW(1,r)	2995.1			1.28	−2.69
Cubic spline	2819.3			4.77	−6.06
BW$_a(1, r)$	3123.8	2.63	0.40	−0.02	−4.33
BW$_a(2, r)$	3120.7	7.98	1.50	0.00	−9.70
UAH ($n = 398$)					
BW(1,r)	1335.9			−0.09	−2.61
Cubic spline	1262.1			1.46	−3.88
BW$_a(1, r)$	1349.2	1.97	2.09	−0.26	−4.50
BW$_a(2, r)$	1356.0	11.75	2.01	−0.25	−14.28

For GISS and CRU, the estimates of r under the $BW_a(2, r)$ are close to one. Hence these two series can be thought of as a non-stationary level plus stationary process where the stationary process is an AR(1) plus measurement error. In each case the AR coefficient is about 0.75. The fact that CRU does not conform to this pattern may be explained by the fact that the measurement volatility appears to depend on time.

The conclusion from this section is that the three considered monthly series in Figure 6.1 separately provide little evidence in support of future global warming in that the $m = 1$ models appear to fit as well as the $m = 2$ models. Only the latter model admits the possibility of future global warming.

6.6 **Joint Butterworth model**

This section proposes, explains, and fits a joint Butterworth model. The model explains the GISS, CRU, and UAH series in terms of a common signal μ_t generated by a Butterworth model (6.6) with $m = 2$. Joint modeling and estimation makes for a more accurate and robust assessment of common trends. Error bounds on the estimated future signal will be more accurate than those based on individual analyses. Taking $m = 2$ permits a permanent non-stationary increase in the signal, consistent with non-transitory global warming.

The measurement equation of the proposed joint temperature model is:

$$\begin{pmatrix} y_{1t} \\ y_{2t} \\ y_{3t} \end{pmatrix} = \begin{pmatrix} 0 \\ \beta_2 \\ \beta_3 \end{pmatrix} + \begin{pmatrix} 1 \\ 1 \\ 1 \end{pmatrix} \mu_t + a \begin{pmatrix} \alpha_t + r_1\varepsilon_{1t} \\ \alpha_t + r_2\varepsilon_{2t} \\ \alpha_t + r_3\varepsilon_{3t} \end{pmatrix}. \qquad (6.8)$$

The indexes 1, 2, 3 correspond to GISS, CRU, and UAH, respectively and row i is deleted from (6.8) if y_{it} is unavailable, $i = 1, 2, 3$. The constructs of this model are as follows:

- Shift parameters β_2 and β_3 measure the differences in overall levels or bases. These differences are estimated using generalized least squares (GLS).
- Level μ_t is common to the three series and $(1 - z)^m \mu_t = \eta_t$ as in (6.6).
- Series i is corrupted by red noise $a(\alpha_t + r_i\varepsilon_{it})$ as in the model of Section 6.5.1. The autoregressive component $\alpha_{t+1} = \rho\alpha_t + \nu_t$ in each red noise is common to the three series.
- Parameter a measures the significance of transient volatility. A generalization of (6.8) is where a depends on i, in which case the AR(1) component α_t has differential impact on each series but this turns out to be difficult to estimate.
- Parameters r_i measure the volatility of the measurement noise in each series.

- The initial value μ_0 has a prior distribution with zero mean with infinite variance. The initial slope δ_0 is modeled as fixed but unknown and hence is an additional parameter to be estimated. The initial AR(1) component α_0 is modeled as zero mean with variance $\sigma^2/(1-\rho^2)$.
- A minor refinement for CRU is to vary r_2 with t since the volatility appears significantly larger in the early years as opposed to the latter. Varying specific volatility is modeled as $\ln r_{2t} = b + ct/m$ where m is the total number of months in the CRU series. The specific CRU log volatility at $t = 0$ is thus b and c is the change in the specific log-volatility over the period from $t = 0$ through $t = m$.

6.6.1 ESTIMATION OF THE JOINT MODEL

The above parametrization is partly motivated by estimation considerations: generally a and the ar_i are of similar magnitude and hence it is convenient to parametrize all these quantities in terms of the "size" parameter a. As a becomes large the transient volatility becomes large in relation to signal volatility. Hence a large a implies a relatively smooth temperature trend. A "common" BW$(2, r)$ type model is derived from the above setup by either setting $\rho = 0$ in which case the mean μ_t is buffeted by a common transient shock $a\alpha_t = a\nu_t$, or by letting $a \to 0$.

The diffuse Kalman filter (de Jong (1991a)), together with a numerical search is used to find the maximum likelihood estimates of the unknown parameters. It pays to conduct the search in terms of carefully chosen scales such as those displayed in Table 6.3. The chosen scales are displayed in Table 6.3 with the results in Table 6.2 used as starting values for the search. The parameters β_2, β_3, and σ^2 parameters are concentrated out of the likelihood and hence their estimates are calculated given the other estimates. All calculations are in terms of the square root version of the Kalman filter ensuring numerical accuracy and stability.

6.6.2 ESTIMATES OF JOINT BUTTERWORTH MODEL FOR GISS, CRU, AND UAH

For these series t runs from January 1850 ($t = 1$) through to February 2010 ($t = 1992$). The CRU ($i = 2$) temperature series is available for the entire

Table 6.3 Estimation results for joint temperature time series model ($n = 3928$)

	AR(1)	GISS		CRU			UAH	
$\ln a$	$\ln \frac{1+\rho}{1-\rho}$	$\ln r_1$	β_2	b	c	β_3	$\ln r_3$	$\ln \sigma$
7.7	0.60	–0.8	–14.1	0.9	–1.7	–14.5	–0.2	–9.6

length. GISS ($i = 1$) starts at January 1880 ($t = 361$). UAH ($i = 3$) starts at December 1978 ($t = 1548$). Indexes $t > 1922$, indicate months beyond February 2010. Maximum likelihood estimates are displayed in Table 6.3.

The estimated specific log–volatilities $\ln ar_i$ are higher than in the separate models of Section 6.5.1. This is expected since the AR component is not tuned to the individual series but rather to all three series. With higher specific volatilities, a smoother temperature trend would be expected, and this is borne out in the results presented in Figure 6.8. The top two panels in Figure 6.8 display CRU, GISS, and UAH series with the smoothed level μ_t obtained under two different noise to signal ratios: $\ln a$ equal to 6.7 and 7.7. The bottom panels display the estimated slope per one hundred years and associated confidence intervals, again for the two values of a. Two values of the signal to noise ratio are used since the log–likelihood is relatively flat in the a direction and hence the a parameter is relatively uncertain. Despite the uncertainty, both estimates of a deliver comparable smoothed estimates of both the level μ_t and the slope δ_t.

The confidence intervals for the level, corresponding to the two top panels of Figure 6.8, are very tight and barely discernible. However, the confidence

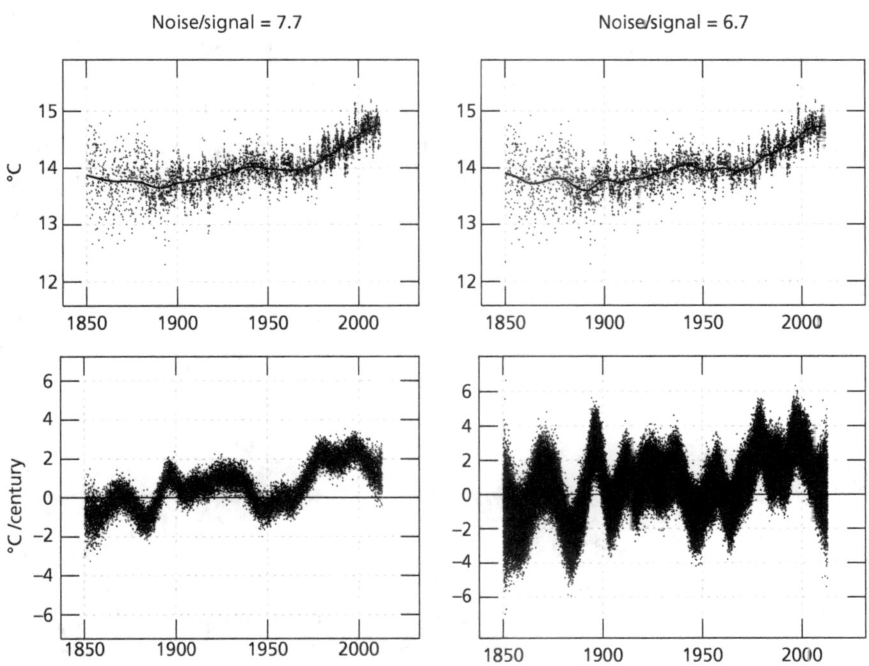

Figure 6.8. Estimated monthly temperature level and slope from joint model

Note: Top two panels display CRU, GISS, and UAH series with the smooth derived from the joint model under two noise to signal ratios: $\ln(a)$ equal to 6.7 and 7.7. Bottom panels display estimated change per one hundred years and simulated estimated confidence bounds.

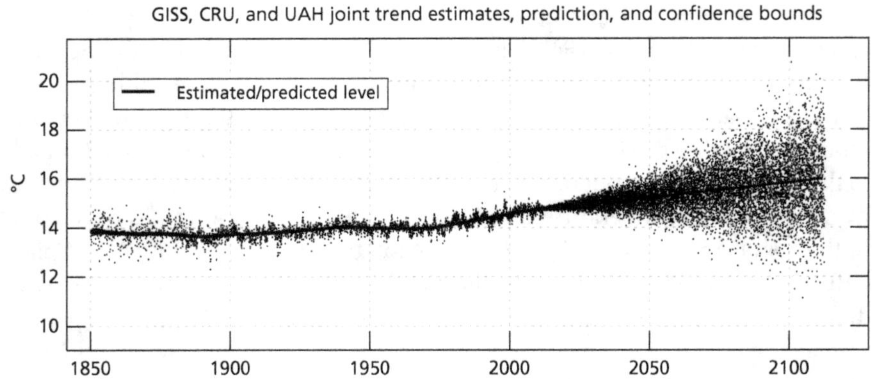

Figure 6.9. Smoothed and forecast average monthly temperature level and simulated 95% confidence band

intervals on the slopes are substantial and slightly wider for the higher noise to signal ratio. The confidence intervals for the latest slope barely encompass zero indicating the positive slope is only marginally significant. Hence the joint model, based on $m = 2$, which admits the possibility of global warming, provides only weak evidence in favor of current global warming. These conclusions are at odds with both sides of an exchange between MIT Professor Richard Lindzen and MIT President Susan Hockfield described in Lindzen (2009). The former claimed "no statistically significant warming since 1995" while the latter claimed climate change was "accelerating."

Figure 6.9 displays the temperature forecasts out to year 2100 and associated confidence intervals based on the joint model with $\ln a = 7.7$, the higher noise to signal ratio. The 95% confidence band, into the future, fan out to quickly encompass 15°C, the approximate current level. Hence the forecast based on these three jointly modeled series gives little support to the contention that temperature signal in one hundred years' time will be significantly higher than today.

6.7 **Conclusions**

This article has discussed the smoothing and extrapolation of climatic temperature series. It is argued that recently proposed methods of smoothing deliver inefficient and non-robust estimates of trend. Classical time series methods are proposed and employed to study trends. These methods find little evidence that studied temperature series contain random walks of order greater than one. Hence global warming is not supported by the analysis of isolated individual time series.

We then fitted a joint model based on a random walk of order two which does not preclude permanent warming. This joint model suggests a marginally significant positive current trend, but with a reducing slope. Extrapolating this joint trend suggests global temperature will be higher in the future but the forecast higher level is not significantly different from the levels today.

7 Forecasting the Boat Race

Geert Mesters and Siem Jan Koopman

7.1 Introduction

The Boat Race between the universities of Cambridge and Oxford is an annual rowing event that started in 1829 and is by now the most popular one day sports event in the world. Each year, a crew from the University of Cambridge rows against a crew from the University of Oxford. The rowing crews consist of eight rowers and one cox. The race takes place on the river Thames between Putney and Mortlake. We consider a class of models for the Boat Race outcomes and investigate the ability of the models to forecast the next outcome of the Boat Race. More specifically, for a variety of dynamic binary model specifications, we assess their ability to forecast the outcome of the race between 1971 and 2010 using information that is available to us just prior to the race. The relative out-of-sample forecasting performance of the models is compared for different loss functions and equal predictive ability tests; see Diebold and Lopez (1996), Diebold and Mariano (1995), and West (1996).

Forecasting the Boat Race is interesting from a number of viewpoints. First, bookmakers and gamblers may increase their expected profits by using the forecasts. Especially if the forecasts from the dynamic models are significantly better than the forecasts resulting from simple procedures, such as flipping a coin or setting the forecast equal to the winner of the previous race. Second, from an econometric forecasting perspective, the historical outcomes of the Boat Race form an interesting binary time series. While the number of observations can be limited, the observations themselves can be spread over a long period of time and with many missing entries. Binary time series occur in many fields of research such as finance, criminology, and computer science. The generation of accurate forecasts for events (such as winning a race, intervening in a market, or closing a computer network) can be of vital importance for many instances.

The presentation of our empirical results for the Boat Race forecasting study in this volume provides us with an opportunity to praise the innovative and empirically motivated time series studies of Andrew Harvey. Although most of his influential work is motivated by statistical and econometric challenges in economics and finance, he also has shown interest in tackling problems in the analysis of time series that have an even more direct impact on society.

A good illustration is his extended study with James Durbin concerning the effect of the seatbelt legislation on the monthly numbers of road passengers killed or seriously injured in Great Britain, see Harvey and Durbin (1986). His enthusiasm for the structural time series approach becomes even more apparent when he illustrates the methodology in fields as diverse as crime ("purses snatched in Hyde Park area of Chicago"), environment ("rainfall in Fortaleza, north-east Brazil" and "ice volumes measured at intervals of 2,000 years"), and many others ("Australian telephone calls to Britain" and "consumption of spirits in the U.K."). His work with Cristiano Fernandes on the exponential family of structural time series models for count data or qualitative observations has also inspired Andrew to work on some more colorful illustrations. For example, in Harvey and Fernandes (1989) a convincing analysis is presented of the results of international football matches played between England and Scotland. The binary time series of the yearly Boat Race has also not escaped their attention. In the years when the second author was just starting to work on his PhD under the superb guidance of Andrew, a full time series analysis of the Boat Race data was already carried out by Fernandes and Harvey (1990).

In our current study we first investigate the in-sample performance of different binary models. The models are summarized by a binary density for the outcome of the race that is defined conditional on a signal that depends on an observable deterministic component and on a latent, stochastic, and dynamic component. The deterministic component includes a constant and a vector of predictors, such as the average difference in weight between the boats, the winner of the toss, and the previous winning distance. The latent stochastic component is modeled by Gaussian dynamic processes such as random walk, autoregressive, autoregressive fractionally integrated, and stochastic cycle processes. The parameters of the models are estimated by the method of Monte Carlo maximum likelihood. In particular, for the stochastic processes that imply a short-memory process we implement the simulation-based methods of Shephard and Pitt (1997), Durbin and Koopman (1997), and Jungbacker and Koopman (2007). These methods are based on the importance sampling technique. When the stochastic component is a long-memory process we alter the estimation method based on the procedures discussed in Mesters et al. (2016).

We further perform an out-of-sample forecasting study. Each model is used to compute a probability forecast for the event that Cambridge wins the next Boat Race. This is done repeatedly for a period of forty years (1971–2010), where we only use information that would have been available just before the race starts. Forecasting in this manner is referred to as "pseudo-out-of-sample" forecasting by Stock and Watson (2003). The relative out-of-sample predictive performance of the different models is compared by different loss functions, or score statistics. We verify whether the models possess significantly different predictive abilities as implied by different loss functions, see Diebold and Mariano (1995) and West (1996).

The remainder of our paper is organized as follows. We continue the introduction with a short outline of the history of the Boat Race and describe the data that is obtained from the history of the race. Section 7.2 discusses the binary models that we consider for forecasting the race. Section 7.3 discusses the forecasting methodology and the predictive ability tests. It further presents and discusses the results of forty years of out-of-sample forecasting. Section 7.4 provides a brief summary of our study.

7.1.1 HISTORY

The first race between the universities of Oxford and Cambridge was organized in 1829. The idea came from two friends; Charles Merivale, a student at Cambridge, and Charles Wordsworth, a student at Oxford. On March 12, 1829 Cambridge sent a challenge to Oxford and the history of the race started. The first race was at Henley-on-Thames and was won by Oxford. In 1839 the race was relocated to London; by now the race had become an annual event taking place between Westminster and Putney. However, the increased crowds at the race made it necessary to move yet again. In 1845 the race was first held on the course between Putney and Mortlake, which is also on the river Thames. This course of the Boat Race is displayed in Figure 7.1 which is reprinted from Drinkwater (1939). In 1836 the Oxford crew selected dark blue to race in and the Cambridge crew selected Eton's school color "duck egg blue" to race in. In 1849, after a gap of two years and for the first and only time, there were two races. The first in March from Putney to Mortlake was won by Cambridge. Oxford challenged Cambridge to a second race in December. This second race in 1849 is the only time in the series to date that the Boat Race was decided by a disqualification, following a foul by Cambridge at the finish.

The race outcomes were about even until 1861, when Oxford started the first winning streak in its history lasting nine years. In 1877 the race was declared a dead heat for the first and only time in its history—although legend has it that the judge, "Honest John" Phelps, was asleep under a bush when the crews reached the finishing line. The 1912 event witnessed another Boat Race first when both boats sank and the race had to be re-run the next day, with Oxford claiming the honors at the second attempt. The race was not held between 1915 and 1919 due to the First World War. When it resumed in 1920, Cambridge embarked on a lengthy period of domination. It would win the race thirteen years running between 1924 and 1936, the longest winning streak in the race's history.

There was another break between 1940 and 1945 because of the Second World War, although four unofficial races were held during this time, all

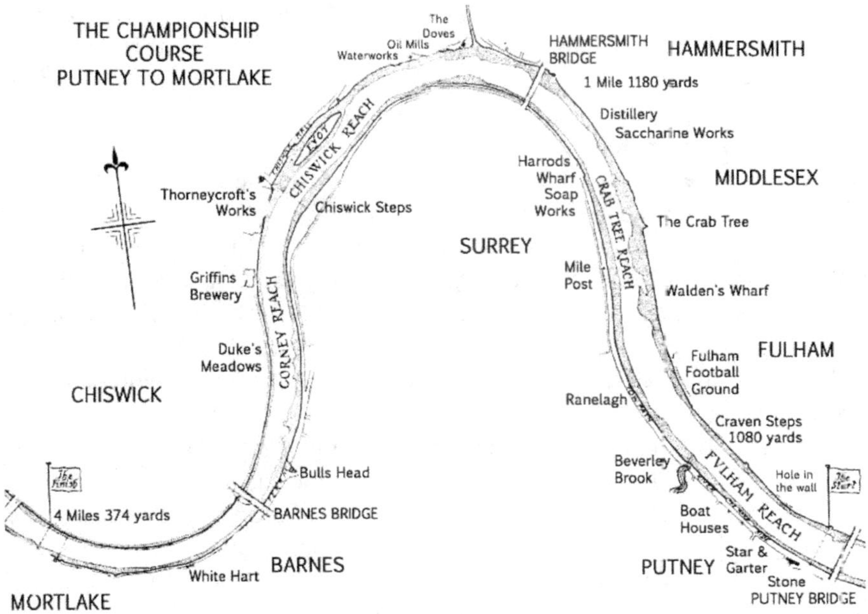

Figure 7.1. The course of the Boat Race

Source: Reprinted from George Carr Drinkwater's book *The Boat Race*

outside London. The 1952 contest witnessed perhaps the most extreme weather in Boat Race history, with Oxford prevailing in the midst of a blizzard. The dark blues also won the one hundreth Boat Race in 1954. In 1981, Sue Brown became the first female to enter the Boat Race, acting as cox for Oxford. The following year, Hugh and Rob Clay of Oxford became the first twins to win the race. The dark blues dominated throughout the eighties, as Cambridge suffered a series of misfortunes. The biggest of these came in 1984, when it managed to write off its boat before the start of the race. Controversy engulfed Oxford at the 1987 race when a section of the crew rose up in mutiny against the president over team selection policy. However, the dispute, which was chronicled in the book and film "True Blue," did not prevent it from winning the race again.

Cambridge regained its pride in 1993 by ending Oxford's domination. It subsequently won the race seven years running, the highlight coming in 1998 when it broke the course record by a massive twenty-six seconds. The last race that we consider in this study was won by Cambridge in 2010.

7.1.2 DATA

The dependent variables in our study are denoted by the elements of the $n \times 1$ vector $\mathbf{y} = (y_1, \ldots, y_n)'$, which indicate the outcome path of the Boat

Race. All data is freely obtained from <http://www.theboatrace.org>. For each year t, where $t = 1$ refers to the year 1829 and $t = n$ to the year 2010, we let $y_t = 1$ indicate that Cambridge has won the race in year t, whereas $y_t = 0$ indicates that Oxford has won the race in year t. For some years the race was not held due to a variety of circumstances; for these years we consider the corresponding outcomes as missing at random. In particular, the years 1830–35, 1837, 1838, 1843, 1844, 1847, 1848, 1850, 1851, 1853, 1855, 1877 (dead heat), and 1915 to 1919 (WWI) and 1940 to 1945 (WWII), are considered missing at random. Only the first race of 1849 (the March race) is included as the second race ended by disqualification. In total we have $n = 182$ observations, of which twenty-eight are missing. By 2010, Cambridge led the series, with eighty compared to seventy-four for Oxford.

During the history of the race the average weight of the rowers has increased substantially. This has led to the suspicion that increasing the weight of the rowers leads to faster boats, due to the presence of more muscle power. The contrary could however also be possible, by reasoning that a lighter boat has less water resistance and would therefore be faster. We include the average log difference in weight between the rowers in the Cambridge and Oxford boats as the first predictor.

The second time-varying predictor that we include is the outcome of the coin toss. The club presidents toss a coin before the race for the right to choose which side of the river (station) they will row on. Their decision is based on the weather conditions and how the various bends in the course might favor the pace of their crew. The north station (Middlesex) has the advantage of the first and last bends, whereas the south (Surrey) station has the longer middle bend see Figure 7.1. It is generally believed that the winner of the coin toss has an improved chance of winning. In most years, the betting odds change severely after the toss outcome has been made public.

The third and final covariate that we include is the distance by which the previous race was won. This winning distance may be viewed as a proxy for the technological and physical gap in ability between the two rowing teams. As the crew members, coaches, and technological advantages may overlap subsequent years, larger winning distances might measure structural advantages. The variable is constructed from the perspective of Cambridge. It is positive when Cambridge has won the previous race and negative when Oxford has won the previous race. The magnitude is measured in terms of boat lengths.

The $n \times 3$ matrix $\mathbf{X}$ includes the time-varying predictors, where element $x_{t,1}$ is the average log difference in weight between the rowers of Cambridge and Oxford in year t, $x_{t,2} = 1$ if Cambridge wins the toss in year t and zero otherwise, and $x_{t,3}$ is the winning distance in boat lengths from the previous race.

7.2 Models and parameter estimation

The models that we use as our basis to forecast the outcome path of the race can be split into two parts: a conditional binary density for the observations and a signal that includes all dynamics and predictors. The conditional binary density is given by

$$p(y_t|\pi_t) = \pi_t^{y_t}(1-\pi_t)^{1-y_t}, \qquad t = 1, \ldots, n, \tag{7.1}$$

where π_t denotes the winning probability for Cambridge in year t. Time-varying probability π_t is unknown to us, but is clearly restricted between zero and one. Therefore we specify the transformed probability as $\theta_t = \log(\pi_t/(1-\pi_t))$ where the transformation is performed by the canonical link function for binary models; see Cox and Snell (1989). We refer to θ_t to as the signal for year t. The conditional density for the observations, given in (7.1), can be rewritten in terms of the latent signal as

$$p(y_t|\theta_t) = \exp[y_t\theta_t - \log(1 + \exp\theta_t)], \qquad t = 1, \ldots, n, \tag{7.2}$$

which is assumed independent over $t = 1, \ldots, n$. It holds that $p(\mathbf{y}|\boldsymbol{\theta}) = \prod_{t=1}^{n} p(y_t|\theta_t)$, where $\boldsymbol{\theta} = (\theta_1, \ldots, \theta_n)'$. We decompose the signal in an observed deterministic part and a latent stochastic part. The deterministic part includes the intercept and three predictors (difference in weight of boat, winner of toss, and the previous winning distance). The latent stochastic part captures the possible dynamic effects which reflect that the events of previous years also affect the outcome of the race this year. The signal decomposition is given by the regression equation

$$\theta_t = \mu + \mathbf{x}_t\boldsymbol{\beta} + u_t, \qquad t = 1, \ldots, n, \tag{7.3}$$

where μ is the intercept, the 3×1 parameter vector $\boldsymbol{\beta}$ measures the effect of the observable predictors $\mathbf{x}_t$, and u_t is the latent stochastic part. The 1×3 vector $\mathbf{x}_t$ is the tth row of matrix $\mathbf{X}$. The models that we consider differ in the way that the latent process u_t is specified. Possible specifications that we include range from setting it to zero, which leads to a logistic regression model, to a complete dynamic specification. In Subsection 7.2.1 we discuss all specifications in some detail.

7.2.1 SIGNAL SPECIFICATIONS

Many dynamic specifications for u_t are possible. In general, we can expect to have a serial correlation structure between outcomes in successive years due to overlapping crews and coaches. For example, during Oxford's first winning streak, between 1862 and 1869, the crew had the same coach George Morrison for six years who was the former oarsman for the crew. Furthermore, technological developments resulting from training methods or boat construction may provide comparative advantages to a team that can last for a number years. We focus on models for u_t that are able to pick up this kind of dynamic effect.

1. The simplest possible specification is the "constant" specification given by
$$u_t = 0 \qquad t = 1, \ldots, n, \tag{7.4}$$
which results in a signal that is completely determined by the intercept and the three observed predictors. This is the baseline specification, which assumes that the entire signal is observable. In effect the model for y_t is reduced to a logistic regression model.
2. The first specification that we consider to capture serial correlation for u_t is the random walk as given by
$$u_{t+1} = u_t + \eta_{t+1}, \qquad \eta_{t+1} \sim N(0, \sigma_\eta^2), \qquad t = 2, \ldots, n, \tag{7.5}$$
where the next value of u_t is equal to the old value plus an error term. The initial state, u_1, has an unknown distribution and is therefore fixed at zero. This corresponds to a 50% winning probability for Cambridge without covariates.
3. The dependence of the outcome on previous realizations of u_t can be tampered by using an autoregressive (AR) specification. The coefficients of the AR(p) polynomial $\phi(B) = 1 - \phi_1 B - \cdots - \phi_p B^p$ enable the adjustment of observations from the past, where B denotes the back shift operator as defined by $B^s y_t = y_{t-s}$ for any integer $s > 0$. The latent process u_t is given by
$$\phi(B) u_{t+1} = \eta_{t+1}, \qquad \eta_{t+1} \sim N(0, \sigma_\eta^2), \qquad t = p+1, \ldots, n. \tag{7.6}$$
When we need to impose a stationary process for u_t, as opposed to the non-stationary random walk process (7.5), we need to assume that the roots of the autoregressive polynomial $\phi(B)$ lie outside the unit circle. The process is initialized for $u_1, \ldots, u_p$ by the unconditional distribution. For example, when $p = 1$ we have
$$u_1 \sim N\left\{0, \sigma_\eta^2/(1-\phi_1^2)\right\}.$$
The resulting model with the AR signal can alternatively be viewed as a dynamic logit model. That is, the model where the signal is given by $u_t = 0$ and $\mathbf{x}_t$ includes $y_{t-1}, \ldots, y_{t-p}$.
4. Since some of the winning streaks are relatively long, we also consider a long memory specification for the latent process u_t. In particular, we investigate the autoregressive fractionally integrated (ARFI) model; see Granger and Joyeux (1980) and Palma (2007). The model for u_t can be expressed by

$$\phi(B)(1-B)^d u_{t+1} = \eta_{t+1}, \quad \eta_{t+1} \sim N(0, \sigma_\eta^2), \quad t = p+1, \ldots, n, \quad (7.7)$$

where the fractional integration part can be expressed by the binomial expansion

$$(1-B)^d = \sum_{k=0}^{\infty} \frac{\Gamma(d+1)}{\Gamma(k+1)\Gamma(d-k+1)} (-1)^k B^k.$$

The fractional parameter d is a real valued constant in the range of $-1 < d < 0.5$. This assumption together with the previous assumption on the autoregressive polynomial ensures that the process is stationary, invertible, and causal; see Palma (2007) for a proof.

5. Another possible component that may be present in some time series is the stochastic cycle process of Harvey (1989). To have a stochastically time-varying cycle component for u_t, we consider the following specification

$$\begin{bmatrix} u_{t+1} \\ u_{t+1}^* \end{bmatrix} = \phi \begin{bmatrix} \cos\lambda & \sin\lambda \\ -\sin\lambda & \cos\lambda \end{bmatrix} \begin{bmatrix} u_t \\ u_t^* \end{bmatrix} + \begin{bmatrix} \eta_{t+1} \\ \eta_{t+1}^* \end{bmatrix}, \quad t = 2, \ldots, n, \quad (7.8)$$

where the errors term η_t and η_t^* are serially and mutually uncorrelated with mean zero and variance σ_η^2. The parameter λ gives the frequency of the cycle and is estimated together with the other parameters. The period of the cycle is $2\pi/\lambda$. Restrictions on the damping term, $0 \le \phi \le 1$, ensure that the process u_t is stationary. The cycle disturbance variance for both η_t and η_t^* is specified as $\sigma_\eta^2 = (1-\phi^2)\sigma_u^2$ where the cycle variance σ_u^2 is for u_t and is treated as the parameter to estimate. The initial distribution of the cycle is then given by

$$u_1 \sim N\left(0, \sigma_u^2\right), \qquad u_1^* \sim N\left(0, \sigma_u^2\right), \qquad E(u_1 u_1^*) = 0.$$

It follows that a constant but stochastic cycle is obtained with $\phi = 1$, which remains properly defined (Harvey and Koopman 1997).

The models that we consider can be summarized by the observational density (7.2), the signal (7.3), and one of the specifications for u_t in (7.4), (7.5), (7.6), (7.7), and (7.8). For each model specification, the unknown parameters are collected in the vector $\boldsymbol{\psi}$.

7.2.2 PARAMETER ESTIMATION

Different estimation procedures may be considered for different model specifications. For exposition and comparison purposes we use the same Monte Carlo maximum likelihood estimation procedure for the estimation of the parameter vector $\boldsymbol{\psi}$ in all cases. In particular, we adopt an estimation procedure that is based on the simulation methods developed in Shephard and Pitt (1997) and Durbin and Koopman (1997). Alternatively, other simulation based estimation methods may be modified to facilitate the estimation of the different models specifications: see, for example, West et al. (1985), Chib and Jeliazkov (2006), Czado and Song (2008), and Richard and Zhang (2007).

The log-likelihood, for all models, is defined as $\ell(\boldsymbol{\psi}) = \log p(\mathbf{y}; \boldsymbol{\psi})$ where $p(\mathbf{y}; \boldsymbol{\psi})$ is the joint density of all observations. The terms after the semi-colon in $p(\cdot; \cdot)$ are the fixed, possibly unknown, arguments of the density function $p(\cdot; \cdot)$. The joint density for models defined by density (7.2) and signal (7.3) is given by

$$p(\mathbf{y}; \boldsymbol{\psi}) = \int_{\boldsymbol{\theta}} p(\mathbf{y}, \boldsymbol{\theta}; \boldsymbol{\psi}) d\boldsymbol{\theta} = \int_{\mathbf{u}} p(\mathbf{y}, \mathbf{u}; \mathbf{X}, \boldsymbol{\psi}) d\mathbf{u} = \int_{\mathbf{u}} p(\mathbf{y}|\mathbf{u}; \mathbf{X}, \boldsymbol{\psi}) p(\mathbf{u}; \boldsymbol{\psi}) d\mathbf{u}, \tag{7.9}$$

where $\mathbf{u} = (u_1, \ldots, u_n)'$ and $p(\cdot)$, $p(\cdot, \cdot)$, and $p(\cdot|\cdot)$ denote marginal, joint, and conditional density functions, respectively. Since $\mathbf{X}$ and $\boldsymbol{\psi}$ are fixed we only need to integrate the stochastic process u_t from the joint density $p(\mathbf{y}, \mathbf{u}; \mathbf{X}, \boldsymbol{\psi})$. We further have

$$p(\mathbf{y}|\mathbf{u}; \mathbf{X}, \boldsymbol{\psi}) = p(\mathbf{y}|\boldsymbol{\theta}; \boldsymbol{\psi}) = \prod_{t=1}^{n} p(y_t|\theta_t; \boldsymbol{\psi}),$$

and

$$p(\mathbf{u}; \boldsymbol{\psi}) = p(u_1, \ldots, u_p; \boldsymbol{\psi}) \prod_{t=p+1}^{n} p(u_t|u_1, \ldots, u_{t-1}; \boldsymbol{\psi}).$$

A closed form solution for integral (7.9) is not available if u_t is stochastic and if $p(y_t|\theta_t; \boldsymbol{\psi})$ is binary. Instead we rely on numerical methods to solve the integral. In particular, we adopt the importance sampling method as it is discussed in, for example, Ripley (1987). It holds that

$$\begin{aligned} p(\mathbf{y}; \boldsymbol{\psi}) &= \int_{\mathbf{u}} \frac{p(\mathbf{y}|\mathbf{u}; \mathbf{X}, \boldsymbol{\psi}) p(\mathbf{u}; \boldsymbol{\psi})}{g(\mathbf{u}|\mathbf{y})} g(\mathbf{u}|\mathbf{y}) d\mathbf{u} \\ &= g(\mathbf{y}) \int_{\mathbf{u}} \frac{p(\mathbf{y}|\mathbf{u}; \mathbf{X}, \boldsymbol{\psi}) p(\mathbf{u}; \boldsymbol{\psi})}{g(\mathbf{y}|\mathbf{u}) g(\mathbf{u})} g(\mathbf{u}|\mathbf{y}) d\mathbf{u}, \end{aligned} \tag{7.10}$$

where $g(\mathbf{u}|\mathbf{y}) = g(\mathbf{y}|\mathbf{u})g(\mathbf{u})/g(\mathbf{y})$ is referred to as the "importance density", which must be proportional to $p(\mathbf{y}, \mathbf{u}; \mathbf{X}, \boldsymbol{\psi})$, easy to sample from and easy to compute. In our approach, we take the importance density as Gaussian with a linear relation between y_t and $\theta_t = \mu + \mathbf{x}_t\boldsymbol{\beta} + u_t$ for $t = 1, \ldots, n$ and for which it holds that $g(\mathbf{y}|\mathbf{u}) = \prod_{t=1}^{n} g(y_t|u_t)$. The mean and variance of the importance density is chosen such that the corresponding first two moments of $p(y_t|\theta_t; \boldsymbol{\psi})$ and $g(y_t|u_t)$ with respect to u_t are set equal. The resulting set of non-linear equations can be solved iteratively using Kalman filter and smoothing methods; see Durbin and Koopman (2012), Section 10.6. A Monte Carlo estimator for (7.10) is given by

$$\hat{p}(y; \psi) = g(y)M^{-1}\sum_{i=1}^{M}\frac{p(y|\mathbf{u}^{(i)}; X, \psi)p(\mathbf{u}^{(i)}; \psi))}{g(y|\mathbf{u}^{(i)})g(\mathbf{u}^{(i)})}, \tag{7.11}$$

where samples $\mathbf{u}^{(i)}$ are drawn from $g(\mathbf{u}|y)$ for $i = 1, \ldots, M$ using simulation smoothing methods.

For any importance density, it holds that $\hat{p}(y; \psi) \to p(y; \psi)$ as $M \to \infty$, which is implied by the law of large numbers. The Lindeberg–Levy central limit theorem implies a $\sqrt{M}$ convergence rate if draws from the importance sampler are independent and if importance weights

$$w(\mathbf{u}^{(i)}, y; X, \psi) = \frac{p(y|\mathbf{u}^{(i)}; X, \psi)p(\mathbf{u}^{(i)}; \psi))}{g(y|\mathbf{u}^{(i)})g(\mathbf{u}^{(i)})} \qquad i = 1, \ldots, M \tag{7.12}$$

have finite mean and variance; see the discussions in Geweke (1989). Failure of this condition to hold can lead to slow and unstable convergence of the estimator.

Further technical details of importance sampling methods for this class of time series models are discussed in Jungbacker and Koopman (2007). All our estimation results are based on this implementation. Mesters et al. (2016) have extended their procedures to allow for the long-memory (ARFI) specification of the signal in (7.7).

Missing values

In the years that the Boat Race was not held the outcomes are considered missing at random. This seems reasonable as there exists a variety of reasons for which the race was not held; see the discussion earlier in Section 7.1. The treatment of missing values within our estimation framework is relatively straightforward. The likelihood estimator (7.11) consists of two components: the likelihood of the approximating model $g(y)$ is rescaled by the mean of the importance sampling weights $w(\mathbf{u}^{(i)}, y; X, \psi)$. When missing values are present in the time series, the likelihood $g(y)$ can still be computed by the Kalman filter as argued by Durbin and Koopman (2012); see Section 4.8. The importance weights are only evaluated for observed elements in y. The realized samples from the importance density $u^{(i)}$ are computed by simulation smoothing methods that also account for missing values; see Durbin and Koopman (2002).

7.2.3 IN-SAMPLE ESTIMATION RESULTS

Our sample of observations y, the Boat Race outcome from 1829 until 2010, is used to optimize the simulated log-likelihood function based on (7.11) but subject to a bias-correction for the log transformation. For each model, we have taken $M = 1000$ draws from the constructed importance density and made use of three antithetic variables to relocate and rescale the weights. The simulated log-likelihood function is evaluated for different values of $\boldsymbol{\psi}$ during the numerical optimization by using the same set of random input values to obtain a smooth log-likelihood function in $\boldsymbol{\psi}$. Most of such standard implementation details are discussed in Durbin and Koopman (1997).

In our discussion we distinguish models that include the two predictors in $\mathbf{x}_t$ and models that depend only on the dynamics in u_t. It allows us to assess whether including dynamic structures, such as the autoregressive structure, improves the model fit when compared to only having the intercept in the model. We include the following model specifications from Section 7.2.1 for the latent stochastic component u_t: constant, random walk, AR(1), ARFI(0,d), ARFI(1,d), and the stochastic cycle.

The parameter estimates of $\boldsymbol{\psi}$ for each specification are presented in Table 7.1. The top panel shows the results for models that include the predictors in $\mathbf{x}_t$. For all models, the estimated long-run mean μ is not significantly different from zero. Hence we can conclude that the models do not significantly predict Cambridge or Oxford as the winner of the Boat Race in the long run. The coefficient estimates for μ are all positive. This finding corresponds to the lead up to 2010 in the series for Cambridge (eighty wins against seventy-four for Oxford).

The difference in weight between the two boats is a strong predictor. The estimated corresponding coefficient is positive and significant for all model specifications. Hence we may conclude that having more muscle power in the boat is more important than having less water resistance. Ideally, we would like to investigate this weight difference in more detail to verify whether the muscle power causes this effect or whether the weight difference presents a proxy for a different set of variables. This requires additional information that is yet unavailable to us.

The outcome of the toss does not come forward as an important predictor. It is insignificant in all models with small coefficient values. This is rather surprising, given that the gambling odds usually change after the toss is made. It may have some impact when we account for certain weather conditions under which it may signify some advantage when a team starts at a specific side of the river. The winning distance by which the previous year's race was won does not provide further predictability. It is insignificant in all models, also with small coefficient values.

The parameters that account for the dynamic effects in the model, including autoregressive coefficients, improve the model fit. For example, the parameter ϕ_1 (or ϕ) is estimated between 0.75 and 0.85 for the AR(1), ARFI(1,d), and stochastic cycle models. The long-memory parameter d is only positive and significant for the ARFI(0,d) model but it cancels out when the ϕ_1 parameter is also included as it is in the case for the ARFI(1,d) model.

The stochastic cycle model shows overall the best in-sample fit according to the estimated log-likelihood value and the AIC criteria. The cycle frequency is estimated at 0.316, which indicates a cycle period of nearly eighteen years. This is a long cycle period and is mainly caused by the long winning streaks in the sample.

The parameter estimates for the model that do not include predictors are shown in the bottom panel of Table 7.1. We have also removed the insignificant mean μ for most of the models. The results are very similar when we compare these for models including predictors. Overall they perform less well in terms of likelihood and AIC values. The estimated dynamic parameters are, however, similar.

Figure 7.2 presents the smoothed estimate of the probabilty π_t which is the logit transformation of the signal θ_t. The estimate is conditional on all data (the smoothed estimate) and is computed using the importance sampling technique, as discussed in Durbin and Koopman (2012), Section 11.4. The top panel shows them for models that include covariates. Here, the estimated signals are mainly determined by the log difference in weight covariate. We have experimented with different transformations for this variable, but the jumps in the weight differences have remained large. The bottom panel for models without predictors presents much smoother estimates of the signal θ_t.

Model diagnostics

To guarantee a $\sqrt{M}$ convergence rate for the Monte Carlo likelihood estimator (7.11), we must ensure that the weight function $w(\mathbf{u}^{(i)}, y; \mathbf{X}, \boldsymbol{\psi})$ in (7.12) has a finite variance as argued by Geweke (1989). To verify whether this is the case for our model specifications, we have implemented the Wald test as discussed by Koopman et al. (2009) and based on extreme value theory. The test statistic is evaluated at the estimated parameters in Table 7.1 and is constructed by the following step-wise procedure:

1. Given a model and its corresponding estimated parameters $\hat{\boldsymbol{\psi}}$, we generate 100 000 weight values $w(\mathbf{u}^{(i)}, y; \mathbf{X}, \hat{\boldsymbol{\psi}})$.
2. Given a threshold value $w^{\min}$, we only consider weights that surpass the threshold.
3. The resulting s exceedences $z_1, \ldots, z_s$ are modeled by the generalized Pareto density $f(w_i; a, b) = -\log b - (1 + a^{-1}) \log(1 + ab^{-1} z_i)$, for $i = 1, \ldots, s$; the coefficients a and b are estimated by maximum likelihood to be straightforward; the estimates are denoted by $\hat{a}$ and $\hat{b}$, respectively. The null hypothesis of variance existence is given by $H_0 : a \leq 0.5$.
4. The test statistic $t_w = \hat{b}^{-1} \sqrt{s/3}(\hat{a} - 0.5)$ is under the null hypothesis asymptotically standard normal. We reject the null for a sufficiently positive value of t_w.

Table 7.1 Parameter estimation results for the Boat Race from 1829 until 2010 ($n = 182$)

	μ	ϕ	σ_η	λ	d	β_1	β_2	β_3	$\log \hat{L}(\psi, y)$	AIC
					Including predictors					
Constant	0.148 $_{0.242}$					0.225 $_{0.063}$	0.099 $_{0.356}$	0.030 $_{0.022}$	–238.967	485.935
RW	0.366 $_{0.876}$		0.161 $_{0.162}$			0.221 $_{0.066}$	0.103 $_{0.387}$	0.019 $_{0.030}$	–238.737	487.474
AR(1)	0.377 $_{0.524}$	0.754 $_{0.096}$	1.264 $_{0.430}$			0.260 $_{0.094}$	–0.115 $_{0.520}$	–0.047 $_{0.045}$	–234.44	480.881
ARFI(0,d,0)	0.424 $_{1.177}$		2.120 $_{0.662}$		0.331 $_{0.127}$	0.335 $_{0.124}$	–0.037 $_{0.618}$	–0.018 $_{0.044}$	–236.837	485.674
ARFI(1,d,0)	0.489 $_{0.619}$	0.778 $_{0.145}$	1.943 $_{0.683}$		–0.079 $_{0.253}$	0.311 $_{0.125}$	–0.173 $_{0.624}$	–0.080 $_{0.059}$	–233.068	480.135
Cycle	0.390 $_{0.573}$	0.858 $_{0.053}$	1.655 $_{0.467}$	0.316 $_{0.082}$		0.304 $_{0.114}$	–0.109 $_{0.637}$	–0.098 $_{0.090}$	–231.273	476.546
					Only dynamics					
Constant	0.078 $_{0.161}$								–248.144	498.289
RW			0.386 $_{0.295}$						–246.894	495.787
AR(1)		0.753 $_{0.101}$	1.075 $_{0.368}$						–240.626	485.253
ARFI(0,d,0)			1.413 $_{0.454}$		0.391 $_{0.104}$				–243.803	491.605
ARFI(1,d,0)		0.771 $_{0.159}$	1.100 $_{0.407}$		–0.038 $_{0.266}$				–240.616	487.232
Cycle		0.872 $_{0.066}$	0.764 $_{0.288}$	0.294 $_{0.086}$					–239.391	484.781

Notes: The standard errors of the estimates are given in small print. The method of Monte Carlo maximum likelihood is based on importance sampling. The number of importance simulations for likelihood evaluation is $M = 1000$. The regression parameters β_1, β_2, and β_3 correspond to the effect of the log average difference in weight between the rowers of the boats of Cambridge and Oxford, the effect of Cambridge winning the toss, and the log winning distance from the previous years race, respectively. The Akaike information criteria is computed as $2\mathbf{k} - 2\log \hat{p}(\mathbf{y}; \hat{\psi})$, where $\mathbf{k}$ denotes the number of parameters present in model k.

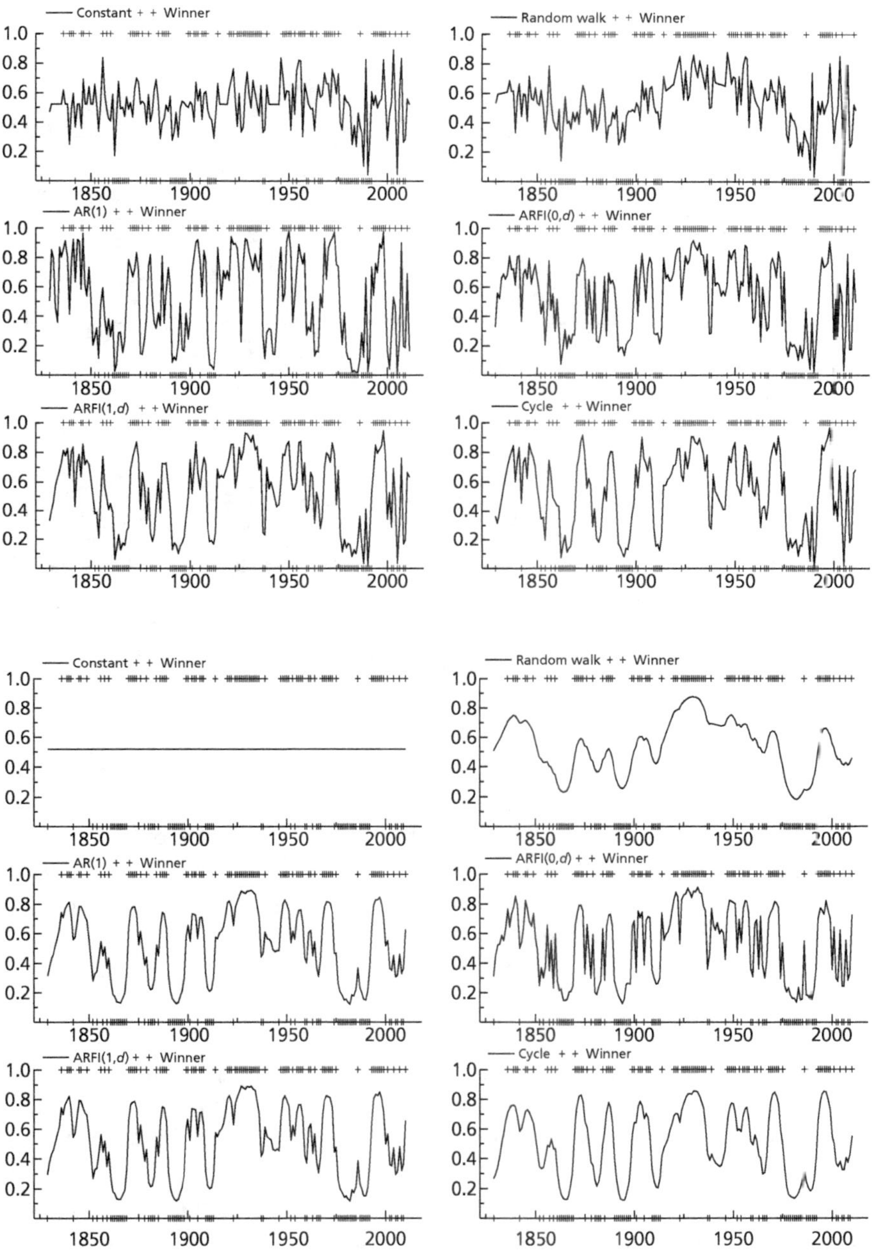

Figure 7.2. Estimated probabilities, $\mathrm{E}_p(\pi_t|\mathbf{y};\mathbf{X},\hat{\boldsymbol{\psi}})$ where $\pi_t = \exp(\theta_t)\,/\,(1+\exp(\theta_t))$, for all models from Table 7.1

Note: The top panel presents the probabilities for models including predictors $\mathbf{X}$ while the bottom panel shows the probabilities for models that only include dynamics.

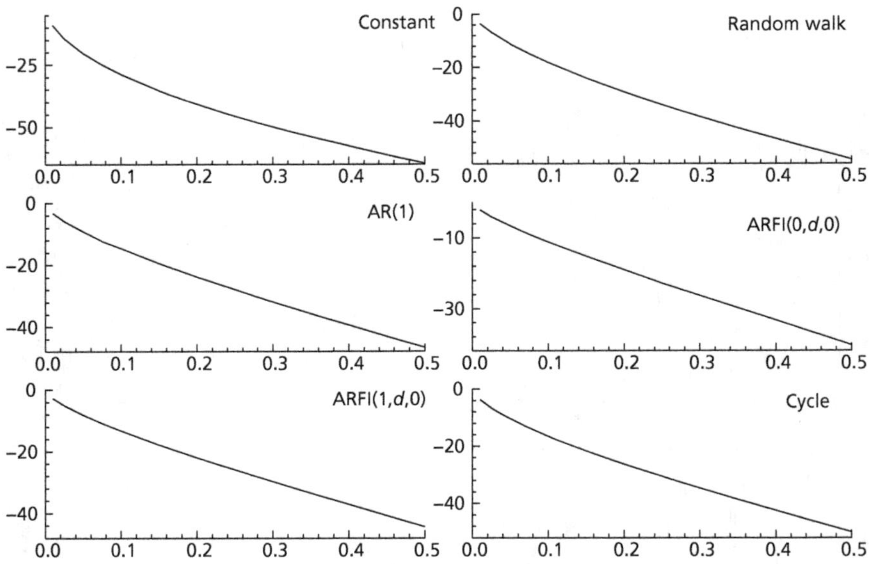

Figure 7.3. Importance sampling test statistics t_w for all models in the top panel of Table 7.1

Note: The test statistics are based on 100 000 simulated weights $w(\mathbf{u}^{(i)}, \mathbf{y}; \mathbf{X}, \psi)$. For each model we compute the test statistics for different thresholds w^{min}. The threshold value is implied by the number of exceedence values s. The x-axis refers to the proportion of exceedences $s/100\,000$ and the y-axis displays t_w. All test statistics are sufficiently negative for variance existence.

In Figure 7.3 we present the test statistics for the models in the top panel of Table 7.1. The test statistics are computed for a multiple of thresholds that result in sets of exceedence weights (with numbers of 1% to 50% of all weights). The test statistics are never rejected, which indicate that the variance exists for many different threshold values. A similar panel of graphics can be produced for models without predictors.

Forecasting outlook

The likelihood values and the AIC criteria in Table 7.1 imply that the in-sample performance of our range of models is quite similar. It does not imply that the out-of-sample forecasting properties are also similar. We can illustrate this by presenting in Figure 7.4 the theoretical autocorrelation functions (ACF) of the estimated latent processes u_t. The ACFs are implied by the model specification with the parameters replaced by their estimates as given in the top panel of Table 7.1.

The constant model takes no serial correlation into account; its ACF values are zero for all lags and is not included in Figure 7.4. All past outcomes of the Boat Race have an equally weighted impact on the forecast of the outcome this

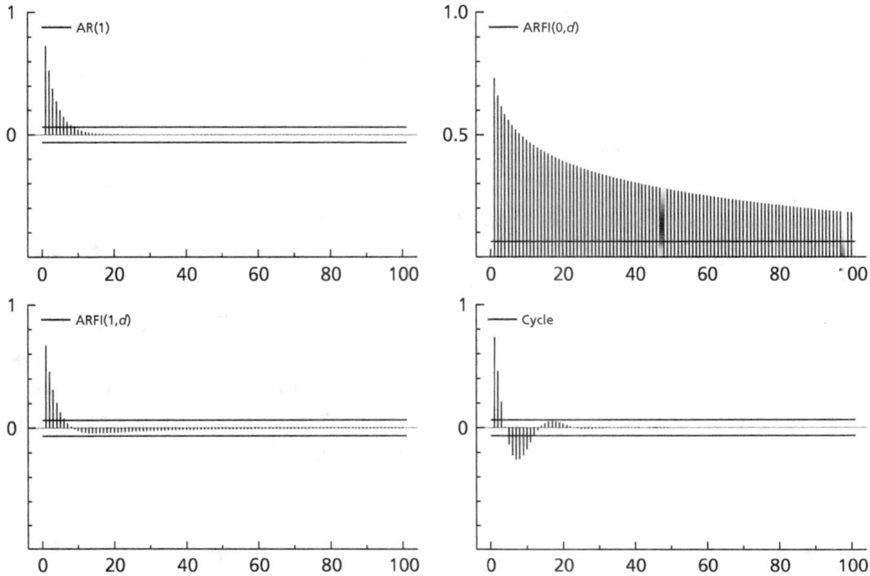

Figure 7.4. Theoretical autocorrelation functions based on the estimated parameters in the top panel of Table 7.1 for a selection of the stochastic processes u_t

year. The random walk model is a non-stationary dynamic process and its ACF is not properly defined. But for the random walk process, all outcomes have no impact on the current forecast except the last, most recent outcome. The AR(1) and ARFI(1,d) models have nearly identical autocorrelation functions. This is not surprising since the d and ϕ parameters in the ARFI(1,d) cancel out against each other to yield the short-memory AR(1) model. In both cases, only the more recent outcomes have an impact on the current forecast. The ACF of the ARFI(0,d) model shows hyperbolic decay, which is characteristic of the long-memory process. The autocorrelation function for the stochastic cycle model is first positive for almost five lags, after which it becomes significantly negative for another five lags. This corresponds to the pattern of winning streaks that we observe in the data.

To summarize the in-sample findings. The overall in-sample performance of the models are similar: no large differences in the fit of the models are obtained. The observed average difference in the weights of the two boats is important as a covariate or predictor. At the time that the forecast is usually made, the weights of the boats are known. The models with an autoregressive structure perform relatively better. The out-of-sample forecasting performance is likely to give different results because the implied ACFs are different for different models, as indicated in Figure 7.4.

7.3 A forty year forecasting assessment

In this section we discuss the out-of-sample forecasting results. All models are used to forecast the Cambridge winning probability repeatedly for a period of forty years. The first forecast is made for the race in 1971 and the final forecast is made for the race in 2010. The forecasts are compared across models based on various loss functions and predictive ability tests, as discussed by Diebold and Mariano (1995). Further details of our design of the forecasting study are discussed below.

Empirical evidence from out-of-sample forecast performance is generally more relevant for bookmakers and gamblers than evidence based on in-sample performances, such as the results presented in Section 7.2. For example, gamblers may wish to know how many times the Boat Race would have been correctly predicted by a particular model for a number of years. A comparison of the models based on these historical forecasting performance yields insight in their comparative advantages. We refer to further discussions by White (2000) and Diebold (2012) where the pros and cons of an in-sample versus out-of-sample model comparison are given in more detail.

7.3.1 FORECASTING PROCEDURE

We perform an out-of-sample forecast study where we repeatedly forecast the probability that Cambridge wins the Boat Race based on information that is available just before the race takes place. Forecasts for years 1971 until 2010 are computed using a rolling forecast window. The choice for the rolling window is motivated by assumptions underlying the test statistics employed.

To illustrate our procedure in detail, consider the first subsample. It consists of the outcomes corresponding to the years 1829 to 1970. Using this subsample, we produce a forecast for 1971. The next subsample contains outcomes from years 1830 to 1971 and produces a forecast for 1972. The final forecast, for 2010, is based on outcomes from 1869 until 2009. In total, we construct $m = 40$ one-year-ahead real-time forecasts for each model.

Let the integer N denote the year 1970, which corresponds to the sample split point and the 114th Boat Race (142–28 missing), respectively. The forecasts are computed for $t = N + 1, \ldots, N + m$ based on subsamples of the observations $y_j, \ldots, y_{N+j-1}$, for $j = 1, \ldots, m$. We estimate the parameter vector ψ for each subsample $y_j, \ldots, y_{N+j-1}$ and denote the estimate by $\hat{\psi}_j$. The predicted probability of Cambridge winning is computed by

$$\hat{\pi}_{N+j|j,\ldots,N+j-1}$$
$$= \frac{\sum_{i=1}^{M} \exp\left(\theta^{(i)}_{N+j|j,\ldots,N+j-1}\right) \;/\; \left(1 + \exp(\theta^{(i)}_{N+j|j,\ldots,N+j-1})\right) w(\mathbf{u}^{(i)}, \mathbf{y}; \mathbf{X}, \hat{\psi}_j)}{\sum_{i=1}^{M} w(\mathbf{u}^{(i)}, \mathbf{y}; \mathbf{X}, \hat{\psi}_j)}, \tag{7.13}$$

where

$$\theta^{(i)}_{N+j|j,\dots,N+j-1} = \hat{\mu} + \mathbf{x}_{N+j}\hat{\beta} + u^{(i)}_{N+j|j,\dots,N+j-1},$$

for $j = 1, \dots, m$. The one-year-ahead prediction for the stochastic component $u^{(i)}_{N+j|j,\dots,N+j-1}$ is based on the linear Gaussian importance model $g(\mathbf{y}|\mathbf{u})g(\mathbf{u})$ and is computed by the Kalman filter; see Durbin and Koopman (2012), Chapter 11, for further details.

Let the set of models $\mathcal{M}$ include all models. The models in $\mathcal{M}$ are indexed by k for $k = 1, \dots, K$, where K is the number of models in set $\mathcal{M}$. We adjust the notation for the one-year-ahead forecast to reflect that $\hat{\pi}^k_{N+j|j,\dots,N+j-1}$ is the forecast for sample $j = 1, \dots, m$ and model $k = 1, \dots, K$.

7.3.2 LOSS FUNCTIONS

To evaluate the predictive accuracy of the different models, we rely on loss functions. We can refer to these as scores when evaluating probability forecasts. In the literature a large variety of loss functions exist and, as discussed by Diebold (1993), appropriate loss functions depend on the situation at hand. From a gamblers perspective, the loss function should have the betting odds in each year as an argument. This would allow us to measure the loss in terms of lost investments in a similar way as when trading rules are evaluated; see Hsu and Kuan (2005). However, this information is not available to us.

In our study we rely on more general loss functions which we have computed as follows. For $j = 1, \dots, m$ and $k = 1, \dots, K$, we compute:

1. Brier loss:

$$L^{1,k}_{N+j} = 2(y_{N+j} - \hat{\pi}^k_{N+j|j,\dots,N+j-1})^2.$$

2. Predictive log-likelihood loss:

$$L^{2,k}_{N+j} = -y_{N+j}\log(\hat{\pi}^k_{N+j|j,\dots,N+j-1}) - (1 - y_{N+j})\log(1 - \hat{\pi}^k_{N+j|j,\dots,N+j-1}).$$

3. Incorrect loss:

$$L^{3,k}_{N+j} = \begin{cases} 1 \text{ if } & \hat{\pi}^k_{N+j|j,\dots,N+j-1} > 0.5 \text{ and } y_{N+j} = 0 \\ 1 \text{ if } & \hat{\pi}^k_{N+j|j,\dots,N+j-1} \leq 0.5 \text{ and } y_{N+j} = 1 \\ 0 \text{ else} & \end{cases}.$$

The Brier loss function, $L^{1,k}_{N+j}$, is proposed by Brier (1950) and is well known for its use in evaluating weather forecasts. It may be viewed as the mean squared error loss function for probabilistic forecasts. The difference is that $\hat{\pi}^k_{N+j|j,\dots,N+j-1}$ is not a forecast for y_{N+j}, but rather a probability statement for the event $y_{N+j} = 1$. The predictive log-likelihood loss function, $L^{2,k}_{N+j}$, uses the negative of the log-likelihood as a measure of accuracy, which has optimal value zero. The incorrect loss function, $L^{3,k}_{N+j}$, measures the loss arising from event forecasting. It takes values of one or zero, indicating loss or success, respectively.

Using these loss functions we define the relative performance of a model l against another model k, with $k, l \in \mathcal{M}$, by

$$d_j^{s,lk} \equiv L^{s,l}_{N+j} - L^{s,k}_{N+j}, \quad s = 1, 2, 3, \quad l \neq k \quad k, l = 1, \ldots, K, \quad j = 1, \ldots, m. \tag{7.14}$$

All predictive ability tests that we consider use functions of the relative performance vectors, $\mathbf{d}^{s,lk} = (d_1^{s,lk}, \ldots, d_m^{s,lk})'$, as test statistics. It is easy to show that the vector series $\mathbf{d}^{s,lk}$ is stationary. Let the first moment of the relative performance indicators be denoted by $\xi^{s,lk} = E[d_j^{s,lk}]$, which we assume independent from j, for all $l, k \in \mathcal{M}$, and $s = 1, 2, 3$. For notational convenience we drop the dependence on the type of loss function s, when discussing the test statistics below.

7.3.3 EQUAL PREDICTIVE ABILITY

Equal predictive ability (EPA) tests are based upon the null hypothesis that their is no difference in accuracy between the competing models. For comparing models l and k this results in the null hypothesis $H_0 : \xi^{lk} = 0$; see Diebold and Mariano (1995) and West (1996).

The widely used EPA t-type test statistic, first proposed in Diebold and Mariano (1995), is given by

$$T_{lk}^{EPA} \equiv \frac{m^{1/2} \bar{d}^{lk}}{\sqrt{\hat{\text{Var}}(d^{lk})}}, \tag{7.15}$$

where $\bar{d}^{lk} = m^{-1} \sum_{j=1}^{m} d_j^{lk}$ and $\hat{\text{Var}}(d^{lk})$ is a consistent estimate to the long-run variance matrix. Several approaches to estimating this variance matrix exist. A convenient choice is given by the Newey and West (1987) estimator. In our empirical work we follow the implementation suggested in Diebold and Mariano (1995).

When the models are nested under the null hypothesis of equal predictive ability, the asymptotic theory from West (1996) may not apply. For our models the nesting relationships are non-trivial. It is unclear given the nonlinearities in the observation density, as well as in some of the signal processes, whether the models are nested in the conventional sense. Clark and McCracken (2001) derive asymptotic properties and critical values for the EPA test statistics when models are nested. We view our EPA test statistic as indicative and we assume that it converges to a standard normal distribution as m goes to infinity, as shown in Clark and McCracken (2001). We report the right-sided p-value for the probability that model l outperforms model k.

7.3.4 RESULTS

We discuss the empirical results for our forty years of out-of-sample forecasting and testing. All models from Section 7.2 with the parameter estimates given in Table 7.1 are used to obtain forecasts for the Boat Race for the years from 1971 until 2010. We do not use the winner of the toss and the winning distance as possible predictors as their values were found insignificant for all models. Further, we do not show the results for the ARFI(1,d) model as its results were found indistinguishable from the AR(1) model results.

We do include a two-state Markov chain model as a simple competitor from a different model class. The Markov chain model is described by the transition probabilities $p_{11} = p(y_t = 1|y_{t-1} = 1)$, $p_{01} = p(y_t = 0|y_{t-1} = 1) = 1 - p_{11}$, $p_{00} = p(y_t = 0|y_{t-1} = 0)$, and $p_{10} = p(y_t = 1|y_{t-1} = 0) = 1 - p_{00}$. These probabilities are estimated from the data. In particular, we estimate μ_i and β_i, where $p_{ii} = \exp(\mu_i + \mathbf{x}_t\beta_i)/[1 + \exp(\mu_i + \mathbf{x}_t\beta_i)]$, for $i = 1, 2$. The likelihood for the Markov chain model is known in closed form.

Also, we include a number of non-parametric forecasting rules for comparison purposes. In particular, we include forecasts that select: last year's race winner (the team is in a flow), last year's race loser (the team is highly motivated), always Cambridge (duck egg blue supporter), and always Oxford (dark blue supporter). The goal is to show that at least a selection of the parametric models of Section 7.2 are able to outperform these simple forecasting procedures. The forecasts for the parametric models of Section 7.2 are computed using a rolling forecasting window. For each model we compute forty forecasts using equation (7.13) with $M = 1000$ number of draws from the importance density.

Forecasting results

The average loss functions $m^{-1}\sum_{j=1}^{m} L_{N+j}^{s,k}$ for models $k = 1, \ldots, K$ are shown in Table 7.2. The non-parametric forecasting rules are not evaluated for the log loss function, $s = 2$, as they do not provide probability forecasts.

First, we discuss the models that include the average log difference in weight between the boats as a predictor. For all loss functions, the binary model with the stochastic cycle component of Harvey (1989) as the signal u_t provides the lowest loss scores. Also, the model that includes the AR(1) component in the signal θ_t yields comparably low scores. The incorrect loss function reveals that our best model (Cycle) is able to predict thirty-one out of forty

Table 7.2 Average loss functions for forty year forecasting for the Boat Race from 1971 until 2010 ($m = 40$ forecasts)

Model	Brier	LOG	Incorrect
		Including predictors	
Constant	0.417	0.610	0.375
Random walk	0.446	0.642	0.375
AR(1)	0.383	0.564	0.275
ARFI(0,d,0)	0.396	0.581	0.325
Cycle	0.371	0.555	0.225
Mchain	0.388	0.563	0.350
		Only dynamics	
Constant	0.522	0.715	0.625
Random walk	0.489	0.706	0.350
AR(1)	0.431	0.623	0.350
ARFI(0,d,0)	0.455	0.648	0.350
Cycle	0.430	0.620	0.350
Mchain	0.479	0.674	0.350
		Non-parametric rules	
Winner	0.700	-	0.350
Loser	1.300	-	0.650
Cambridge	1.200	-	0.600
Oxford	0.800	-	0.400

Note: Each forecast is based on 142 observations. Brier corresponds to loss function $m^{-1}\sum_{j=1}^{m} L_j^{1,k}$, LOG to $m^{-1}\sum_{j=1}^{m} L_j^{2,k}$ and Incorrect to $m^{-1}\sum_{j=1}^{m} L_j^{3,k}$. The highlighted numbers indicate the lowest loss per category per loss function.

races correctly. This is based on predicting Cambridge as the winner if the predicted probability is larger than one half and vice versa. We notice that the forecasting difference between the different parametric signal specification is small. For example, for the Brier loss function, an additional mean squared error of only 0.04 is found between the worst and the best model. The Markov chain model performs comparably to the models with autoregressive dynamics in the state.

For the models that do not include the difference in the weight of the two boats as a predictor, the differences between the models are larger. This is illustrated in Figure 7.5, where we present the predicted probability paths for all parametric models. The difference in weight predictor clearly dominates the forecasts when it is included. When it is not included, the constant and random walk models perform much worse, whereas the other parametric models are less affected. Overall we may conclude that the inclusion of the weight predictor is important. The incorrect loss function is not able to distinguish between the models when the log weight predictor is not included.

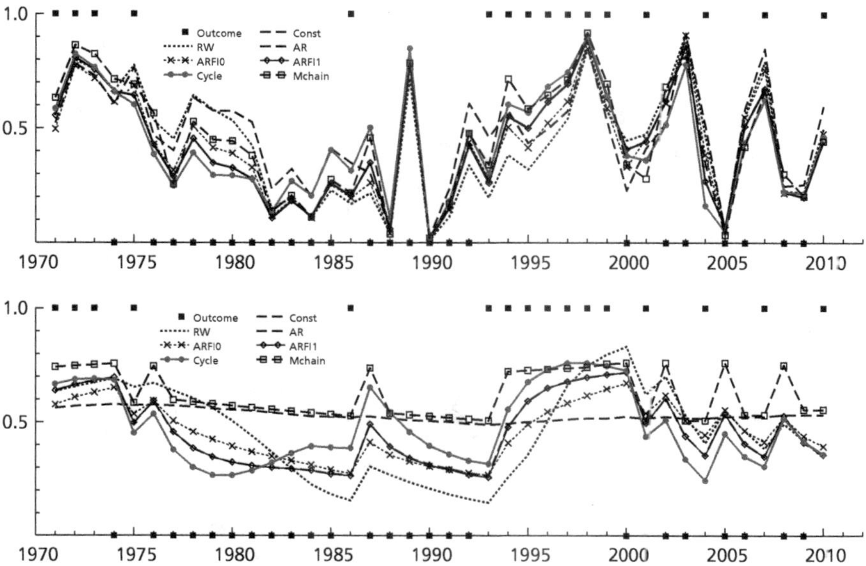

Figure 7.5. Forty year forecasts for the Boat Race from 1971 until 2010 ($m = 40$ forecasts)

Note: Each forecast is based on 142 observations. The top panel shows the forecasts for the parametric models of Section 7.2 including the weight difference of the two boats as a predictor. The bottom panel shows the forecast for models including only dynamics.

The average loss functions for the non-parametric forecasting rules are presented in the lower panel of Table 7.2. The Brier loss function attributes very high losses to these models. It is shown that the "always Cambridge" and "loser" rules predict less than 50% of the races correctly. It seems that always forecasting last year's race winner is a reasonable strategy.

Predictive ability results

In Table 7.3 we present the p-values for the EPA test statistics for the models that include the log difference in weight as a predictor. Each model is compared against all other models for all loss functions. When the p-values are smaller than α, it indicates that the model listed in the row is outperformed by the model listed in the column, with significance level α which we take as $\alpha = 0.1$ in our results. For the Brier loss function, the EPA tests give two main results. First, all non-parametric forecasting rules are significantly outperformed by all parametric models. The rules that predict the "loser" and "always Cambridge" appear to be especially, unsatisfactory when aiming to produce accurate forecasts. Second, the model that includes a random walk process in the signal is outperformed by models that include a more advanced stochastic process in the signal, such as the AR(1), ARFI(0,d), and stochastic cycle processes.

Table 7.3 Equal predictive ability tests for comparing forty year forecasting results for the Boat Race from 1971 until 2010 ($m = 40$ forecasts) for models that include predictors

	Equal predictive ability; including regressors									
Bench. / Alt.	Constant	Random walk	AR(1)	ARFI(0,d)	Cycle	Mchain	Winner	Loser	Cambridge	Oxford
					Brier loss					
Constant	–	0.723	0.174	0.226	0.132	0.212	0.954	1.000	1.000	0.988
Random walk	0.277	–	0.015	0.042	0.050	0.080	0.956	1.000	1.000	0.986
AR(1)	0.826	0.985	–	0.818	0.279	0.596	0.986	1.000	1.000	0.996
ARFI(0,d)	0.774	0.958	0.182	–	0.207	0.393	0.976	1.000	1.000	0.995
Cycle	0.868	0.950	0.721	0.793	–	0.749	0.989	1.000	1.000	0.996
Mchain	0.788	0.920	0.404	0.607	0.251	–	0.988	1.000	1.000	0.994
Winner	0.046	0.044	0.014	0.024	0.011	0.012	–	0.977	0.984	0.692
Loser	0.000	0.000	0.000	0.000	0.000	0.000	0.023	0.000	0.308	0.016
Cambridge	0.000	0.000	0.000	0.000	0.000	0.000	0.016	0.692	–	0.098
Oxford	0.012	0.014	0.004	0.005	0.004	0.006	0.308	0.984	0.902	–
					Log loss					
Constant	–	0.697	0.138	0.185	0.142	0.147				
Random walk	0.303	–	0.017	0.045	0.079	0.063	–	–	–	–
AR(1)	0.862	0.983	–	0.824	0.381	0.484	–	–	–	–
ARFI(0,d)	0.815	0.955	0.176	–	0.272	0.304	–	–	–	–
Cycle	0.858	0.921	0.619	0.728	–	0.588	–	–	–	–
Mchain	0.853	0.937	0.516	0.696	0.412	–	–	–	–	–
Winner	–	–	–	–	–	–	–	–	–	–
Loser	–	–	–	–	–	–	–	–	–	–
Cambridge	–	–	–	–	–	–	–	–	–	–
Oxford	–	–	–	–	–	–	–	–	–	–

	Incorrect loss									
Constant	–	0.500	0.073	0.238	0.023	0.098	0.409	0.995	0.986	0.586
Random walk	0.500	–	0.045	0.205	0.012	0.045	0.391	0.988	0.975	0.596
AR(1)	0.927	0.955	–	0.927	0.073	0.500	0.820	0.999	0.998	0.906
ARFI(0,d)	0.762	0.795	0.073	–	0.018	0.156	0.609	0.997	0.991	0.782
Cycle	0.977	0.988	0.927	0.982	–	0.927	0.958	1.000	1.000	0.970
Mchain	0.902	0.955	0.500	0.844	0.073	–	0.844	0.999	0.999	0.892
Winner	0.591	0.609	0.180	0.391	0.042	0.156	–	0.977	0.984	0.692
Loser	0.005	0.012	0.001	0.003	0.000	0.001	0.023	–	0.308	0.016
Cambridge	0.014	0.025	0.002	0.009	0.000	0.001	0.016	0.692	–	0.098
Oxford	0.414	0.404	0.094	0.218	0.030	0.108	0.308	0.984	0.902	–

Note: The t^{EPA} is constructed for different loss functions, as discussed in Section 7.3.3. The highlighted numbers are significant for $\alpha = 0.1$.

Also, the random walk model is outperformed by the Markov chain model. Perhaps surprisingly, the logistic regression model (constant) is not outperformed by the more advanced dynamic models that take the serial correlation into account. However, the AR(1) and stochastic cycle-based models are most unlikely to be outperformed.

The EPA tests based on the log loss function confirm most of the results that were found for the Brier loss function. Again the random walk model is statistically outperformed by the models that include more advanced stochastic processes. The EPA test based on the log loss function is also not able to distinguish between the forecasting performances of the other parametric models. In other words, no evidence in found in favor of either the logistic regression (constant), AR(1), ARFI(0,d), stochastic cycle, or Markov chain models.

The EPA tests based on the incorrect loss function show somewhat different results. The logistic regression (constant) model is now significantly outperformed by the AR(1), stochastic cycle, and Markov chain models. The random walk model is also outperformed by the AR(1), stochastic cycle, and Markov chain models. Interestingly, the forecasting rule that predicts the last year's race winner seems to perform much better based on the incorrect loss function. Only the stochastic cycle model is able to reject this model. It also rejects the Markov chain model.

In Table 7.4 we show the results for the models that only depend on dynamics and do not include predictors. Here, the differences between the models are more clearly visible for the Brier and log loss functions. The constant, random walk, and all non-parametric forecasting rules are significantly outperformed by the other models for the Brier and log loss functions. For the incorrect loss function the opposite is found. Here, only the constant model and the "always Cambridge" and "loser" prediction rules are outperformed.

When summarizing the testing results for EPA tests, we may conclude that the parametric models predict the Boat Race significantly better when compared to the non-parametric methods based on ad hoc rules. Furthermore, we have provided substantial evidence in favor of models that include an autoregressive component in the signal. These are the models with AR(1) or stochastic cycle process in the signal. There is little evidence against the ARFI(0,d) and Markov chain models but they do not significantly outperform the other models.

7.4 **Conclusion**

The forecasting of the outcomes of the yearly Boat Race between Cambridge and Oxford over the forty year period 1971–2010 is extensively evaluated in a real-time study. The accuracy of the forecasts is measured by different

Table 7.4 Equal predictive ability tests for comparing forty year forecasting results for the Boat Race from 1971 until 2010 ($m = 40$ forecasts) for models that only include dynamics

	Equal predictive ability; only dynamics									
Bench. / Alt.	Constant	Random walk	AR(1)	ARFI(0,d)	Cycle	Mchain	Winner	Loser	Cambridge	Oxford
					Brier loss					
Constant	–	0.328	0.025	0.044	0.027	0.189	0.880	1.000	1.000	0.955
Random walk	0.672	–	0.065	0.205	0.151	0.435	0.965	1.000	1.000	0.976
AR(1)	0.975	0.935	–	0.941	0.495	0.965	0.991	1.000	1.000	0.993
ARFI(0,d)	0.956	0.795	0.059	–	0.241	0.757	0.974	1.000	1.000	0.990
Cycle	0.973	0.849	0.505	0.759	–	0.962	0.991	1.000	1.000	0.990
Mchain	0.811	0.565	0.035	0.243	0.038	–	0.984	1.000	1.000	0.975
Winner	0.120	0.035	0.009	0.026	0.009	0.016	–	0.977	0.984	0.692
Loser	0.000	0.000	0.000	0.000	0.000	0.000	0.023	–	0.308	0.016
Cambridge	0.000	0.000	0.000	0.000	0.000	0.000	0.016	0.692	–	0.098
Oxford	0.045	0.024	0.007	0.010	0.010	0.025	0.308	0.984	0.902	–
					Log loss					
Constant	–	0.461	0.031	0.051	0.031	0.209				
Random walk	0.539	–	0.066	0.164	0.125	0.337				
AR(1)	0.969	0.934	–	0.936	0.470	0.963				
ARFI(0,d)	0.949	0.836	0.064	–	0.236	0.764				
Cycle	0.969	0.875	0.530	0.764	–	0.965				
Mchain	0.791	0.663	0.037	0.236	0.035	–				
Winner							–			
Loser								–		
Cambridge									–	
Oxford										–

(continued)

Table 7.4 Continued

	Equal predictive ability; only dynamics									
Bench. / Alt.	Constant	Random walk	AR(1)	ARFI(0,d)	Cycle	Mchain	Winner	Loser	Cambridge	Oxford
	Incorrect loss									
Constant	–	0.003	0.007	0.003	0.007	0.007	0.007	0.596	0.327	0.059
Random walk	0.997	–	0.500	0.500	0.500	0.500	0.500	0.988	0.984	0.692
AR(1)	0.993	0.500	–	0.500	0.500	0.500	0.500	0.980	0.980	0.704
ARFI(0,d)	0.997	0.500	0.500	–	0.500	0.500	0.500	0.985	0.984	0.692
Cycle	0.993	0.500	0.500	0.500	–	0.500	0.500	0.980	0.984	0.692
Mchain	0.993	0.500	0.500	0.500	0.500	–	0.500	0.977	0.984	0.692
Winner	0.993	0.500	0.500	0.500	0.500	0.500	–	0.977	0.984	0.692
Loser	0.404	0.012	0.020	0.015	0.020	0.023	0.023	–	0.308	0.016
Cambridge	0.673	0.016	0.020	0.016	0.016	0.016	0.016	0.692	–	0.098
Oxford	0.941	0.308	0.296	0.308	0.308	0.308	0.308	0.984	0.902	–

Note: The t^{EPA} is constructed for different loss functions, as discussed in Section 7.3.3. The highlighted numbers are significant for $\alpha = 0.1$.

loss functions and by equal predictive ability tests. The overall finding is that parametric models predict the outcome of the Boat Race significantly better than ad hoc methods. Furthermore, models with a latent autoregressive component in the signal produce the most accurate forecasts. Although this study has been mostly fun for us, we do believe that statistical dynamic models have a serious role to play in event forecasting. While other events may have a more serious impact on us than outcomes of the Boat Race, the ability to forecasts binary time series accurately is important. The formulation of dynamic models, the development of estimation and forecasting procedures, and the assessment of significant outperformance in forecasting accuracy in the context of binary time series may provide many interesting research questions.

8 Tests for serial dependence in static, non-Gaussian factor models

Gabriele Fiorentini and Enrique Sentana

8.1 Introduction

There is a long tradition of factor or multi-index models in finance, where they were originally applied to simplify the computation of the covariance matrix of returns in a mean-variance portfolio allocation framework (see Connor (1984) for a recent monograph). In this context, the common factors usually correspond to unobserved fundamental influences on returns, while the idiosyncratic factors reflect asset-specific risks. In addition, the concept of factors plays a crucial role in two major asset pricing theories: the mutual fund separation theory (see, e.g., Ross (1978)), of which the standard capital asset pricing model (CAPM) is a special case, and the arbitrage pricing theory (see Ross (1976) and Connor (1984) for a unifying approach).

Factor models for low frequency financial returns are routinely estimated by Gaussian maximum likelihood under the assumption that the observations are serially independent using statistical factor analysis routines (see Lawley and Maxwell (1971)). In this context, the EM algorithm of Dempster et al. (1977) and Rubin and Thayer (1982) provides a cheap and reliable procedure for obtaining initial values as close to the optimum as desired, as illustrated by Lehmann and Modest (1988) who successfully employed this algorithm to handle a very large cross-sectional dataset of monthly returns on individual U.S. stocks.

However, there are three empirical characteristics of assets returns which question the adequacy of this estimation procedure. First, there is some evidence of return predictability, which although far from uncontroversial, casts a doubt on the assumption of lack of serial correlation of common and idiosyncratic factors. Second, there is strong evidence that volatilities and correlations vary at high frequencies such as daily. Finally, many empirical studies with financial time series data indicate that the distribution of asset returns is rather leptokurtic, even after controlling for volatility clustering effects. In this context, the lack of normality implies that the Kalman filter prediction equations only provide the best linear least squares predictions and associated mean square errors, as opposed to the first two conditional moments (see Anderson and Moore (1979)), so that one cannot rely on standard results for Gaussian pseudo maximum likelihood estimators and tests, such as those in Bollerslev and Wooldridge (1992).

The objective of our paper is to provide joint diagnostic tests for serial dependence in the common and idiosyncratic factors that take into account the non-normality of asset returns. We will focus on Lagrange multiplier (or score) tests (LM tests), which only require estimation of the static model. As is well known, LM tests are asymptotically equivalent under the null and sequences of local alternatives to both likelihood ratio and Wald tests, and therefore share their optimality properties. In this context, our main contribution is to derive simple algebraic expressions for the score tests of serial correlation in the levels and squares of common and idiosyncratic factors in static factor models when the distribution of the innovations in the latent variables is elliptically symmetric, which can be either parametrically or semiparametrically specified. Elliptical distributions are attractive in this context because they generalize the multivariate normal distribution while retaining its tractability, irrespective of the number of assets. Importantly, our closed form tests are valid even though one must generally resort to simulation methods to approximate the log-likelihood function and its score in non-Gaussian state space models (see, e.g., Durbin and Koopman (2000) and the references therein). In addition, we also explain how to robustify the Gaussian versions of our LM tests when the return distribution is not normal. Finally, we derive tests that focus on either the common factors or the specific factors, or indeed on some of their elements.

We proceed in steps. We initially derive (i) tests against AR/MA-type serial correlation in the latent factors under the maintained assumption that they are conditionally homoskedastic; (ii) tests against ARCH-type effects in those latent variables under the maintained assumption that they are serially uncorrelated; and (iii) joint tests of (i) and (ii). To keep the notation to a minimum, we focus on single factor models throughout, which suffice to illustrate our main results. Extensions to multiple factors are considered in Fiorentini and Sentana (2012). We complement our theoretical results with detailed Monte Carlo exercises to study the finite sample reliability and power of our proposed tests, and to compare them to other existing procedures. Finally, we also apply our methods to monthly stock returns on U.S. broad industry portfolios.

The rest of the paper is organized as follows. First, we study the properties of likelihood-based estimators of the static factor model parameters under the null of serial independence. Then we derive tests against serial correlation in Section 8.3, against conditional heteroskedasticity in Section 8.4, and joint tests in Section 8.5. A Monte Carlo evaluation of all the different tests can be found in Section 8.6, followed by the empirical application to U.S. sectorial stock returns in Section 8.7. Finally, our conclusions, together with several interesting extensions, can be found in Section 8.8. Proofs and auxiliary results are gathered in the Appendices.

8.2 Static factor models

Consider the following traditional (i.e., static, conditionally homoskedastic, and exact) factor model:

$$\left.\begin{array}{c} \mathbf{y}_t = \boldsymbol{\pi} + \mathbf{c}f_t + \boldsymbol{\Gamma}^{1/2}\mathbf{v}_t^*, \\ \begin{pmatrix} f_t \\ \mathbf{v}_t^* \end{pmatrix} | I_{t-1}; \boldsymbol{\phi}_s \sim s\left[\begin{pmatrix} 0 \\ \mathbf{0} \end{pmatrix}, \begin{pmatrix} 1 & \mathbf{0} \\ \mathbf{0} & \mathbf{I}_N \end{pmatrix}; \boldsymbol{\eta}\right] \end{array}\right\} \tag{8.1}$$

where $\mathbf{y}_t$ is an $N \times 1$ vector of observable variables with constant conditional mean $\boldsymbol{\pi}$, f_t is an unobserved common factor, whose constant variance, λ, we have normalized to 1 to avoid the usual scale indeterminacy,[1] $\mathbf{c}$ is the $N \times 1$ vector of factor loadings, $\mathbf{v}_t^*$ is a $N \times 1$ vector of standardized idiosyncratic noises, which are conditionally orthogonal to, but not necessarily independent from, f_t, $\boldsymbol{\Gamma}$ is a $N \times N$ diagonal positive semidefinite (p.s.d.) matrix of constant idiosyncratic variances, I_{t-1} is an information set that contains the values of $\mathbf{y}_t$ and f_t up to, and including, time $t-1$, $\boldsymbol{\theta}_s = (\boldsymbol{\pi}', \mathbf{c}', \boldsymbol{\gamma}')'$, with $\boldsymbol{\gamma} = vecd(\boldsymbol{\Gamma})$ are the mean and variance parameters, $\boldsymbol{\eta}$ are some additional parameters that determine the shape of the conditional distribution of the spherical random vector $(f_t, \mathbf{v}_t^{*\prime})'$, which we assume has a well defined density, and $\boldsymbol{\phi}_s = (\boldsymbol{\theta}_s', \boldsymbol{\eta}')'$.

The most prominent example of spherical distribution is, of course, the standard normal distribution, which we assume corresponds to $\boldsymbol{\eta} = \mathbf{0}$. As in Bollerslev (1986) in a univariate context, and Harvey et al. (1992) in a multivariate one, followed by many others, we shall also consider in some detail a standardized multivariate Student t with ν degrees of freedom, or i.i.d. $t(\mathbf{0}, \mathbf{I}_N, \nu)$ for short, which approaches the multivariate normal as $\nu \to \infty$, or η, its reciprocal, goes to 0. More flexible families of spherical distributions are discrete scale mixtures of normals and Laguerre polynomial expansions of the multivariate normal density (see Amengual et al. (2013)), which could form the basis for a proper semiparametric procedure in which $\boldsymbol{\eta}$ would effectively be regarded as infinite dimensional.

Our assumptions trivially imply that

$$\mathbf{y}_t | I_{t-1}; \boldsymbol{\phi}_s \sim s[\boldsymbol{\pi}, \boldsymbol{\Sigma}(\boldsymbol{\theta}_s), \boldsymbol{\eta}], \tag{8.2}$$

$$\boldsymbol{\Sigma}(\boldsymbol{\theta}_s) = \mathbf{c}\mathbf{c}' + \boldsymbol{\Gamma}, \tag{8.3}$$

[1] To free up the variance of the common factor, we can impose alternative restrictions as, for instance, $c_1 = 1$ or $\mathbf{c}'\mathbf{c} = 1$.

where we have exploited the fact that linear combinations of elliptical random variables are elliptical (see Theorem 2.16 in Fang et al. (1990)). As a result, if we define the standardized innovations

$$\boldsymbol{\varepsilon}_t^* = \boldsymbol{\Sigma}^{-1/2}(\boldsymbol{\theta}_{s0})(\mathbf{y}_t - \boldsymbol{\pi}_0) \tag{8.4}$$

as an $N-$dimensional martingale difference sequence that satisfies $E(\boldsymbol{\varepsilon}_t^*|I_{t-1};\boldsymbol{\phi}_{s0}) = \mathbf{0}$ and $V(\boldsymbol{\varepsilon}_t^*|I_{t-1};\boldsymbol{\phi}_{s0}) = \mathbf{I}_N$, then η fully determines the shape of the conditional density of $\varsigma_t = \boldsymbol{\varepsilon}_t^{*\prime}\boldsymbol{\varepsilon}_t^*$.

The most distinctive feature of these models is that they provide a parsimonious specification of the cross-sectional dependence in the observed variables,[2] which results in a significant reduction in the number of parameters, and allows the estimation of these models with a large number of series (see, e.g., Lehmann and Modest (1988)). For these reasons, Model (8.1) continues to be rather popular in empirical finance applications such as portfolio allocation, asset pricing tests, hedging, and portfolio performance evaluation (see Connor et al. (2010) for details).

The parameters of interest are usually estimated jointly from the log-likelihood function of the observed variables. The ellipticity assumption and the serial independence of the variables involved imply that a modified version of the Kalman filter can still be used to estimate the underlying latent variables even though the innovations are not Gaussian. In particular, we can prove that:

$$E\left(\begin{array}{c} f_t \\ \mathbf{v}_t \end{array}\middle| \mathbf{Y}_t;\boldsymbol{\phi}_s\right) = \begin{bmatrix} \mathbf{c}'\boldsymbol{\Sigma}^{-1}(\boldsymbol{\theta}_s)(\mathbf{y}_t - \boldsymbol{\pi}) \\ \boldsymbol{\Gamma}\boldsymbol{\Sigma}^{-1}(\boldsymbol{\theta}_s)(\mathbf{y}_t - \boldsymbol{\pi}) \end{bmatrix} = \begin{bmatrix} f_{kt}(\boldsymbol{\theta}_s) \\ \mathbf{v}_{kt}(\boldsymbol{\theta}_s) \end{bmatrix}, \tag{8.5}$$

and

$$\begin{aligned} V\left(\begin{array}{c} f_t \\ \mathbf{v}_t \end{array}\middle| \mathbf{Y}_t;\boldsymbol{\phi}_s\right) &= \begin{bmatrix} \upsilon_{kt}(\boldsymbol{\theta}_s,\eta) & -\mathbf{c}'\upsilon_{kt}(\boldsymbol{\theta}_s,\eta) \\ -\mathbf{c}\upsilon_{kt}(\boldsymbol{\theta}_s,\eta) & \mathbf{c}\mathbf{c}'\upsilon_{kt}(\boldsymbol{\theta}_s,\eta) \end{bmatrix} \\ &= \mathfrak{h}[\varsigma_t(\boldsymbol{\theta}_s);\eta]\cdot V\left(\begin{array}{c} f_t - f_{kt}(\boldsymbol{\theta}_s) \\ \mathbf{v}_t - \mathbf{v}_{kt}(\boldsymbol{\theta}_s) \end{array}\middle| \boldsymbol{\phi}_s\right), \end{aligned}$$

where $\mathbf{Y}_t = \{\mathbf{y}_t, \mathbf{y}_{t-1}, \ldots\}$,

$$\mathbf{v}_t = \mathbf{y}_t - \boldsymbol{\pi} - \mathbf{c}f_t = \boldsymbol{\Gamma}^{1/2}\mathbf{v}_t^*,$$

$$\varsigma_t(\boldsymbol{\theta}_s) = \boldsymbol{\varepsilon}_t^{*\prime}(\boldsymbol{\theta}_s)\boldsymbol{\varepsilon}_t^*(\boldsymbol{\theta}_s) = (\mathbf{y}_t - \boldsymbol{\pi})'\boldsymbol{\Sigma}^{-1}(\boldsymbol{\theta}_s)(\mathbf{y}_t - \boldsymbol{\pi})$$

and $\mathfrak{h}(\varsigma_t;\eta)$ is a scalar factor of proportionality that reflects the nonlinear dependence between the elements of a spherical random vector. For example, for the Student t

$$\mathfrak{h}(\varsigma_t;\eta) = \frac{\nu-2}{\nu+N-2}\left(1+\frac{\varsigma_t}{\nu-2}\right),$$

[2] See Sentana (2000) for a random field interpretation of factor models, and their time series and cross-sectional state space representations.

which reduces to 1 under normality (see Harvey et al. (1992)). This scaling factor, whose unconditional mean is 1, multiplies the matrix of unconditional mean square errors

$$V\left(\begin{array}{c} f_t - f_{kt}(\boldsymbol{\theta}_s) \\ \mathbf{v}_t - \mathbf{v}_{kt}(\boldsymbol{\theta}_s) \end{array}\middle| \boldsymbol{\phi}_s\right) = \begin{bmatrix} 1 - \mathbf{c}'\boldsymbol{\Sigma}^{-1}(\boldsymbol{\theta}_s)\mathbf{c} & -\mathbf{c}'\boldsymbol{\Sigma}^{-1}(\boldsymbol{\theta}_s)\boldsymbol{\Gamma} \\ -\boldsymbol{\Gamma}\boldsymbol{\Sigma}^{-1}(\boldsymbol{\theta}_s)\mathbf{c} & \boldsymbol{\Gamma} - \boldsymbol{\Gamma}\boldsymbol{\Sigma}^{-1}(\boldsymbol{\theta}_s)\boldsymbol{\Gamma} \end{bmatrix}$$
$$= \begin{bmatrix} \omega_k(\boldsymbol{\theta}_s) & -\mathbf{c}'\omega_k(\theta_s) \\ -\mathbf{c}\omega_k(\boldsymbol{\theta}_s) & \mathbf{c}\mathbf{c}'\omega_k(\boldsymbol{\theta}_s) \end{bmatrix}, \tag{8.6}$$

which has rank one because we are trying to infer $N + 1$ latent variables from N observed ones. The elements of $f_{kt}(\boldsymbol{\theta}_s)$ and $\mathbf{v}_{kt}(\boldsymbol{\theta}_s)$ are known as the "regression scores" in the factor analysis literature because the weights in (8.5) coincide with the coefficients in the theoretical regression of each unobserved variable onto the observed series, while (8.6) coincides with the unconditional residual covariance matrix from those regressions. As explained in Sentana (2004), the mean squared error (MSE) criterion can be given an intuitive justification in terms of a mean variance investor, since it corresponds to the so-called "tracking error" variability in the finance literature. In that sense, $f_{kt}(\boldsymbol{\theta}_s)$ are the excess returns to the portfolio that best "tracks" f_t, while $\mathbf{v}_{kt}(\boldsymbol{\theta}_s)$ are the excess returns to the original vector of asset returns after we have hedged them against the common source of risk. As we shall see, $f_{kt}(\boldsymbol{\theta}_s)$, $\mathbf{v}_{kt}(\boldsymbol{\theta}_s)$, $\omega_k(\boldsymbol{\theta}_s)$, and $\upsilon_{kt}(\boldsymbol{\theta}_s, \boldsymbol{\eta})$ constitute the basic ingredients of our tests.

In this context, we can formally characterize the asymptotic distribution of three likelihood-based estimators of the static model parameters: the usual maximum likelihood estimator that simultaneously estimates $\boldsymbol{\theta}_s$ and $\boldsymbol{\eta}$; the elliptically symmetric semiparametric estimator of $\boldsymbol{\theta}$ considered by Hodgson and Vorkink (2003), Hafner and Rombouts (2007), and others, which restricts $\boldsymbol{\varepsilon}_t^*$ to have an i.i.d. $s(0, \mathbf{I}_N; \boldsymbol{\eta})$ conditional distribution but does not impose any structure on the distribution of ς_t;[3] and the Gaussian pseudo maximum likelihood estimator of θ, which sets $\boldsymbol{\eta} = \mathbf{0}$ even though the true conditional distribution of $\boldsymbol{\varepsilon}_t^*$ may well be non-normal.

Proposition 1 *Assume that (i) $V(\mathbf{y}_t)$ in (8.3) can be uniquely decomposed into $\mathbf{c}\mathbf{c}'$ and $\boldsymbol{\Gamma}$, and (ii) the matrix*

$$[\boldsymbol{\Gamma} - \mathbf{c}(\mathbf{c}'\boldsymbol{\Gamma}^{-1}\mathbf{c})^{-1}\mathbf{c}'] \odot [\boldsymbol{\Gamma} - \mathbf{c}(\mathbf{c}'\boldsymbol{\Gamma}^{-1}\mathbf{c})^{-1}\mathbf{c}']$$

[3] The main advantage of this estimator over traditional semiparametric estimators is that one can obtain an estimate of the joint density of $\boldsymbol{\varepsilon}_t^*$ from a nonparametric estimate of the univariate density of ς_t, thereby avoiding the curse of dimensionality (see, e.g., appendix B1 in Fiorentini and Sentana (2010b) for details).

has full rank, where $\odot$ denotes the Hadamard product of two matrices of equal orders. Then:

1. *The asymptotic distribution of the maximum likelihood estimators $\hat{\theta}_s$ and $\hat{\eta}$ will be*

$$\sqrt{T}(\hat{\phi}_s - \boldsymbol{\phi}_{s0}) \to N[0, \mathcal{I}^{-1}_{\phi_s\phi_s}(\boldsymbol{\phi}_{s0})],$$

where the information matrix $\mathcal{I}_{\phi_s\phi_s}(\boldsymbol{\phi}_s)$ will be block diagonal between the elements corresponding to $\boldsymbol{\pi}$ and the elements corresponding to $(\mathbf{c}, \boldsymbol{\gamma}, \boldsymbol{\eta})$, with the first block given by $M_{ll}(\eta)\boldsymbol{\Sigma}^{-1}(\boldsymbol{\theta}_s)$ and the second block has the first column given by

$$\begin{bmatrix} M_{ss}(\eta)\{[\mathbf{c}'\boldsymbol{\Sigma}^{-1}(\boldsymbol{\theta}_s)\mathbf{c}]\boldsymbol{\Sigma}^{-1}(\boldsymbol{\theta}_s) + \boldsymbol{\Sigma}^{-1}(\boldsymbol{\theta}_s)\mathbf{c}\mathbf{c}'\boldsymbol{\Sigma}^{-1}(\boldsymbol{\theta}_s)\} + [M_{ss}(\eta) - 1]\boldsymbol{\Sigma}^{-1}(\boldsymbol{\theta}_s)\mathbf{c}\mathbf{c}'\boldsymbol{\Sigma}^{-1}(\boldsymbol{\theta}_s) \\ M_{ss}(\eta)\mathbf{E}'_N[\boldsymbol{\Sigma}^{-1}(\boldsymbol{\theta}_s)\mathbf{c} \otimes \boldsymbol{\Sigma}^{-1}(\boldsymbol{\theta}_s)] + \frac{1}{2}[M_{ss}(\eta) - 1]\boldsymbol{\Sigma}^{-1}(\theta_s)\mathbf{c}\,vecd'[\boldsymbol{\Sigma}^{-1}(\boldsymbol{\theta}_s)] \\ M'_{sr}(\eta)\mathbf{c}'\boldsymbol{\Sigma}^{-1}(\boldsymbol{\theta}_s) \end{bmatrix},$$

the second column given by

$$\begin{bmatrix} M_{ss}(\eta)[\mathbf{c}'\boldsymbol{\Sigma}^{-1}(\boldsymbol{\theta}_s) \otimes \boldsymbol{\Sigma}^{-1}(\boldsymbol{\theta}_s)]\mathbf{E}_N + \frac{1}{2}[M_{ss}(\eta) - 1]vecd[\boldsymbol{\Sigma}^{-1}(\boldsymbol{\theta}_s)]\mathbf{c}'\boldsymbol{\Sigma}^{-1}(\boldsymbol{\theta}_s) \\ \frac{1}{2}M_{ss}(\eta)[\boldsymbol{\Sigma}^{-1}(\boldsymbol{\theta}_s) \odot \boldsymbol{\Sigma}^{-1}(\boldsymbol{\theta}_s)] + \frac{1}{4}[M_{ss}(\eta) - 1]vecd[\boldsymbol{\Sigma}^{-1}(\theta_s)]vecd'[\boldsymbol{\Sigma}^{-1}(\boldsymbol{\theta}_s)] \\ \frac{1}{2}M'_{sr}(\eta)vecd'[\boldsymbol{\Sigma}^{-1}(\boldsymbol{\theta}_s)] \end{bmatrix},$$

and the third column given by

$$\begin{bmatrix} \boldsymbol{\Sigma}^{-1}(\boldsymbol{\theta}_s)\mathbf{c}M_{sr}(\eta) \\ \frac{1}{2}vecd[\boldsymbol{\Sigma}^{-1}(\boldsymbol{\theta}_s)]M_{sr}(\eta) \\ \mathcal{M}_{rr}(\eta) \end{bmatrix},$$

where $\mathbf{E}_n$ is the unique $n^2 \times n$ "diagonalization" matrix which transforms $vec(\mathbf{A})$ into $vecd(\mathbf{A})$ as $vecd(\mathbf{A}) = \mathbf{E}'_n vec(\mathbf{A})$,

$$M_{ll}(\eta) = E\left\{\delta^2[\varsigma_t(\boldsymbol{\theta}), \eta]\frac{\varsigma_t(\boldsymbol{\theta})}{N}\middle|\boldsymbol{\phi}\right\} = E\left\{\frac{2\partial\delta[\varsigma_t(\boldsymbol{\theta}), \eta]}{\partial\varsigma}\frac{\varsigma_t(\boldsymbol{\theta})}{N} + \delta[\varsigma_t(\boldsymbol{\theta}), \eta]\middle|\boldsymbol{\phi}\right\}, \tag{8.7}$$

$$M_{ss}(\eta) = \frac{N}{N+2}\left[1 + V\left\{\delta[\varsigma_t(\boldsymbol{\theta}), \eta]\frac{\varsigma_t}{N}\middle|\boldsymbol{\phi}\right\}\right] = E\left\{\frac{2\partial\delta[\varsigma_t(\theta), \eta]}{\partial\varsigma}\frac{\varsigma_t^2(\theta)}{N(N+2)}\middle|\boldsymbol{\phi}\right\} + 1, \tag{8.8}$$

$$M_{sr}(\eta) = E\left[\left\{\delta[\varsigma_t(\boldsymbol{\theta}), \eta]\frac{\varsigma_t(\theta)}{N} - 1\right\}\mathbf{e}'_{rt}(\boldsymbol{\phi})\middle|\boldsymbol{\phi}\right] = -E\left\{\frac{\varsigma_t(\boldsymbol{\theta})}{N}\frac{\partial\delta[\varsigma_t(\boldsymbol{\theta}), \eta]}{\partial\eta'}\middle|\boldsymbol{\phi}\right\}, \tag{8.9}$$

$$\begin{aligned} \mathcal{M}_{rr}(\eta) &= V[\partial c(\eta)/\partial\eta + \partial g[\varsigma_t(\boldsymbol{\theta}), \eta]/\partial\eta|\phi] \\ &= -E[\partial^2 c(\eta)/\partial\eta\partial\eta' + \partial^2 g[\varsigma_t(\boldsymbol{\theta}), \eta]/\partial\eta\partial\eta'|\boldsymbol{\phi}], \end{aligned} \tag{8.10}$$

$$\delta(\varsigma_t, \eta) = -2\partial g(\varsigma_t, \eta)/\partial\varsigma, \tag{8.11}$$

$c(\boldsymbol{\eta})$ is the constant of integration of the assumed elliptical density and $g(\varsigma_t, \eta)$ its kernel.

2. *Assuming that the population coefficient of multivariate excess kurtosis*

$$\kappa = E(\varsigma_t^2|\boldsymbol{\eta})/[N(N+2)] - 1 \tag{8.12}$$

is such that $-2/(N+2) < \kappa_0 < \infty$, the efficiency bound associated to the elliptically symmetric semiparametric estimator $\mathring{\theta}_s$ will be block diagonal between $\boldsymbol{\pi}$ and $(\mathbf{c}, \boldsymbol{\gamma})$, where the first block coincides with the first block of the information matrix, and the second one with the corresponding block of the information matrix minus

$$\left\{\left[\frac{N+2}{N}\mathrm{M}_{ss}(\boldsymbol{\eta}) - 1\right] - \frac{4}{N\left[(N+2)\kappa + 2\right]}\right\}$$
$$\times\begin{bmatrix} \boldsymbol{\Sigma}^{-1}(\boldsymbol{\theta}_s)\mathbf{cc}'\boldsymbol{\Sigma}^{-1}(\boldsymbol{\theta}_s) & \frac{1}{2}\boldsymbol{\Sigma}^{-1}(\boldsymbol{\theta}_s)\mathbf{c}vecd'[\boldsymbol{\Sigma}^{-1}(\boldsymbol{\theta}_s)] \\ \frac{1}{2}vecd[\boldsymbol{\Sigma}^{-1}(\boldsymbol{\theta})]\mathbf{c}'\boldsymbol{\Sigma}^{-1}(\boldsymbol{\theta}) & vecd[\boldsymbol{\Sigma}^{-1}(\boldsymbol{\theta}_s)]vecd'[\boldsymbol{\Sigma}^{-1}(\boldsymbol{\theta}_s)] \end{bmatrix}.$$

3. *If $\kappa_0 < \infty$, the asymptotic distribution of the Gaussian pseudo maximum likelihood estimator $\bar{\theta}_s$ will be*

$$\sqrt{T}(\bar{\theta}_s - \bar{\phi}_{s0}) \to N[0, \mathcal{A}^{-1}_{\theta_s\theta_s}(\boldsymbol{\theta}_{s0})\mathcal{B}_{\theta_s\theta_s}(\boldsymbol{\phi}_{s0})\mathcal{A}^{-1}_{\theta_s\theta_s}(\boldsymbol{\theta}_{s0})],$$

where

$$\mathcal{A}_{\theta_s\theta_s}(\boldsymbol{\theta}_s) = \mathcal{I}_{\theta_s\theta_s}(\boldsymbol{\theta}_s, \mathbf{0}),$$

$$\mathcal{I}_{\theta_s\theta_s}(\boldsymbol{\theta}_s, \mathbf{0}) = \begin{bmatrix} \boldsymbol{\Sigma}^{-1}(\boldsymbol{\theta}_s) & \mathbf{0} & \mathbf{0} \\ \mathbf{0} & [\mathbf{c}'\boldsymbol{\Sigma}^{-1}(\boldsymbol{\theta}_s)\mathbf{c}]\boldsymbol{\Sigma}^{-1}(\boldsymbol{\theta}_s) + \boldsymbol{\Sigma}^{-1}(\theta_s)\mathbf{cc}'\boldsymbol{\Sigma}^{-1}(\boldsymbol{\theta}_s) & [\mathbf{c}'\boldsymbol{\Sigma}^{-1}(\boldsymbol{\theta}_s) \otimes \boldsymbol{\Sigma}^{-1}(\boldsymbol{\theta}_s)]\mathbf{E}_N \\ \mathbf{0} & \mathbf{E}_N'[\boldsymbol{\Sigma}^{-1}(\boldsymbol{\theta}_s)\mathbf{c} \otimes \boldsymbol{\Sigma}^{-1}(\boldsymbol{\theta}_s)] & \frac{1}{2}\boldsymbol{\Sigma}^{-1}(\boldsymbol{\theta}_s) \odot \boldsymbol{\Sigma}^{-1}(\boldsymbol{\theta}_s) \end{bmatrix}$$

and $\mathcal{B}_{\theta_s\theta_s}(\boldsymbol{\phi})$ has the same expression as $\mathcal{I}_{\theta_s\theta_s}(\boldsymbol{\theta}_s, \boldsymbol{\eta})$ but with $\mathrm{M}_{ll}(\boldsymbol{\eta})$ and $\mathrm{M}_{ss}(\boldsymbol{\eta})$ replaced by 1 and $(\kappa+1)$, respectively.

In the multivariate standardized Student t case, in particular, Fiorentini et al. (2003) show that:

$$\mathrm{M}_{ll}(\eta) = \frac{\nu\,(N+\nu)}{(\nu-2)\,(N+\nu+2)},\ \mathrm{M}_{ss}(\eta) = \frac{(N+\nu)}{(N+\nu+2)},\ \mathrm{M}_{sr}(\eta)$$
$$= -\frac{2\,(N+2)\,\nu^2}{(\nu-2)\,(N+\nu)\,(N+\nu+2)},$$

and

$$M_{rr}(\eta) = \frac{\nu^4}{4}\left[\psi'\left(\frac{\nu}{2}\right) - \psi'\left(\frac{N+\nu}{2}\right)\right] - \frac{N\nu^4\left[\nu^2 + N(\nu-4) - 8\right]}{2(\nu-2)^2(N+\nu)(N+\nu+2)},$$

where $\psi(.)$ is the di-gamma function, which under normality reduce to 1, 1, 0 and $N(N+2)/2$, respectively.

Finally, it is worth mentioning that if we reparametrized the covariance matrix $\mathbf{\Sigma}(\boldsymbol{\theta}_s)$ as $\vartheta_2\mathbf{\Sigma}^\circ(\boldsymbol{\vartheta}_1)$, where ϑ_2 is a scalar and

$$\mathbf{\Sigma}^\circ(\boldsymbol{\vartheta}_1) = \mathbf{c}^*\mathbf{c}^{*\prime} + \mathbf{\Gamma}^*,$$

with $\mathbf{\Gamma} = \vartheta_2\mathbf{\Gamma}^*$ and $\mathbf{c} = \sqrt{\vartheta_2}\mathbf{c}^*$, Proposition 8 in Fiorentini and Sentana (2010b) implies that the maximum likelihood estimator and the elliptically symmetric semiparametric estimator of $\boldsymbol{\vartheta}_1$ would be adaptive (i.e., as efficiently estimated as if we knew η). If we further eliminated the resulting scale indeterminacy by forcing

$$|\mathbf{\Sigma}^\circ(\boldsymbol{\vartheta}_1)| = \left(\prod\nolimits_{i=1}^{N}\gamma_i^*\right)\left(1 + \sum\nolimits_{j=1}^{N} c_j^{*2}/\gamma_j^*\right)$$

to be 1 (or any other fixed value),[4] the same proposition implies that the asymptotic covariance matrices of the three estimators of $\boldsymbol{\vartheta}_1$ and ϑ_2 considered in Proposition 1 would be block diagonal. Moreover, the ML estimator of ϑ_2 could only achieve the asymptotic efficiency of its Gaussian pseudo maximum likelihood estimator, which would be given by the expression:

$$\vartheta_2(\boldsymbol{\vartheta}_1) = \frac{1}{N}\frac{1}{T}\sum_{t=1}^{T}\varsigma_t^\circ(\boldsymbol{\vartheta}_1),$$

$$\varsigma_t^\circ(\boldsymbol{\vartheta}_1) = (\mathbf{y}_t - \pi)'\mathbf{\Sigma}^{\circ-1}(\boldsymbol{\vartheta}_1)(\mathbf{y}_t - \boldsymbol{\pi}),$$

evaluated at the Gaussian pseudo maximum likelihood (PML) estimator $\bar{\vartheta}_1$.

[4] We can solve the resulting determinantal equation for one of the $c^{*\prime}s$, which yields

$$c_i^* = \pm\gamma_i^*\left(\frac{1}{\prod_{i=1}^{N}\gamma_i^*} - 1 - \sum_{j\neq i}^{N} c_j^{*2}/\gamma_j^*\right),$$

or for one of the $\gamma^{*\prime}s$, yielding

$$\gamma_j^* = \left(1 - c_j^2\prod_{i\neq j}\gamma_i^*\right)\Big/\left[\prod_{i\neq j}\gamma_i^*\left(1 + \sum_{i\neq j} c_i^{*2}/\gamma_i^*\right)\right].$$

8.3 Serial correlation tests for common and idiosyncratic factors

8.3.1 BASELINE CASE

The most natural way of introducing serial correlation in Model (8.1) would be to assume that

$$\left.\begin{aligned} \mathbf{y}_t &= \boldsymbol{\pi} + \mathbf{c}x_t + \mathbf{u}_t \\ x_t &= \rho x_{t-1} + f_t \\ \mathbf{u}_t &= diag(\boldsymbol{\rho}^*)\mathbf{u}_{t-1} + \boldsymbol{\Gamma}^{1/2}\mathbf{v}_t^* \end{aligned}\right\} \tag{8.13}$$

and

$$\begin{pmatrix} f_t \\ \mathbf{v}_t^* \end{pmatrix} | I_{t-1}, \boldsymbol{\phi} \sim s\left[\begin{pmatrix} 0 \\ \mathbf{0} \end{pmatrix}, \begin{pmatrix} 1 & \mathbf{0} \\ \mathbf{0} & \mathbf{I}_N \end{pmatrix}, \boldsymbol{\eta}\right], \tag{8.14}$$

where the parameters of interest become $\boldsymbol{\phi} = (\boldsymbol{\theta}', \boldsymbol{\eta}')'$, $\boldsymbol{\theta} = (\theta_s', \boldsymbol{\rho}^{\dagger\prime})'$, with $\boldsymbol{\rho}^\dagger = (\rho, \boldsymbol{\rho}^{*\prime})'$, as this reduces to our baseline specification (8.1) under $H_0 : \boldsymbol{\rho}^\dagger = 0$.

The problem with formulation (8.13) is that unless the true conditional distribution of the latent variables is Gaussian, the conditional distribution of the observed variables given their past values alone is unknown when $\boldsymbol{\rho}^\dagger \neq \mathbf{0}$ and $\boldsymbol{\Gamma}$ has full rank, and can only be approximated by simulation (see, e.g., Durbin and Koopman (2000)). While it is true that the Kalman filter continues to produce the best linear least squares predictions of the underlying state variables in those circumstances (see Anderson and Moore (1979)), its prediction equations do not generally provide the conditional mean vector and covariance matrix of $\mathbf{y}_t$ given $\mathbf{Y}_{t-1}$ (and the parameter values). As a result, we cannot rely on standard results for Gaussian pseudo maximum likelihood estimators and tests, such as those in Bollerslev and Wooldridge (1992). For that reason, in the rest of this section we assume that the mean vector and covariance matrix of $\mathbf{y}_t$ conditional on $\mathbf{Y}_{t-1}$ are given by the usual Kalman filter recursions (see Appendix B in Fiorentini and Sentana (2012)), but the conditional distribution is elliptically symmetric. We will revisit this assumption in Section 8.3.2.

Gaussian versions of dynamic factor models such as (8.13) have become increasingly popular in macroeconomic applications (see, e.g., Bai and Ng (2008b) and the references therein), but they are not widely used for stock returns (see Dungey et al. (2000) or Jegadeesh and Pennacchi (1996) for applications to bonds).

Assuming the stationarity conditions $|\rho| < 1$ and $|\rho_i^*| < 1\ \forall i$ hold, the autocovariance matrices of the observed series will be:

$$\mathbf{G}_\mathbf{y}(j) = \mathbf{c}\mathbf{c}'G_x(j) + \mathbf{G}_\mathbf{u}(j). \tag{8.15}$$

The factor structure applies in particular to $\boldsymbol{\Sigma}$, the unconditional covariance matrix of $\mathbf{y}_t$, even though x_t or $\mathbf{u}_t$ are serially correlated (see Doz and Lenglart

(1999)). It is also easy to see that the autocovariance structure in (8.15) corresponds to a special case of a VARMA(2,1) model since

$$(1-\rho L)[\mathbf{I}_N - diag(\boldsymbol{\rho}^*)L](\mathbf{y}_t - \boldsymbol{\pi}) = [\mathbf{I}_N - diag(\boldsymbol{\rho}^*)L]\mathbf{c}f_t + (1-\rho L)\mathbf{v}_t,$$

whose right hand side has the autocovariance structure of a VMA(1).[5]

As the next proposition shows, however, testing the null of multivariate white noise against such a complex VARMA(2,1) specification is extremely easy. Importantly, we shall distinguish between the optimal score test obtained by exploiting the non-normality of the conditional distribution, and the Gaussian pseudo LM test, which although uses the Gaussian scores, has been robustified against possible non-normality:

Proposition 2 *Let*

$$\bar{G}_f(j;\boldsymbol{\eta}) = \frac{1}{T}\sum_{t=1}^{T} \delta[\varsigma_t(\boldsymbol{\theta}_s),\boldsymbol{\eta}] f_{kt}(\theta_s) f_{kt-j}(\boldsymbol{\theta}_s)$$

denote the sample cross moment of $\delta[\varsigma_t(\theta_s),\boldsymbol{\eta}]f_{kt}(\boldsymbol{\theta}_s)$ *and* $f_{kt-j}(\boldsymbol{\theta}_s)$, *where* $\delta(\varsigma_t,\eta)$ *is defined in (8.11) and* $f_{kt}(\boldsymbol{\theta}_s)$ *is obtained from the updating equations (8.5) of the static factor model (8.1). Similarly, let*

$$\bar{\mathbf{G}}_{\mathbf{v}}(j;\boldsymbol{\eta}) = \frac{1}{T}\sum_{t=1}^{T} \delta[\varsigma_t(\boldsymbol{\theta}_s),\boldsymbol{\eta}]\mathbf{v}_{kt}(\boldsymbol{\theta}_s)\mathbf{v}'_{kt-j}(\boldsymbol{\theta}_s)$$

denote the analogous sample cross moments for the specific factors.

1. *Under the null hypothesis* $H_0 : \boldsymbol{\rho}^\dagger = \mathbf{0}$, *the score test statistic* $LM_{AR(1)}(\boldsymbol{\eta}_0)$ *given by T times*

$$\left(\bar{G}_f(1;\boldsymbol{\eta}_0), vecd'[\boldsymbol{\Gamma}_0^{-1/2}\bar{\mathbf{G}}_{\mathbf{v}}(1;\boldsymbol{\eta}_0)\boldsymbol{\Gamma}_0^{-1/2}]\right)\mathcal{I}^{-1}_{\rho^\dagger\rho^\dagger}(\boldsymbol{\theta}_{s0},\mathbf{0};\boldsymbol{\eta}_0)$$
$$\left(\bar{G}_f(1;\boldsymbol{\eta}_0), vecd'[\boldsymbol{\Gamma}_0^{-1/2}\bar{\mathbf{G}}_{\mathbf{v}}(1;\boldsymbol{\eta}_0)\boldsymbol{\Gamma}_0^{-1/2}]\right)',$$

with

$$\mathcal{I}_{\rho^\dagger\rho^\dagger}(\theta_s,0;\boldsymbol{\eta}) = \mathcal{V}_{\rho^\dagger\rho^\dagger}(\boldsymbol{\theta}_s,0;\boldsymbol{\eta}) \odot \mathcal{V}_{\rho^\dagger\rho^\dagger}(\boldsymbol{\theta}_s,\boldsymbol{\eta};\boldsymbol{\eta}),$$

where

$$\mathcal{V}_{\rho^\dagger\rho^\dagger}(\theta_s,\boldsymbol{\eta};\boldsymbol{\eta}) = V\begin{bmatrix}\delta[\varsigma_t(\boldsymbol{\theta}_s),\boldsymbol{\eta}]f_{kt}(\boldsymbol{\theta}_s)\\ \delta[\varsigma_t(\boldsymbol{\theta}_s),\boldsymbol{\eta}]\boldsymbol{\Gamma}^{-1/2}\mathbf{v}_{kt}(\boldsymbol{\theta}_s)\end{bmatrix} = \mathrm{M}_{ll}(\boldsymbol{\eta})\mathcal{V}_{\rho^\dagger\rho^\dagger}(\boldsymbol{\theta}_s,0;0),$$

$$\mathcal{V}_{\rho^\dagger\rho^\dagger}(\theta_s,0;\boldsymbol{\eta}) = V\begin{bmatrix}f_{kt}(\boldsymbol{\theta}_s)\\ \boldsymbol{\Gamma}^{-1/2}\mathbf{v}_{kt}(\boldsymbol{\theta}_s)\end{bmatrix}$$
$$= \begin{bmatrix}\mathbf{c}'\boldsymbol{\Sigma}^{-1}(\boldsymbol{\theta}_s)\mathbf{c} & \mathbf{c}'\boldsymbol{\Sigma}^{-1}(\boldsymbol{\theta}_s)\boldsymbol{\Gamma}^{1/2}\\ \boldsymbol{\Gamma}^{1/2}\boldsymbol{\Sigma}^{-1}(\boldsymbol{\theta}_s)\mathbf{c} & \boldsymbol{\Gamma}^{1/2}\boldsymbol{\Sigma}^{-1}(\boldsymbol{\theta}_s)\boldsymbol{\Gamma}^{1/2}\end{bmatrix} = \mathcal{V}_{\rho^\dagger\rho^\dagger}(\boldsymbol{\theta}_s,0;0),$$

[5] When $\boldsymbol{\rho}^* = \rho\mathbf{I}_N$, though, the reduced form process becomes a VAR(1) with a scalar companion matrix. As a result, any linear combination of $\mathbf{y}_t$ will have the autocorrelation structure of an AR(1) process with autoregressive coefficient ρ.

will be distributed as a χ^2 with $N+1$ degrees of freedom for N fixed as T goes to infinity. Moreover, this asymptotic null distribution is unaffected if we replace $\boldsymbol{\theta}_{s0}$ and $\boldsymbol{\eta}_0$ by their feasible maximum likelihood estimators in the first part of Proposition 1.

2. *It also remains valid if we replace $\boldsymbol{\theta}_{s0}$ by its elliptically symmetric semiparametric estimator in the second part of Proposition 1, which requires the nonparametric estimation of the density of $\varsigma_t(\boldsymbol{\theta}_s)$.*
3. *Under the same null hypothesis, the Gaussian pseudo score test statistic $LM_{AR(1)}(\mathbf{0})$ given by T times*

$$\left(\bar{G}_f(1;0), vecd'[\boldsymbol{\Gamma}_0^{-1/2}\bar{\mathbf{G}}_{\mathrm{v}}(1;0)\boldsymbol{\Gamma}_0^{-1/2}]\right)\mathcal{I}^{-1}_{\boldsymbol{\rho}^\dagger\boldsymbol{\rho}^\dagger}(\boldsymbol{\theta}_{s0},\mathbf{0};\mathbf{0})$$

$$\left(\bar{G}_f(1;0), vecd'[\boldsymbol{\Gamma}_0^{-1/2}\bar{\mathbf{G}}_{\mathrm{v}}(1;0)\boldsymbol{\Gamma}_0^{-1/2}]\right)',$$

will be distributed as a χ^2 with $N+1$ degrees of freedom for N fixed as T goes to infinity irrespective of the normality of the conditional distribution. This result continues to hold if we replace $\boldsymbol{\theta}_{s0}$ by its Gaussian pseudo maximum likelihood estimator $\bar{\theta}_s$ in the third part of Proposition 1.

Researchers may sometimes be interested in tests that separately assess the serial correlation of either the common factor or the specific factors. In principle, they might even like to focus on a particular v_{it}. By combining the relevant elements of $\bar{G}_f(j;\boldsymbol{\eta})$ and $\bar{\mathbf{G}}_{\mathrm{v}}(1;\boldsymbol{\eta})$ with the corresponding blocks of the information matrix, $\mathcal{I}_{\boldsymbol{\rho}^\dagger\boldsymbol{\rho}^\dagger}(\boldsymbol{\theta}_{s0},\mathbf{0};\boldsymbol{\eta})$, we can easily exploit the results in Proposition 2 to derive the required test statistics for those subcomponents under the maintained hypothesis of serial independence.[6] Intuitively, the reason is that we can interpret $LM_{AR(1)}(\boldsymbol{\eta})$ as a test based on the $N+1$ orthogonality conditions:

$$E\{\delta[\varsigma_t(\boldsymbol{\theta}_s),\boldsymbol{\eta}]f_{kt}(\boldsymbol{\theta}_s)f_{kt-1}(\boldsymbol{\theta}_s)|\theta_s,\mathbf{0},\eta\}=0, \tag{8.16}$$

$$E\{\gamma_i^{-1}\delta[\varsigma_t(\boldsymbol{\theta}_s),\boldsymbol{\eta}]v_{kit}(\boldsymbol{\theta}_s)v_{kit-1}(\boldsymbol{\theta}_s)|\theta_s,\mathbf{0},\eta\}=0 \quad (i=1,\ldots,N). \tag{8.17}$$

Similarly, $LM_{AR(1)}(\mathbf{0})$ is based on

$$E\left[f_{kt}(\boldsymbol{\theta}_s)f_{kt-1}(\boldsymbol{\theta}_s)|\theta_s,\mathbf{0},\boldsymbol{\eta}\right]=0, \tag{8.18}$$

$$E\left[\gamma_i^{-1}v_{kit}(\boldsymbol{\theta}_s)v_{kit-1}(\theta_s)|\theta_s,\mathbf{0},\boldsymbol{\eta}\right]=0 \qquad (i=1,\ldots,N), \tag{8.19}$$

which are the conditions that we would use to test for first order serial correlation if we treated $f_{kt}(\boldsymbol{\theta}_s)$ and $v_{kit}(\boldsymbol{\theta}_s)$ as the series of interest in the Gaussian case (see Breusch and Pagan (1980) or Godfrey (1988)). In that sense, the factor $\delta(\varsigma_t,\eta)$, which is equal to 1 under Gaussianity and to $(N\eta+1)/(1-2\eta+\eta\varsigma_t)$

[6] See Bera and Yoon (1993) for a possible way of orthogonalizing those individual LM tests is under alternatives local to $H_0:\boldsymbol{\rho}^\dagger=\mathbf{0}$.

for the Student t, can be regarded as the type of damping factor for big observations used in the robust estimation literature (see, e.g., Maronna et al. (2006)) because it is a decreasing function of ς_t for fixed $\eta > 0$, the more so the higher η is (see Fiorentini and Sentana (2010a) for a closely related discussion for univariate models).

Given that we have fixed the variance of the innovations in the common factor to 1, the moment conditions (8.18) and (8.19) closely resemble

$$E(f_t f_{t-1}|\theta_s, 0) = 0,$$
$$E(\gamma_i^{-1} v_{it} v_{it-1}|\theta_s, 0) = 0 \qquad (i = 1, \ldots, N),$$

which are the Gaussian-based orthogonality conditions that we could use to test for first order serial correlation if we could observe all the latent variables.

The similarity between these two sets of moment conditions becomes even stronger if we consider individual tests for serial correlation in each latent variable. Let us start with a test of $H_0 : \rho = 0$ under the maintained assumption that $\boldsymbol{\rho}^* = \mathbf{0}$. Part 3 of Proposition 2 implies that the asymptotic variance of $\bar{G}_f(1; \mathbf{0})$ is simply $[\mathbf{c}'\boldsymbol{\Sigma}^{-1}(\boldsymbol{\theta}_s)\mathbf{c}]^2$. But we can use (8.6) to interpret $\mathbf{c}'\boldsymbol{\Sigma}^{-1}(\boldsymbol{\theta}_s)\mathbf{c}$ as the R^2 in the theoretical least squares projection of f_t on a constant and $\mathbf{y}_t$. Therefore, the higher the degree of observability of the common factor, the closer the asymptotic variance of $\bar{G}_f(1; \mathbf{0})$ will be to 1, which is the asymptotic variance of the first sample autocorrelation of f_t. Intuitively, this convergence result simply reflects the fact that the common factor becomes observable in the limit, which implies that our Gaussian test of $H_0 : \rho = 0$ will become arbitrarily close to a Gaussian first order serial correlation test for the common factor as the "signal to noise" ratio $\mathbf{c}'\boldsymbol{\Sigma}^{-1}(\boldsymbol{\theta}_s)\mathbf{c}$ approaches 1. Before the limit, though, our test takes into account the unobservability of f_t. A particularly interesting situation arises if we consider models in which N is large. Since $\mathbf{c}'\boldsymbol{\Sigma}^{-1}(\boldsymbol{\theta}_s)\mathbf{c} = (\mathbf{c}'\boldsymbol{\Gamma}^{-1}\mathbf{c})/[1 + (\mathbf{c}'\boldsymbol{\Gamma}^{-1}\mathbf{c})]$ under the assumption that $\boldsymbol{\Gamma}$ has full rank, the aforementioned R^2 converges to 1 as N grows because $(\mathbf{c}'\boldsymbol{\Gamma}^{-1}\mathbf{c}) \to \infty$ in those circumstances due to the pervasive nature of the common factor (see, e.g., Sentana (2004)).

Likewise, Part 3 of Proposition 2 implies that the asymptotic variance of $\bar{G}_{v_i}(1; \mathbf{0})$ is $[\gamma_i \sigma^{ii}(\boldsymbol{\theta}_s)]^2$, where $\sigma^{ii}(\boldsymbol{\theta}_s)$ denotes the ith diagonal element of $\boldsymbol{\Sigma}^{-1}(\theta_s)$. But we can again use (8.6) to interpret $\gamma_i \sigma^{ii}(\vartheta_s)$ as the R^2 in the theoretical least squares projection of v_{it} on a constant and $\mathbf{y}_t$. Therefore, we can apply a similar line of reasoning to a Gaussian test of $H_0 : \rho_i^* = 0$ under the maintained assumption that both ρ and the remaining elements of $\boldsymbol{\rho}^*$ are 0. In this respect, note that $\sigma^{ii}(\boldsymbol{\theta}_s) = \gamma_i^{-1} - \gamma_i^{-2} c_i^2/[1 + (\mathbf{c}'\boldsymbol{\Gamma}^{-1}\mathbf{c})]$ when $\boldsymbol{\Gamma}$ has full rank, which means that $\gamma_i \sigma^{ii}(\boldsymbol{\theta}_s)$ also converges to 1 as N increases for fixed c_i and γ_i.

Nevertheless, it is important to emphasize that our joint tests take into account the covariance between the Kalman filter estimators of common and specific factors, even though the latent variables themselves are uncorrelated. In fact, $\mathcal{V}_{\rho^\dagger\rho^\dagger}(\boldsymbol{\theta}, \eta; \eta)$ has rank N instead of $N+1$ because of the negative relationship $\mathbf{v}_{kt}(\boldsymbol{\theta}) = \mathbf{y}_t - \boldsymbol{\pi} - \mathbf{c}f_{kt}(\theta)$, which rules out the direct application of the multivariate serial correlation test, discussed in Section 8.3.3, to the vector process $[f_{kt}(\boldsymbol{\theta}_s), \mathbf{v}'_{kt}(\theta_s)]'$.

Part 3 of Proposition 2 also implies that the asymptotic distribution of the Gaussian tests does not depend on normality or indeed ellipticity. Effectively, this result mimics the fact that under conditional homoskedasticity, standard score tests for serial correlation in observed series are also robust to non-normality. In fact, we can strengthen this intuition as follows. Since $V[f_{kt}(\boldsymbol{\theta}_s)|\boldsymbol{\theta}_s, 0, \eta] = \mathbf{c}'\boldsymbol{\Sigma}^{-1}(\boldsymbol{\theta}_s)\mathbf{c}$, we can obtain an asymptotically equivalent test of $H_0 : \rho = 0$ by computing the F test of the regression of $f_{kt}(\boldsymbol{\theta}_s)$ on a constant and $f_{kt-1}(\boldsymbol{\theta}_s)$, whose asymptotic null distribution does not depend on Gaussianity.

Finally, it is worth mentioning that the orthogonality conditions (8.16) and (8.17) remain valid when $\mathbf{y}_t$ is serially uncorrelated irrespective of $V(\mathbf{y}_t)$ having an exact single factor structure. Therefore, one could also use them to derive a standard moment test (see, e.g., Newey and McFadden (1994), Newey (1985) and Tauchen (1985)), which will continue to have non-trivial power even though it will no longer be an LM test (see Sentana and Shah (1993) for an interpretation of $\boldsymbol{\theta}_s$ when $\boldsymbol{\Sigma}(\boldsymbol{\theta}_s)$ is misspecified). Naturally, the same applies to (8.18) and (8.19).

8.3.2 EXTENSIONS

Unobservable conditional means

The assumption that the distribution of $\mathbf{y}_t$ conditional on its past alone is elliptically symmetric but with a mean vector and covariance matrix given by the usual Gaussian Kalman filter recursions may be regarded as a way of constructing a convenient auxiliary model that coincides with the model of interest for $\boldsymbol{\rho}^\dagger = 0$ or $\eta = 0$, but whose log-likelihood function and score we can obtain in closed form for every possible value of $\boldsymbol{\rho}^\dagger$ when $\eta \neq \mathbf{0}$. In this regard, it is important to bear in mind that the fact that we can compute the true log-likelihood function of $\mathbf{y}_t$ under the null of $\boldsymbol{\rho}^\dagger = \mathbf{0}$ is not sufficient to compute its derivative with respect to $\boldsymbol{\rho}^\dagger$. Nevertheless, it is possible to use the EM principle to obtain this score. Remarkably, it turns out that the score of the potentially non-Gaussian state space model (8.13) and the approximating model used in the Section 8.3.1 coincide under the null, even though the Kalman filter prediction equations do not provide the true conditional mean and covariance matrix under the alternative. As a result, the test statistics we have derived in Proposition 2 remain valid for Model (8.13) too. The following proposition formalizes our claim for the multivariate Student t, but we conjecture it applies to most other elliptical distributions:

Proposition 3 *Let* $\mathbf{s}_t(\boldsymbol{\phi}) = \partial l_t(\boldsymbol{\phi})/\partial\boldsymbol{\phi}$ *denote the log-likelihood score of the conditionally elliptical model for* $\mathbf{y}_t|\mathbf{Y}_{t-1};\boldsymbol{\phi}$ *in Section 8.3.1. Similarly, let* $\mathbf{q}_t(\phi) = \partial p(\mathbf{y}_t|\mathbf{Y}_{t-1};\boldsymbol{\phi})/\partial\boldsymbol{\phi}$ *denote the exact log-likelihood score of model (8.13). If (8.14) is a (standardized) multivariate Student t with* $0 \leq \eta < .5$ *then* $l_t(\boldsymbol{\phi}) = p(\mathbf{y}_t|\mathbf{Y}_{t-1};\boldsymbol{\phi})$ *and* $\mathbf{s}_t(\boldsymbol{\phi}) = \mathbf{q}_t(\phi)$ *when evaluated at* $\boldsymbol{\rho}^\dagger = 0$.

In other words, the approximating model "smoothly embeds" (in the sense used by Gallant and Tauchen (1996) in their Theorem 2) the original model in those circumstances.

Moving average processes

Specification (8.13) assumes that common and specific factors follow AR(1) processes. However, recent macroeconomic applications of dynamic factor models have often considered moving average processes instead, sometimes treating the lagged latent variables as additional factors (see, again, Bai and Ng (2008b)). Thus, we could alternatively assume that

$$\begin{aligned} x_t &= f_t + \varphi f_{t-1}, \\ \mathbf{u}_t &= \mathbf{v}_t + diag(\boldsymbol{\varphi}^*)\mathbf{v}_{t-1}. \end{aligned} \tag{8.20}$$

In this case, the autocorrelation structure of $\mathbf{y}_t$ corresponds to a restricted VMA(1) process. Although the Kalman filter recursions for this dynamic model change, we can show that the scores corresponding to $\boldsymbol{\varphi}^\dagger = (\varphi, \varphi^{*\prime})'$ evaluated at $\boldsymbol{\varphi}^\dagger = \mathbf{0}$ numerically coincide with the scores corresponding to $\boldsymbol{\rho}^\dagger$ in model (8.13) evaluated at $\rho^\dagger = \mathbf{0}$. Hence, we can also interpret $LM_{AR(1)}(\boldsymbol{\eta})$ in Proposition 2 as the LM test of $H_0 : \boldsymbol{\varphi}^\dagger = \mathbf{0}$. This result mimics the well known fact that MA(1) and AR(1) processes provide locally equivalent alternatives in univariate tests for serial correlation (see, e.g., Godfrey (1988)).

Higher order processes

Consider now the following alternative:

$$\begin{aligned} x_t &= \sum_{l=1}^{h} \rho_l x_{t-l} + f_t, \\ u_{it} &= \sum_{l=1}^{h_i^*} \rho_{il}^* u_{it-l} + v_{it}, \qquad (i = 1, \ldots, N). \end{aligned}$$

In view of the discussion in Section 8.3.1, it is perhaps not surprising that the score test of $\rho_l = 0$ will be based on a modified version of (8.18) with $f_{kt-l}(\boldsymbol{\theta}_s)$ replacing $f_{kt-1}(\boldsymbol{\theta}_s)$, while the test of $\rho_{il}^* = 0$ will be based on the analogue version of (8.19). Given that $\mathbf{y}_t$ is i.i.d. under the null, it is not difficult to show that the joint test for higher order dynamics will be given by T times the sum of terms of the form

$$\begin{aligned} &\left(\bar{G}_f(l;\boldsymbol{\eta})\ vecd'[\boldsymbol{\Gamma}^{-1/2}\bar{\mathbf{G}}_{\mathbf{v}}(l;\boldsymbol{\eta})\boldsymbol{\Gamma}^{-1/2}\right)\mathcal{I}^{-1}_{\boldsymbol{\rho}^\dagger\boldsymbol{\rho}^\dagger}(\boldsymbol{\theta}_s, 0;\boldsymbol{\eta}) \\ &\left(\bar{G}_f(l;\boldsymbol{\eta})\ vecd'[\boldsymbol{\Gamma}^{-1/2}\bar{\mathbf{G}}_{\mathbf{v}}(l;\boldsymbol{\eta})\boldsymbol{\Gamma}^{-1/2}\right)'. \end{aligned}$$

As expected, these statistics are also LM tests against MA(h) structures in the factors. And if for some reason we wanted to test for different orders of serial correlation in different latent variables, then we should eliminate the irrelevant autocovariances from the above expression.

Similarly, we could be interested either in models in which the autoregressive structure of the latent variable follows some restricted distributed lag, or in panel data type structures in which $\rho_{il}^* = \rho_l^* \ \forall i, l$ to alleviate the incidental parameter problems for large N. In those cases, we can use the usual chain rule to obtain the relevant moment conditions and their asymptotic covariance matrix. For example, if we assume that $\rho_l = \rho \ \forall l$, the relevant orthogonality condition of the Gaussian tests will become

$$E\left[f_{kt}(\boldsymbol{\theta}_s)\sum_{l=1}^{h} f_{kt-l}(\theta_s)|\theta_s, \mathbf{0}\right] = 0,$$

with $h \cdot [\mathbf{c}'\boldsymbol{\Sigma}^{-1}(\theta_s)\mathbf{c}]^2$ being the corresponding asymptotic variance. Interestingly, this expression is entirely analogous to the so-called Hodrick (1992) standard errors used in LM tests for long-run return predictability in univariate regressions with overlapping observations.

8.3.3 THE RELATIVE POWER OF AR TESTS IN MULTIVARIATE CONTEXTS

Although we investigate the finite sample properties of our proposed tests of serial correlation in common and specific factors in Section 8.6, it is illustrative to theoretically compare their power to other possibilities, such as the multivariate generalization of the Box and Pierce (1970) test proposed by Hosking (1981),[7] a standard univariate AR(1) test applied to the equally weighted portfolio (EWP), and a joint test of univariate first-order autocorrelation in all N series ($H_0 : vecd[\mathbf{G}_\mathbf{y}(1)] = \mathbf{0}$), which takes into account that the $y_{it}'s$ are contemporaneously correlated even when they are serially uncorrelated.[8] We consider a single factor model of the form:

$$\begin{aligned} y_{it} &= \pi_i + c_i x_t + u_{it} \qquad (i = 1, \ldots, 5) \\ x_t &= \rho x_{t-1} + \sqrt{1-\rho^2} f_t \\ u_{it} &= \rho_i^* u_{it-1} + \sqrt{1-\rho_i^{*2}} v_{it} \end{aligned}$$

where $\boldsymbol{\pi} = (.5, .4, .5, .4, .5)$, $\mathbf{c} = (5, 4, 5, 4, 5)$, $\boldsymbol{\gamma} = (5, 9, 5, 9, 5)$, and $\rho_i^* = \rho^* \forall i$. Such a design is motivated by the empirical application in Section 8.7 and our desire to avoid exchangeable models, in which unusual simplifications

[7] In the first order case, one can reinterpret his proposal as a test of the null hypothesis of lack of serial correlation against an unrestricted VAR(1) model, as in Hendry (1971), Guilkey (1974), and Harvey (1982).

[8] Given the single factor structure of $\boldsymbol{\Sigma}$, this test differs from Test 2 in Harvey (1982), which tests the null hypothesis $H_0 : vecd(\mathbf{G}_y(1)) = \mathbf{0}$ under the maintained assumption that $\boldsymbol{\Sigma}$ is diagonal.

occur. We evaluate asymptotic power against *compatible* sequences of local alternatives of the form $\boldsymbol{\rho}_{0T}^{\dagger} = \bar{\boldsymbol{\rho}}^{\dagger}/\sqrt{T}$ (see Appendix 8.10 for details). To avoid penalizing Hosking's test, in this section we only consider the Gaussian versions of our tests. In any case, all the Gaussian tests that we compare will be robust to the presence of non-normality.

In view of the discussion following Proposition 2, it is worth looking at the first two unconditional moments of $\mathbf{y}_t$. In this sense, note that by construction $E(x_t) = 0$, $V(x_t) = 1$, $E(u_{it}) = 0$, $V(u_{it}) = \gamma_i$, and $cov(x_t, u_{it}) = 0$ both under the null and the different alternatives, which implies that $E(\mathbf{y}_t) = \boldsymbol{\pi}$ and $V(\mathbf{y}_t) = \mathbf{cc}' + \boldsymbol{\Gamma}$. Thus, the unconditional standard deviations will be $\sqrt{30}$ for the first, third, and fifth series, and 5 for the second and fourth ones, while the unconditional correlations will be $.8\mathring{3}$ (odd with odd), .73 (odd with even) or .64 (even with even). Finally, the "signal to noise" ratio $\mathbf{c}'\boldsymbol{\Sigma}^{-1}\mathbf{c}$, which coincides with the R^2 in the theoretical least squares projection of f_t on a constant and $\mathbf{y}_t$, is .95.[9] As for the means, note that we have implicitly imposed that linear factor pricing holds because $\boldsymbol{\pi} = .1\mathbf{c}$. Although this restriction is inconsequential for our econometric results, it implies an a priorily realistic unconditional mean variance frontier, with a maximum Sharpe ratio of .34 on an annual basis.[10]

Figure 8.1(a) shows that when $\rho^* = 1.5\rho$ our proposed test of $H_0 : \boldsymbol{\rho}^{\dagger} = \mathbf{0}$ is the most powerful at the usual 5% significance level, closely followed by the test of $H_0 : \boldsymbol{\rho}^* = \mathbf{0}$. Next, we find the pormenteau test of $H_0 : vec[\mathbf{G}_\mathbf{y}(1)] = \mathbf{0}$ and the univariate test applied to EWP, which is barely distinguishable from the test of serial correlation in the common factor and very close to the "diagonal" serial correlation test of $H_0 : vecd[\mathbf{G}_\mathbf{y}(1)] = \mathbf{0}$. However, the results depend on the "signal to noise" ratio $\mathbf{c}'\boldsymbol{\Sigma}^{-1}\mathbf{c}$. Figure 8.1(b) shows the equivalent picture when we multiply all the elements of $\boldsymbol{\gamma}$ by 10, so that the R^2 in the regression of f_t on $\mathbf{y}_t$ reduces to .65. In this case, the power of our test of serial correlation in f_t, the univariate test on EWP, and especially the diagonal test increases substantially. In contrast, Figure 8.1(c) illustrates the effects of dividing the elements of $\boldsymbol{\gamma}$ by 5, so that the aforementioned R^2 reaches .99. In this context, the diagonal test becomes the least powerful.

The other crucial determinant of the power of the different tests is the relative magnitudes of ρ and ρ^*. Figure 8.2(a) shows the effect of setting $\rho^* = 0$ for our baseline signal to noise ratio, while Figure 8.2(b) illustrates the effects of $\rho = 0$. In the first case, the test of serial correlation in the common factor becomes the most powerful, with the test of serial correlation in the specific factors having power virtually equal to size, while exactly the opposite happens in the second case.[11]

[9] A more common measure of the importance of commonalities is the R^2 in the theoretical regression of each series on the common factor, which is $.8\mathring{3}$ for the odd numbered series and .64 for the even numbered ones.

[10] The ex-ante optimal mean variance portfolio percentage weights are (25.7, 11.4, 25.7, 11.4, 25.7).

[11] Although the test of $H_0 : \rho = 0$ has non-trivial local power when $\rho = 0$ but $\boldsymbol{\rho}^* \neq \mathbf{0}$, and the same is true of the test of $H_0 : \boldsymbol{\rho}^* = \mathbf{0}$ when $\boldsymbol{\rho}^* = \mathbf{0}$ but $\rho \neq 0$, a much larger horizontal axis would be necessary to appreciate those effects.

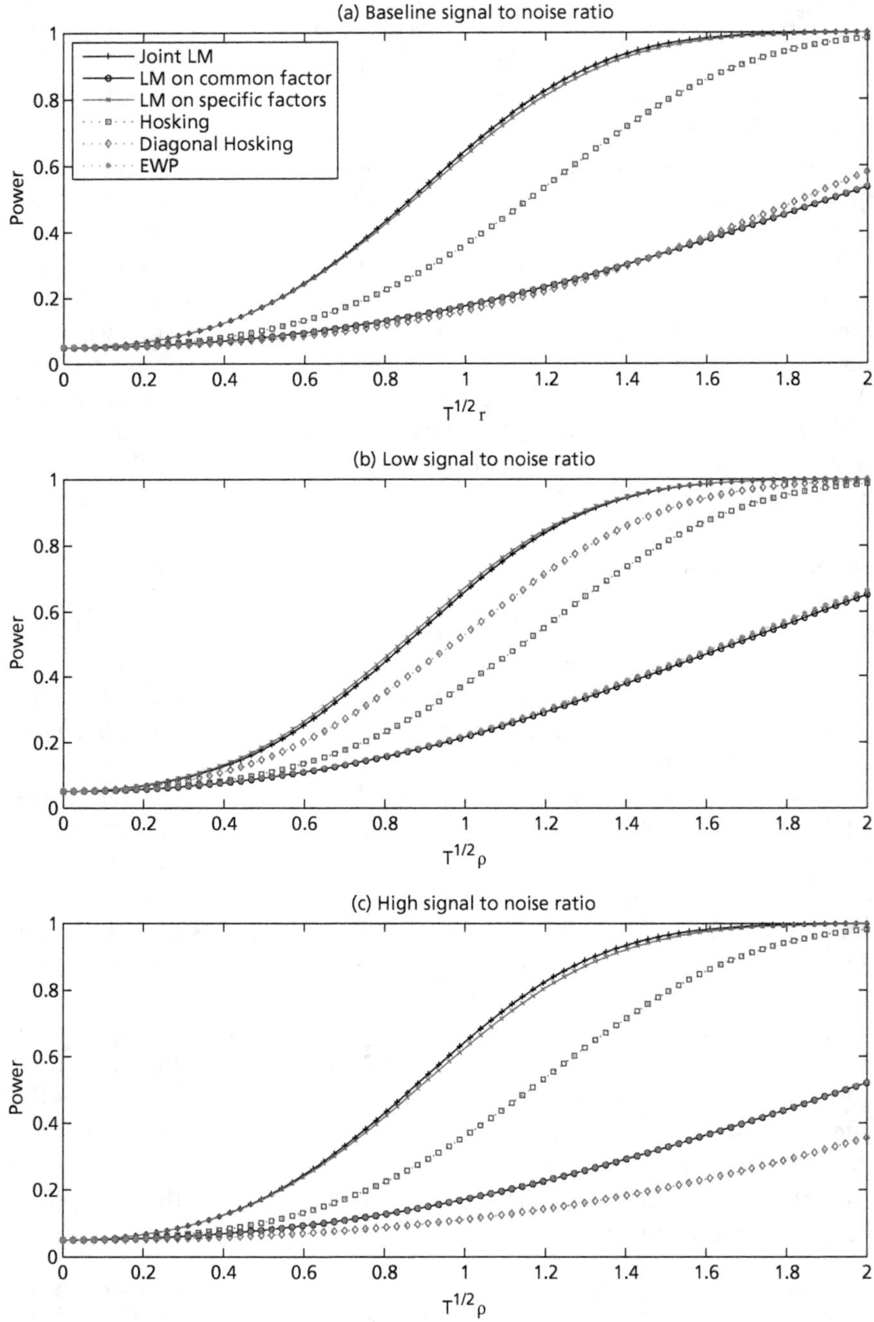

Figure 8.1. Power of mean dependence tests at 5% level against local alternatives at baseline, low, and high signal to noise ratios

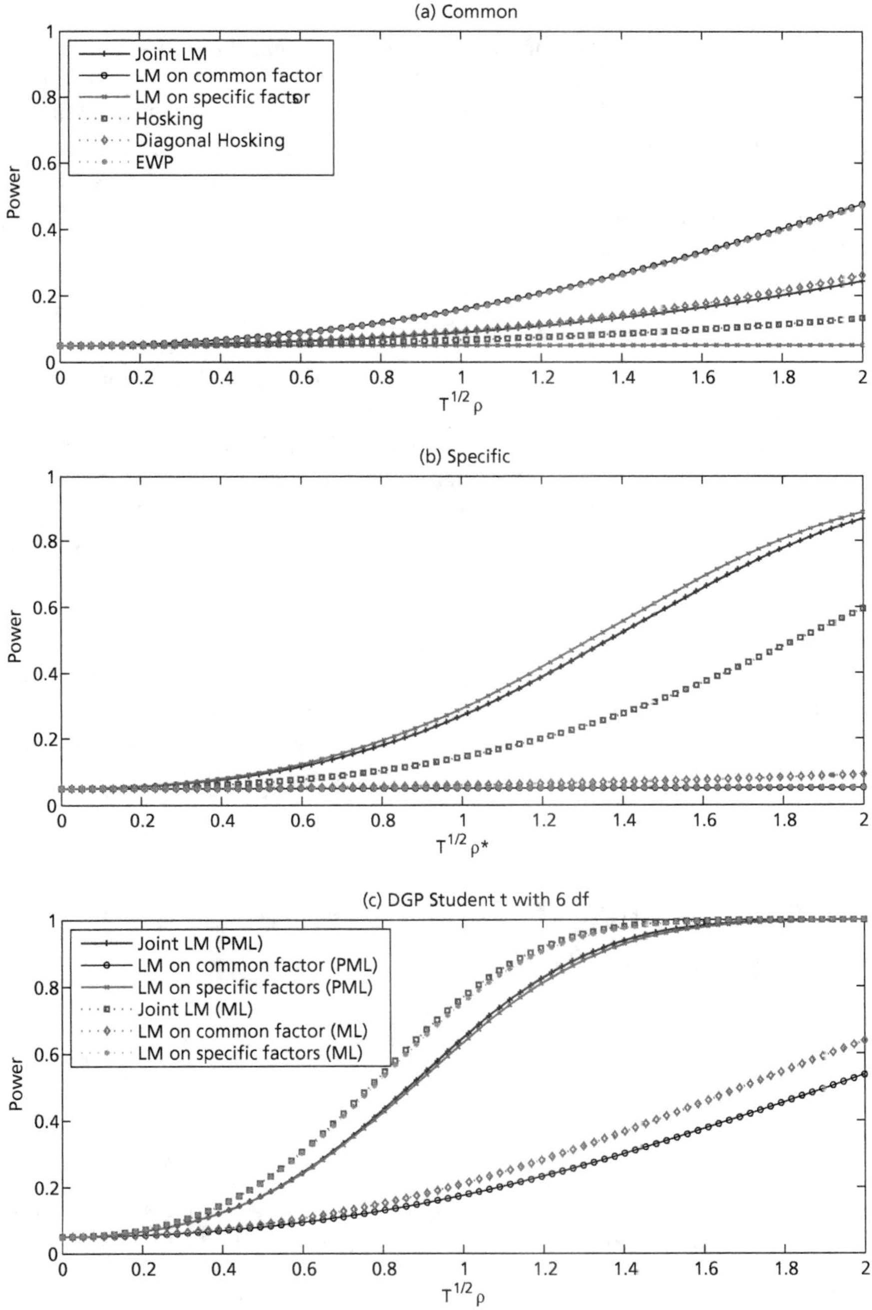

Figure 8.2. Power of mean dependence tests at 5% level against local alternatives for common, specific, and DGP Student t with 6 df

8.3.4 THE RELATIVE POWER OF THE NORMALITY TESTS

Let us now assess the gains that accrue from exploiting the non-normality in the distribution of returns. It is not difficult to show that the ratio of non-centrality parameters of the normality test $LM_{AR(1)}(\mathbf{0})$ and the elliptical likelihood test $LM_{AR(1)}(\boldsymbol{\eta})$ is $\mathrm{M}_{ll}^{-1}(\boldsymbol{\eta}_0)$, which reflects the fact that the non-centrality parameter of the Gaussian tests is invariant to the true conditional distribution of the data. In the multivariate Student t case with $\nu_0 > 4$, in particular, this asymptotic efficiency ratio becomes

$$\frac{(\nu_0 - 2)(\nu_0 + N + 2)}{\nu_0(\nu_0 + N)}. \tag{8.21}$$

For any given N, this ratio is monotonically increasing in ν_0, and approaches 1 from below as $\nu_0 \to \infty$, and 0 from above as $\nu_0 \to 2^+$. For instance, for $N = 1$ it takes the values of $.9\mathring{3}$ and .8 for $\nu_0 = 9$ and $\nu_0 = 5$, respectively. At the same time, this ratio is decreasing in N for a given ν_0, which reflects the fact that Fisher's information for the mean is "increasing" in N in the Student t case (see Fiorentini and Sentana (2010b)). For $N = 3$ and $\nu_0 = 9$, for instance, it takes the value of .907, while for $\nu_0 = 5$, its value is only .75.

It is also straightforward to map those efficiency ratios into power gains by considering sequences of local alternatives. For illustrative purposes, we look at the baseline design in Section 8.3.3 under the assumption that the true conditional distribution of ε_t^* is a multivariate t_6. Figure 8.2(c) shows that the power gains that accrue to our proposed serial correlation tests by exploiting the leptokurtosis of the Student t distribution are far from trivial.

8.4 Tests for ARCH effects in common and idiosyncratic factors

8.4.1 BASELINE CASE

The alternative that we consider next is the following conditionally heteroskedastic factor model:

$$\left.\begin{array}{c} \mathbf{y}_t = \boldsymbol{\pi} + \mathbf{c}f_t + \mathbf{v}_t \\ \begin{pmatrix} f_t \\ \mathbf{v}_t \end{pmatrix} | I_{t-1}; \theta, \eta \sim s\left[\begin{pmatrix} 0 \\ \mathbf{0} \end{pmatrix}, \begin{pmatrix} \lambda_t(\boldsymbol{\theta}) & \mathbf{0} \\ \mathbf{0} & \boldsymbol{\Gamma}_t(\boldsymbol{\theta}) \end{pmatrix}, \boldsymbol{\eta}\right] \end{array}\right\}, \tag{8.22}$$

with

$$\left.\begin{array}{r} \lambda_t(\boldsymbol{\theta}) = 1 + \alpha[E(f_{t-1}^2|\mathbf{Y}_{t-1}; \boldsymbol{\theta}, 0) - 1], \\ \gamma_{it}(\boldsymbol{\theta}) = \gamma_i + \alpha_i^*[E(v_{it-1}^2|\mathbf{Y}_{t-1}; \boldsymbol{\theta}, 0) - \gamma_i], \qquad (i = 1, \ldots, N) \end{array}\right\} \tag{8.23}$$

where $E(f_{t-1}^2|\mathbf{Y}_{t-1};\boldsymbol{\theta},\mathbf{0})$ and $E(v_{it-1}^2|\mathbf{Y}_{t-1};\boldsymbol{\theta},\mathbf{0})$ are the Kalman filter estimators of the squares of the underlying common and idiosyncratic factors obtained from this model (see Appendix B in Fiorentini and Sentana (2012) for details). In this case, the parameters of interest become $\boldsymbol{\phi} = (\theta_s', \boldsymbol{\eta}')'$, $\boldsymbol{\theta} = (\boldsymbol{\theta}_s', \alpha^{\dagger\prime})'$, where $\boldsymbol{\alpha}^{\dagger} = (\alpha, \alpha^*)$ and $\boldsymbol{\alpha}^* = (\alpha_1, \ldots, \alpha_N)$. Although it is in principle very important to distinguish between $I_{t-1} = \{\mathbf{y}_{t-1}, \mathbf{f}_{t-1}, \mathbf{y}_{t-2}, \mathbf{f}_{t-2}, \ldots\}$, and the econometrician's information set $\mathbf{Y}_{t-1}$, which only includes lagged values of $\mathbf{y}_t$, (see Harvey et al. (1992)), for ease of exposition we postpone the discussion of those cases in which $\lambda_t(\boldsymbol{\theta}) \notin \mathbf{Y}_{t-1}$ until Section 8.4.2.

Given (8.22) and (8.23), the distribution of $\mathbf{y}_t$ conditional on $\mathbf{Y}_{t-1}$ is $N(\mathbf{0}, \boldsymbol{\Sigma}_t)$, where $\boldsymbol{\Sigma}_t = \mathbf{cc}'\lambda_t + \boldsymbol{\Gamma}_t$ has the usual exact factor structure. For this reason, we shall refer to the data generation process specified by (8.22) as a multivariate conditionally heteroskedastic exact factor model, which reduces to our baseline specification (8.1) under the null hypothesis that $H_0 : \boldsymbol{\alpha}^{\dagger} = \mathbf{0}$. But even if f_t or $\mathbf{v}_t$ are conditionally heteroskedastic, provided that they are covariance stationary, model (8.22) also implies an unconditional exact factor structure for $\mathbf{y}_t$. That is, the unconditional covariance matrix, $\boldsymbol{\Sigma}$, can be written as:

$$\boldsymbol{\Sigma} = E(\boldsymbol{\Sigma}_t|\boldsymbol{\theta}) = \mathbf{cc}' + \boldsymbol{\Gamma}, \tag{8.24}$$

because we have set the unconditional variance of the common factor to 1 to eliminate the usual scale indeterminacy.[12]

The above model has very interesting implications for correlations. A stylized fact that has been noted before is that periods when markets are increasingly correlated are also times when markets are volatile (see King et al. (1994)). Since the empirical evidence typically suggests that changes in the unobservable factor lead to individual stocks moving in the same direction, Model (8.22) implies that periods when the volatility of the unobservable factor rises are also those when, *ceteris paribus*, individual stocks appear to exhibit greater inter-correlation. Specifically, the conditional correlation coefficient between any two elements of $\mathbf{y}_t$ is given by

$$\rho_{12t} = \frac{c_1c_2\lambda_t}{\sqrt{c_1^2\lambda_t + \gamma_{1t}}\sqrt{c_2^2\lambda_t + \gamma_{2t}}}.$$

Hence, ρ_{12t} will be increasing in λ_t if $c_1c_2 > 0$ and decreasing in γ_{1t} and γ_{2t}.

A more precise way to characterize the serial dependence structure implied by Model (8.22) is to consider the autocovariance structure of

$$vec[(\mathbf{y}_t - \boldsymbol{\pi})(\mathbf{y}_t - \boldsymbol{\pi})'] = (\mathbf{c} \otimes \mathbf{c})f_t^2 + vec(\mathbf{v}_t\mathbf{v}_t') + (\mathbf{I}_{N_2} + \mathbf{K}_{NN})(\mathbf{c} \otimes \mathbf{I}_N)vec(f_t\mathbf{v}_t),$$

[12] See Fiorentini et al. (2004) for symmetric scaling assumptions for integrated ARCH models.

where $\mathbf{K}_{mn}$ is the commutation matrix of orders m and n (see Magnus and Neudecker (1988)). Given that $vec(f_t\mathbf{v}_t)$ is a martingale difference sequence, $\mathbf{y}_t$ follows a weak ARCH model (see Nijman and Sentana (1996)) which shares the factor structure in (8.15) not for the levels but for the squares and cross-products of the observed variables $\mathbf{y}_t$ (see Appendix 8.10 for further details). In this sense, another empirically appealing feature of (8.22) is that all linear combinations of $\mathbf{y}_t$ will follow weak ARCH processes as long as α and $\boldsymbol{\alpha}^*$ are strictly positive.

Sentana and Fiorentini (2001) develop tests of the null hypothesis $H_0: \alpha = 0$ under the maintained hypotheses that $\boldsymbol{\alpha}^* = \mathbf{0}$ and the conditional distribution is Gaussian. The following proposition extends their results to joint tests of ARCH effects in common and specific factors in elliptical contexts.

Proposition 4 *Let*

$$\bar{S}_f(j;\boldsymbol{\eta}) = \frac{1}{T}\sum\nolimits_{t=1}^{T}\{\delta[\varsigma_t(\boldsymbol{\theta}_s),\boldsymbol{\eta}]f_{kt}^2(\boldsymbol{\theta}_s)+\omega_k(\boldsymbol{\theta}_s)-1\}\,[f_{kt-j}^2(\boldsymbol{\theta}_s)+\omega_k(\boldsymbol{\theta}_s)-1]$$

denote the sample cross moment of $\delta[\varsigma_t(\theta_s),\boldsymbol{\eta}]f_{kt}^2(\boldsymbol{\theta}_s)+\omega_k(\boldsymbol{\theta}_s)-1$ *and* $E(f_{t-j}^2|Y_{t-j};\boldsymbol{\theta}_s,0,0) = f_{kt-j}^2(\boldsymbol{\theta}_s)+\omega_k(\boldsymbol{\theta}_s)-1$, *where* $\delta(\varsigma_t,\eta)$ *is defined in (8.11) and* $f_{kt}(\boldsymbol{\theta}_s)$ *and* $\omega_k(\boldsymbol{\theta}_s)$ *are obtained from the updating equations (8.5) of the static factor model (8.1). Similarly, let*

$$\bar{\mathbf{S}}_{\mathbf{v}}(j;\boldsymbol{\eta}) = \frac{1}{T}\sum\nolimits_{t=1}^{T} vecd\{\delta[\varsigma_t(\boldsymbol{\theta}_s),\boldsymbol{\eta}]\mathbf{v}_{kt}(\boldsymbol{\theta}_s)\mathbf{v}_{kt}'(\boldsymbol{\theta}_s)+\mathbf{c}\mathbf{c}'\omega_k(\boldsymbol{\theta}_s)-\boldsymbol{\Gamma}\}$$
$$\times vecd[\mathbf{v}_{kt-j}(\boldsymbol{\theta}_s)\mathbf{v}_{kt-j}'(\boldsymbol{\theta}_s)+\mathbf{c}\mathbf{c}'\omega_k(\boldsymbol{\theta}_s)-\boldsymbol{\Gamma}]$$

denote the analogous sample cross moments for the specific factors.

1. *Under the null hypothesis* $H_0: \boldsymbol{\alpha}^\dagger = \mathbf{0}$, *the score test statistic* $LM_{ARCH(1)}(\boldsymbol{\eta})$ *given by*

$$\tfrac{T}{4}\left(\bar{S}_f(1;\boldsymbol{\eta}_0), vecd'[\boldsymbol{\Gamma}_0^{-1}\bar{\mathbf{S}}_{\mathbf{v}}(1;\boldsymbol{\eta})\boldsymbol{\Gamma}_0^{-1}]\right)\mathcal{I}_{\alpha^\dagger\boldsymbol{\alpha}^\dagger}^{-1}(\boldsymbol{\theta}_{s0},0;\boldsymbol{\eta}_0)\Big(\bar{S}_f(1;\boldsymbol{\eta}_0),$$
$$vecd'[\boldsymbol{\Gamma}_0^{-1}\bar{\mathbf{S}}_{\mathbf{v}}(1;\boldsymbol{\eta}_0)\boldsymbol{\Gamma}_0^{-1}]\Big)',$$

is distributed as a χ^2 *with* $N+1$ *degrees of freedom for* N *fixed as* T *goes to infinity, where*

$$\mathcal{I}_{\boldsymbol{\alpha}^\dagger\boldsymbol{\alpha}^\dagger}(\theta_s,0;\boldsymbol{\eta}) = \mathcal{V}_{\boldsymbol{\alpha}^\dagger\boldsymbol{\alpha}^\dagger}(\boldsymbol{\theta}_s,0;\boldsymbol{\eta})\odot\mathcal{V}_{\boldsymbol{\alpha}^\dagger\boldsymbol{\alpha}^\dagger}(\boldsymbol{\theta}_s,\boldsymbol{\eta};\boldsymbol{\eta}),$$

$$\mathcal{V}_{\boldsymbol{\alpha}^\dagger\boldsymbol{\alpha}^\dagger}(\theta_s,\eta;\eta) = V\begin{bmatrix}\frac{1}{\sqrt{2}}\{\delta[\varsigma_t(\boldsymbol{\theta}_s),\boldsymbol{\eta}]f_{kt}^2(\boldsymbol{\theta}_s)+\omega_k(\boldsymbol{\theta}_s)-1\}\\ \frac{1}{\sqrt{2}}\boldsymbol{\Gamma}^{-1}vecd\{\delta[\varsigma_t(\boldsymbol{\theta}_s),\boldsymbol{\eta}]\mathbf{v}_{kt}(\boldsymbol{\theta}_s)\mathbf{v}_{kt}'(\boldsymbol{\theta}_s)+\mathbf{c}\mathbf{c}'\omega_k(\boldsymbol{\theta}_s)-\boldsymbol{\Gamma}\}\end{bmatrix} =$$

$$M_{ss}(\eta)\begin{bmatrix}[\mathbf{c}'\boldsymbol{\Sigma}^{-1}(\boldsymbol{\theta}_s)\mathbf{c}]^2 & \mathbf{c}'\boldsymbol{\Sigma}^{-1}(\boldsymbol{\theta}_s)\boldsymbol{\Gamma}^{1/2}\odot\mathbf{c}'\boldsymbol{\Sigma}^{-1}(\boldsymbol{\theta}_s)\boldsymbol{\Gamma}^{1/2}\\ \boldsymbol{\Gamma}^{1/2}\boldsymbol{\Sigma}^{-1}(\boldsymbol{\theta}_s)\mathbf{c}\odot\boldsymbol{\Gamma}^{1/2}\boldsymbol{\Sigma}^{-1}(\boldsymbol{\theta}_s)\mathbf{c} & \boldsymbol{\Gamma}^{1/2}\boldsymbol{\Sigma}^{-1}(\boldsymbol{\theta}_s)\boldsymbol{\Gamma}^{1/2}\odot\boldsymbol{\Gamma}^{1/2}\boldsymbol{\Sigma}^{-1}(\theta_s)\boldsymbol{\Gamma}^{1/2}\end{bmatrix}+$$

$$\frac{[\text{M}_{ss}(\eta)-1]}{2}\begin{bmatrix} [\mathbf{c}'\boldsymbol{\Sigma}^{-1}(\boldsymbol{\theta}_s)\mathbf{c}]^2 & [\mathbf{c}'\boldsymbol{\Sigma}^{-1}(\boldsymbol{\theta}_s)\mathbf{c}]vecd'[\boldsymbol{\Gamma}^{1/2}\boldsymbol{\Sigma}^{-1}(\boldsymbol{\theta}_s)\boldsymbol{\Gamma}^{1/2}] \\ [\mathbf{c}\Sigma^{-1}(\boldsymbol{\theta}_s)\mathbf{c}]vecd[\boldsymbol{\Gamma}^{1/2}\boldsymbol{\Sigma}^{-1}(\boldsymbol{\theta}_s)\boldsymbol{\Gamma}^{1/2}] & vecd[\boldsymbol{\Gamma}^{1/2}\boldsymbol{\Sigma}^{-1}(\boldsymbol{\theta}_s)\boldsymbol{\Gamma}^{1/2}]vecd'[\boldsymbol{\Gamma}^{1/2}\boldsymbol{\Sigma}^{-1}(\boldsymbol{\theta}_s)\boldsymbol{\Gamma}^{1/2}] \end{bmatrix}.$$

and $\mathcal{V}_{\boldsymbol{\alpha}^\dagger\boldsymbol{\alpha}^\dagger}(\boldsymbol{\theta}_s, 0; \boldsymbol{\eta})$ mimics $\mathcal{V}_{\boldsymbol{\alpha}^\dagger\boldsymbol{\alpha}^\dagger}(\boldsymbol{\theta}_s, \boldsymbol{\eta}; \boldsymbol{\eta})$ after replacing $\text{M}_{ss}(\eta)$ by $\kappa + 1$. Moreover, this asymptotic null distribution is unaffected if we replace $\boldsymbol{\theta}_{s0}$ and $\boldsymbol{\eta}_0$ by their feasible maximum likelihood estimators in Proposition 1.

2. *It also remains valid if we replace $\boldsymbol{\theta}_{s0}$ by its elliptically symmetric semiparametric estimator in Proposition 1, which requires the nonparametric estimation of the density of $\varsigma_t(\boldsymbol{\theta}_s)$.*
3. *Under the same null hypothesis, the Gaussian pseudo score test statistic $LM_{ARCH(1)}(\mathbf{0})$ given by*

$$\tfrac{T}{4}\left(\bar{S}_f(1;\mathbf{0}), vecd'[\boldsymbol{\Gamma}_0^{-1}\bar{\mathbf{S}}_{\mathrm{v}}(1;\mathbf{0})\boldsymbol{\Gamma}_0^{-1}]\right)\mathcal{B}^{-1}_{\boldsymbol{\alpha}^\dagger\alpha^\dagger}(\boldsymbol{\phi}_0)\left(\bar{S}_f(1;\mathbf{0}), vecd'[\boldsymbol{\Gamma}_0^{-1}\bar{\mathbf{S}}_{\mathrm{v}}(1;\mathbf{0})\boldsymbol{\Gamma}_0^{-1}]\right)',$$

with

$$\mathcal{B}_{\boldsymbol{\alpha}^\dagger\boldsymbol{\alpha}^\dagger}(\phi) = \mathcal{V}_{\boldsymbol{\alpha}^\dagger\boldsymbol{\alpha}^\dagger}(\boldsymbol{\theta}_s, 0; \boldsymbol{\eta}) \odot \mathcal{V}_{\alpha^\dagger\boldsymbol{\alpha}^\dagger}(\boldsymbol{\theta}_s, 0; \boldsymbol{\eta}), \tag{8.25}$$

will be distributed as a χ^2 with $N+1$ degrees of freedom for N fixed as T goes to infinity, irrespective of whether the elliptical conditional distribution is normal. This result continues to hold if we replace $\boldsymbol{\theta}_{s0}$ by its Gaussian pseudo maximum likelihood estimator $\bar{\theta}_s$ in Proposition 1.

Researchers may once more be interested in tests that separately assess the conditional heteroskedasticity of either the common factor or the specific factors. Indeed, they might even like to focus on a particular v_{it}. By combining the relevant elements of $\bar{S}_f(j;\boldsymbol{\eta})$ and $\bar{\mathbf{S}}_{\mathrm{v}}(1;\boldsymbol{\eta})$ with the corresponding blocks of the information matrix, $\mathcal{I}_{\boldsymbol{\alpha}^\dagger\boldsymbol{\alpha}^\dagger}(\boldsymbol{\theta}_s, 0; \boldsymbol{\eta})$, we can easily exploit the results in Proposition 4 to derive the required test statistics for those subcomponents under the maintained hypothesis of serial independence. Intuitively, the reason is that we can interpret $LM_{ARCH(1)}(\boldsymbol{\eta})$ as a test based on the $N+1$ orthogonality conditions:

$$E\left\{\begin{array}{c} \frac{1}{2}\{\delta[\varsigma_t(\boldsymbol{\theta}_s), \eta]f_{kt}^2(\boldsymbol{\theta}_s) + \omega_k(\boldsymbol{\theta}_s) - 1\} \\ \cdot[f_{kt-1}^2(\boldsymbol{\theta}_s) + \omega_k(\boldsymbol{\theta}_s) - 1]|\theta_s, 0, \boldsymbol{\eta} \end{array}\right\} = 0, \tag{8.26}$$

$$E\left\{\begin{array}{c} \frac{1}{2}\gamma_i^{-2}\{\delta[\varsigma_t(\boldsymbol{\theta}_s), \boldsymbol{\eta}]v_{kit}^2(\boldsymbol{\theta}_s) + c_i^2\omega_k(\boldsymbol{\theta}_s) - \gamma_i\} \\ \cdot[v_{kit-1}^2(\boldsymbol{\theta}_s) + c_i^2\omega_k(\boldsymbol{\theta}_s) - \gamma_i]|\theta_s, \mathbf{0}, \eta \end{array}\right\} = 0 \quad (i = 1, \ldots, N). \tag{8.27}$$

Similarly, $LM_{ARCH(1)}(\mathbf{0})$ is based on

$$E\left\{\begin{array}{c} \frac{1}{2}[f_{kt}^2(\boldsymbol{\theta}_s) + \omega_k(\boldsymbol{\theta}_s) - 1] \\ \cdot[f_{kt-1}^2(\boldsymbol{\theta}_s) + \omega_k(\boldsymbol{\theta}_s) - 1]|\theta_s, \mathbf{0}, \boldsymbol{\eta} \end{array}\right\} = 0, \tag{8.28}$$

$$E\left\{\begin{array}{c} \frac{1}{2}\gamma_i^{-2}[v_{kit}^2(\boldsymbol{\theta}_s) + c_i^2\omega_k(\boldsymbol{\theta}_s) - \gamma_i] \\ \cdot[v_{kit-1}^2(\boldsymbol{\theta}_s) + c_i^2\omega_k(\boldsymbol{\theta}_s) - \gamma_i]|\theta_s, \mathbf{0}, \eta \end{array}\right\} = 0 \qquad (i = 1, \ldots, N). \tag{8.29}$$

As in the serial correlation tests, $\delta(\varsigma_t, \eta)$ acts as a damping factor for big observations (see Fiorentini and Sentana (2010a) for a closely related discussion on univariate models).[13]

Once again, given that we normalize $V(f_t)$ to 1, the moment conditions (8.28) and (8.29) closely resemble

$$E[(f_t^2 - 1)(f_{t-1}^2 - 1)|\theta_s, \mathbf{0}, \boldsymbol{\eta}] = 0,$$
$$E[\gamma_i^{-2}(v_{it}^2 - \gamma_i)(v_{it-1}^2 - \gamma_i)|\theta_s, \mathbf{0}, \boldsymbol{\eta}] = 0 \qquad (i = 1, \ldots, N),$$

which are the Gaussian-based orthogonality conditions that we would use to test for first order ARCH effects if we could observe the latent variables (see, e.g., Engle (1982)).

The similarity between these two sets of moment conditions becomes even stronger if we consider individual tests for ARCH in each latent variable. Let us start with a test of $H_0 : \alpha = 0$ under the maintained assumption that $\boldsymbol{\alpha}^* = \mathbf{0}$. Part 3 of Proposition 4 implies that the asymptotic variance of $\bar{S}_f(1; \mathbf{0})$ is simply $\frac{1}{2}(3\kappa + 2)^2[\mathbf{c}'\boldsymbol{\Sigma}^{-1}(\boldsymbol{\theta}_s)\mathbf{c}]^4$. But as we saw in Section 8.3.1, we can interpret $\mathbf{c}'\boldsymbol{\Sigma}^{-1}(\boldsymbol{\theta}_s)\mathbf{c}$ as the R^2 in the theoretical least squares projection of f_t on a constant and $\mathbf{y}_t$. Therefore, the higher the degree of observability of the common factor, the closer the asymptotic variance of $\bar{S}_f(1; 0)$ will be to $\frac{1}{2}(3\kappa + 2)^2$, which is the asymptotic variance of the first sample autocovariance of f_t^2 under the null. Intuitively, this convergence result simply reflects the fact that the common factor becomes observable in the limit, which implies that our test of $H_0 : \alpha = 0$ will become arbitrarily close to a first order ARCH test for the common factor as the "signal to noise" ratio $\mathbf{c}'\boldsymbol{\Sigma}^{-1}(\boldsymbol{\theta}_s)\mathbf{c}$ approaches 1. Before the limit, though, our test takes into account the unobservability of f_t.

Also, Part 3 of Proposition 4 implies that the asymptotic variance of $\bar{S}_{v_{ki}v_{ki}}(1, 0)$ is $\frac{1}{2}(3\kappa + 2)^2[\gamma_i\sigma^{ii}(\boldsymbol{\theta}_s)]^4$, where $\sigma^{ii}(\boldsymbol{\theta}_s)$ denotes the ith diagonal element of $\boldsymbol{\Sigma}^{-1}(\boldsymbol{\theta}_s)$. But since we can again interpret $\gamma_i\sigma^{ii}(\theta_s)$ as the R^2 in the theoretical least squares projection of v_{it} on a constant and $\mathbf{y}_t$, we can apply a similar line of reasoning to a test of $H_0 : \alpha_i^* = 0$ under the maintained assumption that $\alpha = 0$ and the remaining elements of $\boldsymbol{\alpha}^*$ are 0. Once again, though, it is important to emphasize that our joint tests take into account the covariance between the Kalman filter estimators of the underlying factors, even though the latent variables themselves are uncorrelated.

Part 3 of Proposition 4 also implies that the asymptotic distribution of the Gaussian tests does not depend on normality, although if the conditional distribution of $\mathbf{y}_t$ given $\mathbf{Y}_{t-1}$ were not elliptical, then one would have to replace $\mathcal{V}_{\alpha^\dagger\alpha^\dagger}(\boldsymbol{\theta}_s, 0; \boldsymbol{\eta})$ in (8.25) by the joint unconditional covariance matrix

[13] This factor also plays an important role in the beta-t-ARCH models proposed by Harvey and Chakravarty (2008), although if one derived an LM test for conditional homoskedasticity against their models, $\delta(\varsigma_t, \eta)$ would appear not only in the regressand but also in the regressor.

of $\frac{1}{\sqrt{2}}[f_{kt}^2(\boldsymbol{\theta}_s) + \omega_k(\boldsymbol{\theta}_s) - 1]$ and $\frac{1}{\sqrt{2}}\boldsymbol{\Gamma}^{-1}vecd[\mathbf{v}_{kt}(\boldsymbol{\theta}_s)\mathbf{v}_{kt}'(\boldsymbol{\theta}_s)+\mathbf{c}\mathbf{c}'\omega_k(\boldsymbol{\theta}_s) - \boldsymbol{\Gamma}]$ under the null of $H_0 : \alpha^\dagger = \mathbf{0}$. The advantage of using the theoretical expressions in the elliptical case is that they should improve the finite sample reliability of the Gaussian tests.

Interestingly, such robust versions of the test for ARCH effects in common and idiosyncratic factors can be regarded as the factor analytic analogues to the suggestion that Koenker (1981) made to robustify tests of conditional homoskedasticity based on Gaussian scores, such as the original univariate ARCH test in Engle (1982), whose information matrix version is only valid under conditional normality. In fact, we can obtain an asymptotically equivalent test of $H_0 : \alpha = 0$ by computing the F test of the regression of $f_{kt}^2(\boldsymbol{\theta}_s)$ on a constant and $f_{kt-1}^2(\boldsymbol{\theta}_s)$, whose asymptotic null distribution remains valid irrespective of the normality of $f_{kt}(\boldsymbol{\theta}_s)$ because it is effectively using a consistent estimator of $V[f_{kt}^2(\boldsymbol{\theta}_s)]$ as the residual variance of the regression under the null. But if we impose that the residual variance is $2[\mathbf{c}'\boldsymbol{\Sigma}^{-1}(\boldsymbol{\theta}_s)\mathbf{c}]^2$ instead, which is its value under normality because $V[f_{kt}(\boldsymbol{\theta}_s)|\theta_s, 0, \eta] = \mathbf{c}'\boldsymbol{\Sigma}^{-1}(\boldsymbol{\theta}_s)\mathbf{c}$, then our F test will be incorrectly sized when the conditional distribution is not Gaussian.

Again, it would be straightforward to adapt Proposition 4 to handle large N panel data restrictions such as $\alpha_i^* = \alpha^* \; \forall i$, as in Sentana et al. (2008). Further, given that the orthogonality conditions (8.26) and (8.27) remain valid when $\mathbf{y}_t$ is serially independent irrespective of $V(\mathbf{y}_t)$ having an exact single factor structure, one could also use them to derive a standard moment test that will still have non-trivial power even though it will no longer be an LM test.

8.4.2 EXTENSIONS

Unobservable conditional variances

Following the discussion at the beginning of Section 5 in Harvey et al. (1992), Specification (8.22) assumes that the conditional variances of common and specific factors are a function of lagged observable variables. But it may seem more natural to assume that those variances are in fact functions of the lagged latent variables. Specifically,

$$\left.\begin{aligned} \lambda_t(\boldsymbol{\theta}) &= 1 + \alpha(f_{t-j}^2 - 1), \\ \gamma_{it}(\boldsymbol{\theta}) &= \gamma_i + \alpha_i^*(v_{it-1}^2 - \gamma_i), \qquad (i = 1, \ldots, N). \end{aligned}\right\} \tag{8.30}$$

The problem with this formulation is that even in the Gaussian case the log-likelihood function can no longer be written in closed form except when $\boldsymbol{\alpha}^\dagger = \mathbf{0}$, and one has to resort to simulation methods, such as the MCMC procedures put forward by Fiorentini et al. (2004). As explained by Sentana et al. (2008), the combination of (8.22) with (8.23) may be regarded as a convenient auxiliary model that coincides with the model of interest for $\boldsymbol{\alpha}^\dagger = \mathbf{0}$, but whose log-likelihood function and score we can obtain in closed form for every possible value of $\alpha^\dagger$. In this regard, it is important to bear in mind the fact that we can compute the true log-likelihood function of $\mathbf{y}_t$ under the null of $\boldsymbol{\alpha}^\dagger = \mathbf{0}$ is not sufficient to compute its derivative with respect to $\boldsymbol{\alpha}^\dagger$. Fortunately, it is once again possible to use the EM principle to obtain this score. Remarkably, it turns out that the score of the model with latent variances (8.30) is virtually identical to the score of the approximating model under the null of conditionally homoskedasticity despite both the non-measurability of λ_t and $\boldsymbol{\Gamma}_t$ and the potential non-normality of the conditional distribution. In fact, they would coincide if we had followed Section 5.2 of Harvey et al. (1992) instead, and replaced the conditional variances of common and specific factors in (8.23) by

$$\left.\begin{aligned} \lambda_t(\boldsymbol{\theta}) &= 1 + \alpha[E(f_{t-1}^2|\mathrm{Y}_{t-1};\boldsymbol{\theta},\boldsymbol{\eta}) - 1], \\ \gamma_{it}(\boldsymbol{\theta}) &= \gamma_i + \alpha_i^*[E(v_{it-1}^2|\mathrm{Y}_{t-1};\boldsymbol{\theta},\boldsymbol{\eta}) - \gamma_i], \qquad (i = 1, \ldots, N) \end{aligned}\right\} \quad (8.31)$$

where

$$E(f_{t-1}^2|\mathrm{Y}_{t-1};\boldsymbol{\theta}_s, 0, \boldsymbol{\eta}) = f_{kt-1}^2(\boldsymbol{\theta}_s) + \upsilon_{kt-1}(\boldsymbol{\theta}_s, \boldsymbol{\eta}) - 1,$$

$$E(v_{it-1}^2|\mathrm{Y}_{t-1};\boldsymbol{\theta}_s, 0, \eta) = v_{ikt-1}^2(\boldsymbol{\theta}_s) + c_i^2\upsilon_{kt-1}(\theta_s, \boldsymbol{\eta}) - \gamma_i,$$

with $f_{kt}(\boldsymbol{\theta}_s)$, $\omega_k(\boldsymbol{\theta}_s)$, and $\upsilon_{kt-1}(\boldsymbol{\theta}_s, \boldsymbol{\eta})$ defined in (8.5). The following result, which generalizes Proposition 1 in Sentana et al. (2008), formalizes our claim for the multivariate Student t, but we conjecture it applies to most other elliptical distributions:

Proposition 5 *Let* $\mathbf{s}_t(\boldsymbol{\phi}) = \partial l_t(\boldsymbol{\phi})/\partial\boldsymbol{\phi}$ *denote the log-likelihood score of the conditionally heteroskedastic model for* $\mathbf{y}_t|\mathrm{Y}_{t-1};\boldsymbol{\phi}$ *in (8.22) when the variances of the latent factors are given by (8.31). Similarly, let* $\mathbf{q}_t(\boldsymbol{\phi}) = \partial p(\mathbf{y}_t|\mathrm{Y}_{t-1};\boldsymbol{\phi})/\partial\boldsymbol{\phi}$ *denote the exact log-likelihood score of the same model when the variances of the latent factors are given by (8.30) instead. If the conditional distribution is a (standardized) multivariate Student t with* $0 \leq \eta < .5$ *then* $l_t(\boldsymbol{\phi}) = p(\mathbf{y}_t|\mathrm{Y}_{t-1};\boldsymbol{\phi})$ *and* $\mathbf{s}_t(\boldsymbol{\phi}) = \mathbf{q}_t(\boldsymbol{\phi})$ *when evaluated at* $\boldsymbol{\alpha}^\dagger = \mathbf{0}$.

Therefore, the approximating model that uses (8.31) "smoothly embeds" the original model in those circumstances.

Higher order processes

Consider the following alternative specification:

$$\lambda_t(\boldsymbol{\theta}) = 1 + \sum_{j=1}^{q} \alpha_j[E(f_{t-j}^2|Y_{t-j};\boldsymbol{\theta}, 0) - 1],$$

$$\gamma_{it}(\boldsymbol{\theta}) = \gamma_i + \sum_{j=1}^{q_i^*} \alpha_{ij}^*[E(v_{it-j}^2|Y_{t-j};\boldsymbol{\theta}, 0) - \gamma_i]. \qquad (i = 1, \ldots, N).$$

In view of the discussion in Section 8.4.1, it is perhaps not surprising that the score test of $\alpha_j = 0$ will be based on a modified version of (8.28) with

$f^2_{kt-j}(\boldsymbol{\theta}_s)$ replacing $f^2_{kt-1}(\boldsymbol{\theta}_s)$, while the test of $\alpha^*_{ij} = 0$ will be based on the analogue version of (8.29). Given that $\mathbf{y}_t$ is i.i.d. under the null hypothesis, it is not difficult to show that the joint test for higher order dynamics will be given by $\frac{1}{4}T$ times the sum of terms of the form

$$\left(\bar{S}_f(j;\boldsymbol{\eta}), vecd'[\boldsymbol{\Gamma}^{-1}\bar{\mathbf{S}}_v(j;\boldsymbol{\eta})\boldsymbol{\Gamma}^{-1}]\right)\mathcal{I}^{-1}_{\alpha^\dagger\alpha^\dagger}(\boldsymbol{\theta}_s,0;\boldsymbol{\eta})\left(\bar{S}_f(j;\eta), vecd'[\boldsymbol{\Gamma}^{-1}\bar{\mathbf{S}}_v(j;\boldsymbol{\eta})\boldsymbol{\Gamma}^{-1}]\right)'.$$

Once again, we could eliminate the irrelevant autocovariances from the above expression to test for different orders of serial correlation in the squares of different latent variables.

The univariate empirical evidence, though, suggests that GARCH(1,1) specifications of the form

$$\begin{aligned}
\lambda_t(\boldsymbol{\theta}) &= 1-\alpha-\beta+\alpha E(f^2_{t-j}|Y_{t-1};\boldsymbol{\theta},0)+\beta\lambda_{t-1}(\boldsymbol{\theta})\\
&= 1+\alpha\textstyle\sum_{j=1}^{\infty}\beta^{j-1}[E(f^2_{t-j}|Y_{t-j};\boldsymbol{\theta},0)-1],\\
\gamma_{it}(\boldsymbol{\theta}) &= \gamma_i(1-\alpha^*_i-\beta^*_i)+\alpha^*_i E(v^2_{it-j}|Y_{t-1};\boldsymbol{\theta},0)+\beta^*_i\gamma_{it-1}(\boldsymbol{\theta})\\
&= \gamma_i+\alpha^*_i\textstyle\sum_{j=1}^{\infty}(\beta^*_i)^{j-1}[E(v^2_{it-j}|Y_{t-j};\boldsymbol{\theta},0)-\gamma_i]
\end{aligned}$$

should be more realistic than unrestricted ARCH(q) ones. As Bollerslev (1986) noted in a univariate context, however, one cannot derive a score test for conditional homoskedasticity versus these GARCH(1,1) specifications in the usual way, because β and β^*_i are only identified under the alternative. A possible solution to testing situations such as this one involves computing the test statistic for many values of β and β^*_i in the range [0,1), which are then combined to construct an overall statistic, as initially suggested by Davies (1977, 1987). Andrews (2001) discusses ways of obtaining critical values for such tests by regarding the different LM statistics as continuous stochastic processes indexed with respect to the parameters β and β^*_i $(i = 1,\ldots,N)$. Unfortunately, his procedure is difficult to apply in our context because $\dim(\boldsymbol{\beta}^\dagger) = N+1$. An alternative solution involves choosing arbitrary values of the underidentified parameters to carry out a score test of $\alpha = 0$ and $\boldsymbol{\alpha}^* = \mathbf{0}$ based on the moment conditions

$$E\left\{[\delta[\varsigma_t(\boldsymbol{\theta}_s),\boldsymbol{\eta}]f^2_{kt}(\boldsymbol{\theta}_s)+\omega_k(\boldsymbol{\theta}_s)-1]\sum_{l=1}^{\infty}\beta^{l-1}[f^2_{kt-l}(\boldsymbol{\theta}_s)+\omega_k(\boldsymbol{\theta}_s)-1]|\theta_s,0,\mathbf{0}\right\}=0,$$

$$E\left\{[\delta[\varsigma_t(\boldsymbol{\theta}_s),\boldsymbol{\eta}]v^2_{kit}(\boldsymbol{\theta}_s)+c_i^2\omega_k(\boldsymbol{\theta}_s)-\gamma_i]\sum_{l=1}^{\infty}(\beta^*_i)^{l-1}[v^2_{kit-l}(\boldsymbol{\theta}_s)+c_i^2\omega_k(\theta_s)-\gamma_i]|\theta_s,0,\mathbf{0}\right\}=0,$$

whose asymptotic covariance matrix would be

$$\sum_{l=0}^{\infty} diag^l[\beta,\boldsymbol{\beta}^{*\prime}]\mathcal{I}_{\alpha^\dagger\alpha^\dagger}(\theta_s,0;\boldsymbol{\eta})diag^l[\beta,\boldsymbol{\beta}^{*\prime}],$$

which can be obtained in closed form. The values of β and β^* influence the small sample power of these tests, achieving maximum power when they coincide with their true values (see Demos and Sentana (1998)), but the advantage is that the resulting tests have standard distributions under H_0. An attractive possibility is to set β and $\boldsymbol{\beta}^*$ to the decay factor recommended by Group (1996) to obtain exponentially weighted volatility estimates for f_{kt} and v_{ikt}.

8.4.3 THE RELATIVE POWER OF ARCH TESTS IN MULTIVARIATE CONTEXTS

We compare the power of our LM tests, Hosking's test applied to $vech[(\mathbf{y}_t - \boldsymbol{\pi})(\mathbf{y}_t - \boldsymbol{\pi})']$ as in Duchesne and Lalancette (2003), a standard univariate ARCH(1) test applied to the EW portfolio, a joint test of univariate first-order autocorrelation in all $N(N+1)/2$ squares and cross-products of the (demeaned) observed series, and an analogous test that only focuses on their squares. Note that our joint LM test can also be understood as test of univariate first-order autocorrelation in the squares of $[f_{kt}(\theta_s), \mathbf{v}'_{kt}(\boldsymbol{\theta}_s)]$.[14] We consider another non-exchangeable single factor model of the form:

$$\begin{aligned} y_{it} &= \pi_i + c_i f_t + v_{it} \qquad (i = 1, \ldots, 5) \\ \lambda_t &= (1-\alpha) + \alpha f_{t-1}^2 \\ \gamma_{it} &= \gamma_i(1-\alpha_i^*) + \alpha_i^* v_{it-1}^2 \end{aligned}$$

where $\boldsymbol{\pi} = (.5, .4, .5, .4, .5)$, $\mathbf{c} = (5, 4, 5, 4, 5)$, $\boldsymbol{\gamma} \propto (5, 9, 5, 9, 5)$, and $\alpha_i^* = \alpha^*$ $\forall i$, whose first two unconditional moments are also empirically motivated, as they coincide with those of the model considered in Section 8.3.3. We evaluate power against *compatible* sequences of local alternatives of the form $\boldsymbol{\alpha}_{0T}^{\dagger} = \bar{\boldsymbol{\alpha}}^{\dagger}/\sqrt{T}$ (see Appendix 8.10 for details). To avoid penalizing Hosking's test, in this section we only consider the Gaussian versions of our tests.[15]

For the baseline case in which $\boldsymbol{\gamma} = (5, 9, 5, 9, 5)$, and $\alpha^* = \alpha$, Figure 8.3(a) shows that our proposed test of $H_0 : \boldsymbol{\alpha}^{\dagger} = \mathbf{0}$ is the most powerful at the usual 5% significance level, followed by our test of $H_0 : \boldsymbol{\alpha}^* = \mathbf{0}$. Next we find our test of ARCH effects in the common factor and the univariate ARCH test applied to EWP, the diagonal serial correlation tests of $vecd[(\mathbf{y}_t - \boldsymbol{\pi})(\mathbf{y}_t - \boldsymbol{\pi})']$, and $vech[(\mathbf{y}_t - \boldsymbol{\pi})(\mathbf{y}_t - \boldsymbol{\pi})']$, and finally the pormanteau test of unrestricted first-order serial dependence, which suffers from having a very large number of degrees of freedom. Once again, though, this ranking crucially depends on the "signal to noise" ratio $\mathbf{c}'\boldsymbol{\Sigma}^{-1}\mathbf{c}$. Figure 8.3(b) shows the equivalent picture when we multiply all the elements of $\boldsymbol{\gamma}$ by ten, so that the R^2 in the regression of f_t on $\mathbf{y}_t$ reduces to .65. In this case, the power of the two univariate tests decreases substantially, while the power of the diagonal tests increases. In contrast, Figure 8.3(c) illustrates the effects of dividing the elements of $\boldsymbol{\gamma}$ by five, so that the aforementioned R^2 reaches .99. Not surprisingly, the power of the two univariate tests almost coincides because EWP and $f_{kt}(\boldsymbol{\theta}_0)$ become very highly correlated.

[14] Another implication of the single factor structure of $\boldsymbol{\Sigma}$ is that our proposed LM test differs from the multivariate ARCH test considered by Dufour et al. (2010), who apply Hosking's test to the vech of the outer product of standardized values of $\mathbf{y}_t$ obtained from a Cholesky decomposition of $\bar{\boldsymbol{\Sigma}}$.

[15] See Footnote 12 in Fiorentini and Sentana (2012) for ways of making Hosking's tests for squares and cross-products robust to non-normality.

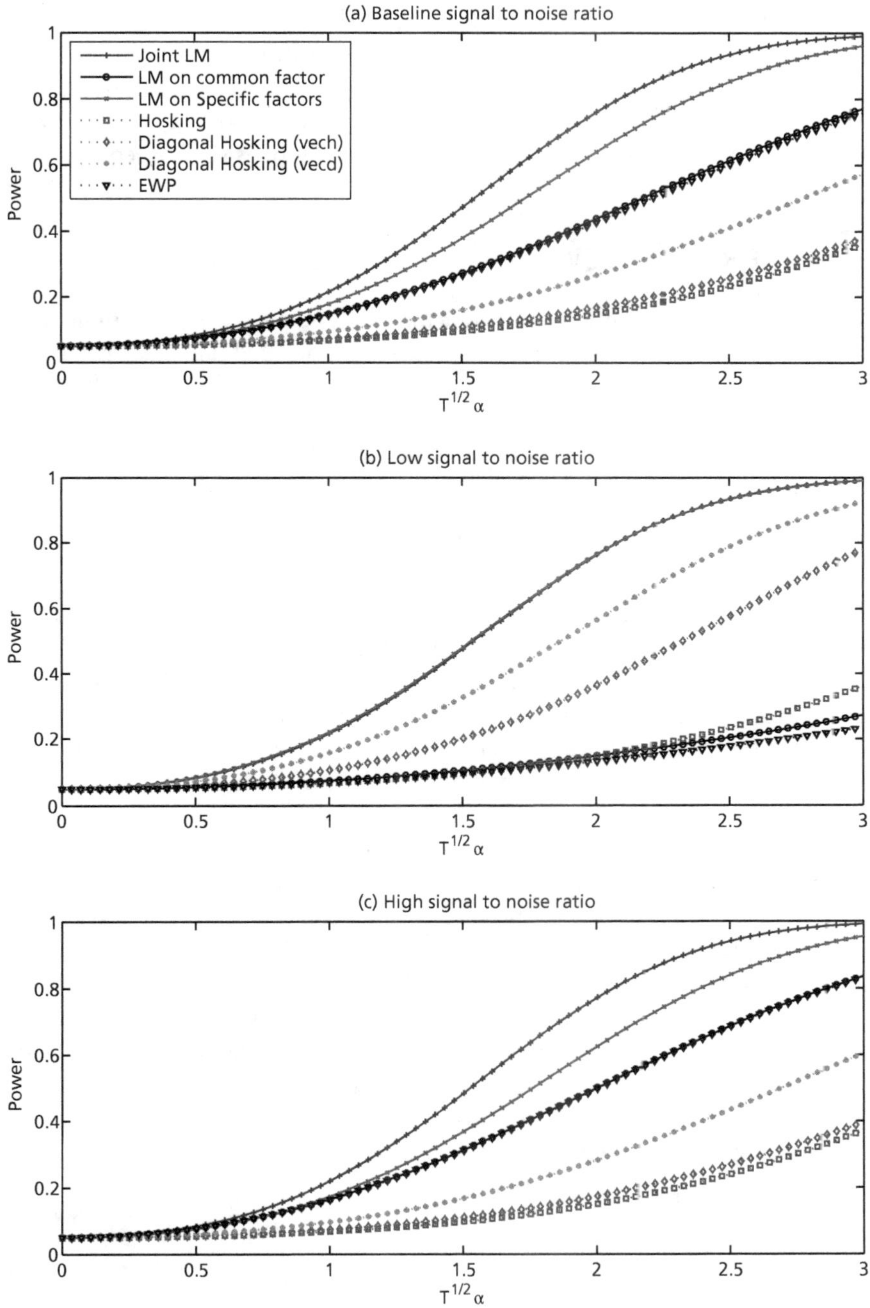

Figure 8.3. Power of variance dependence tests at 5% level against local alternative for baseline, low, and high signal to noise ratios

The other crucial determinant of the power of the different tests is the relative magnitudes of α and α^*. Figure 8.4(a) shows the effect of setting $\alpha^* = 0$ for our baseline signal to noise ratio, while Figure 8.4(b) illustrates the effects of $\alpha = 0$. In the first case, the test of conditional homoskedasticity in the common factor becomes the most powerful, with the specific factors test having power virtually equal to size, while exactly the opposite happens in the second case.

8.4.4 THE RELATIVE POWER OF THE NORMALITY TESTS

To keep the algebra simple, we shall initially compare the individual tests of $H_0 : \alpha = 0$ under the maintained assumption that all the remaining ARCH parameters in $\boldsymbol{\alpha}^*$ are 0. In this context, we can show that the ratio of non-centrality parameters of the Gaussian test and the elliptical test is $4/\{[3\text{M}_{ss}(\boldsymbol{\eta}_0) - 1](3\kappa_0 + 2)\}$. In the multivariate Student t case with $\nu_0 > 4$, in particular, this asymptotic efficiency ratio becomes

$$\frac{(\nu_0 + N + 2)\,(\nu_0 - 4)}{(\nu_0 - 1)(\nu_0 + N - 1)}.$$

For any given N, this ratio is monotonically increasing in ν_0, and approaches 1 from below as $\nu_0 \to \infty$, and 0 from above as $\nu_0 \to 4^+$. For instance, for $N = 1$, it takes the values of $.8\mathring{3}$ and .4 for $\nu_0 = 9$ and $\nu_0 = 5$, respectively. At the same time, this ratio is decreasing in N for a given ν_0. For $N = 3$ and $\nu_0 = 9$, for instance, it takes the value of .795, while for $\nu_0 = 5$, its value is only .75. Exactly the same results apply to tests of $H_0 : \alpha_i^* = 0$.

More generally, we can combine the asymptotic distribution of the different estimators of $\boldsymbol{\alpha}^\dagger$ under the null derived in Proposition 4 with the expressions in Appendix 8.10 to obtain the non-centrality parameters of joint tests of $\boldsymbol{\alpha}^* = \mathbf{0}$ or $\boldsymbol{\alpha}^\dagger = \mathbf{0}$. Unlike in the case of the mean parameters, though, the asymptotic relative efficiency of the different tests depends on the values of the static factor analysis parameters $\boldsymbol{\theta}_s$. In any case, it is straightforward to map those efficiency ratios into power gains by considering sequences of local alternatives. For illustrative purposes, we look at the baseline design in Section 8.4.3 under the assumption that the true conditional distribution of $\boldsymbol{\varepsilon}_t^*$ is a multivariate t_6. Figure 8.4(c) shows that the power gains are even bigger for our proposed ARCH tests, which is in line with the asymptotic relative efficiency results derived above.

8.5 **Joint tests for serial dependence**

In this section we shall consider joint tests of AR(1)–ARCH(1) effects in common and specific factors. Therefore, our alternative will be a single factor

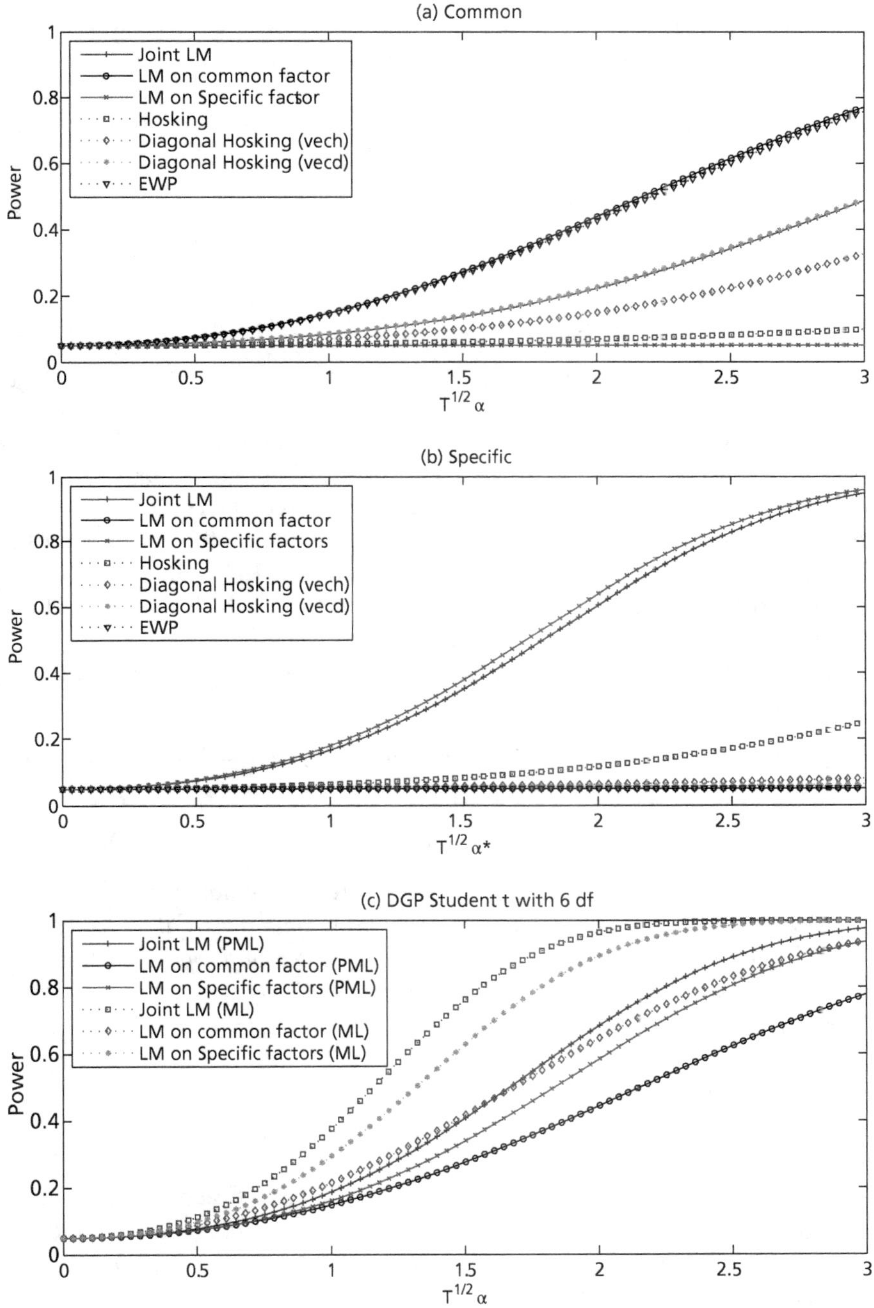

Figure 8.4. Power of variance dependence tests at 5% level against local alternative for common, specific, and DGP Student t with 6 df

version of a dynamic, conditionally heteroskedastic exact factor model in which both common and idiosyncratic factors follow covariance stationary AR(1)–ARCH(1) type processes. Specifically,

$$\left.\begin{array}{c} \mathbf{y}_t = \boldsymbol{\pi} + \mathbf{c}x_t + \mathbf{u}_t \\ x_t = \rho x_{t-l} + f_t \\ \mathbf{u}_t = diag(\boldsymbol{\rho}^*)\mathbf{u}_{t-1} + \mathbf{v}_t \\ \begin{pmatrix} f_t \\ \mathbf{v}_t \end{pmatrix} |I_{t-1};\boldsymbol{\theta} \sim s\left[\begin{pmatrix} 0 \\ \mathbf{0} \end{pmatrix}, \begin{pmatrix} \lambda_t(\boldsymbol{\theta}) & \mathbf{0} \\ \mathbf{0} & \boldsymbol{\Gamma}_t(\boldsymbol{\theta}) \end{pmatrix}, \boldsymbol{\eta}\right], \\ V(f_t|I_{t-1};\boldsymbol{\theta}) = \lambda_t(\boldsymbol{\theta}) = 1 + \alpha[E(f_{t-1}^2|Y_{t-1};\boldsymbol{\theta},\mathbf{0}) - 1], \\ V(v_{it}|I_{t-1};\boldsymbol{\theta}) = \gamma_{it}(\boldsymbol{\theta}) = \gamma_i + \alpha_i^*[E(v_{it-1}^2|Y_{t-1};\boldsymbol{\theta},\mathbf{0}) - \gamma_i], \quad (i = 1, \ldots, N) \end{array}\right\} . \tag{8.32}$$

When the conditional variances of the common and idiosyncratic factors are constant (i.e., $\alpha = 0$ and $\boldsymbol{\alpha}^* = \mathbf{0}$), the above formulation reduces to (8.13). Similarly, when the levels of the latent variables are unpredictable (i.e., $\rho = 0$ and $\boldsymbol{\rho}^* = \mathbf{0}$), the above model simplifies to (8.22). Finally, under the null hypothesis of lack of predictability in mean ($\boldsymbol{\rho}^\dagger = \mathbf{0}$) and variance ($\boldsymbol{\alpha}^\dagger = \mathbf{0}$), Model (8.32) reduces to the traditional (static) factor Model (8.1), which is our baseline specification.

It turns out that the joint tests of AR(1)–ARCH(1) in Propositions 2 and 4 is simply the sum of the separate tests:

Proposition 6 1. *Under the joint null hypothesis $H_0 : \boldsymbol{\rho}^\dagger = \mathbf{0}, \boldsymbol{\alpha}^\dagger = \mathbf{0}$ the score test statistic*

$$LM_{AR(1)-ARCH(1)}(\boldsymbol{\eta}_0) = LM_{AR(1)}(\boldsymbol{\eta}_0) + LM_{ARCH(1)}(\boldsymbol{\eta}_0),$$

will be distributed as a χ^2 with $2(N+1)$ degrees of freedom for N fixed as T goes to infinity. This asymptotic null distribution is unaffected if we replace $\boldsymbol{\theta}_s$ and $\boldsymbol{\eta}_0$ by their joint maximum likelihood estimators in Proposition 1.

2. *It also remains valid if we replace $\boldsymbol{\theta}_{s0}$ by its elliptically symmetric semiparametric estimator, which requires the nonparametric estimation of the density of $\varsigma_t(\boldsymbol{\theta}_s)$.*
3. *Under the same null hypothesis*

$$LM_{AR(1)-ARCH(1)}(\mathbf{0}) = LM_{AR(1)}(\mathbf{0}) + LM_{ARCH(1)}(\mathbf{0})$$

will also be distributed as a χ^2 with $2(N+1)$ degrees of freedom for N fixed as T goes to infinity irrespective of whether the elliptical conditional distribution is normal. This result continues to hold if we replace $\boldsymbol{\theta}_{s0}$ by its Gaussian pseudo maximum likelihood estimator $\bar{\theta}_s$ in Proposition 1.

Intuitively, the reason is that the serial correlation orthogonality conditions (8.16)–(8.17) are asymptotically orthogonal to the ARCH orthogonality conditions (8.26)–(8.27) because all odd order moments of multivariate spherical distributions are 0, which means that the joint test is simply the sum of its two components.

This additivity, though, no longer holds for non-spherical distributions, in which case one could robustify the Gaussian tests by using as weighting matrix

$$\begin{bmatrix} \mathcal{B}_{\rho^\dagger\rho^\dagger}(\phi) & \mathcal{B}_{\rho^\dagger\alpha^\dagger}(\phi) \\ \mathcal{B}'_{\rho^\dagger\alpha^\dagger}(\phi) & \mathcal{B}_{\alpha^\dagger\alpha^\dagger}(\phi) \end{bmatrix} = \begin{bmatrix} \mathcal{V}_{\rho^\dagger\rho^\dagger}(\theta_s,0;\varrho) & \mathcal{V}_{\rho^\dagger\alpha^\dagger}(\theta_s,0;\varrho) \\ \mathcal{V}'_{\rho^\dagger\alpha^\dagger}(\theta_s,0;\varrho) & \mathcal{V}_{\alpha^\dagger\alpha^\dagger}(\theta_s,0;\varrho) \end{bmatrix} \odot \begin{bmatrix} \mathcal{V}_{\rho^\dagger\rho^\dagger}(\theta_s,0;\varrho) & \mathcal{V}_{\rho^\dagger\alpha^\dagger}(\theta_s,0;\varrho) \\ \mathcal{V}'_{\rho^\dagger\alpha^\dagger}(\theta_s,0;\varrho) & \mathcal{V}_{\alpha^\dagger\alpha^\dagger}(\theta_s,0;\varrho) \end{bmatrix},$$

where

$$\begin{bmatrix} \mathcal{V}_{\rho^\dagger\rho^\dagger}(\theta_s,0;\varrho) & \mathcal{V}_{\rho^\dagger\alpha^\dagger}(\theta_s,0;\varrho) \\ \mathcal{V}'_{\rho^\dagger\alpha^\dagger}(\theta_s,0;\varrho) & \mathcal{V}_{\alpha^\dagger\alpha^\dagger}(\theta_s,0;\varrho) \end{bmatrix} = V \begin{bmatrix} f_{kt}(\theta_s) \\ \Gamma^{-1/2}\mathbf{v}_{kt}(\theta_s) \\ \frac{1}{\sqrt{2}}[f_{kt}^2(\theta_s)+\omega_k(\theta_s)-1] \\ \frac{1}{\sqrt{2}}\Gamma^{-1} vecd[\mathbf{v}_{kt}(\theta_s)\mathbf{v}'_{kt}(\theta_s)+\mathbf{c}\mathbf{c}'\omega_k(\theta_s)-\Gamma] \end{bmatrix} \tag{8.33}$$

has to be computed taking into account the third and fourth multivariate moments of the distribution of $\mathbf{y}_t$, except for $\mathcal{V}_{\rho^\dagger\rho^\dagger}(\theta_s,0;\varrho)$, whose Gaussian expression remains valid.

8.6 **Monte Carlo analysis**

8.6.1 DESIGN

We assess the finite sample performance of the different testing procedures discussed above by means of an extensive Monte Carlo exercise, with an experimental design that nests those in Sections 8.3.3 and 8.4.3, and is thereby adapted to the empirical application in Section 8.7. For that reason, we only report the results for samples of 720 observations each (plus another one hundred for initialization) in which the cross-sectional dimension is $N = 5$. This sample size corresponds to sixty years of monthly data, roughly the same as in our empirical analysis. In this sense, the main reason for looking at a small cross-sectional dimension is to handicap our proposed tests relative to the existing multivariate serial dependence tests, which in the case of Hosking's test applied to $vech[(\mathbf{y}_t-\boldsymbol{\pi})(\mathbf{y}_t-\boldsymbol{\pi})']$ already involves 784 degrees of freedom for $N = 7$. We carry out 20 000 replications for the purposes of estimating actual sizes and powers with high precision.[16] All the examples of the DGP in (8.32)

[16] For instance, the 95% confidence interval for a nominal size of 5% would be (4.7%, 5.3%).

considered can be written as nonexchangeable single factor models of the form:

$$\begin{aligned}
y_{it} &= \pi_i + c_i x_t + u_{it} \qquad (i = 1, \ldots, 5)\\
x_t &= \rho x_{t-1} + f_t\\
u_{it} &= \rho_i^* u_{it-1} + v_{it} \qquad (i = 1, \ldots, 5)\\
\lambda_t &= (1 - \alpha - \beta)(1 - \rho^2) + \alpha(f_{kt-1}^2 + \omega_k - 1) + \beta\lambda_{t-1}\\
\gamma_{it} &= \gamma_i(1 - \alpha_i^* - \beta_i^*)(1 - \rho_i^*)^2 +\\
&\quad \alpha_i^*(v_{it-1}^2 + c_i^2\omega_k - \gamma_i) + \rho_i^*\gamma_{it-1} \qquad (i = 1, \ldots, 5)
\end{aligned}$$

with $\boldsymbol{\pi} = (.5, .4, .5, .4, .5)$, $\mathbf{c} = (5, 4, 5, 4, 5)$, $\gamma = (5, 9, 5, 9, 5)$, $\rho_i^* = \rho^*$, $\alpha_i^* = \alpha^*$ and $\beta_i^* = \beta^*$ $\forall i$. Thus, the values of ρ, ρ^*, α, α^*, β, β^* fully explain the differences between our designs.

We generate samples from a Gaussian distribution, a Student t with 6 degrees of freedom, a discrete scale mixture of normals (DSMN) with the same kurtosis but finite higher order moments, and an asymmetric Student t such that the marginal distribution of an equally-weighted portfolio of $\mathbf{y}_t$ has the maximum negative skewness compatible with the kurtosis of a univariate t_6 (see Mencía and Sentana (2009, 2012) for details). These distributions allow us to assess the reliability of the robust Gaussian tests, and to shed some light on the "efficiency consistency" trade-offs of those tests that exploit the leptokurtosis of financial returns.

We draw spherical Gaussian random vectors using the NAG library Fortran G05FDF routine after initialization by G05CBF. To sample standardized Student t vectors, we simply divide those Gaussian random vectors by the square root of an independent univariate Gamma (3,2) random variable, and scale the result by 2. Similarly, we generate a standardized version of a two-component scale mixture of multivariate normals as

$$\boldsymbol{\varepsilon}_t^* = \frac{s_t + (1 - s_t)\sqrt{\kappa^*}}{\sqrt{\pi + (1 - \pi)\kappa^*}} \cdot \boldsymbol{\varepsilon}_t^\circ,$$

where $\boldsymbol{\varepsilon}_t^\circ$ is a spherical multivariate normal, κ^* the variance ratio of the two components, and s_t is an independent Bernoulli variate with $P(s_t = 1) = \pi$, which we draw by comparing π with a uniform from G05CAC. Specifically, we choose $\pi = .05$ and $\kappa^* = .1438$. Finally, following Mencía and Sentana (2012), we generate a standardized asymmetric multivariate t by choosing

$$\boldsymbol{\varepsilon}_t^* = \boldsymbol{\beta}\left[\xi_t^{-1} - c(\beta,\eta)\right] + \sqrt{\frac{\zeta_t}{\xi_t}}\,\Xi^{1/2}\boldsymbol{\varepsilon}_t^\circ, \tag{8.34}$$

where ξ_t is Gamma random variable with parameters $(2\eta)^{-1}$ and $\delta^2/2$ with $\delta = (1 - 2\eta)\eta^{-1}c(\beta,\eta)$, $\boldsymbol{\beta}$ is a $N \times 1$ parameter vector, and Ξ is the $N \times N$ positive definite matrix

$$\Xi = \frac{1}{c(\beta,\eta)}\left[I_N + \frac{c(\beta,\eta) - 1}{\boldsymbol{\beta}'\boldsymbol{\beta}}\beta\beta'\right],$$

with

$$c(\beta,\eta) = \frac{-(1-4\eta) + \sqrt{(1-4\eta)^2 + 8\boldsymbol{\beta}'\boldsymbol{\beta}\,(1-4\eta)\,\eta}}{4\boldsymbol{\beta}'\boldsymbol{\beta}\eta}.$$

In this sense, note that $\lim_{\beta'\beta \longrightarrow 0} c(\beta,\eta) = 1$, so that the above distribution collapses to the usual multivariate symmetric t when $\beta = \mathbf{0}$. In the asymmetric t case, though, we use $\beta = -10^6 \boldsymbol{\iota}_N$.

Importantly, we use the same underlying pseudo-random numbers in all designs to minimize experimental error. In particular, we make sure that the standard Gaussian random vectors are the same for all four distributions. Given that the usual routines for simulating gamma random variables involve some degree of rejection, which unfortunately can change for different values of η, we use the slower but smooth inversion method based on the NAG G01FFF gamma quantile function so that we can keep the underlying uniform variates fixed across simulations. Those uniform random variables are also used to generate the DSMN random vectors.

Finally, we combine the underlying random numbers with the vector of conditional means $\boldsymbol{\mu}_t(\boldsymbol{\theta}_0)$ and Cholesky decomposition of the covariance matrix $\boldsymbol{\Sigma}_t(\boldsymbol{\theta}_0)$ provided by the relevant Kalman filter recursions, which we describe in Appendix B of Fiorentini and Sentana (2012).[17] We start up the recursions by exploiting covariance stationarity with $x_{-100|-100} = u_{i,-100|-100} = 0$, $\lambda_{-100} = 1 - \rho^2$, $\gamma_{i,-100} = (1 - \rho_i^{*2})\gamma_i$, $\boldsymbol{\Omega}_{11,-100|-100} = diag(1, \boldsymbol{\gamma}')$, and $\boldsymbol{\Omega}_{12,-100|-100} = \boldsymbol{\Omega}_{22,-100|-100} = diag(1 - \rho^2, 1 - \rho^{*2}\boldsymbol{\iota}_5)$.

For each Monte Carlo sample thus generated, our ML estimation procedure employs the following numerical strategy. First, we estimate the static mean and variance parameters $\boldsymbol{\theta}_s$ under normality with a scoring algorithm that combines the E04LBF routine with the analytical expressions for the score and the $\mathcal{A}(\boldsymbol{\phi}_0)$ matrix appearing in the proof of Proposition 1. For this purpose, the EM algorithm of Rubin and Thayer (1982) provides very good initial values. Then we compute Mardia's (1970) sample coefficient of multivariate kurtosis κ, on the basis of which we obtain the sequential method of moments estimator of η suggested by Fiorentini et al. (2003), which exploits the theoretical relationship $\eta = \kappa/(4\kappa + 2)$. Next, we could use this estimator as an initial value for a univariate optimization procedure that uses the E04ABF routine to obtain the sequential ML estimator of η discussed by Amengual et al. (2013), which keeps $\boldsymbol{\pi}$, $\mathbf{c}$, and $\boldsymbol{\gamma}$ fixed at their Gaussian PML estimators. The resulting estimates of η, together with the PMLE of $\boldsymbol{\theta}_s$, become the initial values for the t-based ML estimators, which are obtained with the same scoring algorithm as the Gaussian PML estimator, but this time using the analytical expressions for the information matrix $\mathcal{I}(\boldsymbol{\phi}_0)$ in Proposition 1. We rule out numerically problematic solutions by imposing the inequality constraint $0 \leq \eta \leq .499$.

[17] The choice of a Cholesky factor is inconsequential for the all estimators of the static factor model parameters that we consider, and for all simulated distributions except the asymmetric t.

Computational details for the elliptically symmetric semiparametric procedure can be found in Appendix B of Fiorentini and Sentana (2010b). Given that a proper cross-validation procedure is extremely costly to implement in a Monte Carlo exercise, we have chosen the "optimal" bandwidth in Silverman (1986).

8.6.2 FINITE SAMPLE SIZE

The size properties under the null of our proposed LM tests, Hosking's test, the univariate first-order serial correlation test of EWP, and the joint test of univariate first-order autocorrelation in all N series introduced in Section 8.3.3 are summarized in Figures 8.5(a) to (d) using Davidson and MacKinnon's (1998) p-value discrepancy plots, which show the difference between actual and nominal test sizes for every possible nominal size. When the distribution is Gaussian, all tests are very accurate. The same conclusion is obtained when the distribution is a Student t, although in this case the elliptically symmetric semiparametric (SSP) tests show some very minor distortions. In contrast, when the true distribution is a DSMN, the tests based on the Student t PMLE's also show some size distortions, but they are very small. Finally, all tests are remarkably reliable when the conditional distribution is an asymmetric Student t, which partly reflects the fact that the elliptically symmetric estimators of the autocorrelation coefficients remain consistent in this case (see Proposition 10 in Fiorentini and Sentana (2010b)).

In turn, Figures 8.6(a) to (d) show the size of the two-sided versions of our ARCH(1) LM tests, Hosking's test applied to $vech[(\mathbf{y}_t - \boldsymbol{\pi})(\mathbf{y}_t - \boldsymbol{\pi})']$, a univariate first-order ARCH test applied to EWP, the joint test of univariate first-order autocorrelation in all $N(N+1)/2$ squares and cross-products of the (demeaned) observed series introduced in Section 8.4.3, and the analogous test that only focuses on their squares. In the Gaussian case, all tests are fairly accurate, except the SSP tests, which are rather conservative, and Hosking's test, which is rather liberal. This liberality is exacerbated when the true distribution is a Student t, and is shared to some extent by the diagonal version that looks at all $N(N+1)/2$ squares and cross-products, which reflects the imprecision in unrestrictedly estimating higher order moments in this case. As expected, the non-robust version of the normal test rejects far too often, while all the other tests follow a similar pattern: they are liberal for low significance values, and conservative for large ones. Not surprisingly, the sizes of the Student t tests also become highly distorted when the distribution is a DSMN, but the two robust versions of the normal tests are also somewhat unreliable in that context. Finally, those versions of the Gaussian tests that are only robust to kurtosis also suffer substantial size distortions when the conditional distribution is an asymmetric Student t, but the ones that are also robust to asymmetries are not very reliable either.

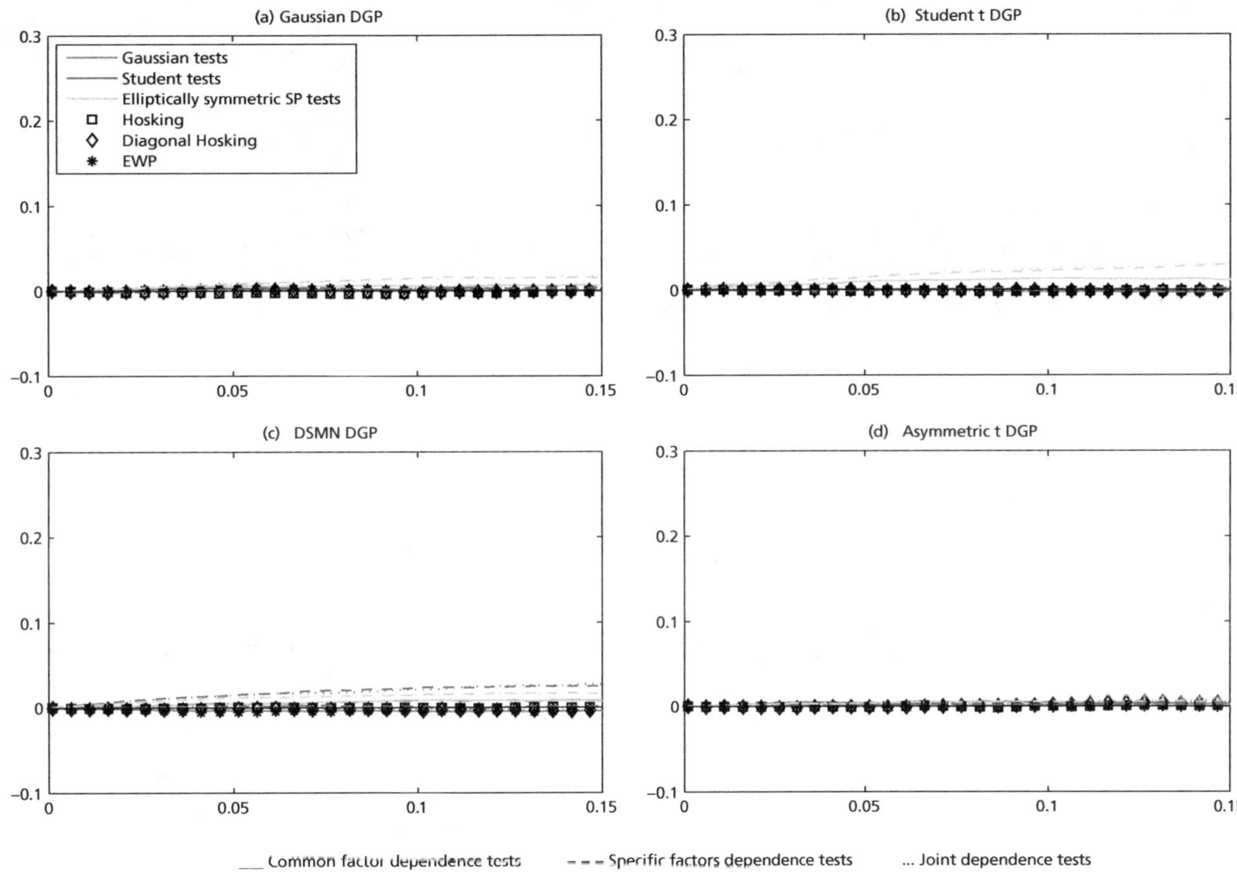

Figure 8.5. P-value discrepancy plot: tests against AR(1) alternatives

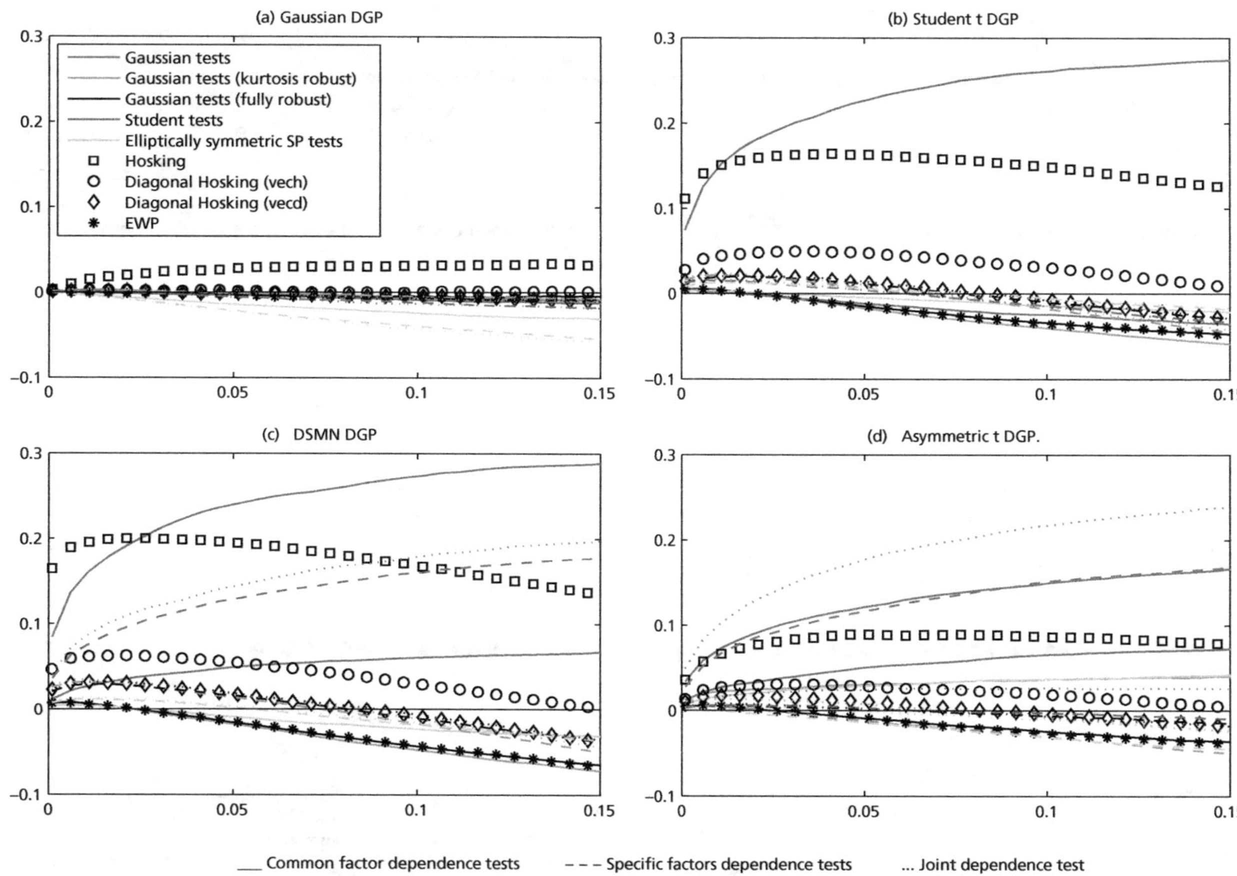

Figure 8.6. P-value discrepancy plot: tests against ARCH(1) alternatives

Figures 8.7(a) to (d) show the size of all our two-sided LM tests for GARCH(1,1) effects calculated with the discount factors $\bar{\beta} = \bar{\beta}^* = .94$ suggested in Group (1996). The behavior of these tests is fairly similar to that of the ARCH(1) tests, although in this case the asymptotically valid tests show a stronger tendency to underreject in finite samples.

8.6.3 FINITE SAMPLE POWER

In order to gauge the power of the serial correlation tests we look at a design in which $\rho = .03$ and $\rho_i^* = .045$ but $\alpha = \alpha^* = \beta = \beta^* = 0$. The evidence at the 5% significance level is presented in panels (a) and (b) of Table 8.1, which include raw rejection rates, as well as size adjusted ones based on the empirical distribution obtained under the null, which in this case provides the closest match because the Gaussian PML estimators of $\boldsymbol{\theta}_s$ that ignore the dynamics in $\mathbf{y}_t$ remain consistent in the presence of serial correlation or conditional heteroskedasticity, as shown by Doz and Lenglart (1999) and Sentana and Fiorentini (2001), respectively.

As expected from our theoretical analysis, the power of the normal tests does not depend much on the actual distribution of the data, while the tests that exploit the leptokurtosis of $\mathbf{y}_t$ offer noticeable power gains in the case of the multivariate t, especially the parametric versions. Another result that we saw in Section 8.3.3 is that in this design the joint test of $H_0 : \boldsymbol{\rho}^\dagger = \mathbf{0}$ is only marginally more powerful than the joint test of $H_0 : \boldsymbol{\rho}^* = \mathbf{0}$, which in turn is substantially more powerful than the individual test of $H_0 : \rho = 0$. Standard serial correlation tests also behave very much in line with the theoretical analysis in that section.

We also look at a design with $\rho = \rho^* = 0$ but $\alpha = \alpha^* = .05$ and $\beta = \beta^* = 0.75$ to assess the power of the ARCH(1) and GARCH(1,1) tests. A comparison of panels (c)–(e) and (d)–(f) confirms that GARCH(1,1) tests are more powerful than their ARCH(1) counterparts, even though the riskmetrics values for $\bar{\beta}$ and $\bar{\beta}^*$are much higher than the true values of these parameters. We also find that the power of the fully robust versions of the normal tests is slightly reduced when the distribution of the simulated data is leptokurtic. In contrast, the tests that exploit the leptokurtosis of $\mathbf{y}_t$ clearly become more powerful. Another result that we saw in Section 8.4.3 is that in this design the joint tests of $H_0 : \boldsymbol{\alpha}^\dagger = \mathbf{0}$ are more powerful than the joint tests of $H_0 : \boldsymbol{\alpha}^* = \mathbf{0}$, which in turn are substantially more powerful than tests of $H_0 : \alpha = 0$. Finally, standard first-order serial correlation tests applied to the squares and cross-products of $\mathbf{y}_t$ do not have much power once we take into account their substantial size distortions under the null, except for the ARCH test applied to the EWP, which is almost as powerful as the analogous test for the common factor.

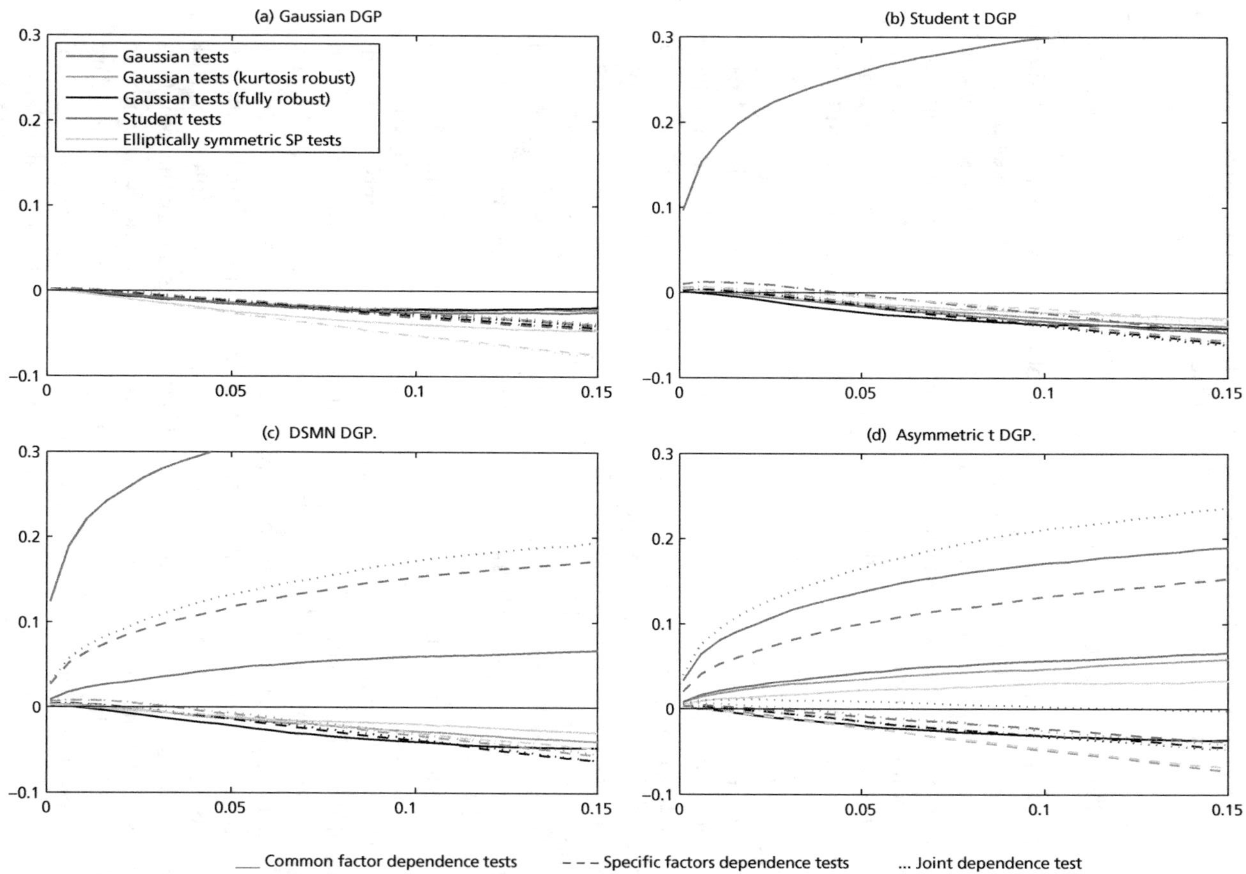

Figure 8.7. P-value discrepancy plot: tests against GARCH(1,1) alternatives

Table 8.1 Test power

(a) AR(1) tests. DGP: Gaussian (ρ=.03,ρ_i^*=.045,α=α^*=β=β^*=0)

	Common			Specific			Joint			Hosking		
	PML	ML	SSP	PML	ML	SSP	PML	ML	SSP	Gen	Vecd	EWP
Rejection rate	0.121	0.121	0.126	0.395	0.396	0.401	0.402	0.402	0.411	0.203	0.110	0.121
Size adjusted	0.116	0.115	0.117	0.390	0.391	0.376	0.398	0.399	0.381	0.209	0.109	0.117

(b) AR(1) tests. DGP: Student t_6 (ρ=.03,ρ_i^*=.045,α=α^*=β=β^*=0)

	Common			Specific			Joint			Hosking		
	PML	ML	SSP	PML	ML	SSP	PML	ML	SSP	Gen	Vecd	EWP
Rejection rate	0.120	0.143	0.155	0.391	0.500	0.524	0.397	0.509	0.539	0.202	0.110	0.120
Size adjusted	0.119	0.143	0.138	0.394	0.502	0.479	0.399	0.511	0.439	0.206	0.110	0.118

(c) ARCH(1) tests. DGP: Gaussian (ρ=ρ^*=0,α=α^*=.05,β=β^*=.75)

	Common			Specific			Joint			Hosking			
	PML	ML	SSP	PML	ML	SSP	PML	ML	SSP	Gen	Vech	Vecd	EWP
Rejection rate	0.263	0.261	0.228	0.391	0.391	0.315	0.469	0.473	0.389	0.279	0.197	0.219	0.259
Size adjusted	0.270	0.270	0.264	0.401	0.405	0.391	0.480	0.487	0.475	0.215	0.192	0.222	0.265

(d) ARCH(1) tests. DGP: Student t_6 (ρ=ρ^*=0,α=α^*=.05,β=β^*=.75)

	Common			Specific			Joint			Hosking			
	PML	ML	SSP	PML	ML	SSP	PML	ML	SSP	Gen	Vech	Vecd	EWP
Rejection rate	0.229	0.238	0.259	0.377	0.397	0.444	0.438	0.484	0.543	0.510	0.293	0.258	0.226
Size adjusted	0.265	0.267	0.268	0.339	0.384	0.423	0.390	0.467	0.517	0.196	0.189	0.223	0.265

(e) GARCH(1,1) tests ($\bar{\beta}$=$\bar{\beta}^*$=.94). DGP: Gaussian (ρ=ρ^*=0,α=α^*=.05,β=β^*=.75)

	Common			Specific			Joint		
	PML	ML	SSP	PML	ML	SSP	PML	ML	SSP
Rejection rate	0.321	0.321	0.292	0.499	0.499	0.437	0.592	0.594	0.525
Size adjusted	0.358	0.355	0.350	0.538	0.540	0.533	0.631	0.635	0.522

(f) GARCH(1,1) tests ($\bar{\beta}$=$\bar{\beta}^*$=.94). DGP: Student t_6 (ρ=ρ^*=0,α=α^*=.05,β=β^*=.75)

	Common			Specific			Joint		
	PML	ML	SSP	PML	ML	SSP	PML	ML	SSP
Rejection rate	0.286	0.330	0.352	0.456	0.545	0.600	0.530	0.652	0.714
Size adjusted	0.337	0.372	0.380	0.511	0.554	0.612	0.574	0.662	0.726

Table 8.2 Descriptive statistics

Industry portfolios							
			Correlations				
Sector	Means	Std.dev.	Cnsmr	Manuf	HiTec	Hlth	Other
Cnsmr	.566	4.481	*1*				
Manuf	.543	4.178	.804	*1*			
HiTec	.497	5.320	.734	.718	*1*		
Hlth	.733	4.995	.710	.668	.634	*1*	
Other	.500	4.998	.878	.848	.739	.708	*1*

Notes: Sample: January 1953 to December 2008. Industry definitions: Cnsmr: Consumer durables, nondurables, wholesale, retail, and some services (laundries, repair shops). Manuf: Manufacturing, energy, and utilities. HiTec: Business equipment, telephone and television transmission. Hlth: Healthcare, medical equipment, and drugs. Other: Other—mines, construction, building material, transport, hotels, business services, entertainment, finance.

8.7 **Empirical application**

In this section we initially apply the procedures previously developed to the returns on five portfolios of U.S. stocks grouped by industry in excess of the one-month treasury bill rate (from Ibbotson Associates), which we have obtained from Ken French's Data Library. Specifically, each NYSE, AMEX, and NASDAQ stock is assigned to an industry portfolio at the end of June of year t based on its four-digit SIC code at the time[18] (see <http://mba.tuck.dartmouth.edu/pages/faculty/ken.french/data_library.html> for further details). We use monthly data from 1952 to 2008, so that our sample starts soon after the March 1951 Treasury Federal Reserve Accord whereby the Federal Reserve stopped its wartime pegging of interest rates. Nevertheless, since we reserve 1952 to compute pre-sample values, we effectively work with 672 observations.

Table 8.2 contains the sample means, standard deviation, and contemporaneous correlations for the excess returns on those portfolios. For our purposes, the two most relevant empirical characteristics are the strong degree of contemporaneous correlation between the series, and their leptokurtosis. Regarding the first aspect, it is customary to look at the ratio of the largest eigenvalue of the sample covariance matrix to its trace in order to judge the representativeness of the first principal component of $\mathbf{y}_t$. However, this measure, which is .79 in our case, fails to take into account the fact that unlike principal components, factor models fully explain the variances of all the y_{it}'s thanks to the inclusion of idiosyncratic components. For that reason, we prefer

[18] Industry definitions: Cnsmr: Consumer durables, nondurables, wholesale, retail, and some services (laundries, repair shops). Manuf: Manufacturing, energy, and utilities. HiTec: Business equipment, telephone, and television transmission. Hlth: Healthcare, medical equipment, and drugs. Other: Other—mines, constr, bldmt, trans, hotels, bus serv, entertainment, finance.

to look at the fraction of the (square) Frobenius norm of the sample covariance matrix accounted for by a single factor model, which is 99.47%.[19]

As for the Gaussianity of the data, the Kuhn-Tucker test of normality against the alternative of multivariate Student t proposed by Fiorentini et al. (2003), which tests the restriction on the first two moments of $\varsigma_t(\boldsymbol{\theta}_0)$ implicit in the single condition

$$E\left[\frac{N(N+2)}{4} - \frac{N+2}{2}\varsigma_t(\boldsymbol{\theta}_0) + \frac{1}{4}\varsigma_t^2(\boldsymbol{\theta}_0)\right] = E[m_{kt}(\boldsymbol{\theta}_0)] = 0,$$

yields a value of 1478.9 despite having one degree of freedom. In contrast, the test of multivariate normal against asymmetric alternatives in Mencía and Sentana (2012), which assesses whether

$$E\left\{\boldsymbol{\varepsilon}_t(\boldsymbol{\theta}_0)\left[\varsigma_t(\boldsymbol{\theta}_0) - (N+2)\right]\right\} = E[\mathbf{m}_{st}(\boldsymbol{\theta}_0, 0)] = \mathbf{0}, \qquad (8.35)$$

yields 7.01, whose p-value is 22%. On this basis, we decided to estimate a multivariate t distribution. The ML estimator of the Student tail parameter η is .189, which corresponds to 5.3 degrees of freedom. This confirms our empirical motivation for developing testing procedures that exploit such a prevalent feature of the data.

Nevertheless, both parametric and semiparametric elliptically-based procedures are sensitive to the assumption of elliptical symmetry. For that reason, we follow Mencía and Sentana (2012), and test the null hypothesis of multivariate Student t innovations against the multivariate asymmetric t distribution in (8.34). Their statistic checks the following moment conditions:

$$E\left[\frac{N\eta+1}{1-2\eta+\eta\varsigma_t(\boldsymbol{\theta})}\boldsymbol{\varepsilon}_t(\boldsymbol{\theta})\left[\varsigma_t(\boldsymbol{\theta}) - (N+2)\right]\right] = E[\mathbf{m}_{st}(\boldsymbol{\theta}_0, \eta_0)] = \mathbf{0},$$

which reduce to (8.35) when $\eta = 0$. The asymptotic distribution that takes into account the fact that $\boldsymbol{\theta}$ and η have to be replaced by their t-based ML estimators $\hat{\boldsymbol{\theta}}_T$ and $\hat{\eta}_T$ is

$$\frac{\sqrt{T}}{T}\sum_{t=1}^{T}\mathbf{m}_{st}(\hat{\boldsymbol{\theta}}_T, \hat{\eta}_T) \to N\left[0, 2(N+2)(N\eta_0+1)\boldsymbol{\Sigma}_0\right].$$

The test statistic is 3.83 with a p-value of 57%, so we cannot reject the null hypothesis that the distribution of $\mathbf{y}_t$ is multivariate Student t at conventional levels.

Table 8.3 presents the three different estimates of the unconditional covariance parameters, namely Gaussian PMLE, Student t ML, and SSP. As can be seen, the discrepancies are fairly minor, especially in the case of estimators that exploit the leptokurtosis of the data. Consequently, the time series evolution

[19] The Frobenius norm of a general matrix $\mathbf{A}$, $||\mathbf{A}||$ say, is the Euclidean norm of $vec(\mathbf{A})$, which can be easily computed as the square root of the sum of its square singular values since $vec'(\mathbf{A})vec(\mathbf{A}) = tr(\mathbf{A}^2)$. Given that $V(\mathbf{y}_t)$ is a real, square symmetric matrix with spectral decomposition $\boldsymbol{U\Delta U}'$, with U orthonormal, it is easy to see $||V(\mathbf{y}_t)||^2$ can be additively decomposed as the sum of the square eigenvalues of $V(\mathbf{y}_t)$.

Table 8.3 Estimates of $\Sigma = cc' + \Gamma$

	Industry portfolios					
	Factor loadings			Specific variances		
Sector	PML	ML	SSP	PML	ML	SSP
Cnsmr	4.130	4.309	4.292	3.024	3.263	3.215
Manuf	3.708	3.840	3.847	3.710	3.683	3.705
HiTec	4.223	4.337	4.342	10.465	8.453	8.997
Hlth	3.791	4.120	4.075	10.574	10.915	10.870
Other	4.740	4.900	4.909	2.518	3.105	3.062

Notes: Sample: January 1953 to December 2008. Industry definitions: Cnsmr: Consumer durables, nondurables, wholesale, retail, and some services (laundries, repair shops). Manuf: Manufacturing, energy, and utilities. HiTec: Business equipment, telephone, and television transmission. Hlth: Healthcare, medical equipment, and drugs. Other: Other—mines, construction building material, transport, hotels, business services, entertainment, finance. PML refers to the Gaussian-based ML estimators, ML to the Student t ones, and SSP to the elliptically symmetric semiparametric estimators.

Table 8.4a Serial correlation tests (p-values, %)

	Ar(1)			Ar(3)			Ar(12)		
	PML	ML	SSP	PML	ML	SSP	PML	ML	SSP
Common factor	0.35	2.64	1.35	19.75	35.49	24.04	39.59	53.85	59.63
Specific factors	1.46	2.70	1.45	1.40	8.84	4.11	0.06	0.00	0.00
Joint	0.11	0.87	0.30	1.52	11.31	4.71	0.11	0.00	0.00

of the corresponding Kalman filter estimates of the common factor are very highly correlated with each other (> .999), and also with the excess returns on the Fama and French market portfolio ($\simeq$.978), which corresponds to the value weighted return on all NYSE, AMEX, and NASDAQ stocks in CRSP.

Table 8.4a reports the results of the two multivariate serial correlation tests discussed in Section 8.3.3. As can be seen, there is evidence of first order serial correlation in the industry return series. Nevertheless, it is interesting to understand whether the dependence is due to the common factor or the specific ones. In this sense, note that we have considered not only tests against AR(1) dynamics in common and specific factors, but also tests against restricted AR(3) and AR(12) specifications in which the autoregressive coefficients are all assumed to be the same. The motivation for such tests is twofold. First, there is a substantial body of empirical evidence which suggests that expected returns are smooth processes, while observed returns have a small first order autocorrelation. Second, a rather interesting example of persistent expected returns is an AR(h) model in which $\rho = \rho\iota$, where ι is a vector of h 1's. The results in Section 8.3.2 imply that a test of $\rho = 0$ in this context essentially involves assessing the significance of the sum of the first h autocorrelations of f_{kt}. In this sense, our procedure is entirely analogous to the one recommended by Jegadeesh (1989) to test for the long run predictability of individual asset returns without introducing overlapping observations (see, also, Cochrane (1991) and Hodrick

Table 8.4b Conditional heteroskedasticity tests (p-values, %)

	Arch(1)			Garch(·,1)		
	PML	ML	SSP	PML	ML	SSP
Common factor	0.36	6.12	1.79	0.00	0.26	0.01
Specific factors	0.00	0.00	0.00	0.00	0.00	0.00
Joint	0.00	0.00	0.00	0.00	0.00	0.00

Notes: Sample: July:1962 to June:2007. Industry definitions: Cnsmr: Consumer durables, non-durables, wholesale, retail, and some services (laundries, repair shops). Manuf: Manufacturing, energy, and utilities. HiTec: Business equipment, telephone, and television transmission. Hlth: Healthcare, medical equipment, and drugs. Other: Other—mines, construction building material, transport, hotels, business services, entertainment, finance. PML refers to the (fully robust) tests based on the Gaussian ML estimators, ML to the Student t ones, SSP to the elliptically symmetric semiparametric estimators.

(1992)). The intuition is that if returns contain a persistent but mean reverting predictable component, a persistent right hand side variable may pick it up.

The results reported in Table 8.4a show clear evidence of first order serial correlation in both common and specific factors. There is also some evidence that the idiosyncratic factors may have persistent mean-reverting components. In contrast, there is no evidence that such a component is present in the common factor. This interesting divergence could be due to the market being more closely followed by investors than the hedged components of the industry portfolios.

Table 8.4b presents our tests for conditional heteroskedasticity. Given the strong evidence for leptokurtosis, we only report the values of the fully robust versions of the different Gaussian tests. Not surprisingly, the multivariate serial dependence tests reject conditional homoskedasticity. We also find very strong evidence of ARCH effects in the idiosyncratic factors. In contrast, the ARCH(1) tests do not provide such a clear evidence in the case of the common factor. Nevertheless, the GARCH(1,1) tests strongly reject the null of conditionally homoskedasticity.

Our conclusions do not seem to be very sensitive to the degree of aggregation of our data. When we repeat exactly the same exercise with the excess returns of the ten portfolios of U.S. stocks grouped by industry in Ken French's Data Library, we obtain rather similar results.

8.8 **Conclusions and extensions**

We obtain simple algebraic expressions for the score tests of serial correlation in the levels and squares of common and idiosyncratic factors in static factor models. The implicit orthogonality conditions resemble the orthogonality conditions of models with observed factors but the weighting matrices reflect their unobservability. We robustify our Gaussian tests against non-normality, and derive more powerful versions when the conditional distribution is elliptically symmetric, which can be either parametrically or semipametrically specified. We also explain how to derive tests that focus on either the common factors or the specific factors, or indeed on some of their elements.

Importantly, we show that despite the non-Gaussian nature of the state-space models that we consider, which makes it generally impossible to compute the log-likelihood function and its score without resorting to simulation methods, our tests coincide with the correct score tests.

We conduct Monte Carlo exercises to study the finite sample reliability and power of our proposed tests, and to compare them to existing multivariate serial dependence tests. Our simulation results suggest that the serial correlation tests have fairly accurate finite sample sizes, while the tests for conditional homoskedasticity show some size distortions. Given that $\mathbf{y}_t$ is i.i.d. under the null, it would be useful to explore bootstrap procedures, which could also exploit the fact that elliptical distributions are parametric in $N - 1$ dimensions, and non-parametric in only one (see Dufour et al. (2010) for alternative finite-sample refinements of existing multivariate serial dependence tests). We also confirm that there are clear power gains from exploiting the cross-sectional dependence structure implicit in factor models, the leptokurtosis of financial returns, as well as the persistent behavior of conditional variances.

Finally, we apply our methods to monthly stock returns on U.S. broad industry portfolios. We find clear evidence in favor of first order serial correlation in common and specific factors, weaker evidence for persistent components in the idiosyncratic terms, and no evidence that such a component appears in the common factor. We also find strong evidence for persistent serial correlation in the volatility of common and specific terms.

It should be possible to robustify the serial dependence tests which assume that the return distribution is a Student t when in fact it is not along the lines described by Amengual and Sentana (2010) for mean-variance efficiency tests, and study their relative power in those circumstances. It should also be feasible to develop semiparametric tests that do not impose the assumption of elliptical symmetry. Another interesting extension would be to consider non-parametric alternatives such as the ones studied by Hong and Shehadeh (1999) and Duchesne and Lalancette (2003) among others, in which the lag length is implicitly determined by the choice of bandwidth parameter in a kernel-based estimator of a spectral density matrix. In addition, we could test for the effect of exogenous regressors in either the conditional mean vector or the conditional covariance matrix of returns. Moreover, we could use the test statistics that we have derived to obtain easy to compute indirect estimators of the dynamic models that define our alternative hypothesis along the lines suggested by Calzolari et al. (2004).

The extension of our methods to models in which N/T is non-negligible would also constitute a very valuable addition with many potentially interesting empirical finance applications. In those circumstances, though, we would expect our proposed tests to be more reliable in finite samples and more powerful asymptotically than the Hosking-type multivariate serial correlation tests for the levels, squares, and cross products of the variables under consideration, which involve orders of magnitude more degrees of freedom for fixed N.

Another particularly interesting extension would be to allow for serial dependence under the null. Specifically, suppose that we take as our new null hypothesis the factor model with AR(1) dynamics in the latent variables that we considered as the alternative in Section 8.3.1, and as our new alternative a model with a common factor that follows an AR(2) process instead. Although a Lagrange multiplier test of the new null hypothesis in the time domain is conceptually straightforward (see, e.g., Engle and Watson (1981)), the algebra is incredibly tedious, and the recursive scores difficult to interpret. In contrast, the frequency domain procedures in Harvey (1989) and Fernandes (1990) yield scores which are once again entirely analogous to the univariate frequency domain score obtained if we treated the smoothed estimator of the common factor, $x_{kt|T}$, as if it were observed. We explore this interesting research avenue in Fiorentini and Sentana (2013).

APPENDICES

8.9 Proofs

PROPOSITION 1

Given that the conditional density of $\boldsymbol{\varepsilon}_t^*|I_{t-1};\boldsymbol{\phi}$ is $\exp[c(\boldsymbol{\eta})+g(\varsigma_t,\boldsymbol{\eta})]$ under ellipticity, the log-likelihood function of a sample of size T will take the form $L_T(\boldsymbol{\phi})=\sum_{t=1}^T l_t(\boldsymbol{\phi})$, with $l_t(\boldsymbol{\phi})=d_t(\boldsymbol{\theta})+c(\boldsymbol{\eta})+g\left[\varsigma_t(\boldsymbol{\theta}),\boldsymbol{\eta}\right]$, where $d_t(\boldsymbol{\theta})=-1/2\ln|\boldsymbol{\Sigma}_t(\boldsymbol{\theta})|$ is the Jacobian, $\varsigma_t(\boldsymbol{\theta})=\boldsymbol{\varepsilon}_t^{*\prime}(\boldsymbol{\theta})\boldsymbol{\varepsilon}_t^*(\boldsymbol{\theta})$, $\boldsymbol{\varepsilon}_t^*(\boldsymbol{\theta})=\boldsymbol{\Sigma}_t^{-1/2}(\boldsymbol{\theta})\boldsymbol{\varepsilon}_t(\boldsymbol{\theta})$, and $\boldsymbol{\varepsilon}_t(\boldsymbol{\theta})=\mathbf{y}_t-\boldsymbol{\mu}_t(\boldsymbol{\theta})$. Let $\mathbf{s}_t(\boldsymbol{\phi})$ denote the score function $\partial l_t(\boldsymbol{\phi})/\partial\boldsymbol{\phi}$, and partition it into two blocks, $\mathbf{s}_{\boldsymbol{\theta}t}(\boldsymbol{\phi})$ and $s_{\boldsymbol{\eta}t}(\boldsymbol{\phi})$, whose dimensions conform to those of $\boldsymbol{\theta}$ and $\boldsymbol{\eta}$, respectively. Fiorentini and Sentana (2010b) show that if $\boldsymbol{\Sigma}_t(\boldsymbol{\theta})$ has full rank and $\boldsymbol{\mu}_t(\boldsymbol{\theta})$, $\boldsymbol{\Sigma}_t(\boldsymbol{\theta})$, $c(\boldsymbol{\eta})$, and $g\left[\varsigma_t(\boldsymbol{\theta}),\boldsymbol{\eta}\right]$ are differentiable, then

$$\mathbf{s}_{\boldsymbol{\theta}t}(\boldsymbol{\phi})=\frac{\partial d_t(\boldsymbol{\theta})}{\partial\boldsymbol{\theta}}+\frac{\partial g_t\left[\varsigma_t(\boldsymbol{\theta}),\boldsymbol{\eta}\right]}{\partial\boldsymbol{\theta}}=[\mathbf{Z}_{lt}(\boldsymbol{\theta}),\mathbf{Z}_{st}(\boldsymbol{\theta})]\begin{bmatrix}\mathbf{e}_{lt}(\boldsymbol{\phi})\\ \mathbf{e}_{st}(\boldsymbol{\phi})\end{bmatrix}=\mathbf{Z}_{dt}(\boldsymbol{\theta})\mathbf{e}_{dt}(\boldsymbol{\phi}),$$

$$\mathbf{s}_{\boldsymbol{\eta}t}(\boldsymbol{\phi})=\partial c(\boldsymbol{\eta})/\partial\boldsymbol{\eta}+\partial g\left[\varsigma_t(\boldsymbol{\theta}),\boldsymbol{\eta}\right]/\partial\boldsymbol{\eta}=\mathbf{e}_{rt}(\boldsymbol{\phi}),$$

where

$$\begin{aligned}
\mathbf{Z}_{lt}(\boldsymbol{\theta})&=\partial\boldsymbol{\mu}_t'(\boldsymbol{\theta})/\partial\theta\cdot\Sigma_t^{-1/2\prime}(\boldsymbol{\theta}),\\
\mathbf{Z}_{st}(\boldsymbol{\theta})&=\frac{1}{2}\partial vec'\left[\boldsymbol{\Sigma}_t(\boldsymbol{\theta})\right]/\partial\theta\cdot[\boldsymbol{\Sigma}_t^{-1/2\prime}(\theta)\otimes\Sigma_t^{-1/2\prime}(\boldsymbol{\theta})],\\
\mathbf{e}_{lt}(\boldsymbol{\theta},\boldsymbol{\eta})&=\delta\left[\varsigma_t(\boldsymbol{\theta}),\boldsymbol{\eta}\right]\boldsymbol{\varepsilon}_t^*(\boldsymbol{\theta}),\\
\mathbf{e}_{st}(\theta,\eta)&=vec\left\{\delta\left[\varsigma_t(\boldsymbol{\theta}),\boldsymbol{\eta}\right]\boldsymbol{\varepsilon}_t^*(\boldsymbol{\theta})\boldsymbol{\varepsilon}_t^{*\prime}(\boldsymbol{\theta})-\mathbf{I}_N\right\},
\end{aligned}$$

and $\delta(\varsigma_t, \eta)$ is defined in (8.11). Given correct specification, the results in Crowder (1976) imply that $\mathbf{e}_t(\boldsymbol{\phi}) = [\mathbf{e}'_{dt}(\boldsymbol{\phi}), \mathbf{e}_{rt}(\boldsymbol{\phi})]'$ evaluated at the true parameter values follows a vector martingale difference, and therefore, the same is true of the score vector $\mathbf{s}_t(\boldsymbol{\phi})$. His results also imply that, under suitable regularity conditions, the asymptotic distribution of the feasible ML estimator will be $\sqrt{T}(\hat{\boldsymbol{\phi}}_T - \boldsymbol{\phi}_0) \to N[\mathbf{0}, \mathcal{I}^{-1}(\boldsymbol{\phi}_0)]$, where $\mathcal{I}(\boldsymbol{\phi}_0) = E[\mathcal{I}_t(\boldsymbol{\phi}_0)|\boldsymbol{\phi}_0]$,

$$\mathcal{I}_t(\boldsymbol{\phi}) = -E\left[\mathbf{h}_t(\boldsymbol{\phi})|\mathbf{z}_t, I_{t-1}; \boldsymbol{\phi}\right] = V\left[\mathbf{s}_t(\boldsymbol{\phi})|\mathbf{z}_t, I_{t-1}; \boldsymbol{\phi}\right] = \mathbf{Z}_t(\boldsymbol{\theta})\mathcal{M}(\boldsymbol{\eta})\mathbf{Z}'_t(\theta),$$

$$\mathbf{h}_t(\boldsymbol{\phi}) = \frac{\partial \mathbf{s}_t(\boldsymbol{\phi})}{\partial \boldsymbol{\phi}'} = \frac{\partial^2 l_t(\boldsymbol{\phi})}{\partial \boldsymbol{\phi} \partial \boldsymbol{\phi}'},$$

$$\mathbf{Z}_t(\boldsymbol{\theta}) = \begin{pmatrix} \mathbf{Z}_{dt}(\boldsymbol{\theta}) & 0 \\ 0 & \mathbf{I}_q \end{pmatrix} = \begin{pmatrix} \mathbf{Z}_{lt}(\boldsymbol{\theta}) & \mathbf{Z}_{st}(\boldsymbol{\theta}) & 0 \\ 0 & 0 & \mathbf{I}_q \end{pmatrix},$$

and $\mathcal{M}(\boldsymbol{\eta}) = V\left[\mathbf{e}_t(\boldsymbol{\phi})|\boldsymbol{\phi}\right]$. In particular, Crowder (1976) requires that: (a) $\boldsymbol{\phi}_0$ is locally identified and belongs to the interior of the admissible parameter space, which is a compact subset of $\mathbb{R}^{p+r}$; (b) the Hessian matrix is non-singular and continuous throughout some neighborhood of $\boldsymbol{\phi}_0$; (c) there is uniform convergence to the integrals involved in the computation of the mean vector and covariance matrix of $\mathbf{s}_t(\boldsymbol{\phi})$; and (d) $-E^{-1}\left[-T^{-1}\sum_t \mathbf{h}_t(\boldsymbol{\phi})\right] T^{-1}\sum_t \mathbf{h}_t(\boldsymbol{\phi}) \xrightarrow{p} \mathbf{I}_{p+r}$, where $E^{-1}\left[-T^{-1}\sum_t \mathbf{h}_t(\boldsymbol{\phi})\right]$ is positive definite on a neighborhood of $\boldsymbol{\phi}_0$. These regularity conditions are easy to verify in our i.i.d. context. In particular, the conditions in the theorem statement guarantee the identification of the factor model parameters and the positive definiteness of the Hessian matrix (see Theorem 12.1 in Anderson and Rubin (1956) and Theorem 2 in Kano (1983)). So the only remaining task is to find out the expression for the unconditional information matrix. In this context, Proposition 2 in Fiorentini and Sentana (2010b) states that:

$$\mathcal{M}(\boldsymbol{\eta}) = \begin{pmatrix} \mathcal{M}_{ll}(\boldsymbol{\eta}) & 0 & 0 \\ 0 & \mathcal{M}_{ss}(\boldsymbol{\eta}) & \mathcal{M}_{sr}(\eta) \\ 0 & \mathcal{M}'_{sr}(\boldsymbol{\eta}) & \mathcal{M}_{rr}(\boldsymbol{\eta}) \end{pmatrix},$$

$$\mathcal{M}_{ll}(\boldsymbol{\eta}) = V[\mathbf{e}_{lt}(\boldsymbol{\phi})|\phi] = \mathrm{M}_{ll}(\boldsymbol{\eta})\mathbf{I}_N,$$

$$\mathcal{M}_{ss}(\boldsymbol{\eta}) = V[\,\mathbf{e}_{st}(\boldsymbol{\phi})|\,\phi]$$
$$= \mathrm{M}_{ss}(\boldsymbol{\eta})\left(\mathbf{I}_{N^2} + \mathbf{K}_{NN}\right) + [\mathrm{M}_{ss}(\eta) - 1]vec(\mathbf{I}_N)vec'(\mathbf{I}_N),$$

$$\mathcal{M}_{sr}(\boldsymbol{\eta}) = E[\,\mathbf{e}_{st}(\boldsymbol{\phi})\mathbf{e}'_{rt}(\boldsymbol{\phi})\big|\,\boldsymbol{\phi}] = -E\left\{\partial \mathbf{e}_{st}(\boldsymbol{\phi})/\partial \eta'\big|\,\boldsymbol{\phi}\right\} = vec(\mathbf{I}_N)\mathrm{M}_{sr}(\boldsymbol{\eta}),$$

where $\mathrm{M}_{ll}(\boldsymbol{\eta})$, $\mathrm{M}_{ss}(\boldsymbol{\eta})$, $\mathrm{M}_{sr}(\boldsymbol{\eta})$, and $\mathcal{M}_{rr}(\boldsymbol{\eta})$ are defined in (8.7), (8.8), (8.9), and (8.10), respectively. Therefore, all we need is the matrix $\mathbf{Z}_{dt}(\theta_s)$, which in turn requires the Jacobian of the conditional mean and covariance functions. Differentiating the Kalman filter prediction equations we obtain $d\boldsymbol{\mu}_t(\boldsymbol{\theta}) = d\boldsymbol{\pi}$ and

$$d\boldsymbol{\Sigma}_t(\boldsymbol{\theta}_s) = d(\mathbf{cc}' + \boldsymbol{\Gamma}) = (d\mathbf{c})\mathbf{c}' + \mathbf{c}(d\mathbf{c}') + d\boldsymbol{\Gamma}$$

(see Magnus and Neudecker (1988)). Hence, the only three nonzero terms of the Jacobian will be:

$$\frac{\partial \boldsymbol{\mu}_t(\boldsymbol{\theta}_s)}{\partial \pi'} = \mathrm{I}_N; \quad \frac{\partial vec\,[\boldsymbol{\Sigma}_t(\boldsymbol{\theta}_s)]}{\partial \mathrm{c}'} = (\mathrm{I}_{N^2} + \mathrm{K}_{NN})(\mathrm{c} \otimes \mathrm{I}_N); \quad \frac{\partial vec\,[\boldsymbol{\Sigma}_t(\boldsymbol{\theta}_s)]}{\partial \boldsymbol{\gamma}'} = \mathrm{E}_N.$$

As a result,

$$\mathrm{Z}_{dt}(\boldsymbol{\theta}_s) = \begin{bmatrix} \boldsymbol{\Sigma}^{-1/2\prime}(\boldsymbol{\theta}_s) & 0 \\ 0 & [\mathrm{c}'\boldsymbol{\Sigma}^{-1/2\prime}(\theta_s) \otimes \boldsymbol{\Sigma}^{-1/2\prime}(\boldsymbol{\theta}_s)] \\ 0 & \frac{1}{2}\mathrm{E}'_N[\boldsymbol{\Sigma}^{-1/2\prime}(\boldsymbol{\theta}_s) \otimes \boldsymbol{\Sigma}^{-1/2\prime}(\boldsymbol{\theta}_s)] \end{bmatrix} = \mathrm{Z}_d(\boldsymbol{\phi}).$$

After some straightforward algebraic manipulations, we get that the elliptically symmetric score is

$$\begin{aligned} \mathrm{s}_{\pi t}(\boldsymbol{\theta}_s, \boldsymbol{\eta}) &= \delta[\varsigma_t(\boldsymbol{\theta}_s); \boldsymbol{\eta}]\boldsymbol{\Sigma}^{-1}(\boldsymbol{\theta}_s)(\mathrm{y}_t - \boldsymbol{\pi}) \\ \mathrm{s}_{ct}(\boldsymbol{\theta}_s, \boldsymbol{\eta}) &= \delta[\varsigma_t(\boldsymbol{\theta}_s); \boldsymbol{\eta}]\boldsymbol{\Sigma}^{-1}(\boldsymbol{\theta}_s)(\mathrm{y}_t - \boldsymbol{\pi})(\mathrm{y}_t - \boldsymbol{\pi})'\boldsymbol{\Sigma}^{-1}(\boldsymbol{\theta}_s)\mathrm{c} - \boldsymbol{\Sigma}^{-1}(\boldsymbol{\theta}_s)\mathrm{c} \\ \mathrm{s}_{\gamma t}(\boldsymbol{\theta}_s, \boldsymbol{\eta}) &= \tfrac{1}{2} vecd\{\delta[\varsigma_t(\boldsymbol{\theta}_s); \boldsymbol{\eta}]\boldsymbol{\Sigma}^{-1}(\boldsymbol{\theta}_s)(\mathrm{y}_t - \boldsymbol{\pi})(\mathrm{y}_t - \boldsymbol{\pi})'\boldsymbol{\Sigma}^{-1}(\boldsymbol{\theta}_s) - \boldsymbol{\Sigma}^{-1}(\boldsymbol{\theta}_s)\}. \end{aligned} \tag{8.36}$$

Assuming that $\Gamma > \mathbf{0}$ we can use the Woodbury formula

$$\boldsymbol{\Sigma}^{-1}(\boldsymbol{\theta}_s) = \boldsymbol{\Gamma}^{-1} - (1 + \mathrm{c}'\boldsymbol{\Gamma}^{-1}\mathrm{c})^{-1}\boldsymbol{\Gamma}^{-1}\mathrm{cc}'\boldsymbol{\Gamma}^{-1} = \boldsymbol{\Gamma}^{-1}[\boldsymbol{\Gamma} - (1 + \mathrm{c}'\boldsymbol{\Gamma}^{-1}\mathrm{c})^{-1}\mathrm{cc}']\boldsymbol{\Gamma}^{-1} \tag{8.37}$$

to write

$$\begin{aligned} &\delta[\varsigma_t(\boldsymbol{\theta}_s); \boldsymbol{\eta}]\boldsymbol{\Sigma}^{-1}(\boldsymbol{\theta}_s)(\mathrm{y}_t - \boldsymbol{\pi})(\mathrm{y}_t - \boldsymbol{\pi})'\boldsymbol{\Sigma}^{-1}(\boldsymbol{\theta}_s)\mathrm{c} - \boldsymbol{\Sigma}^{-1}(\boldsymbol{\theta}_s)\mathrm{c} \\ &= \boldsymbol{\Gamma}^{-1}\{\delta[\varsigma_t(\boldsymbol{\theta}_s); \boldsymbol{\eta}]\mathrm{v}_{kt}(\boldsymbol{\theta}_s)f_{kt}(\theta_s) - \mathrm{c}\omega_k(\boldsymbol{\theta}_s)\}, \\ &\boldsymbol{\Sigma}^{-1}(\boldsymbol{\theta}_s)(\mathrm{y}_t - \boldsymbol{\pi})(\mathrm{y}_t - \boldsymbol{\pi})'\boldsymbol{\Sigma}^{-1}(\boldsymbol{\theta}_s) - \boldsymbol{\Sigma}^{-1}(\boldsymbol{\theta}_s) \\ &= \boldsymbol{\Gamma}^{-1}\{\delta[\varsigma_t(\boldsymbol{\theta}_s); \boldsymbol{\eta}]\mathrm{v}_{kt}(\boldsymbol{\theta}_s)\mathrm{v}'_{kt}(\boldsymbol{\theta}_s) + \mathrm{cc}'\omega_k(\boldsymbol{\theta}_s) - \boldsymbol{\Gamma}\}\boldsymbol{\Gamma}^{-1}, \end{aligned}$$

and

$$\varsigma_t(\boldsymbol{\theta}_s) = (\mathrm{y}_t - \boldsymbol{\pi})'\boldsymbol{\Sigma}^{-1}(\boldsymbol{\theta}_s)(\mathrm{y}_t - \boldsymbol{\pi}) = (\mathrm{y}_t - \boldsymbol{\pi})'\boldsymbol{\Gamma}^{-1}(\mathrm{y}_t - \boldsymbol{\pi}) - f_{kt}^2(\boldsymbol{\theta}_s)/\omega_k(\boldsymbol{\theta}_s),$$

which greatly simplifies the computation of all the elements of $\mathrm{s}_{\theta t}(\boldsymbol{\theta}_s, \boldsymbol{\eta})$, as well as $\mathrm{s}_{\eta t}(\mathrm{y}_t|\mathrm{Y}_{t-1}; \boldsymbol{\theta})$ (see Sentana (2000)).

If we put all the previous elements together, we can finally obtain the conditional (and unconditional) information matrix, which in view of the expression for $\mathrm{Z}_{dt}(\boldsymbol{\theta}_s)$ will be block diagonal between the elements corresponding to $\boldsymbol{\pi}$, and the elements corresponding to $(\mathrm{c}, \boldsymbol{\gamma}, \boldsymbol{\eta})$, with the diagonal blocks given in the statement of the first part of the proposition. Once again, the Woodbury formula simplifies considerably the computation of the information matrix when $\Gamma > \mathbf{0}$ because $\boldsymbol{\Sigma}^{-1}(\boldsymbol{\theta}_s)\mathrm{c} = (1 + \mathrm{c}'\boldsymbol{\Gamma}^{-1}\mathrm{c})^{-1}\boldsymbol{\Gamma}^{-1}\mathrm{c}$ and $\mathrm{c}'\boldsymbol{\Sigma}^{-1}(\boldsymbol{\theta}_s)\mathrm{c} = (1 + \mathrm{c}'\boldsymbol{\Gamma}^{-1}\mathrm{c})^{-1}\mathrm{c}'\boldsymbol{\Gamma}^{-1}\mathrm{c}$. Expression (8.37) also makes clear that the block of the information matrix corresponding to γ is guaranteed to be full rank.

Next, we can use Proposition 7 in Fiorentini and Sentana (2010b) to obtain the elliptically symmetric semiparametric score and corresponding efficiency bound. Specifically, they will be given by:

$$\begin{aligned} \mathring{\mathrm{s}}_{\theta t}(\boldsymbol{\phi}_0) = \mathrm{Z}_{dt}(\boldsymbol{\theta}_0)\mathrm{e}_{dt}(\boldsymbol{\phi}_0) - \mathrm{W}_d(\boldsymbol{\phi}_0) &\left\{\left[\delta[\varsigma_t(\boldsymbol{\theta}_0), \boldsymbol{\eta}_0]\frac{\varsigma_t(\boldsymbol{\theta}_0)}{N} - 1\right]\right. \\ &\left. - \frac{2}{(N+2)\kappa_0 + 2}\left[\frac{\varsigma_t(\boldsymbol{\theta}_0)}{N} - 1\right]\right\}, \end{aligned} \tag{8.38}$$

and

$$\mathring{\mathcal{S}}(\boldsymbol{\phi}_0) = \mathcal{I}_{\theta\theta}(\boldsymbol{\phi}_0) - \mathbf{W}_d(\boldsymbol{\phi}_0)\mathbf{W}_d'(\boldsymbol{\phi}_0) \cdot \left\{ \left[\frac{N+2}{N} \mathrm{M}_{ss}(\boldsymbol{\eta}_0) - 1 \right] - \frac{4}{N\left[(N+2)\kappa_0 + 2\right]} \right\}, \tag{8.39}$$

respectively, where

$$\begin{aligned}\mathbf{W}_d(\boldsymbol{\phi}) &= \mathbf{Z}_d(\boldsymbol{\phi})[0', vec'(\mathbf{I}_N)]' = E[\mathbf{Z}_{dt}(\boldsymbol{\theta})|\boldsymbol{\phi}][0', vec'(\mathbf{I}_N)]' \\ &= E\left\{ \frac{1}{2} \partial vec' \left[\boldsymbol{\Sigma}_t(\boldsymbol{\theta})\right] / \partial\theta \cdot vec[\boldsymbol{\Sigma}_t^{-1}(\theta)] \middle| \boldsymbol{\phi} \right\} \\ &= -E\left\{ \partial d_t(\theta)/\partial\boldsymbol{\theta} | \boldsymbol{\phi} \right\} = E[\mathbf{W}_{dt}(\boldsymbol{\phi})|\boldsymbol{\phi}]. \end{aligned} \tag{8.40}$$

But since in the case of a static factor model

$$\mathbf{W}_{dt}'(\boldsymbol{\phi}) = \left[0 \ \mathbf{c}'\boldsymbol{\Sigma}^{-1}(\boldsymbol{\theta}_s) \ \tfrac{1}{2} vecd'[\boldsymbol{\Sigma}^{-1}(\boldsymbol{\theta}_s)] \right] = \mathbf{W}_d'(\boldsymbol{\phi}), \tag{8.41}$$

we will have that:

$$\begin{aligned} \mathring{\mathbf{s}}_{\pi t}(\boldsymbol{\theta}_s, \boldsymbol{\eta}) &= \mathbf{s}_{\pi t}(\boldsymbol{\theta}_s, \boldsymbol{\eta}), \\ \mathring{\mathbf{s}}_{\mathbf{c}t}(\boldsymbol{\theta}_s, \boldsymbol{\eta}) &= \mathbf{s}_{\mathbf{c}t}(\boldsymbol{\theta}_s, \boldsymbol{\eta}) - \boldsymbol{\Sigma}^{-1}(\boldsymbol{\theta}_s)\mathbf{c}\Big[\{\delta[\varsigma_t(\boldsymbol{\theta}_s), \boldsymbol{\eta}]\varsigma_t(\theta_s)/N - 1\} \\ &\quad - \tfrac{2}{(N+2)\kappa+2} (\varsigma_t(\boldsymbol{\theta}_s)/N - 1) \Big], \\ \mathring{\mathbf{s}}_{\gamma t}(\boldsymbol{\theta}_s, \boldsymbol{\eta}) &= \mathbf{s}_{\gamma t}(\boldsymbol{\theta}_s, \boldsymbol{\eta}) - \tfrac{1}{2} vecd[\boldsymbol{\Sigma}^{-1}(\boldsymbol{\theta})]\Big[\{\delta[\varsigma_t(\boldsymbol{\theta}_s), \boldsymbol{\eta}]\varsigma_t(\theta_s)/N - 1\} \\ &\quad - \tfrac{2}{(N+2)\kappa+2} (\varsigma_t(\boldsymbol{\theta}_s)/N - 1) \Big]. \end{aligned}$$

Expression (8.41) also implies that the elliptically symmetric semiparametric efficiency bound will be block diagonal between $\boldsymbol{\pi}$ and $(\mathbf{c}, \boldsymbol{\gamma})$, where the first block coincides with the first block of the information matrix, and the second one with the expression given in the second part of the proposition.

Finally, the Gaussian PML estimator of the conditional mean and variance parameters $\boldsymbol{\theta}$ sets to 0 the average value of $\mathbf{s}_{\boldsymbol{\theta}_s t}(\boldsymbol{\theta}, 0)$, which is trivially obtained from (8.36) by noting that $\delta(\varsigma_t, 0) = 1$. Given that the regularity conditions A.1 in Bollerslev and Wooldridge (1992) are satisfied in our context, then we know from Proposition 3 in Fiorentini and Sentana (2010b) that $\sqrt{T}(\bar{\theta}_T - \boldsymbol{\theta}_0) \to N\left[0, \mathcal{C}(\boldsymbol{\phi}_0)\right]$, where

$$\begin{aligned} \mathcal{C}(\boldsymbol{\phi}) = \mathcal{A}^{-1}(\boldsymbol{\phi})\mathcal{B}(\boldsymbol{\phi})\mathcal{A}^{-1}(\boldsymbol{\phi}), \quad \mathcal{A}(\boldsymbol{\phi}) &= -E\left[\mathbf{h}_{\theta\theta t}(\boldsymbol{\theta}, 0)|\phi\right] = E\left[\mathcal{A}_t(\boldsymbol{\phi})|\phi\right], \\ \mathcal{A}_t(\boldsymbol{\phi}) &= -E[\mathbf{h}_{\theta\theta t}(\boldsymbol{\theta}; 0)|\, \mathbf{z}_t, I_{t-1}; \phi] = \mathbf{Z}_{dt}(\boldsymbol{\theta})\mathcal{K}(0)\mathbf{Z}_{dt}'(\boldsymbol{\theta}), \\ \mathcal{B}(\boldsymbol{\phi}) &= V\left[\mathbf{s}_{\boldsymbol{\theta} t}(\theta, 0)|\phi\right] = E\left[\mathcal{B}_t(\boldsymbol{\phi})|\phi\right], \\ \mathcal{B}_t(\boldsymbol{\phi}) &= V[\mathbf{s}_{\theta t}(\boldsymbol{\theta}; 0)|\, \mathbf{z}_t, I_{t-1}; \boldsymbol{\phi}] = \mathbf{Z}_{dt}(\boldsymbol{\theta})\mathcal{K}(\kappa)\mathbf{Z}_{dt}'(\boldsymbol{\theta}), \end{aligned}$$

$$\text{and } \mathcal{K}(\kappa) = V[\mathbf{e}_{dt}(\boldsymbol{\theta}, 0)|\, \mathbf{z}_t, I_{t-1}; \boldsymbol{\phi}] = \begin{bmatrix} \mathbf{I}_N & 0 \\ 0 & (\kappa+1)\left(\mathbf{I}_{N^2} + \mathbf{K}_{NN}\right) + \kappa\, vec(\mathbf{I}_N)vec'(\mathbf{I}_N) \end{bmatrix}, \tag{8.42}$$

which only depends on $\boldsymbol{\eta}$ through κ. Hence, we can easily see that $\mathcal{A}_{\boldsymbol{\theta}_s\boldsymbol{\theta}_s}(\boldsymbol{\phi})$ coincides with $\mathcal{I}_{\boldsymbol{\theta}_s\boldsymbol{\theta}_s}(\boldsymbol{\theta}_s, 0)$ irrespective of the distribution of $\mathbf{y}_t$ because the model is static and $\mathcal{A}_{\boldsymbol{\theta}_s\boldsymbol{\theta}_s t}(\boldsymbol{\phi}) = -E[\mathbf{h}_{\boldsymbol{\theta}_s\boldsymbol{\theta}_s t}(\theta, 0)|I_{t-1}; \boldsymbol{\phi}]$ is equal to $\mathcal{I}_{\boldsymbol{\theta}_s\boldsymbol{\theta}_s t}(\boldsymbol{\theta}_s, 0)$. A closely related argument shows that $\mathcal{B}_{\boldsymbol{\theta}_s\boldsymbol{\theta}_s}(\boldsymbol{\phi})$ also mimics the expression for the information matrix if we replace $\mathrm{M}_{ll}(\boldsymbol{\eta})$ by 1 and $\mathrm{M}_{ss}(\boldsymbol{\eta})$ by $(\kappa + 1)$.

More generally, if $\boldsymbol{\varepsilon}_t^*|I_{t-1};\theta_0,\boldsymbol{\varrho}_0$ is i.i.d. $(0,\mathbf{I}_N)$ with density function $f(\boldsymbol{\varepsilon}_t^*;\boldsymbol{\varrho})$, where $\boldsymbol{\varrho}$ are some shape parameters and $\varrho=0$ denotes normality, then Proposition 1 in Bollerslev and Wooldridge (1992) coupled with the static nature of the model implies that:

$$\mathcal{B}_t(\boldsymbol{\phi}) = \mathbf{Z}_{dt}(\boldsymbol{\theta}_s)\mathcal{K}(\boldsymbol{\varrho})\,\mathbf{Z}'_{dt}(\theta_s),$$

where

$$\mathcal{K}(\boldsymbol{\varrho}) = V[\mathbf{e}_{dt}(\theta,0)|\,I_{t-1};\boldsymbol{\theta},\boldsymbol{\varrho}] \tag{8.43}$$

is the matrix of third and fourth order central moments of ε_t^*, whose first block is the identity matrix of order N. This means that the block diagonality between $\boldsymbol{\pi}$ and $(\mathbf{c},\boldsymbol{\gamma})$ disappears if the true distribution is asymmetric even though $\mathcal{B}_{\pi\pi}(\boldsymbol{\phi})$ continues to equal $\mathcal{I}_{\pi\pi}(\boldsymbol{\theta}_s,0)$. In view of $\mathbf{s}_{\boldsymbol{\theta}t}(\boldsymbol{\theta}_s,0)$, an alternative expression will be

$$\mathcal{B}_{\boldsymbol{\theta}_s\boldsymbol{\theta}_s}(\boldsymbol{\phi}) = V\begin{bmatrix} \boldsymbol{\Sigma}^{-1}(\boldsymbol{\theta}_s)(\mathbf{y}_t-\boldsymbol{\pi}) \\ \boldsymbol{\Gamma}^{-1}[\mathbf{v}_{kt}(\boldsymbol{\theta}_s)f_{kt}(\theta_s)-\mathbf{c}\omega_k(\boldsymbol{\theta}_s)] \\ \frac{1}{2}vecd\{\boldsymbol{\Gamma}^{-1}[\mathbf{v}_{kt}(\boldsymbol{\theta}_s)\mathbf{v}'_{kt}(\boldsymbol{\theta}_s)+\mathbf{c}\mathbf{c}'\omega_k(\boldsymbol{\theta}_s)-\boldsymbol{\Gamma}]\boldsymbol{\Gamma}^{-1}\} \end{bmatrix},$$

which is more amenable for empirical applications.

PROPOSITION 2

Initially, the proof follows the same steps as the proof of Proposition 1. Therefore, we need expressions for $\partial\boldsymbol{\mu}_t(\boldsymbol{\theta})/\partial\boldsymbol{\theta}$ and $\partial vec[\boldsymbol{\Sigma}_t(\boldsymbol{\theta})]/\partial\boldsymbol{\theta}$ to obtain $\mathbf{Z}_{dt}(\boldsymbol{\theta})$. Given our maintained assumption about the coincidence of the first two conditional moments with the Kalman filter prediction equations, we will have that

$$d\boldsymbol{\mu}_t(\boldsymbol{\theta}) = d\boldsymbol{\pi} + d(\,\mathbf{c}\;\mathbf{I}_N\,)\begin{pmatrix} x_{t|t-1}(\boldsymbol{\theta}) \\ \mathbf{u}_{t|t-1}(\boldsymbol{\theta}) \end{pmatrix} + (\,\mathbf{c}\;\mathbf{I}_N\,)d\begin{pmatrix} x_{t|t-1}(\boldsymbol{\theta}) \\ \mathbf{u}_{t|t-1}(\boldsymbol{\theta}) \end{pmatrix}$$

and

$$d\boldsymbol{\Sigma}_t(\boldsymbol{\theta}) = d(\,\mathbf{c}\;\mathbf{I}_N\,)\boldsymbol{\Omega}_{t|t-1}(\boldsymbol{\theta})\begin{pmatrix} \mathbf{c}' \\ \mathbf{I}_N \end{pmatrix} + (\,\mathbf{c}\;\mathbf{I}_N\,)d\boldsymbol{\Omega}_{t|t-1}(\boldsymbol{\theta})\begin{pmatrix} \mathbf{c}' \\ \mathbf{I}_N \end{pmatrix}$$
$$+(\,\mathbf{c}\;\mathbf{I}_N\,)\boldsymbol{\Omega}_{t|t-1}(\boldsymbol{\theta})d\begin{pmatrix} \mathbf{c}' \\ \mathbf{I}_N \end{pmatrix},$$

whence

$$\frac{\partial\boldsymbol{\mu}_t(\boldsymbol{\theta})}{\partial\boldsymbol{\theta}'} = \frac{\partial\boldsymbol{\pi}}{\partial\boldsymbol{\theta}'} + [x_{t|t-1}(\boldsymbol{\theta})\otimes\mathbf{I}_N]\frac{\partial\mathbf{c}}{\partial\boldsymbol{\theta}'} + \mathbf{c}\frac{\partial x_{t|t-1}(\boldsymbol{\theta})}{\partial\boldsymbol{\theta}'} + \frac{\partial\mathbf{u}_{t|t-1}(\boldsymbol{\theta})}{\partial\boldsymbol{\theta}'}$$

and

$$\frac{\partial vec[\boldsymbol{\Sigma}_t(\boldsymbol{\theta})]}{\partial\boldsymbol{\theta}'} = (\mathbf{I}_{N^2}+\mathbf{K}_{NN})[(\,\mathbf{c}\;\mathbf{I}_N\,)\boldsymbol{\Omega}_{t|t-1}(\boldsymbol{\theta})\otimes\mathbf{I}_N]\begin{pmatrix} \partial\mathbf{c}/\partial\boldsymbol{\theta}' \\ \mathbf{0} \end{pmatrix}$$
$$+[(\,\mathbf{c}\;\mathbf{I}_N\,)\otimes(\,\mathbf{c}\;\mathbf{I}_N\,)]\frac{\partial vec[\boldsymbol{\Omega}_{t|t-1}(\boldsymbol{\theta})]}{\partial\boldsymbol{\theta}'}.$$

Similarly,

$$\frac{\partial x_{t|t-1}(\boldsymbol{\theta})}{\partial\boldsymbol{\theta}'} = x_{t|t-1}(\boldsymbol{\theta})\frac{\partial\rho}{\partial\boldsymbol{\theta}'} + \rho\frac{\partial x_{t-1|t-1}(\boldsymbol{\theta})}{\partial\boldsymbol{\theta}'},$$

and

$$\frac{\partial \mathbf{u}_{t|t-1}(\boldsymbol{\theta})}{\partial \theta'} = [\mathbf{u}'_{t|t-1}(\boldsymbol{\theta}) \otimes \mathbf{I}_N]\mathbf{E}_N \frac{\partial \boldsymbol{\rho}^*}{\partial \theta'} + diag(\boldsymbol{\rho}^*) \frac{\partial \mathbf{u}_{t-1|t-1}(\boldsymbol{\theta})}{\partial \theta'}.$$

In fact, it is easy to see that this last expression reduces to

$$\frac{\partial u_{it|t-1}(\boldsymbol{\theta})}{\partial \boldsymbol{\theta}'} = u_{it|t-1}(\boldsymbol{\theta}) \frac{\partial \rho_i^*}{\partial \boldsymbol{\theta}'} + \rho_i^* \frac{\partial u_{it-1|t-1}(\boldsymbol{\theta})}{\partial \boldsymbol{\theta}'}.$$

In addition, if we differentiate the updating equation we obtain

$$\frac{\partial vec[\boldsymbol{\Omega}_{t|t-1}(\boldsymbol{\theta})]}{\partial \boldsymbol{\theta}'} = (\mathbf{I}_{(N+1)^2} + \mathbf{K}_{N+1,N+1}) \left\{ \begin{bmatrix} \rho & 0 \\ 0 & diag(\boldsymbol{\rho}^*) \end{bmatrix} \otimes \mathbf{I}_{N+1} \right\} \mathbf{E}_{N+1} \begin{pmatrix} \partial \rho / \partial \boldsymbol{\theta}' \\ \partial \boldsymbol{\rho}^* / \partial \boldsymbol{\theta}' \end{pmatrix}$$

$$+ \mathbf{E}_{N+1} \begin{pmatrix} 0 \\ \partial \boldsymbol{\gamma} / \partial \boldsymbol{\theta}' \end{pmatrix} + \left\{ \begin{bmatrix} \rho & 0 \\ 0 & diag(\boldsymbol{\rho}^*) \end{bmatrix} \otimes \begin{bmatrix} \rho & 0 \\ 0 & diag(\boldsymbol{\rho}^*) \end{bmatrix} \right\} \frac{\partial vec[\boldsymbol{\Omega}_{t-1|t-1}(\theta)]}{\partial \boldsymbol{\theta}'}.$$

In principle, we would need expressions for $\partial x_{t-1|t-1}(\boldsymbol{\theta})/\partial \boldsymbol{\theta}'$, $\partial u_{it-1|t-1}(\boldsymbol{\theta})/\partial \boldsymbol{\theta}'$ and $\partial vec[\boldsymbol{\Omega}_{t-1|t-1}(\boldsymbol{\theta})]/\partial \theta'$. However, since we are only interested in evaluating the score at $\rho = 0$ and $\boldsymbol{\rho}^* = \mathbf{0}$, those expressions become unnecessary.

In addition, it is worth noting that under the null $x_{t|t-1}(\theta_s, \mathbf{0}) = 0$, $\mathbf{u}_{t|t-1}(\boldsymbol{\theta}_s, \mathbf{0}) = \mathbf{0}$, $\boldsymbol{\Omega}_{t|t-1}(\boldsymbol{\theta}_s, \mathbf{0}) = diag(1, \boldsymbol{\gamma})$, $\boldsymbol{\Sigma}_t(\boldsymbol{\theta}_s, \mathbf{0}) = \mathbf{cc}' + \boldsymbol{\Gamma} = \boldsymbol{\Sigma}(\boldsymbol{\theta}_s)$, $x_{t|t}(\boldsymbol{\theta}_s, \mathbf{0}) = f_{kt}(\boldsymbol{\theta}_s)$ and $\mathbf{u}_{t|t}(\boldsymbol{\theta}_s, \mathbf{0}) = \mathbf{v}_{kt}(\boldsymbol{\theta}_s)$, so that

$$\frac{\partial \boldsymbol{\mu}_t(\boldsymbol{\theta}_s, \mathbf{0})}{\partial \boldsymbol{\theta}'} = \frac{\partial \boldsymbol{\pi}}{\partial \theta'} + \mathbf{c} f_{kt}(\boldsymbol{\theta}_s) \frac{\partial \rho}{\partial \boldsymbol{\theta}'} + diag[\mathbf{v}_{kt}(\theta_s)] \frac{\partial \boldsymbol{\rho}^*}{\partial \theta'}$$

and

$$\frac{\partial vec[\boldsymbol{\Sigma}_t(\boldsymbol{\theta}_s, \mathbf{0})]}{\partial \boldsymbol{\theta}'} = (\mathbf{I}_{N^2} + \mathbf{K}_{NN})(\mathbf{c} \otimes \mathbf{I}_N) \frac{\partial \mathbf{c}}{\partial \theta'} + \mathbf{E}_N \frac{\partial \boldsymbol{\gamma}}{\partial \boldsymbol{\theta}'}.$$

Hence

$$\mathbf{Z}_{dt}(\boldsymbol{\theta}_s, \mathbf{0}) = \begin{bmatrix} \boldsymbol{\Sigma}^{-1/2\prime}(\boldsymbol{\theta}_s) & \mathbf{0} \\ \mathbf{0} & \frac{1}{2}(\mathbf{c}' \otimes \mathbf{I}_N)(\mathbf{I}_{N^2} + \mathbf{K}_{NN})[\boldsymbol{\Sigma}^{-1/2\prime}(\boldsymbol{\theta}_s) \otimes \boldsymbol{\Sigma}^{-1/2\prime}(\boldsymbol{\theta}_s)] \\ \mathbf{0} & \frac{1}{2}\mathbf{E}'_N[\boldsymbol{\Sigma}^{-1/2\prime}(\boldsymbol{\theta}_s) \otimes \boldsymbol{\Sigma}^{-1/2\prime}(\boldsymbol{\theta}_s)] \\ f_{kt-1}(\boldsymbol{\theta}_s)\mathbf{c}'\boldsymbol{\Sigma}^{-1/2\prime}(\boldsymbol{\theta}_s) & \mathbf{0} \\ diag[\mathbf{v}_{kt-1}(\boldsymbol{\theta}_s)]\boldsymbol{\Sigma}^{-1/2\prime}(\boldsymbol{\theta}_s) & \mathbf{0} \end{bmatrix},$$

$$\mathbf{Z}_d(\boldsymbol{\phi}) = \begin{bmatrix} \boldsymbol{\Sigma}^{-1/2\prime}(\boldsymbol{\theta}_s) & \mathbf{0} \\ \mathbf{0} & \frac{1}{2}(\mathbf{c}' \otimes \mathbf{I}_N)(\mathbf{I}_{N^2} + \mathbf{K}_{NN})[\boldsymbol{\Sigma}^{-1/2\prime}(\boldsymbol{\theta}_s) \otimes \boldsymbol{\Sigma}^{-1/2\prime}(\boldsymbol{\theta}_s)] \\ \mathbf{0} & \frac{1}{2}\mathbf{E}'_N[\boldsymbol{\Sigma}^{-1/2\prime}(\boldsymbol{\theta}_s) \otimes \boldsymbol{\Sigma}^{-1/2\prime}(\boldsymbol{\theta}_s)] \\ \mathbf{0} & \mathbf{0} \\ \mathbf{0} & \mathbf{0} \end{bmatrix}$$

and

$$\mathbf{W}_d(\boldsymbol{\phi}) = \left[\mathbf{0}' \;\; \mathbf{c}'\boldsymbol{\Sigma}^{-1}(\theta_s) \;\; \tfrac{1}{2} vecd'[\boldsymbol{\Sigma}^{-1}(\boldsymbol{\theta}_s)] \;\; 0 \;\; \mathbf{0}' \right]', \tag{8.44}$$

where we have used the fact that

$$\left. \begin{aligned} E[f_{kt}(\boldsymbol{\theta}_s)|\boldsymbol{\theta}_s, \mathbf{0}, 0] = E[\mathbf{c}'\boldsymbol{\Sigma}^{-1}(\boldsymbol{\theta}_s)(\mathbf{y}_t - \boldsymbol{\pi})|\boldsymbol{\theta}_s, \mathbf{0}, 0] = 0 \\ E[\mathbf{v}_{kt}(\boldsymbol{\theta}_s)|\boldsymbol{\theta}_s, \mathbf{0}, 0] = E[\boldsymbol{\Gamma}\boldsymbol{\Sigma}^{-1}(\boldsymbol{\theta}_s)(\mathbf{y}_t - \boldsymbol{\pi})|\boldsymbol{\theta}_s, \mathbf{0}, 0] = 0 \end{aligned} \right\} \tag{8.45}$$

irrespective of the distribution of $\mathbf{y}_t$.

As a result, the elliptically symmetric score under the null will be

$$\begin{bmatrix} s_{\pi t}(\boldsymbol{\theta}_s, 0, \boldsymbol{\eta}) \\ s_{ct}(\boldsymbol{\theta}_s, 0, \boldsymbol{\eta}) \\ s_{\gamma t}(\boldsymbol{\theta}_s, 0, \boldsymbol{\eta}) \\ s_{\rho t}(\boldsymbol{\theta}_s, 0, \boldsymbol{\eta}) \\ s_{\rho^* t}(\boldsymbol{\theta}_s, 0, \boldsymbol{\eta}) \end{bmatrix} = \begin{bmatrix} \delta[\varsigma_t(\boldsymbol{\theta}_s); \boldsymbol{\eta}]\boldsymbol{\Sigma}^{-1}(\boldsymbol{\theta}_s)(\mathbf{y}_t - \boldsymbol{\pi}) \\ \delta[\varsigma_t(\boldsymbol{\theta}_s); \boldsymbol{\eta}]\boldsymbol{\Sigma}^{-1}(\boldsymbol{\theta}_s)(\mathbf{y}_t - \boldsymbol{\pi})(\mathbf{y}_t - \boldsymbol{\pi})'\boldsymbol{\Sigma}^{-1}(\boldsymbol{\theta}_s)\mathbf{c} - \boldsymbol{\Sigma}^{-1}(\boldsymbol{\theta}_s)\mathbf{c} \\ \frac{1}{2}vecd\{\delta[\varsigma_t(\boldsymbol{\theta}_s); \boldsymbol{\eta}]\boldsymbol{\Sigma}^{-1}(\boldsymbol{\theta}_s)(\mathbf{y}_t - \boldsymbol{\pi})(\mathbf{y}_t - \boldsymbol{\pi})'\boldsymbol{\Sigma}^{-1}(\boldsymbol{\theta}_s) - \boldsymbol{\Sigma}^{-1}(\boldsymbol{\theta}_s)\} \\ f_{kt-1}(\boldsymbol{\theta}_s)\delta[\varsigma_t(\boldsymbol{\theta}_s); \boldsymbol{\eta}]\mathbf{c}'\boldsymbol{\Sigma}^{-1}(\theta_s)(\mathbf{y}_t - \boldsymbol{\pi}) \\ diag[\mathbf{v}_{kt-1}(\boldsymbol{\theta}_s)]\delta[\varsigma_t(\boldsymbol{\theta}_s); \boldsymbol{\eta}]\boldsymbol{\Sigma}^{-1}(\boldsymbol{\theta}_s)(\mathbf{y}_t - \boldsymbol{\pi}) \end{bmatrix}.$$

Therefore, the only difference relative to the static factor model are the scores $s_{\rho t}(\boldsymbol{\theta}_s, 0, \boldsymbol{\eta})$ and $s_{\rho^* t}(\boldsymbol{\theta}_s, 0, \boldsymbol{\eta})$. In this sense, if we assume that $\boldsymbol{\Gamma} > \mathbf{0}$, then we can use the Woodbury formula once again to show that

$$\begin{bmatrix} s_{\rho t}(\boldsymbol{\theta}_s, 0, \boldsymbol{\eta}) \\ s_{\rho^* t}(\boldsymbol{\theta}_s, 0, \boldsymbol{\eta}) \end{bmatrix} = \begin{bmatrix} \delta[\varsigma_t(\boldsymbol{\theta}_s); \boldsymbol{\eta}]f_{kt-1}(\boldsymbol{\theta}_s)f_{kt}(\boldsymbol{\theta}_s) \\ \delta[\varsigma_t(\boldsymbol{\theta}_s); \boldsymbol{\eta}]diag[\mathbf{v}_{kt-1}(\boldsymbol{\theta}_s)]\boldsymbol{\Gamma}^{-1}\mathbf{v}_{kt}(\boldsymbol{\theta}_s) \end{bmatrix}.$$

Using the expression for $\mathbf{Z}_{dt}(\boldsymbol{\theta}_s, 0)$, together with (8.45), it is easy to show that the unconditional information matrix $\mathcal{I}_{\theta\theta}(\boldsymbol{\theta}_s, 0, \boldsymbol{\eta})$ will be block diagonal between π, $(\mathbf{c}, \gamma, \eta)$, and $\boldsymbol{\rho}^\dagger$, with the first two blocks as in the static case. Consequently, in computing our ML-based tests we can safely ignore the sampling uncertainty in estimating $\boldsymbol{\theta}_s$ and $\boldsymbol{\eta}$. In addition, we can write

$$\mathcal{I}_{\rho^\dagger\rho^\dagger t}(\theta, 0, \boldsymbol{\eta}) = diag\begin{bmatrix} f_{kt-1}(\boldsymbol{\theta}_s) \\ \boldsymbol{\Gamma}^{-1/2}\mathbf{v}_{kt-1}(\boldsymbol{\theta}_s) \end{bmatrix} \mathcal{V}_{\rho^\dagger\rho^\dagger}(\boldsymbol{\theta}_s, \eta; \eta) diag\begin{bmatrix} f_{kt-1}(\boldsymbol{\theta}_s) \\ \boldsymbol{\Gamma}^{-1/2}\mathbf{v}_{kt-1}(\boldsymbol{\theta}_s) \end{bmatrix},$$

where

$$\begin{aligned} \mathcal{V}_{\rho^\dagger\rho^\dagger}(\theta_s, \eta; \eta) &= V\begin{bmatrix} \delta[\varsigma_t(\boldsymbol{\theta}_s), \boldsymbol{\eta}]f_{kt}(\boldsymbol{\theta}_s) \\ \boldsymbol{\Gamma}^{-1/2}\delta[\varsigma_t(\boldsymbol{\theta}_s), \boldsymbol{\eta}]\mathbf{v}_{kt}(\boldsymbol{\theta}_s) \end{bmatrix} \\ &= \text{M}_{ll}(\boldsymbol{\eta})\begin{bmatrix} \mathbf{c}'\boldsymbol{\Sigma}^{-1}(\boldsymbol{\theta}_s)\mathbf{c} & \mathbf{c}'\boldsymbol{\Sigma}^{-1}(\boldsymbol{\theta}_s)\boldsymbol{\Gamma}^{1/2} \\ \boldsymbol{\Gamma}^{1/2}\boldsymbol{\Sigma}^{-1}(\boldsymbol{\theta}_s)\mathbf{c} & \boldsymbol{\Gamma}^{1/2}\boldsymbol{\Sigma}^{-1}(\theta_s)\boldsymbol{\Gamma}^{1/2} \end{bmatrix} \\ &= \text{M}_{ll}(\boldsymbol{\eta})\begin{bmatrix} (\mathbf{c}'\boldsymbol{\Gamma}^{-1}\mathbf{c})/(1+\mathbf{c}'\boldsymbol{\Gamma}^{-1}\mathbf{c}) & \mathbf{c}'\boldsymbol{\Gamma}^{-1/2}/(1+\mathbf{c}'\boldsymbol{\Gamma}^{-1}\mathbf{c}) \\ \boldsymbol{\Gamma}^{-1/2}\mathbf{c}/(1+\mathbf{c}'\boldsymbol{\Gamma}^{-1}\mathbf{c}) & \mathbf{I}_N - \boldsymbol{\Gamma}^{-1/2}\mathbf{c}\mathbf{c}'\boldsymbol{\Gamma}^{-1/2}/(1+\mathbf{c}'\boldsymbol{\Gamma}^{-1}\mathbf{c}) \end{bmatrix}. \end{aligned}$$

Thus, the only remaining item is the calculation of the second moments appearing in $\mathcal{V}_{\alpha^\dagger\alpha^\dagger}(\boldsymbol{\theta}_s, 0; \eta)$. But since

$$E[f_{kt}^2(\boldsymbol{\theta}_s)|\boldsymbol{\theta}_s, 0, \eta] = E[\mathbf{c}'\boldsymbol{\Sigma}^{-1}(\boldsymbol{\theta}_s)(\mathbf{y}_t - \boldsymbol{\pi})(\mathbf{y}_t - \boldsymbol{\pi})'\boldsymbol{\Sigma}^{-1}(\boldsymbol{\theta}_s)\mathbf{c}|\boldsymbol{\theta}_s, 0, \eta]$$
$$= \mathbf{c}'\boldsymbol{\Sigma}^{-1}(\boldsymbol{\theta}_s)\mathbf{c} = \mathbf{c}'\boldsymbol{\Gamma}^{-1}\mathbf{c}/(1+\mathbf{c}'\boldsymbol{\Gamma}^{-1}\mathbf{c}),$$
$$E\{\mathbf{v}_{kt}(\boldsymbol{\theta}_s)f_{kt}(\boldsymbol{\theta}_s)|\boldsymbol{\theta}_s, 0, \boldsymbol{\eta}\} = E\{[\boldsymbol{\Gamma}\boldsymbol{\Sigma}^{-1}(\boldsymbol{\theta}_s)(\mathbf{y}_t - \boldsymbol{\pi})(\mathbf{y}_t - \boldsymbol{\pi})'\boldsymbol{\Sigma}^{-1}(\boldsymbol{\theta}_s)\mathbf{c}|\boldsymbol{\theta}_s, 0, \boldsymbol{\eta}\}$$
$$= \boldsymbol{\Gamma}\boldsymbol{\Sigma}^{-1}(\boldsymbol{\theta}_s)\mathbf{c} = \mathbf{c}/(1+\mathbf{c}'\boldsymbol{\Gamma}^{-1}\mathbf{c})$$

and

$$E\{\mathbf{v}_{kt}(\boldsymbol{\theta}_s)\mathbf{v}_{kt}(\boldsymbol{\theta}_s)']|\boldsymbol{\theta}_s, 0, \eta] = E[\boldsymbol{\Gamma}\boldsymbol{\Sigma}^{-1}(\boldsymbol{\theta}_s)(\mathbf{y}_t - \pi)(\mathbf{y}_t - \boldsymbol{\pi})'\boldsymbol{\Sigma}^{-1}(\boldsymbol{\theta}_s)\boldsymbol{\Gamma}]|\boldsymbol{\theta}_s, \mathbf{0}, \boldsymbol{\eta}\}$$
$$= \boldsymbol{\Gamma}\boldsymbol{\Sigma}^{-1}(\boldsymbol{\theta}_s)\boldsymbol{\Gamma} = \boldsymbol{\Gamma} - cc'/(1+\mathbf{c}'\boldsymbol{\Gamma}^{-1}\mathbf{c}),$$

we finally obtain that $\mathcal{V}_{\rho^\dagger\rho^\dagger}(\boldsymbol{\theta}_s, 0; \eta)$ mimics $\mathcal{V}_{\rho^\dagger\rho^\dagger}(\boldsymbol{\theta}_s, \eta; \eta)$ if we replace $\text{M}_{ll}(\boldsymbol{\eta})$ by 1.

In addition, it follows from (8.44) that the elliptically symmetric semiparametric scores for ρ and $\boldsymbol{\rho}^*$ coincide with the parametric ones, and that the elliptically symmetric semiparametric efficiency bound will be block diagonal between $\boldsymbol{\pi}$, $(\rho, \boldsymbol{\rho}^*)$, and $(\mathbf{c}, \boldsymbol{\gamma})$, where the first two blocks coincide with the first two blocks of the information matrix, and the third one with the corresponding bound in the static factor model.

Finally, let us consider the tests based on the Gaussian PML scores $s_{\rho t}(\boldsymbol{\theta}_s, \mathbf{0}, \mathbf{0})$ and $s_{\rho^* t}(\boldsymbol{\theta}_s, \mathbf{0}, \mathbf{0})$ when $\mathbf{y}_t|I_{t-1}; \boldsymbol{\phi}$ is i.i.d. $D(\boldsymbol{\pi}, \boldsymbol{\Sigma}(\theta_s); \boldsymbol{\varrho})$ but not necessarily normal or elliptical. To do so, let us partition the parameter vector $\boldsymbol{\theta}$ as $(\boldsymbol{\theta}_s, \boldsymbol{\rho}^\dagger)$. It is well known (see, e.g., Engle (1984)) that a robust Gaussian pseudo score test of the null hypothesis $H_0 : \boldsymbol{\rho}^\dagger = \mathbf{0}$ can be computed as

$$\left[\frac{\sqrt{T}}{T}\sum_{t=1}^{T} \mathbf{s}'_{\rho^\dagger t}(\tilde{\theta}_s, 0, 0)\right] \mathcal{A}^{\rho^\dagger \rho^\dagger}(\boldsymbol{\phi}_0)\, \mathcal{C}^{-1}_{\rho^\dagger \rho^\dagger}(\boldsymbol{\phi}_0)\, \mathcal{A}^{\rho^\dagger \rho^\dagger}(\boldsymbol{\phi}_0) \left[\frac{\sqrt{T}}{T}\sum_{t=1}^{T} \mathbf{s}_{\rho^\dagger t}(\tilde{\theta}_s, 0, 0)\right],$$

where $\mathbf{s}_{\rho^\dagger t}(\tilde{\theta}_s, 0, 0)$ is the Gaussian score evaluated at the restricted PML estimator $\tilde{\theta}_s$, $\mathcal{A}^{\rho^\dagger \rho^\dagger}(\boldsymbol{\phi}_0)$ is the relevant block of the inverse of the expected Hessian matrix $\mathcal{A}(\boldsymbol{\phi}) = -E[\mathbf{h}_{\theta\theta t}(\theta, 0)|\phi]$, and $\mathcal{C}_{\rho^\dagger \rho^\dagger}(\boldsymbol{\phi}_0)$ is the corresponding block of the usual sandwich expression $\mathcal{C}(\boldsymbol{\phi}) = \mathcal{A}^{-1}(\boldsymbol{\phi})\mathcal{B}(\boldsymbol{\phi})\mathcal{A}^{-1}(\boldsymbol{\phi})$, with $\mathcal{B}(\boldsymbol{\phi}) = V[\mathbf{s}_{\theta t}(\boldsymbol{\theta}, 0)|\phi]$. Once again, the structure of $\mathbf{Z}_{dt}(\boldsymbol{\theta})$, together with (8.45), implies that $\mathcal{A}(\boldsymbol{\phi})$ will be block diagonal between $(\rho, \boldsymbol{\rho}^*)$ and $(\boldsymbol{\pi}, \mathbf{c}, \boldsymbol{\gamma})$ irrespective of the true distribution of $\mathbf{y}_t$. In addition, $\mathcal{A}_{\rho^\dagger \rho^\dagger}(\boldsymbol{\phi})$ will coincide with $\mathcal{I}_{\rho^\dagger \rho^\dagger}(\theta_s, 0, 0)$. A closely related argument shows that $\mathcal{B}(\boldsymbol{\phi})$ will also be block diagonal between $(\rho, \boldsymbol{\rho}^*)$ and $(\boldsymbol{\pi}, \mathbf{c}, \boldsymbol{\gamma})$, and that $\mathcal{B}_{\rho^\dagger \rho^\dagger}(\boldsymbol{\phi}) = \mathcal{A}_{\rho^\dagger \rho^\dagger}(\boldsymbol{\phi})$, which validates the expression for $LM_{AR(1)}(\mathbf{0})$.

PROPOSITION 3

For the sake of brevity, the proof will be developed for the following univariate model:

$$\begin{aligned} y_t &= \pi + x_t + \gamma^{1/2} v_t^*, \\ x_t &= \rho x_{t-1} + f_t, \end{aligned}$$
$$\begin{pmatrix} f_t \\ v_t^* \end{pmatrix} |I_{t-1} \sim t\left[\begin{pmatrix} 0 \\ 0 \end{pmatrix}, \begin{pmatrix} 1 & 0 \\ 0 & 1 \end{pmatrix}, \eta\right].$$

Nevertheless, it can be tediously extended to cover the general case. Given that when $\rho = 0$ the log-likelihood function of this model coincides with the log-likelihood function of the model considered in Section 8.2, we only need to look at the score associated to this parameter.

It is easy to see that the joint distribution of y_t and x_t given the past of both variables will be

$$\begin{pmatrix} y_t \\ x_t \end{pmatrix} |I_{t-1} \sim t\left[\begin{pmatrix} \rho x_{t-1} \\ \rho x_{t-1} \end{pmatrix}, \begin{pmatrix} 1+\gamma & 1 \\ 1 & 1 \end{pmatrix}, \eta\right].$$

Hence, we can write down the joint log-likelihood as

$$c_2(\eta) - \frac{1}{2}\ln \gamma + g[\varsigma_t(\rho, \gamma); \eta],$$

where

$$c_2(\eta) = \ln\left[\Gamma\left(\frac{2\eta+1}{2\eta}\right)\right] - \ln\left[\Gamma\left(\frac{1}{2\eta}\right)\right] - \ln\left(\frac{1-2\eta}{\eta}\right) - \ln \pi$$

is the (log) constant of integration,

$$\gamma = \left| \begin{pmatrix} 1+\gamma & 1 \\ 1 & 1 \end{pmatrix} \right|$$

the Jacobian, and

$$g[\varsigma_t(\rho,\gamma);\eta] = -\left(\frac{2\eta+1}{2\eta}\right) \ln\left[1 + \frac{\eta}{1-2\eta}\varsigma_t(\rho,\gamma)\right],$$

with

$$\begin{aligned} \varsigma_t(\rho,\gamma) &= \begin{pmatrix} y_t - \rho x_{t-1} & x_t - \rho x_{t-1} \end{pmatrix} \begin{pmatrix} 1+\gamma & 1 \\ 1 & 1 \end{pmatrix}^{-1} \begin{pmatrix} y_t - \rho x_{t-1} \\ x_t - \rho x_{t-1} \end{pmatrix} \\ &= \gamma^{-1}(y_t - x_t)^2 + (x_t - \rho x_{t-1})^2, \end{aligned}$$

the (log) kernel of the bivariate Student t density.

Given that we can write the standardized residuals as

$$\begin{aligned} \begin{pmatrix} 1+\gamma & 1 \\ 1 & 1 \end{pmatrix}^{-1/2} \begin{pmatrix} y_t - \rho x_{t-1} \\ x_t - \rho x_{t-1} \end{pmatrix} &= \begin{pmatrix} \frac{1}{\gamma} & -\frac{1}{\gamma} \\ 0 & 1 \end{pmatrix} \begin{pmatrix} y_t - \rho x_{t-1} \\ x_t - \rho x_{t-1} \end{pmatrix} \\ &= \begin{pmatrix} \gamma^{-1/2}(y_t - x_t) \\ x_t - \rho x_{t-1} \end{pmatrix} \end{aligned}$$

and the gradient of the conditional mean vector with respect to ρ will be x_{t-1} times the vector $(1,1)'$, we will have that the score of the joint log-likelihood function with respect to ρ will be given by

$$\begin{aligned} &-\frac{2\eta+1}{1-2\eta+\eta\varsigma_t(\rho,\gamma)} x_{t-1} \begin{pmatrix} 1 & 1 \end{pmatrix} \begin{pmatrix} \gamma^{-1/2} & 0 \\ -\gamma^{-1/2} & 1 \end{pmatrix} \begin{pmatrix} \gamma^{-1/2}(y_t - x_t) \\ x_t - \rho x_{t-1} \end{pmatrix} \\ &= -\frac{2\eta+1}{1-2\eta+\eta\varsigma_t(\rho,\gamma)} (x_t - \rho x_{t-1}) x_{t-1}. \end{aligned}$$

The Kullback inequality then implies that score of the marginal log-likelihood function of y_t with respect to ρ will be given by

$$E\left[-\frac{2\eta+1}{1-2\eta+\eta\varsigma_t(\rho,\gamma)} (x_t - \rho x_{t-1}) x_{t-1} \middle| \mathrm{Y}_T, \rho \right].$$

This expected value becomes analytically tractable when $\rho = 0$. First of all, the expression inside the expectation simplifies to

$$E\left[\frac{2\eta+1}{1-2\eta+\eta\left[\gamma^{-1}(y_t - x_t)^2 + x_t^2\right]} x_t x_{t-1} \middle| \mathrm{Y}_T, \rho = 0 \right].$$

Second, the joint distribution of y_t and x_t is i.i.d. over time, which means that the expected value of this product is equal to

$$E\left[\frac{2\eta+1}{1-2\eta+\eta\left[\gamma^{-1}(y_t - x_t)^2 + x_t^2\right]} x_t \middle| y_t, \rho = 0 \right] E\left[x_{t-1} | y_{t-1}, \rho = 0\right].$$

But the distribution of $x_t = f_t$ given y_t will also be t with mean

$$f_{kt}(\gamma) = \frac{1}{1+\gamma} y_t, \tag{8.46}$$

variance

$$\begin{aligned} \upsilon_{kt}(\gamma, \eta) &= \frac{1-2\eta}{1-\eta}\left(1+\frac{\eta}{1-2\eta}\frac{y_t^2}{1+\gamma}\right)\left(1-\frac{1}{1+\gamma}\right) \\ &= \frac{1-2\eta}{1-\eta}\left(1+\frac{\eta}{1-2\eta}\frac{y_t^2}{1+\gamma}\right)\omega_k(\gamma) \end{aligned} \tag{8.47}$$

and shape parameter

$$\frac{\eta}{1+\eta}, \tag{8.48}$$

since the degrees of freedom of the conditional distribution of x_t given y_t are 1 plus the degrees of freedom of the joint distribution. Therefore, the second term is simply given by the lagged value of (8.46). The first term is trickier, as we need to find the expected value of

$$\frac{2\eta+1}{1-2\eta+\eta\left[\gamma^{-1}(y_t-x_t)^2+x_t^2\right]}x_t. \tag{8.49}$$

To do so, it is convenient to follow Fiorentini et al. (2003) and write x_t in terms of a conditionally standardized Student t component x_t^* as follows:

$$x_t = \frac{1}{1+\gamma}y_t + \sqrt{\frac{1-2\eta}{1-\eta}\left(1+\frac{\eta}{1-2\eta}\frac{y_t^2}{1+\gamma}\right)\frac{\gamma}{1+\gamma}}x_t^*,$$

$$x_t^* = \sqrt{\frac{1-\eta}{\eta}} \times \sqrt{\varsigma_t/\xi_t}u_t,$$

where u_t is either 1 or -1 with probability 1/2, ζ_t is a chi-square random variable with 1 degree of freedom and ξ_t is a gamma random variable with mean $1+\eta^{-1}$ and variance $2(1+\eta^{-1})$, with u_t, ζ_t, and ξ_t mutually independent and independent of y_t and I_{t-1}.

In turn, this decomposition implies that

$$\begin{aligned} \varsigma_t(0,\gamma) = \gamma^{-1}(y_t-x_t)^2+x_t^2 &= \frac{y_t^2}{1+\gamma}+\left(\frac{1+\gamma}{\gamma}\right)\left(x_t-\frac{1}{1+\gamma}y_t\right)^2 \\ &= \frac{y_t^2}{1+\gamma}\left(1+\frac{\zeta_t}{\xi_t}\right)+\frac{1-2\eta}{\eta}\frac{\zeta_t}{\xi_t}, \end{aligned}$$

so that the denominator of (8.49) can be written as

$$1-2\eta+\eta\varsigma_t(0,\gamma) = \left(1-2\eta+\frac{\eta y_t^2}{1+\gamma}\right)\left(\frac{\xi_t+\zeta_t}{\xi_t}\right).$$

As a result, (8.49) becomes

$$\left(1-2\eta+\frac{\eta y_t^2}{1+\gamma}\right)^{-1}\left(\frac{\xi_t}{\xi_t+\zeta_t}\right)\frac{2\eta+1}{1+\gamma}y_t +$$

$$\left(1-2\eta+\frac{\eta y_t^2}{1+\gamma}\right)^{-1}\sqrt{\frac{1-2\eta}{1-\eta}\left(1+\frac{\eta}{1-2\eta}\frac{y_t^2}{1+\gamma}\right)\frac{\gamma}{1+\gamma}}\sqrt{\frac{1-\eta}{\eta}}\left(\frac{\xi_t}{\xi_t+\zeta_t}\right)\sqrt{\frac{\zeta_t}{\xi_t}}u_t.$$

The expected value of the second summand conditional on y_t is 0 because of the symmetry of u_t. In contrast, we can use the properties of the beta distribution to prove that

$$E\left(\frac{\xi_t}{\xi_t+\zeta_t}\right)=\frac{1+\eta}{1+2\eta}$$

and consequently, that

$$E\left[\left(1-2\eta+\frac{\eta y_t^2}{1+\gamma}\right)^{-1}\left(\frac{\xi_t}{\xi_t+\zeta_t}\right)\frac{2\eta+1}{1+\gamma}y_t\middle|y_t\right]$$

$$=\left(1-2\eta+\frac{\eta y_t^2}{1+\gamma}\right)^{-1}\frac{1+\eta}{1+\gamma}y_t.$$

Therefore, we have proved that

$$E\left[\frac{2\eta+1}{1-2\eta+\eta\left[\gamma^{-1}(y_t-x_t)^2+x_t^2\right]}x_tx_{t-1}\middle|Y_T,\rho=0\right]$$

$$=\left(1-2\eta+\frac{\eta y_t^2}{1+\gamma}\right)^{-1}\frac{1+\eta}{1+\gamma}y_t\frac{1}{1+\gamma}y_{t-1}.$$

Finally, using the general expressions for the score of the approximating model obtained in the proof of Proposition 2, we will have that the score with respect to ρ of such a univariate log-likelihood function under the null of $\rho=0$ will be given

$$\frac{\eta+1}{1-2\eta+\eta(1+\gamma)^{-1}y_t^2}(1+\gamma)^{-1/2}y_t(1+\gamma)^{-1/2}(1+\gamma)^{-1}y_{t-1},$$

as required.

PROPOSITION 4

We start again by differentiating the prediction equations, which yield $d\boldsymbol{\mu}_t(\boldsymbol{\theta})=d\boldsymbol{\pi}$ and

$$d\boldsymbol{\Sigma}_t(\boldsymbol{\theta})=(d\mathbf{c})\lambda_t(\theta)\mathbf{c}+\mathbf{c}[d\lambda_t(\boldsymbol{\theta})]\mathbf{c}'+\mathbf{c}\lambda_t(\boldsymbol{\theta})d\mathbf{c}'+d\boldsymbol{\Gamma}_t(\boldsymbol{\theta}),$$

whence

$$\frac{\partial\boldsymbol{\mu}_t(\boldsymbol{\theta})}{\partial\boldsymbol{\theta}'}=\frac{\partial\boldsymbol{\pi}}{\partial\boldsymbol{\theta}'}$$

and

$$\frac{\partial vec[\boldsymbol{\Sigma}_t(\boldsymbol{\theta})]}{\partial\theta'}=(\mathbf{I}_{N^2}+\mathbf{K}_{NN})[\mathbf{c}\lambda_{t|t-1}(\boldsymbol{\theta})\otimes\mathbf{I}_N]\frac{\partial\mathbf{c}}{\partial\boldsymbol{\theta}'}+(\mathbf{c}\otimes\mathbf{c})\frac{\partial\lambda_t(\boldsymbol{\theta})}{\partial\theta'}+\mathbf{E}_N\frac{\partial\boldsymbol{\gamma}_t(\boldsymbol{\theta})}{\partial\boldsymbol{\theta}'}.$$

But since

$$\lambda_t(\boldsymbol{\theta})=1+\alpha[E(f_{t-1}^2|Y_{t-1};\boldsymbol{\theta},0)-1],$$

$$\gamma_{it}(\boldsymbol{\theta})=\gamma_i+\alpha_i^*[E(v_{it-1}^2|Y_{t-1};\boldsymbol{\theta},0)-\gamma_i],$$

we will have that:

$$\frac{\partial\lambda_t(\boldsymbol{\theta})}{\partial\boldsymbol{\theta}} = \alpha\frac{\partial E(f_{t-1}^2|Y_{t-1};\boldsymbol{\theta},0)}{\partial\boldsymbol{\theta}} + \frac{\partial\alpha}{\partial\boldsymbol{\theta}}[E(f_{t-1}^2|Y_{t-1};\boldsymbol{\theta},0)-1],$$

$$\frac{\partial\gamma_{it}(\boldsymbol{\theta})}{\partial\boldsymbol{\theta}} = \frac{\partial\gamma_i}{\partial\boldsymbol{\theta}} + \alpha_i^*\frac{\partial E(v_{it-1}^2|Y_{t-1};\boldsymbol{\theta},0)}{\partial\boldsymbol{\theta}} + \frac{\partial\alpha_i^*}{\partial\theta}[E(v_{it-1}^2|Y_{t-1};\boldsymbol{\theta},0)-\gamma_i].$$

This implies that under the null hypothesis of $\boldsymbol{\alpha}^\dagger = \mathbf{0}$,

$$\frac{\partial\lambda_t(\boldsymbol{\theta}_s,0)}{\partial\boldsymbol{\theta}} = \frac{\partial\alpha}{\partial\boldsymbol{\theta}}[f_{kt-1}^2(\boldsymbol{\theta}_s)+\omega_k(\boldsymbol{\theta}_s)-1],$$

$$\frac{\partial\gamma_{it}(\boldsymbol{\theta}_s,0)}{\partial\boldsymbol{\theta}} = \frac{\partial\gamma_i}{\partial\boldsymbol{\theta}} + \frac{\partial\alpha_i^*}{\partial\boldsymbol{\theta}}[v_{kit-1}^2(\boldsymbol{\theta}_s)+c_i^2\omega_k(\boldsymbol{\theta}_s)-\gamma_i],$$

where we have used the fact that $\boldsymbol{\Sigma}_t(\boldsymbol{\theta}_s,0) = \mathbf{cc}'+\boldsymbol{\Gamma} = \boldsymbol{\Sigma}(\theta_s)\ \forall t$.

As a result,

$$\mathbf{Z}_{dt}(\boldsymbol{\theta}_s,0) = \begin{bmatrix} \boldsymbol{\Sigma}^{-1/2\prime}(\boldsymbol{\theta}_s) & \mathbf{0} \\ \mathbf{0} & \frac{1}{2}(\mathbf{c}'\otimes\mathbf{I}_N)(\mathbf{I}_{N^2}+\mathbf{K}_{NN})[\boldsymbol{\Sigma}^{-1/2\prime}(\boldsymbol{\theta}_s)\otimes\boldsymbol{\Sigma}^{-1/2}(\boldsymbol{\theta}_s)] \\ \mathbf{0} & \frac{1}{2}\mathbf{E}_N'[\boldsymbol{\Sigma}^{-1/2\prime}(\boldsymbol{\theta}_s)\otimes\boldsymbol{\Sigma}^{-1/2}(\boldsymbol{\theta}_s)] \\ \mathbf{0} & \frac{1}{2}[f_{kt-1}^2(\boldsymbol{\theta}_s)+\omega_k(\boldsymbol{\theta}_s)-1][\mathbf{c}'\boldsymbol{\Sigma}^{-1/2\prime}(\boldsymbol{\theta}_s)\otimes\mathbf{c}'\boldsymbol{\Sigma}^{-1/2}(\boldsymbol{\theta}_s)] \\ \mathbf{0} & \frac{1}{2}dg[\mathbf{v}_{kt-1}(\boldsymbol{\theta}_s)\mathbf{v}_{kt-1}'(\boldsymbol{\theta}_s)+\mathbf{cc}'\omega_k(\boldsymbol{\theta}_s)-\boldsymbol{\Gamma}]\mathbf{E}_N'[\boldsymbol{\Sigma}^{-1/2\prime}(\boldsymbol{\theta}_s)\otimes\boldsymbol{\Sigma}^{-1/2}(\boldsymbol{\theta}_s)] \end{bmatrix},$$

whence it is easy to see that

$$\mathbf{Z}_d(\boldsymbol{\phi}) = \begin{bmatrix} \boldsymbol{\Sigma}^{-1/2\prime}(\boldsymbol{\theta}_s) & \mathbf{0} \\ \mathbf{0} & \frac{1}{2}(\mathbf{c}'\otimes\mathbf{I}_N)(\mathbf{I}_{N^2}+\mathbf{K}_{NN})[\boldsymbol{\Sigma}^{-1/2\prime}(\boldsymbol{\theta}_s)\otimes\boldsymbol{\Sigma}^{-1/2}(\boldsymbol{\theta}_s)] \\ \mathbf{0} & \frac{1}{2}\mathbf{E}_N'[\boldsymbol{\Sigma}^{-1/2\prime}(\boldsymbol{\theta}_s)\otimes\boldsymbol{\Sigma}^{-1/2}(\boldsymbol{\theta}_s)] \\ \mathbf{0} & \mathbf{0} \\ \mathbf{0} & \mathbf{0} \end{bmatrix}$$

and

$$\mathbf{W}_d(\boldsymbol{\phi}) = \left[\mathbf{0}\ \mathbf{c}'\boldsymbol{\Sigma}^{-1}(\boldsymbol{\theta}_s)\ \tfrac{1}{2}vecd'[\boldsymbol{\Sigma}^{-1}(\boldsymbol{\theta}_s)]\ \mathbf{0}\ \mathbf{0}\right]', \tag{8.50}$$

where we have used the fact that

$$\left.\begin{aligned} E[f_{kt-1}^2(\boldsymbol{\theta}_s)+\omega_k(\boldsymbol{\theta}_s)-1|\boldsymbol{\theta}_s,0] &= 0 \\ E[v_{kit-1}^2(\boldsymbol{\theta}_s)+c_i^2\omega_k(\boldsymbol{\theta}_s)-\gamma_i|\boldsymbol{\theta}_s,0] &= 0 \end{aligned}\right\} \tag{8.51}$$

irrespective of the true distribution of $\mathbf{y}_t$.

In addition, it follows that the elliptical score under the null will be:

$$\begin{bmatrix} s_{\pi t}(\boldsymbol{\theta}_s,0,\boldsymbol{\eta}) \\ s_{ct}(\boldsymbol{\theta}_s,0,\boldsymbol{\eta}) \\ s_{\gamma t}(\boldsymbol{\theta}_s,0,\boldsymbol{\eta}) \\ s_{\alpha t}(\boldsymbol{\theta}_s,0,\boldsymbol{\eta}) \\ \\ s_{\alpha^* t}(\boldsymbol{\theta}_s,0,\boldsymbol{\eta}) \end{bmatrix} = \begin{bmatrix} \delta[\varsigma_t(\boldsymbol{\theta}_s),\boldsymbol{\eta}]\boldsymbol{\Sigma}^{-1}(\boldsymbol{\theta}_s)(\mathbf{y}_t-\boldsymbol{\pi}) \\ \delta[\varsigma_t(\boldsymbol{\theta}_s),\boldsymbol{\eta}]\boldsymbol{\Sigma}^{-1}(\boldsymbol{\theta}_s)(\mathbf{y}_t-\pi)(\mathbf{y}_t-\boldsymbol{\pi})'\Sigma^{-1}(\boldsymbol{\theta}_s)\mathbf{c}-\boldsymbol{\Sigma}^{-1}(\theta_s)\mathbf{c} \\ \frac{1}{2}vecd[\delta[\varsigma_t(\boldsymbol{\theta}_s),\boldsymbol{\eta}]\boldsymbol{\Sigma}^{-1}(\boldsymbol{\theta}_s)(\mathbf{y}_t-\boldsymbol{\pi})(\mathbf{y}_t-\boldsymbol{\pi})'\boldsymbol{\Sigma}^{-1}(\boldsymbol{\theta}_s)-\boldsymbol{\Sigma}^{-1}(\theta_s)] \\ \frac{1}{2}[f_{kt-1}^2(\boldsymbol{\theta}_s)+\omega_k(\boldsymbol{\theta}_s)-1] \\ \{\delta[\varsigma_t(\boldsymbol{\theta}_s),\boldsymbol{\eta}]\mathbf{c}'\boldsymbol{\Sigma}^{-1}(\boldsymbol{\theta})(\mathbf{y}_t-\boldsymbol{\pi})(\mathbf{y}_t-\boldsymbol{\pi})'\boldsymbol{\Sigma}^{-1}(\boldsymbol{\theta}_s)\mathbf{c}-\mathbf{c}'\boldsymbol{\Sigma}^{-1}(\boldsymbol{\theta}_s)\mathbf{c}\} \\ \frac{1}{2}dg[\mathbf{v}_{kt-1}(\boldsymbol{\theta}_s)\mathbf{v}_{kt-1}'(\boldsymbol{\theta}_s)+\mathbf{cc}'\omega_k(\boldsymbol{\theta}_s)-\boldsymbol{\Gamma}] \\ \times vecd\{\delta[\varsigma_t(\boldsymbol{\theta}_s),\boldsymbol{\eta}]\boldsymbol{\Sigma}^{-1}(\boldsymbol{\theta}_s)(\mathbf{y}_t-\boldsymbol{\pi})(\mathbf{y}_t-\boldsymbol{\pi})'\boldsymbol{\Sigma}^{-1}(\boldsymbol{\theta}_s)-\boldsymbol{\Sigma}^{-1}(\boldsymbol{\theta}_s)\} \end{bmatrix}.$$

Therefore, the only difference relative to the static factor model are the scores $s_{\alpha t}(\boldsymbol{\theta}_s,0,\boldsymbol{\eta})$ and $s_{\alpha^* t}(\boldsymbol{\theta}_s,0,\boldsymbol{\eta})$. In this sense, if we assume that $\boldsymbol{\Gamma} > \mathbf{0}$ we can use the

Woodbury formula to show that

$$\begin{aligned}&\delta[\varsigma_t(\boldsymbol{\theta}_s);\eta]\mathbf{c}'\boldsymbol{\Sigma}^{-1}(\boldsymbol{\theta}_s)(\mathbf{y}_t-\boldsymbol{\pi})(\mathbf{y}_t-\boldsymbol{\pi})'\boldsymbol{\Sigma}^{-1}(\boldsymbol{\theta}_s)\mathbf{c}-\mathbf{c}'\boldsymbol{\Sigma}^{-1}(\boldsymbol{\theta}_s)\mathbf{c}\\ &=\delta[\varsigma_t(\boldsymbol{\theta}_s);\eta]f_{kt}^2(\boldsymbol{\theta}_s)+\omega_k(\boldsymbol{\theta}_s)-1,\end{aligned}$$

so that

$$\begin{bmatrix} s_{\alpha t}(\boldsymbol{\theta}_s,0,\eta)\\ s_{\alpha^* t}(\boldsymbol{\theta}_s,0,\eta)\end{bmatrix}=\begin{bmatrix}\frac{1}{2}[f_{kt-1}^2(\boldsymbol{\theta}_s)+\omega_k(\boldsymbol{\theta}_s)-1]\{\delta[\varsigma_t(\boldsymbol{\theta}_s),\eta]f_{kt}^2(\boldsymbol{\theta}_s)+\omega_k(\boldsymbol{\theta}_s)-1\}\\ \frac{1}{2}dg[\mathbf{v}_{kt-1}(\boldsymbol{\theta}_s)\mathbf{v}_{kt-1}'(\boldsymbol{\theta}_s)+\mathbf{c}\mathbf{c}'\omega_k(\boldsymbol{\theta}_s)-\boldsymbol{\Gamma}]\\ \times vecd\{\boldsymbol{\Gamma}^{-1}[\delta[\varsigma_t(\theta_s),\eta]\mathbf{v}_{kt}(\boldsymbol{\theta}_s)\mathbf{v}_{kt}'(\boldsymbol{\theta}_s)+\mathbf{c}\mathbf{c}'\omega_k(\boldsymbol{\theta}_s)-\boldsymbol{\Gamma}]\boldsymbol{\Gamma}^{-1}\}\end{bmatrix}.$$

Using the expression for $\mathbf{Z}_{dt}(\boldsymbol{\theta}_s,\mathbf{0})$, together with (8.51), it is easy to show that the unconditional information matrix $\mathcal{I}_{\theta\theta}(\boldsymbol{\theta},\mathbf{0},\eta)$ will be block diagonal between $\boldsymbol{\pi}$, $(\mathbf{c},\gamma,\eta)$, and $\boldsymbol{\alpha}^\dagger$, with the first two blocks as in the static case. Consequently, in computing our ML-based tests we can safely ignore the sampling uncertainty in estimating $\boldsymbol{\theta}_s$ and $\boldsymbol{\eta}$. In addition, we can write

$$\begin{aligned}\mathcal{I}_{\alpha^\dagger\alpha^\dagger t}(\boldsymbol{\theta},\mathbf{0},\boldsymbol{\eta})&=diag\begin{bmatrix}\frac{1}{\sqrt{2}}[f_{kt-1}^2(\boldsymbol{\theta}_s)+\omega_k(\theta_s)-1]\\ \frac{1}{\sqrt{2}}\boldsymbol{\Gamma}^{-1}vecd[\mathbf{v}_{kt-1}(\theta_s)\mathbf{v}_{kt-1}'(\boldsymbol{\theta}_s)+\mathbf{c}\mathbf{c}'\omega_k(\boldsymbol{\theta}_s)-\boldsymbol{\Gamma}]\end{bmatrix}\\ &\times\mathcal{V}_{\boldsymbol{\alpha}^\dagger\boldsymbol{\alpha}^\dagger}(\boldsymbol{\theta}_s,\eta;\eta)\\ &\times diag\begin{bmatrix}\frac{1}{\sqrt{2}}[f_{kt-1}^2(\boldsymbol{\theta}_s)+\omega_k(\theta_s)-1]\\ \frac{1}{\sqrt{2}}\boldsymbol{\Gamma}^{-1}vecd[\mathbf{v}_{kt-1}(\theta_s)\mathbf{v}_{kt-1}'(\boldsymbol{\theta}_s)+\mathbf{c}\mathbf{c}'\omega_k(\boldsymbol{\theta}_s)-\boldsymbol{\Gamma}]\end{bmatrix}\end{aligned}$$

where

$$\mathcal{V}_{\boldsymbol{\alpha}^\dagger\boldsymbol{\alpha}^\dagger}(\theta_s,\eta;\eta)=V\begin{bmatrix}\frac{1}{\sqrt{2}}\{\delta[\varsigma_t(\boldsymbol{\theta}_s),\eta]f_{kt}^2(\boldsymbol{\theta}_s)+\omega_k(\boldsymbol{\theta}_s)-1\}\\ \frac{1}{\sqrt{2}}\boldsymbol{\Gamma}^{-1}vecd\{\delta[\varsigma_t(\boldsymbol{\theta}_s),\eta]\mathbf{v}_{kt}(\boldsymbol{\theta}_s)\mathbf{v}_{kt}'(\boldsymbol{\theta}_s)+\mathbf{c}\mathbf{c}'\omega_k(\boldsymbol{\theta}_s)-\boldsymbol{\Gamma}\}\end{bmatrix}$$

$$=\mathrm{M}_{ss}(\boldsymbol{\eta})\begin{bmatrix}[\mathbf{c}'\boldsymbol{\Sigma}^{-1}(\boldsymbol{\theta}_s)\mathbf{c}]^2 & \mathbf{c}'\boldsymbol{\Sigma}^{-1}(\boldsymbol{\theta}_s)\boldsymbol{\Gamma}^{1/2}\odot\mathbf{c}'\boldsymbol{\Sigma}^{-1}(\boldsymbol{\theta}_s)\boldsymbol{\Gamma}^{1/2}\\ \boldsymbol{\Gamma}^{1/2}\boldsymbol{\Sigma}^{-1}(\boldsymbol{\theta}_s)\mathbf{c}\odot\boldsymbol{\Gamma}^{1/2}\boldsymbol{\Sigma}^{-1}(\boldsymbol{\theta}_s)\mathbf{c} & \boldsymbol{\Gamma}^{1/2}\boldsymbol{\Sigma}^{-1}(\boldsymbol{\theta}_s)\boldsymbol{\Gamma}^{1/2}\odot\boldsymbol{\Gamma}^{1/2}\boldsymbol{\Sigma}^{-1}(\theta_s)\boldsymbol{\Gamma}^{1/2}\end{bmatrix}\tag{8.52}$$

$$+\frac{[\mathrm{M}_{ss}(\boldsymbol{\eta})-1]}{2}\begin{bmatrix}[\mathbf{c}'\boldsymbol{\Sigma}^{-1}(\boldsymbol{\theta}_s)\mathbf{c}]^2 & [\mathbf{c}'\boldsymbol{\Sigma}^{-1}(\boldsymbol{\theta}_s)\mathbf{c}]vecd'[\boldsymbol{\Gamma}^{1/2}\boldsymbol{\Sigma}^{-1}(\boldsymbol{\theta}_s)\boldsymbol{\Gamma}^{1/2}]\\ [\mathbf{c}\boldsymbol{\Sigma}^{-1}(\boldsymbol{\theta}_s)\mathbf{c}]vecd[\boldsymbol{\Gamma}^{1/2}\boldsymbol{\Sigma}^{-1}(\boldsymbol{\theta}_s)\boldsymbol{\Gamma}^{1/2}] & vecd[\boldsymbol{\Gamma}^{1/2}\boldsymbol{\Sigma}^{-1}(\boldsymbol{\theta}_s)\boldsymbol{\Gamma}^{1/2}]vecd'[\boldsymbol{\Gamma}^{1/2}\boldsymbol{\Sigma}^{-1}(\boldsymbol{\theta}_s)\boldsymbol{\Gamma}^{1/2}]\end{bmatrix}\tag{8.53}$$

Thus, the only remaining item is the calculation of fourth order terms appearing in $\mathcal{V}_{\boldsymbol{\alpha}^\dagger\boldsymbol{\alpha}^\dagger}(\boldsymbol{\theta}_s,0;\eta)$. But if we write

$$f_{kt}^2(\boldsymbol{\theta}_s)+\omega_k(\boldsymbol{\theta}_s)-1=\mathbf{c}'\boldsymbol{\Sigma}^{-1}(\boldsymbol{\theta}_s)(\mathbf{y}_t-\boldsymbol{\pi})(\mathbf{y}_t-\boldsymbol{\pi})'\boldsymbol{\Sigma}^{-1}(\boldsymbol{\theta}_s)\mathbf{c}-[1-\omega_k(\boldsymbol{\theta}_s)],$$

then it is easy to see that

$$\begin{aligned}&E[f_{kt}^2(\boldsymbol{\theta}_s)+\omega_k(\boldsymbol{\theta}_s)-1]^2\\ &=E\{vec[\mathbf{c}'\boldsymbol{\Sigma}^{-1}(\boldsymbol{\theta}_s)(\mathbf{y}_t-\boldsymbol{\pi})(\mathbf{y}_t-\boldsymbol{\pi})'\boldsymbol{\Sigma}^{-1}(\boldsymbol{\theta}_s)\mathbf{c}]vec'[\mathbf{c}'\boldsymbol{\Sigma}^{-1}(\boldsymbol{\theta}_s)(\mathbf{y}_t-\boldsymbol{\pi})(\mathbf{y}_t-\boldsymbol{\pi})'\boldsymbol{\Sigma}^{-1}(\boldsymbol{\theta}_s)\mathbf{c}]\}\\ &\quad-[1-\omega_k(\boldsymbol{\theta}_s)]^2\\ &=[\mathbf{c}'\boldsymbol{\Sigma}^{-1/2\prime}(\boldsymbol{\theta}_s)\otimes\mathbf{c}'\boldsymbol{\Sigma}^{-1/2\prime}(\theta_s)]E[vec(\boldsymbol{\varepsilon}_t^*\boldsymbol{\varepsilon}_t^{*\prime})vec'(\boldsymbol{\varepsilon}_t^*\varepsilon_t^{*\prime})][\boldsymbol{\Sigma}^{-1/2}(\boldsymbol{\theta}_s)\mathbf{c}\otimes\boldsymbol{\Sigma}^{-1/2}(\boldsymbol{\theta}_s)\mathbf{c}]\\ &\quad-[1-\omega_k(\boldsymbol{\theta}_s)]^2\end{aligned}$$

$$= [c'\boldsymbol{\Sigma}^{-1/2\prime}(\boldsymbol{\theta}_s) \otimes c'\boldsymbol{\Sigma}^{-1/2\prime}(\theta_s)](\kappa+1)[(\mathbf{I}_{N^2}+\mathbf{K}_{NN}) + vec(\mathbf{I}_N)\, vec'(\mathbf{I}_N)]$$
$$[\boldsymbol{\Sigma}^{-1/2}(\boldsymbol{\theta}_s)c \otimes \boldsymbol{\Sigma}^{-1/2}(\boldsymbol{\theta}_s)c] - [1-\omega_k(\theta_s)]^2$$
$$= (\kappa+1)\{2[c'\boldsymbol{\Sigma}^{-1}(\boldsymbol{\theta}_s)c]^2 + [c'\boldsymbol{\Sigma}^{-1}(\theta_s)c]^2\} - [c'\boldsymbol{\Sigma}^{-1}(\boldsymbol{\theta}_s)c]^2 = (3\kappa+2)[c'\boldsymbol{\Sigma}^{-1}(\boldsymbol{\theta}_s)c]^2.$$

Similarly, since

$$vecd[\mathbf{v}_{kt}(\boldsymbol{\theta}_s)\mathbf{v}'_{kt}(\boldsymbol{\theta}_s)+cc'\omega_k(\boldsymbol{\theta}_s) - \boldsymbol{\Gamma}]$$
$$= \mathbf{E}'_N\{vec[\boldsymbol{\Gamma}\boldsymbol{\Sigma}^{-1}(\theta_s)(\mathbf{y}_t-\boldsymbol{\pi})(\mathbf{y}_t-\boldsymbol{\pi})'\boldsymbol{\Sigma}^{-1}(\boldsymbol{\theta}_s)\boldsymbol{\Gamma}] - vec[\boldsymbol{\Gamma} - cc'\omega_k(\boldsymbol{\theta}_s)]\},$$

we will have that

$$E\{vecd[\mathbf{v}_{kt}(\boldsymbol{\theta}_s)\mathbf{v}'_{kt}(\boldsymbol{\theta}_s)+cc'\omega_k(\boldsymbol{\theta}_s) - \boldsymbol{\Gamma}]vecd'[\mathbf{v}_{kt}(\boldsymbol{\theta}_s)\mathbf{v}'_{kt}(\boldsymbol{\theta}_s)+cc'\omega_k(\boldsymbol{\theta}_s) - \boldsymbol{\Gamma}]\}$$
$$= \mathbf{E}'_N E\{vec[\boldsymbol{\Gamma}\boldsymbol{\Sigma}^{-1}(\theta_s)(\mathbf{y}_t-\boldsymbol{\pi})(\mathbf{y}_t-\boldsymbol{\pi})'\Sigma^{-1}(\boldsymbol{\theta}_s)\boldsymbol{\Gamma}]vec'[\boldsymbol{\Gamma}\boldsymbol{\Sigma}^{-1}(\theta_s)(\mathbf{y}_t-\boldsymbol{\pi})(\mathbf{y}_t-\boldsymbol{\pi})'\boldsymbol{\Sigma}^{-1}(\boldsymbol{\theta}_s)\boldsymbol{\Gamma}]\}\mathbf{E}_N$$
$$-vecd[\boldsymbol{\Gamma} - cc'\omega_k(\boldsymbol{\theta}_s)]vecd'[\boldsymbol{\Gamma} - cc'\omega_k(\boldsymbol{\theta}_s)]$$
$$= \mathbf{E}'_N[\boldsymbol{\Gamma}\boldsymbol{\Sigma}^{-1/2\prime}(\theta_s) \otimes \boldsymbol{\Gamma}\boldsymbol{\Sigma}^{-1/2\prime}(\boldsymbol{\theta}_s)]E[vec(\boldsymbol{\varepsilon}_t^*\boldsymbol{\varepsilon}_t^{*\prime})vec'(\boldsymbol{\varepsilon}_t^*\boldsymbol{\varepsilon}_t^{*\prime})][\boldsymbol{\Sigma}^{-1/2}(\boldsymbol{\theta}_s)\boldsymbol{\Gamma} \otimes \boldsymbol{\Sigma}^{-1/2}(\boldsymbol{\theta}_s)\boldsymbol{\Gamma}]$$
$$-vecd[\boldsymbol{\Gamma} - cc'\omega_k(\boldsymbol{\theta}_s)]vecd'[\boldsymbol{\Gamma} - cc'\omega_k(\boldsymbol{\theta}_s)]$$
$$= \mathbf{E}'_N[\boldsymbol{\Gamma}\boldsymbol{\Sigma}^{-1/2\prime}(\theta_s) \otimes \boldsymbol{\Gamma}\boldsymbol{\Sigma}^{-1/2\prime}(\boldsymbol{\theta}_s)](\kappa+1)[(\mathbf{I}_{N^2}+\mathbf{K}_{NN}) + vec(\mathbf{I}_N)\, vec'(\mathbf{I}_N)]$$
$$\times[\boldsymbol{\Sigma}^{-1/2}(\boldsymbol{\theta}_s)\boldsymbol{\Gamma} \otimes \boldsymbol{\Sigma}^{-1/2}(\boldsymbol{\theta}_s)\boldsymbol{\Gamma}] - vecd[\boldsymbol{\Gamma} - cc'\omega_k(\boldsymbol{\theta}_s)]vecd'[\boldsymbol{\Gamma} - cc'\omega_k(\boldsymbol{\theta}_s)]$$
$$= (\kappa+1)\{2[\boldsymbol{\Gamma}\boldsymbol{\Sigma}^{-1}(\boldsymbol{\theta}_s)\boldsymbol{\Gamma} \odot \boldsymbol{\Gamma}\boldsymbol{\Sigma}^{-1}(\boldsymbol{\theta})\boldsymbol{\Gamma}] + vecd[\boldsymbol{\Gamma}\boldsymbol{\Sigma}^{-1}(\boldsymbol{\theta}_s)\boldsymbol{\Gamma}]vecd'[\boldsymbol{\Gamma}\boldsymbol{\Sigma}^{-1}(\boldsymbol{\theta}_s)\boldsymbol{\Gamma}]\}$$
$$-vecd[\boldsymbol{\Gamma} - cc'\omega_k(\boldsymbol{\theta}_s)]vecd[\boldsymbol{\Gamma} - cc'\omega_k(\boldsymbol{\theta}_s)]\mathbf{E}_N$$
$$= 2(\kappa+1)[\boldsymbol{\Gamma}\boldsymbol{\Sigma}^{-1}(\boldsymbol{\theta}_s)\boldsymbol{\Gamma} \odot \boldsymbol{\Gamma}\boldsymbol{\Sigma}^{-1}(\boldsymbol{\theta}_s)\boldsymbol{\Gamma}] + \kappa vecd[\boldsymbol{\Gamma}\boldsymbol{\Sigma}^{-1}(\boldsymbol{\theta}_s)\boldsymbol{\Gamma}]vecd'[\boldsymbol{\Gamma}\boldsymbol{\Sigma}^{-1}(\boldsymbol{\theta}_s)\boldsymbol{\Gamma}]\}.$$

Finally,

$$E\{vecd[\mathbf{v}_{kt}(\boldsymbol{\theta}_s)\mathbf{v}'_{kt}(\boldsymbol{\theta}_s)+cc'\omega_k(\boldsymbol{\theta}_s) - \boldsymbol{\Gamma}][f_{kt}^2(\boldsymbol{\theta}_s) + \omega_k(\theta_s) - 1]\}$$
$$= \mathbf{E}'_N E\{vec[\boldsymbol{\Gamma}\boldsymbol{\Sigma}^{-1}(\theta_s)(\mathbf{y}_t-\boldsymbol{\pi})(\mathbf{y}_t-\boldsymbol{\pi})'\Sigma^{-1}(\boldsymbol{\theta}_s)\boldsymbol{\Gamma}]vec'[c'\Sigma^{-1}(\boldsymbol{\theta}_s)(\mathbf{y}_t-\boldsymbol{\pi})(\mathbf{y}_t-\boldsymbol{\pi})'\boldsymbol{\Sigma}^{-1}(\boldsymbol{\theta}_s)c]\}$$
$$-vecd[\boldsymbol{\Gamma} - cc'\omega_k(\boldsymbol{\theta}_s)][1-\omega_k(\boldsymbol{\theta}_s)]$$
$$= \mathbf{E}'_N[\boldsymbol{\Gamma}\boldsymbol{\Sigma}^{-1/2\prime}(\theta_s) \otimes \boldsymbol{\Gamma}\boldsymbol{\Sigma}^{-1/2\prime}(\boldsymbol{\theta}_s)]E[vec(\boldsymbol{\varepsilon}_t^*\boldsymbol{\varepsilon}_t^{*\prime})vec'(\boldsymbol{\varepsilon}_t^*\boldsymbol{\varepsilon}_t^{*\prime})][\boldsymbol{\Sigma}^{-1/2}(\boldsymbol{\theta}_s)c \otimes \boldsymbol{\Sigma}^{-1/2}(\boldsymbol{\theta}_s)c]$$
$$-vecd[\boldsymbol{\Gamma} - cc'\omega_k(\boldsymbol{\theta}_s)][1-\omega_k(\boldsymbol{\theta}_s)]$$
$$= \mathbf{E}'_N[\boldsymbol{\Gamma}\boldsymbol{\Sigma}^{-1/2\prime}(\theta_s) \otimes \boldsymbol{\Gamma}\boldsymbol{\Sigma}^{-1/2\prime}(\boldsymbol{\theta}_s)](\kappa+1)[(\mathbf{I}_{N^2}+\mathbf{K}_{NN}) + vec(\mathbf{I}_N)\, vec'(\mathbf{I}_N)]$$
$$\times[\boldsymbol{\Sigma}^{-1/2}(\boldsymbol{\theta}_s)c \otimes \boldsymbol{\Sigma}^{-1/2}(\boldsymbol{\theta}_s)c] - vecd[\boldsymbol{\Gamma} - cc'\omega_k(\boldsymbol{\theta}_s)][1-\omega_k(\boldsymbol{\theta}_s)]$$
$$= 2(\kappa+1)[\boldsymbol{\Gamma}\boldsymbol{\Sigma}^{-1}(\boldsymbol{\theta}_s)c \odot \boldsymbol{\Gamma}\boldsymbol{\Sigma}^{-1}(\boldsymbol{\theta}_s)c] + \kappa vecd[\boldsymbol{\Gamma}\boldsymbol{\Sigma}^{-1}(\boldsymbol{\theta}_s)\boldsymbol{\Gamma}][c'\boldsymbol{\Sigma}^{-1}(\boldsymbol{\theta}_s)c].$$

Therefore, $\mathcal{V}_{\alpha^\dagger\alpha^\dagger}(\boldsymbol{\theta}_s, 0; \eta)$ mimics $\mathcal{V}_{\alpha^\dagger\alpha^\dagger}(\boldsymbol{\theta}_s, \eta; \eta)$ if we replace $\mathrm{M}_{ss}(\boldsymbol{\eta})$ by $\kappa+1$.

In addition, it follows from (8.50) that the elliptically symmetric semiparametric scores for $\boldsymbol{\alpha}^\dagger$ coincide with the parametric ones, and that the elliptically symmetric semiparametric efficiency bound will be block diagonal between $\boldsymbol{\pi}$, $(\mathbf{c}, \boldsymbol{\gamma})$, and $\boldsymbol{\alpha}^\dagger$, where the first and last blocks coincide with the corresponding blocks of the information matrix, and the second one with the corresponding bound in the static factor model.

Finally, let us consider the tests based on the Gaussian PML scores $s_{\alpha t}(\boldsymbol{\theta}_s, 0, 0)$ and $s_{\alpha^* t}(\boldsymbol{\theta}_s, 0, 0)$ when $\mathbf{y}_t|I_{t-1};\boldsymbol{\phi}$ is i.i.d. $D(\boldsymbol{\pi}, \boldsymbol{\Sigma}(\boldsymbol{\theta}_s); \boldsymbol{\varrho})$ but not necessarily normal or elliptical. Once again, a robust Gaussian pseudo score test of the null hypothesis $H_0 : \boldsymbol{\alpha}_1^\dagger = \mathbf{0}$ can be computed as

$$\left[\frac{\sqrt{T}}{T}\sum_{t=1}^{T}\mathbf{s}'_{\alpha^\dagger t}(\tilde{\theta}_s, 0, 0)\right]\mathcal{A}^{\alpha^\dagger\alpha^\dagger}(\boldsymbol{\phi}_0)\,\mathcal{C}^{-1}_{\alpha^\dagger\alpha^\dagger}(\boldsymbol{\phi}_0)\,\mathcal{A}^{\alpha^\dagger\alpha^\dagger}(\boldsymbol{\phi}_0)\left[\frac{\sqrt{T}}{T}\sum_{t=1}^{T}\mathbf{s}'_{\alpha^\dagger t}(\tilde{\theta}_s, 0, 0)\right],$$

where $\mathbf{s}'_{\alpha^\dagger t}(\tilde{\theta}_s, 0, 0)$ is the Gaussian score evaluated at the restricted PML estimator $\tilde{\theta}_s$, $\mathcal{A}^{\alpha^\dagger\alpha^\dagger}(\boldsymbol{\phi}_0)$ is the relevant block of the inverse of the expected Hessian matrix $\mathcal{A}(\boldsymbol{\phi}) = -E\left[\mathbf{h}_{\theta\theta t}(\boldsymbol{\theta}, 0)|\phi\right]$ and $\mathcal{C}_{\alpha^\dagger\alpha^\dagger}(\boldsymbol{\phi}_0)$ is the corresponding block of the usual sandwich expression $\mathcal{C}(\boldsymbol{\phi}) = \mathcal{A}^{-1}(\boldsymbol{\phi})\mathcal{B}(\boldsymbol{\phi})\mathcal{A}^{-1}(\boldsymbol{\phi})$, with $\mathcal{B}(\boldsymbol{\phi}) = V\left[\mathbf{s}_{\theta t}(\boldsymbol{\theta}, 0)|\phi\right]$ (see, e.g., Engle (1984)). The structure of $\mathbf{Z}_{dt}(\boldsymbol{\theta})$, together with (8.51) and the fact that $\mathcal{A}_{\theta\theta t}(\boldsymbol{\phi})$ equals $\mathcal{I}_{\theta\theta t}(\boldsymbol{\theta}_s, 0, 0)$, implies that $\mathcal{A}(\boldsymbol{\phi})$ will be block diagonal between $(\alpha, \boldsymbol{\alpha}^*)$ and $(\boldsymbol{\pi}, \mathbf{c}, \boldsymbol{\gamma})$ irrespective of the true distribution of $\mathbf{y}_t$. In addition, it is easy to see that

$$\mathcal{A}_{\alpha^\dagger\alpha^\dagger}(\phi) = E[\mathcal{A}_{\alpha^\dagger\alpha^\dagger t}(\boldsymbol{\phi})|\boldsymbol{\theta}_s, 0; \varrho] = \mathcal{V}_{\alpha^\dagger\alpha^\dagger}(\boldsymbol{\theta}_s, 0; \varrho) \odot \mathcal{V}_{\alpha^\dagger\alpha^\dagger}(\boldsymbol{\theta}_s, 0; 0),$$

where

$$\mathcal{V}_{\alpha^\dagger\alpha^\dagger}(\theta_s, 0; \varrho) = V\left[\begin{array}{c}\frac{1}{\sqrt{2}}[f^2_{kt}(\boldsymbol{\theta}_s) + \omega_k(\theta_s) - 1] \\ \frac{1}{\sqrt{2}}\boldsymbol{\Gamma}^{-1}vecd[\mathbf{v}_{kt}(\boldsymbol{\theta}_s)\mathbf{v}'_{kt}(\boldsymbol{\theta}_s) + \mathbf{c}\mathbf{c}'\omega_k(\boldsymbol{\theta}_s) - \boldsymbol{\Gamma}]\end{array}\right|\boldsymbol{\theta}_s, 0; \varrho\right].$$

A closely related argument shows that $\mathcal{B}_t(\boldsymbol{\phi})$ will also be block diagonal between $(\alpha, \boldsymbol{\alpha}^*)$ and $(\boldsymbol{\pi}, \mathbf{c}, \boldsymbol{\gamma})$. Further, the stationarity of $\mathbf{y}_t$ implies that

$$\mathcal{B}_{\alpha^\dagger\alpha^\dagger}(\phi) = E[\mathcal{B}_{\alpha^\dagger\alpha^\dagger t}(\boldsymbol{\phi})|\boldsymbol{\theta}_s, 0; \varrho] = \mathcal{V}_{\alpha^\dagger\alpha^\dagger}(\boldsymbol{\theta}_s, 0; \varrho) \odot \mathcal{V}_{\alpha^\dagger\alpha^\dagger}(\boldsymbol{\theta}_s, 0; \varrho),$$

which is generally different from $\mathcal{A}_{\alpha^\dagger\alpha^\dagger}(\boldsymbol{\phi})$. As we have seen in (8.53) above, $\mathcal{V}_{\alpha^\dagger\alpha^\dagger}(\boldsymbol{\theta}_s, 0; \varrho)$ will simplify considerably when $\boldsymbol{\varepsilon}^*_t$ is spherical. In any case, the block diagonality of $\mathcal{A}(\boldsymbol{\phi})$ and $\mathcal{B}(\phi)$ implies that

$$\mathcal{A}^{\alpha^\dagger\alpha^\dagger}(\boldsymbol{\phi}_0)\,\mathcal{C}^{-1}_{\alpha^\dagger\alpha^\dagger}(\boldsymbol{\phi}_0)\,\mathcal{A}^{\alpha^\dagger\alpha^\dagger}(\boldsymbol{\phi}_0) = \mathcal{B}^{-1}_{\alpha^\dagger\alpha^\dagger t}(\boldsymbol{\phi}),$$

which proves the last part of the proposition.

PROPOSITION 5

For the sake of brevity, the proof will be developed for the following univariate model:

$$y_t = f_t + v_t,$$

$$\begin{pmatrix} f_t \\ v_t \end{pmatrix}|I_{t-1} \sim t\left\{\begin{pmatrix} 0 \\ 0 \end{pmatrix}, \begin{bmatrix} 1 + \alpha(f^2_{t-1} - 1) & 0 \\ 0 & \gamma \end{bmatrix}, \eta\right\},$$

where $\alpha \geq 0$ and $\gamma \geq 0$. Nevertheless, it can be tediously extended to cover the general case. Given that when $\alpha = 0$ the log-likelihood function of this model coincides with the log-likelihood function of the model considered in Section 8.2, we only need to look at the score associated to this parameter.

It is easy to see that the joint distribution of y_t and f_t give the past of both variables will be

$$\begin{pmatrix} y_t \\ f_t \end{pmatrix} |I_{t-1} \sim t\left[\begin{pmatrix} 0 \\ 0 \end{pmatrix}, \begin{pmatrix} 1+\alpha(f_{t-1}^2-1)+\gamma & 1+\alpha(f_{t-1}^2-1) \\ 1+\alpha(f_{t-1}^2-1) & 1+\alpha(f_{t-1}^2-1) \end{pmatrix}, \eta\right].$$

Hence, we can write down the joint log-likelihood as

$$c_2(\eta) - \frac{1}{2}\ln\gamma - \frac{1}{2}\ln[1+\alpha(f_{t-1}^2-1)] + g[\varsigma_t(\rho,\gamma);\eta],$$

where

$$c_2(\eta) = \ln\left[\Gamma\left(\frac{2\eta+1}{2\eta}\right)\right] - \ln\left[\Gamma\left(\frac{1}{2\eta}\right)\right] - \ln\left(\frac{1-2\eta}{\eta}\right) - \ln\pi$$

is the (log) constant of integration,

$$\gamma[1+\alpha(f_{t-1}^2-1)] = \left|\begin{pmatrix} 1+\alpha(f_{t-1}^2-1)+\gamma & 1+\alpha(f_{t-1}^2-1) \\ 1+\alpha(f_{t-1}^2-1) & 1+\alpha(f_{t-1}^2-1) \end{pmatrix}\right|$$

the Jacobian and

$$g[\varsigma_t(\alpha,\gamma);\eta] = -\left(\frac{2\eta+1}{2\eta}\right)\ln\left[1+\frac{\eta}{1-2\eta}\varsigma_t(\rho,\gamma)\right],$$

with

$$\begin{aligned}\varsigma_t(\alpha,\gamma) &= \begin{pmatrix} y_t & f_t \end{pmatrix}\begin{pmatrix} 1+\alpha(f_{t-1}^2-1)+\gamma & 1+\alpha(f_{t-1}^2-1) \\ 1+\alpha(f_{t-1}^2-1) & 1+\alpha(f_{t-1}^2-1) \end{pmatrix}^{-1}\begin{pmatrix} y_t \\ f_t \end{pmatrix} \\ &= \gamma^{-1}(y_t-f_t)^2 + [1+\alpha(f_{t-1}^2-1)]^{-1}f_t^2,\end{aligned}$$

the (log) kernel of the bivariate Student t density.

Given that we can write the standardized residuals as

$$\begin{aligned}&\begin{pmatrix} 1+\alpha(f_{t-1}^2-1)+\gamma & 1+\alpha(f_{t-1}^2-1) \\ 1+\alpha(f_{t-1}^2-1) & 1+\alpha(f_{t-1}^2-1) \end{pmatrix}^{-1/2}\begin{pmatrix} y_t \\ f_t \end{pmatrix} \\ &= \begin{pmatrix} \gamma^{-1/2} & -\gamma^{-1/2} \\ 0 & [1+\alpha(f_{t-1}^2-1)]^{-1/2} \end{pmatrix}\begin{pmatrix} y_t \\ f_t \end{pmatrix} \\ &= \begin{pmatrix} \gamma^{-1/2}(y_t-f_t) \\ [1+\alpha(f_{t-1}^2-1)]^{-1/2}f_t \end{pmatrix}\end{aligned}$$

and the gradient of the *vec* of the conditional covariance matrix with respect to α will be f_{t-1}^2-1 times the vector $(1,1,1,1)'$, we will have that the score of the joint log-likelihood function with respect to α will be given by

$$\frac{1}{2}(f_{t-1}^2-1)(1,1,1,1)\times$$

$$\left[\begin{pmatrix} \gamma^{-1/2} & 0 \\ -\gamma^{-1/2} & [1+\alpha(f_{t-1}^2-1)]^{-1/2} \end{pmatrix} \otimes \begin{pmatrix} \gamma^{-1/2} & 0 \\ -\gamma^{-1/2} & [1+\alpha(f_{t-1}^2-1)]^{-1/2} \end{pmatrix}\right]\times$$

$$vec\begin{pmatrix} \frac{2\eta+1}{1-2\eta+\eta\varsigma_t(\alpha,\gamma)}\frac{(y_t-f_t)^2}{\gamma}-1 & \frac{2\eta+1}{1-2\eta+\eta\varsigma_t(\alpha,\gamma)}\frac{(y_t-f_t)}{\sqrt{\gamma}}\frac{f_t}{\sqrt{1+\alpha(f_{t-1}^2-1)}} \\ \frac{2\eta+1}{1-2\eta+\eta\varsigma_t(\alpha,\gamma)}\frac{(y_t-f_t)}{\sqrt{\gamma}}\frac{f_t}{\sqrt{1+\alpha(f_{t-1}^2-1)}} & \frac{2\eta+1}{1-2\eta+\eta\varsigma_t(\alpha,\gamma)}\frac{f_t^2}{1+\alpha(f_{t-1}^2-1)}-1 \end{pmatrix}=$$

$$\frac{1}{2}\left(\frac{2\eta+1}{1-2\eta+\eta\varsigma_t(\alpha,\gamma)}\frac{f_t^2}{1+\alpha(f_{t-1}^2-1)}-1\right)\frac{f_{t-1}^2-1}{1+\alpha(f_{t-1}^2-1)},$$

where we have used the fact that

$$\begin{pmatrix}\gamma^{-1/2} & -\gamma^{-1/2} \\ 0 & [1+\alpha(f_{t-1}^2-1)]^{-1/2}\end{pmatrix}\begin{pmatrix}1 & 1\\ 1 & 1\end{pmatrix}\begin{pmatrix}\gamma^{-1/2} & 0 \\ -\gamma^{-1/2} & [1+\alpha(f_{t-1}^2-1)]^{-1/2}\end{pmatrix}$$

$$=\begin{pmatrix}0 & 0 \\ 0 & [1+\alpha(f_{t-1}^2-1)]^{-1}\end{pmatrix}.$$

The Kullback inequality implies that score of the marginal log-likelihood function of y_t with respect to α will be given by

$$\frac{1}{2}E\left[\left(\frac{2\eta+1}{1-2\eta+\eta\varsigma_t(\alpha,\gamma)}\frac{f_t^2}{1+\alpha(f_{t-1}^2-1)}-1\right)\frac{f_{t-1}^2-1}{1+\alpha(f_{t-1}^2-1)}\middle|\mathbf{Y}_T,\alpha\right].$$

This expected value becomes analytically tractable when $\alpha=0$. First of all, the expression inside the expectation simplifies to

$$E\left[\left(\frac{(2\eta+1)f_t^2}{1-2\eta+\eta\left[\gamma^{-1}(y_t-f_t)^2+f_t^2\right]}-1\right)(f_{t-1}^2-1)\middle|\mathbf{Y}_T,\alpha=0\right].$$

Second, the joint distribution of y_t and f_t is i.i.d. over time, which means that the expected value of this product should be equal to

$$E\left[\left(\frac{(2\eta+1)f_t^2}{1-2\eta+\eta\left[\gamma^{-1}(y_t-f_t)^2+f_t^2\right]}-1\right)\middle|y_t,\alpha=0\right]E\left[f_{t-1}^2-1\middle|y_{t-1},\alpha=0\right].$$

But since f_t given y_t has a Student t distribution with (conditional) mean, variance, and shape parameter given by (8.46), (8.47), and (8.48), respectively, the second term is simply given by

$$f_{kt-1}^2(\gamma)+\upsilon_{kt-1}(\gamma,\eta)-1.$$

The first term is trickier, as we need to find the expected value of

$$\frac{(2\eta+1)f_t^2}{1-2\eta+\eta\left[\gamma^{-1}(y_t-f_t)^2+f_t^2\right]}-1. \tag{8.54}$$

To do so, it is convenient to follow Fiorentini et al. (2003) and write f_t in terms of a conditionally standardized Student t component f_t^* as follows:

$$f_t=\frac{1}{1+\gamma}y_t+\sqrt{\frac{1-2\eta}{1-\eta}\left(1+\frac{\eta}{1-2\eta}\frac{y_t^2}{1+\gamma}\right)\frac{\gamma}{1+\gamma}}f_t^*,$$

$$f_t^* = \sqrt{\frac{1-\eta}{\eta}} \times \sqrt{\zeta_t/\xi_t} u_t,$$

where u_t is either 1 or -1 with probability 1/2, ζ_t is a chi-square random variable with 1 degree of freedom and ξ_t is a gamma random variable with mean $1+\eta^{-1}$ and variance $2(1+\eta^{-1})$, with u_t, ζ_t, and ξ_t mutually independent and independent of y_t and I_{t-1}.

This decomposition allows us to express

$$\varsigma_t(0,\gamma) = \gamma^{-1}(y_t - f_t)^2 + f_t^2 = \frac{y_t^2}{1+\gamma} + \left(\frac{1+\gamma}{\gamma}\right)\left(f_t - \frac{1}{1+\gamma}y_t\right)^2$$
$$= \frac{y_t^2}{1+\gamma} + \frac{1-2\eta}{\eta}\left(1 + \frac{\eta}{1-2\eta}\frac{y_t^2}{1+\gamma}\right)\frac{\zeta_t}{\xi_t}$$
$$= \frac{y_t^2}{1+\gamma}\left(1+\frac{\zeta_t}{\xi_t}\right) + \frac{1-2\eta}{\eta}\frac{\zeta_t}{\xi_t},$$

so that the denominator of (8.49) can be written as

$$1 - 2\eta + \eta\varsigma_t(0,\gamma) = 1 - 2\eta + \eta\frac{y_t^2}{1+\gamma}\left(1+\frac{\zeta_t}{\xi_t}\right) + (1-2\eta)\frac{\zeta_t}{\xi_t} =$$

$$\left(1 - 2\eta + \frac{\eta y_t^2}{1+\gamma}\right)\left(\frac{\xi_t+\zeta_t}{\xi_t}\right) = (1-2\eta)\left(1 + \frac{\eta}{1-2\eta}\frac{y_t^2}{1+\gamma}\right)\left(\frac{\xi_t+\zeta_t}{\xi_t}\right).$$

As for the numerator, we are left with $2\eta + 1$ times

$$f_t^2 = \frac{1}{(1+\gamma)^2}y_t^2 + \frac{1-2\eta}{\eta}\left(1 + \frac{\eta}{1-2\eta}\frac{y_t^2}{1+\gamma}\right)\frac{\gamma}{1+\gamma}\frac{\zeta_t}{\xi_t}$$
$$+ \frac{2}{1+\gamma}y_t\sqrt{\frac{1-2\eta}{\eta}\left(1 + \frac{\eta}{1-2\eta}\frac{y_t^2}{1+\gamma}\right)\frac{\gamma}{1+\gamma}}\sqrt{\frac{\zeta_t}{\xi_t}}u_t.$$

Therefore, we can re-write (8.54) as -1 plus $2\eta + 1$ times

$$\frac{\frac{1}{(1+\gamma)^2}y_t^2}{(1-2\eta)\left(1+\frac{\eta}{1-2\eta}\frac{y_t^2}{1+\gamma}\right)}\left(\frac{\xi_t}{\xi_t+\zeta_t}\right) + \frac{\gamma}{\eta(1+\gamma)}\left(\frac{\zeta_t}{\xi_t+\zeta_t}\right)$$
$$+ \frac{\frac{2}{1+\gamma}y_t\sqrt{\frac{1-2\eta}{\eta}\left(1+\frac{\eta}{1-2\eta}\frac{y_t^2}{1+\gamma}\right)\frac{\gamma}{1+\gamma}}\sqrt{\frac{\zeta_t}{\xi_t}}u_t}{(1-2\eta)\left(1+\frac{\eta}{1-2\eta}\frac{y_t^2}{1+\gamma}\right)\left(\frac{\xi_t+\zeta_t}{\xi_t}\right)}.$$

The expected value of the last summand is clearly 0 because of the symmetry of u_t. In contrast, we can use the properties of the beta distribution to prove that

$$E\left(\frac{\xi_t}{\xi_t+\zeta_t}\right) = \frac{1+\eta}{1+2\eta}$$

and

$$E\left(\frac{\zeta_t}{\xi_t+\zeta_t}\right)=\frac{\eta}{1+2\eta}.$$

If we put all the pieces together we end up with

$$\frac{1+\eta}{\left(1-2\eta+\eta\frac{y^2}{1+\gamma}\right)}\frac{y_t^2}{(1+\gamma)^2}+\frac{\gamma}{1+\gamma}-1=\frac{1+\eta}{\left(1-2\eta+\eta\frac{y^2}{1+\gamma}\right)}f_{kt}^2(\gamma)+\upsilon_{kt}(\gamma,0)$$

$$-1=\frac{1+\eta}{\left(1-2\eta+\eta\frac{y^2}{1+\gamma}\right)}\left[f_{kt}^2(\gamma)+\frac{1-\eta}{1+\eta}\upsilon_{kt}(\gamma,\eta)\right]-1.$$

As a result, the score of the true log-likelihood at $\alpha=0$ is

$$\frac{1}{2}\left[\frac{1+\eta}{\left(1-2\eta+\eta\frac{y^2}{1+\gamma}\right)}f_{kt}^2(\gamma)+\upsilon_{kt}(\gamma,0)-1\right]\left[f_{kt-1}^2(\gamma)+\upsilon_{kt-1}(\gamma,\eta)-1\right].$$

Interestingly, note that $\upsilon_{kt}(\gamma,\eta)$ is evaluated in the regressand at its Gaussian value $(=\omega_k(\gamma))$, while in the regressor it is evaluated at the true value of η.

Consider now the following HRS-style auxiliary model

$$y_t=f_t+v_t,$$

$$\begin{pmatrix}f_t\\v_t\end{pmatrix}|I_{t-1}\sim t\left\{\begin{pmatrix}0\\0\end{pmatrix},\begin{bmatrix}1+\alpha[f_{t-1|t-1}^2(\theta)+\omega_{t-1|t-1}(\theta,\eta)-1] & 0\\0 & \gamma\end{bmatrix},\eta\right\},$$

$$f_{t|t}(\theta)=\frac{1+\alpha[f_{t-1|t-1}^2(\theta)+\omega_{t-1|t-1}(\boldsymbol{\theta},\eta)-1]}{1+\alpha[f_{t-1|t-1}^2(\theta)+\omega_{t-1|t-1}(\boldsymbol{\theta},\eta)-1]+\gamma}\cdot y_t,$$

$$\omega_{t|t}(\boldsymbol{\theta},\eta)=\frac{1-2\eta}{1-\eta}\left(1+\frac{\eta}{1-2\eta}\frac{y_t^2}{1+\alpha[f_{t-1|t-1}^2(\theta)+\omega_{t-1,t-1}(\boldsymbol{\theta},\eta)-1]+\gamma}\right)$$

$$\times\frac{1+\alpha[f_{t-1|t-1}^2(\theta)+\omega_{t-1|t-1}(\boldsymbol{\theta},\eta)-1]}{1+\alpha[f_{t-1|t-1}^2(\theta)+\omega_{t-1|t-1}(\boldsymbol{\theta},\eta)-1]+\gamma}\cdot\gamma,$$

and $\alpha\geq 0,\gamma\geq 0$. In order to compute the score of this model with respect to α, we need the derivative of the conditional variance of y_t with respect to this parameter This derivative will be

$$[f_{t-1|t-1}^2(\theta)+\omega_{t-1|t-1}(\boldsymbol{\theta},\eta)-1]+\alpha\left[2f_{t-1|t-1}(\theta)\frac{\partial f_{t-1|t-1}(\theta)}{\partial\alpha}+\frac{\partial\omega_{t-1|t-1}(\boldsymbol{\theta},\eta)}{\partial\alpha}\right].$$

However, since we are only interested in evaluating it at $\alpha=0$ we do not need to compute the second term.

The other component of the derivative will be given by the expression

$$\frac{1+\eta}{1-2\eta+\eta\varepsilon_t^{*2}(\boldsymbol{\theta})}\varepsilon_t^{*2}(\boldsymbol{\theta})-1,$$

where

$$\varepsilon_t^*(\boldsymbol{\theta}) = \frac{y_t}{\sqrt{1+\alpha[f_{t-1|t-1}^2(\theta) + \omega_{t-1|t-1}(\boldsymbol{\theta},\eta) - 1] + \gamma}}.$$

Hence, under the null of $\alpha = 0$ the score with respect to α will be

$$\frac{1}{2}\left(\frac{1+\eta}{1-2\eta+\eta\frac{y_t^2}{1+\gamma}}\frac{y_t^2}{1+\gamma} - 1\right)\frac{1}{1+\gamma}[f_{t-1|t-1}^2(\gamma,0) + \omega_{t-1|t-1}(\gamma,0,\eta) - 1]$$

But since

$$\left(\frac{1+\eta}{1-2\eta+\eta\frac{y_t^2}{1+\gamma}}\frac{y_t^2}{1+\gamma} - 1\right)\frac{1}{1+\gamma} = \frac{1+\eta}{\left(1-2\eta+\eta\frac{y_t^2}{1+\gamma}\right)}\frac{1}{(1+\gamma)^2}y_t^2 + \frac{\gamma}{1+\gamma} - 1,$$

the pseudo log-likelihood score of the auxiliary model coincides with the score of the true model when we evaluate them at α=0. Hence, the Student t version of HRS auxiliary model smoothly embeds the true model at those parameter values.

PROPOSITION 6

The proof of this proposition combines many elements of the proofs of Propositions 2 and 4. Given that Model (8.32) reduces to Model (8.13) when $\alpha = 0$ and $\boldsymbol{\alpha}^* = \mathbf{0}$ for every possible value of the parameters $\boldsymbol{\pi}, \rho, \boldsymbol{\rho}^*, \mathbf{c}$, and $\boldsymbol{\gamma}$, while it reduces to Model (8.22) when $\rho = 0$ and $\boldsymbol{\rho}^* = \mathbf{0}$ for every possible value of the parameters $\boldsymbol{\pi}, \mathbf{c}, \boldsymbol{\gamma}, \alpha$, and $\boldsymbol{\alpha}^*$, then it trivially follows that under the joint null of $\boldsymbol{\rho}^\dagger = \mathbf{0}$ and $\boldsymbol{\alpha}^\dagger = \mathbf{0}$ we will have that

$$\mathbf{Z}_{dt}(\boldsymbol{\theta}_s, 0, 0) =$$

$$\begin{bmatrix} \boldsymbol{\Sigma}^{-1/2\prime}(\boldsymbol{\theta}_s) & 0 \\ 0 & \frac{1}{2}(\mathbf{c}'\otimes\mathbf{I}_N)(\mathbf{I}_{N^2}+\mathbf{K}_{NN})[\boldsymbol{\Sigma}^{-1/2\prime}(\boldsymbol{\theta}_s)\otimes\boldsymbol{\Sigma}^{-1/2\prime}(\boldsymbol{\theta}_s)]0 \\ 0 & \frac{1}{2}\mathbf{E}_N'[\boldsymbol{\Sigma}^{-1/2\prime}(\boldsymbol{\theta}_s)\otimes\boldsymbol{\Sigma}^{-1/2}(\boldsymbol{\theta}_s)] \\ f_{kt-1}(\boldsymbol{\theta}_s)\mathbf{c}'\boldsymbol{\Sigma}^{-1/2\prime}(\boldsymbol{\theta}_s) & 0 \\ diag[\mathbf{v}_{kt-1}(\boldsymbol{\theta}_s)]\boldsymbol{\Sigma}^{-1/2\prime}(\boldsymbol{\theta}_s) & 0 \\ 0 & \frac{1}{2}[f_{kt-1}^2(\boldsymbol{\theta}_s)+\omega_k(\boldsymbol{\theta}_s)-1][\mathbf{c}'\boldsymbol{\Sigma}^{-1/2\prime}(\boldsymbol{\theta}_s)\otimes\mathbf{c}'\boldsymbol{\Sigma}^{-1/2}(\boldsymbol{\theta}_s)] \\ 0 & \frac{1}{2}dg[\mathbf{v}_{kt-1}(\boldsymbol{\theta}_s)\mathbf{v}_{kt-1}'(\boldsymbol{\theta}_s)+\mathbf{c}\mathbf{c}'\omega_k(\boldsymbol{\theta}_s)-\boldsymbol{\Gamma}]\mathbf{E}_N'[\boldsymbol{\Sigma}^{-1/2\prime}(\boldsymbol{\theta}_s)\otimes\boldsymbol{\Sigma}^{-1/2}(\boldsymbol{\theta}_s)] \end{bmatrix},$$

whence

$$\mathbf{Z}_d(\boldsymbol{\phi}) = \begin{bmatrix} \boldsymbol{\Sigma}^{-1/2\prime}(\boldsymbol{\theta}) & 0 \\ 0 & \frac{1}{2}(\mathbf{c}'\otimes\mathbf{I}_N)(\mathbf{I}_{N^2}+\mathbf{K}_{NN})[\boldsymbol{\Sigma}^{-1/2\prime}(\boldsymbol{\theta}_s)\otimes\boldsymbol{\Sigma}^{-1/2\prime}(\boldsymbol{\theta}_s)] \\ 0 & \frac{1}{2}\mathbf{E}_N'[\boldsymbol{\Sigma}^{-1/2\prime}(\boldsymbol{\theta}_s)\otimes\boldsymbol{\Sigma}^{-1/2}(\boldsymbol{\theta}_s)] \\ 0 & 0 \\ 0 & 0 \\ 0 & 0 \\ 0 & 0 \end{bmatrix}$$

and

$$\mathbf{W}_d(\boldsymbol{\phi}) = \left[0 \;\; \mathbf{c}'\boldsymbol{\Sigma}^{-1}(\boldsymbol{\theta}_s) \;\; \tfrac{1}{2}vecd'[\boldsymbol{\Sigma}^{-1}(\boldsymbol{\theta}_s)] \;\; 0 \;\; 0 \;\; 0 \;\; 0\right]'. \quad (8.55)$$

As a result, the score vector under the null will be

$$\begin{bmatrix} s_{\pi t}(\boldsymbol{\theta}_s,0,0,\boldsymbol{\eta}) \\ s_{ct}(\boldsymbol{\theta}_s,0,0,\boldsymbol{\eta}) \\ s_{\gamma t}(\boldsymbol{\theta}_s,0,0,\eta) \\ s_{\rho t}(\boldsymbol{\theta}_s,0,0,\boldsymbol{\eta}) \\ s_{\rho^* t}(\boldsymbol{\theta}_s,0,0,\boldsymbol{\eta}) \\ s_{\alpha t}(\boldsymbol{\theta}_s,0,0,\boldsymbol{\eta}) \\ s_{\alpha^* t}(\boldsymbol{\theta}_s,0,0,\eta) \end{bmatrix} = \begin{bmatrix} \delta[\varsigma_t(\boldsymbol{\theta}_s),\eta]\boldsymbol{\Sigma}^{-1}(\boldsymbol{\theta}_s)(\mathbf{y}_t-\boldsymbol{\pi}) \\ \delta[\varsigma_t(\boldsymbol{\theta}_s),\eta]\boldsymbol{\Sigma}^{-1}(\boldsymbol{\theta}_s)(\mathbf{y}_t-\boldsymbol{\pi})(\mathbf{y}_t-\boldsymbol{\pi})'\boldsymbol{\Sigma}^{-1}(\boldsymbol{\theta}_s)\mathbf{c}-\boldsymbol{\Sigma}^{-1}(\theta_s)\mathbf{c} \\ \frac{1}{2}vecd[\delta[\varsigma_t(\boldsymbol{\theta}_s),\eta]\boldsymbol{\Sigma}^{-1}(\boldsymbol{\theta}_s)(\mathbf{y}_t-\boldsymbol{\pi})(\mathbf{y}_t-\boldsymbol{\pi})'\boldsymbol{\Sigma}^{-1}(\vartheta_s)-\boldsymbol{\Sigma}^{-1}(\theta_s)] \\ f_{kt-1}(\boldsymbol{\theta}_s)\delta[\varsigma_t(\boldsymbol{\theta}_s);\eta]\mathbf{c}'\boldsymbol{\Sigma}^{-1}(\theta_s)(\mathbf{y}_t-\boldsymbol{\pi}) \\ diag[\mathbf{v}_{kt-1}(\boldsymbol{\theta}_s)]\delta[\varsigma_t(\boldsymbol{\theta}_s);\eta]\boldsymbol{\Sigma}^{-1}(\theta_s)(\mathbf{y}_t-\boldsymbol{\pi}) \\ \frac{1}{2}[f^2_{kt-1}(\boldsymbol{\theta}_s)+\omega_k(\boldsymbol{\theta}_s)-1]\{\delta[\varsigma(\boldsymbol{\theta}_s),\eta]\mathbf{c}'\boldsymbol{\Sigma}^{-1}(\boldsymbol{\theta}_s)(\mathbf{y}_t-\boldsymbol{\pi})(\mathbf{y}_t-\boldsymbol{\pi})'\boldsymbol{\Sigma}^{-1}(\boldsymbol{\theta}_s)\mathbf{c} \\ -\mathbf{c}'\boldsymbol{\Sigma}^{-1}(\boldsymbol{\theta}_s)\mathbf{c}\} \\ \frac{1}{2}dg[\mathbf{v}_{kt-1}(\boldsymbol{\theta}_s)\mathbf{v}'_{kt-1}(\boldsymbol{\theta}_s)+\mathbf{c}\mathbf{c}'\omega_k(\boldsymbol{\theta}_s)-\boldsymbol{\Gamma}] \\ \times vecd[\delta[\varsigma_t(\boldsymbol{\theta}_s),\eta]\boldsymbol{\Sigma}^{-1}(\boldsymbol{\theta}_s)(\mathbf{y}_t-\boldsymbol{\pi})(\mathbf{y}_t-\boldsymbol{\pi})'\boldsymbol{\Sigma}^{-1}(\boldsymbol{\theta}_s)-\boldsymbol{\Sigma}^{-1}(\boldsymbol{\theta}_s)] \end{bmatrix}.$$

But this score is simply made up of the components of the different special cases that we have already studied, so the only thing left to do is to study the blocks of the information matrix and the other efficiency bounds that corresponds to the cross product of

$$[s_{\rho t}(\boldsymbol{\theta}_s,0,0,\boldsymbol{\eta}),\mathbf{s}'_{\rho^* t}(\boldsymbol{\theta}_s,0,0,\boldsymbol{\eta})]$$

with

$$[s_{\alpha t}(\boldsymbol{\theta}_s,0,0,\eta),\mathbf{s}'_{\alpha^* t}(\boldsymbol{\theta}_s,0,0,\boldsymbol{\eta})].$$

When the observed variables are elliptically distributed, the vector

$$[f_{kt-1}(\boldsymbol{\theta}_s),\mathbf{v}'_{kt-1}(\theta_s)]$$

is unconditionally orthogonal to the vector

$$\{[f^2_{kt-1}(\boldsymbol{\theta}_s)+\omega_k(\boldsymbol{\theta}_s)-1],vecd'[\mathbf{v}_{kt-1}(\boldsymbol{\theta}_s)\mathbf{v}'_{kt-1}(\boldsymbol{\theta}_s)+\mathbf{c}\mathbf{c}'\omega_k(\boldsymbol{\theta}_s)-\boldsymbol{\Gamma}]\},$$

so all the relevant off-diagonal blocks of $\mathcal{I}_{\theta\theta}(\boldsymbol{\phi}_0)$, $\mathring{\mathcal{S}}(\boldsymbol{\phi}_0)$, $\mathcal{A}(\boldsymbol{\phi}_0)$, and $\mathcal{B}(\boldsymbol{\phi}_0)$ will be 0, which confirms the additive decomposition of the different joint tests under elliptical symmetry.

For general distributions, though, the expressions for $\mathcal{A}(\phi_0)$ and $\mathcal{B}(\boldsymbol{\phi}_0)$ are more involved. Specifically, while it is still true that these matrices will remain block diagonal between $(\boldsymbol{\rho}^\dagger,\boldsymbol{\alpha}^\dagger)$ and $\boldsymbol{\theta}_s$ regardless of the true distribution of $\mathbf{y}_t$ in view of (8.45) and (8.51), and that $\mathcal{A}(\boldsymbol{\phi}_0)$ will also be block diagonal between $\boldsymbol{\rho}^\dagger$ and $\boldsymbol{\alpha}^\dagger$, with the relevant expressions for $\mathcal{A}_{\rho^\dagger\rho^\dagger}(\boldsymbol{\phi}_0)$ and $\mathcal{A}_{\alpha^\dagger\alpha^\dagger}(\boldsymbol{\phi}_0)$ as in the proofs of Propositions 2 and 4, respectively, it will no longer be true that $\mathcal{B}(\boldsymbol{\phi}_0)$ will be block diagonal between AR and ARCH parameters, even though $\mathcal{B}_{\rho^\dagger\rho^\dagger}(\boldsymbol{\phi}_0)=\mathcal{A}_{\rho^\dagger\rho^\dagger}(\boldsymbol{\phi}_0)$. Nevertheless, straightforward calculations show that the blocks of $\mathcal{B}_t(\boldsymbol{\phi}_0)$ corresponding to $(\boldsymbol{\rho}^\dagger,\alpha^\dagger)$ will be given by

$$diag\begin{bmatrix} f_{kt-1}(\boldsymbol{\theta}_s) \\ \boldsymbol{\Gamma}^{-1/2}\mathbf{v}_{kt-1}(\boldsymbol{\theta}_s) \\ \frac{1}{\sqrt{2}}[f^2_{kt-1}(\boldsymbol{\theta}_s)+\omega_k(\theta_s)-1] \\ \frac{1}{\sqrt{2}}\boldsymbol{\Gamma}^{-1}vecd[\mathbf{v}_{kt-1}(\theta_s)\mathbf{v}'_{kt-1}(\boldsymbol{\theta}_s)+\mathbf{c}\mathbf{c}'\omega_k(\boldsymbol{\theta}_s)-\boldsymbol{\Gamma}] \end{bmatrix}$$

$$\times\begin{bmatrix} \mathcal{V}_{\rho^\dagger\rho^\dagger}(\theta_s,0;\boldsymbol{\varrho}) & \mathcal{V}_{\rho^\dagger\alpha^\dagger}(\boldsymbol{\theta}_s,0;\boldsymbol{\varrho}) \\ \mathcal{V}'_{\rho^\dagger\alpha^\dagger}(\boldsymbol{\theta}_s,0;\boldsymbol{\varrho}) & \mathcal{V}_{\alpha^\dagger\alpha^\dagger}(\boldsymbol{\theta}_s,0;\boldsymbol{\varrho}) \end{bmatrix}\times$$

$$diag \begin{bmatrix} f_{kt-1}(\boldsymbol{\theta}_s) \\ \boldsymbol{\Gamma}^{-1/2}\mathbf{v}_{kt-1}(\boldsymbol{\theta}_s) \\ \frac{1}{\sqrt{2}}[f_{kt-1}^2(\boldsymbol{\theta}_s) + \omega_k(\theta_s) - 1] \\ \frac{1}{\sqrt{2}}\boldsymbol{\Gamma}^{-1} vecd[\mathbf{v}_{kt-1}(\theta_s)\mathbf{v}'_{kt-1}(\boldsymbol{\theta}_s) + \mathbf{cc}'\omega_k(\boldsymbol{\theta}_s) - \boldsymbol{\Gamma}] \end{bmatrix},$$

which confirms (8.33) in view of the stationarity of $\mathbf{y}_t$.

8.10 Local power calculations

Let $\mathbf{m}_t(\boldsymbol{\theta}_1, \boldsymbol{\theta}_2)$ denote the h influence functions used to develop the following moment test of $H_0 : \boldsymbol{\theta}_2 = \mathbf{0}$:

$$M_T = T\bar{\mathbf{m}}'_T(\boldsymbol{\theta}_{10}, \mathbf{0})\boldsymbol{\Psi}^{-1}\bar{\mathbf{m}}_T(\boldsymbol{\theta}_{10}, \mathbf{0}), \tag{8.56}$$

where $\bar{\mathbf{m}}_T(\boldsymbol{\theta}_{10}, \mathbf{0})$ is the sample average of $\mathbf{m}_t(\boldsymbol{\theta})$ evaluated under the null, $\boldsymbol{\Psi}$ is the corresponding asymptotic covariance matrix, and $\boldsymbol{\theta}_{10}$ the true values of the remaining model parameters. In order to obtain the non-centrality parameter of this test under Pitman sequences of local alternatives of the form $H_{la} : \boldsymbol{\theta}_{2T} = \bar{\theta}_2/\sqrt{T}$, it is convenient to linearize $\mathbf{m}_t(\boldsymbol{\theta}_{10}, \mathbf{0})$ with respect to θ_2 around its true value $\boldsymbol{\theta}_{2T}$. This linearization yields

$$\sqrt{T}\bar{\mathbf{m}}_T(\boldsymbol{\theta}_{10}, \mathbf{0}) = \sqrt{T}\bar{\mathbf{m}}_T(\boldsymbol{\theta}_{10}, \boldsymbol{\theta}_{2T}) - \frac{1}{T}\sum\nolimits_{t=1}^{T} \frac{\partial \mathbf{m}_t(\boldsymbol{\theta}_{10}, \boldsymbol{\theta}^*_{2T})}{\partial \boldsymbol{\theta}'_2}\bar{\theta}_2,$$

where $\boldsymbol{\theta}^*_{2T}$ is some "intermediate" value between $\boldsymbol{\theta}_{2T}$ and $\mathbf{0}$. As a result,

$$\sqrt{T}\bar{\mathbf{m}}_T(\boldsymbol{\theta}_{10}, \mathbf{0}) \to N[\mathbf{M}(\boldsymbol{\theta}_{10}, \mathbf{0})\bar{\theta}_2, \boldsymbol{\Psi}],$$

under standard regularity conditions, where

$$\mathbf{M}(\boldsymbol{\theta}_{10}, \mathbf{0}) = E[-\partial \mathbf{m}_t(\boldsymbol{\theta}_{10}, \mathbf{0})/\partial \boldsymbol{\theta}'_2],$$

so that the non-centrality parameter of the moment test (8.56) will be

$$\bar{\theta}'_2\mathbf{M}'(\boldsymbol{\theta}_{10}, \mathbf{0})\boldsymbol{\Psi}^{-1}\mathbf{M}(\boldsymbol{\theta}_{10}, \mathbf{0})\bar{\theta}_2. \tag{8.57}$$

On this basis, we can easily obtain the limiting probability of M_T exceeding some pre-specified quantile of a central χ^2_h distribution from the cdf of a non-central χ^2 distribution with h degrees of freedom and non-centrality parameter (8.57). When $\mathbf{m}_t(\boldsymbol{\theta}_1, \boldsymbol{\theta}_2)$ coincides with a subset of the true scores with respect to $\boldsymbol{\theta}_2$, $\mathbf{M}(\boldsymbol{\theta}_{10}, \mathbf{0})$ and $\boldsymbol{\Psi}$ can be readily obtained from the relevant blocks of the information matrix. Similarly, they can be obtained from the $\mathcal{A}(\boldsymbol{\phi})$ and $\mathcal{B}(\boldsymbol{\phi})$ matrices, respectively, when $\mathbf{m}_t(\theta_1, \boldsymbol{\theta}_2)$ coincides with a subset of the Gaussian scores.

Importantly, (8.57) remains valid when we replace θ_{10} by its ML estimator under the null if $\mathbf{m}_t(\theta_1, \mathbf{0})$ and the scores corresponding to $\boldsymbol{\theta}_1$, $\mathbf{s}_{\theta_1 t}(\boldsymbol{\theta}_1, \mathbf{0})$ say, are asymptotically uncorrelated when H_0 is true, as in all our tests. The same applies to the Gaussian PMLEs because the matrices involved in the asymptotic expansions are block diagonal too. More generally, it would be convenient to work with the alternative influence functions

$$\mathbf{m}_t^{\perp}(\boldsymbol{\theta}_1, 0) = \mathbf{m}_t(\boldsymbol{\theta}_1, 0) - cov[\mathbf{m}_t(\boldsymbol{\theta}_1, 0), \mathbf{s}_{\boldsymbol{\theta}_1 t}(\boldsymbol{\theta}_1, 0)] V^{-1}[\mathbf{s}_{\boldsymbol{\theta}_1 t}(\boldsymbol{\theta}_1, 0)] \mathbf{s}_{\boldsymbol{\theta}_1 t}(\boldsymbol{\theta}_1, 0),$$

which can be interpreted as the residual in the regression of $\mathbf{m}_t(\boldsymbol{\theta}_1, 0)$ onto $\mathbf{s}_{\boldsymbol{\theta}_1 t}(\boldsymbol{\theta}_1, 0)$.

8.10.1 SERIAL CORRELATION TESTS

Let us assume without loss of generality that $\boldsymbol{\pi} = \mathbf{0}$. Hosking's test is effectively based on the influence functions

$$\mathbf{m}_{lt}(\boldsymbol{\theta}_s, \boldsymbol{\rho}^{\dagger}) = vec[\mathbf{y}_t \mathbf{y}'_{t-1} - \mathbf{G}_{\mathbf{y}}(1)]$$

evaluated at $\boldsymbol{\rho}^{\dagger} = \mathbf{0}$. But since

$$\mathbf{G}_{\mathbf{y}}(1) = \mathbf{c}\mathbf{c}'\rho + diag(\boldsymbol{\gamma} \odot \boldsymbol{\rho}^*)$$

for the model considered in Section 8.3.3 in view of (8.15), and

$$vec[\mathbf{G}_{\mathbf{y}}(1)] = (\mathbf{c} \otimes \mathbf{c})\rho + vec[diag(\boldsymbol{\gamma} \odot \boldsymbol{\rho}^*)],$$

it trivially follows that

$$\mathbf{M}_l(\boldsymbol{\theta}_s, 0) = E[\partial \mathbf{m}_{lt}(\boldsymbol{\theta}_s, 0)/\partial \boldsymbol{\rho}^{\dagger\prime}] = -[\,(\mathbf{c} \otimes \mathbf{c})\ \mathbf{E}_N \boldsymbol{\Gamma}\,].$$

Hence, we will have that

$$\mathbf{M}_l(\boldsymbol{\theta}_s, 0)\bar{\rho}^{\dagger} = -[(\mathbf{c} \otimes \mathbf{c})\rho + \mathbf{E}_N \boldsymbol{\gamma} \rho^*]$$

when

$$\bar{\rho}^{\dagger\prime} = (\rho\ \rho^* \boldsymbol{\iota}'_N).$$

As for the asymptotic covariance matrix, the proof of Proposition 4 in Fiorentini and Sentana (2012) implies that if $\boldsymbol{\rho}^{\dagger} = \mathbf{0}$, then

$$\sqrt{T}\mathbf{m}_{lt}(\boldsymbol{\theta}_s, 0) = \sqrt{T} vec(\mathbf{y}_t \mathbf{y}'_{t-1}) \to N(0, \boldsymbol{\Sigma} \otimes \boldsymbol{\Sigma})$$

irrespective of the distribution of $\mathbf{y}_t$.

Since the diagonal serial correlation test uses the influence functions

$$vecd[\mathbf{y}_t \mathbf{y}'_{t-1} - \mathbf{G}_{\mathbf{y}}(1)] = \mathbf{E}'_N vec[\mathbf{y}_t \mathbf{y}'_{t-1} - \mathbf{G}_y(1)],$$

it is easy to obtain the corresponding Jacobian matrix by premultiplying $\mathbf{M}_l(\boldsymbol{\theta}_s, 0)$ by $\mathbf{E}'_N$. Specifically,

$$\mathbf{E}'_N \mathbf{M}_l(\boldsymbol{\theta}_s, 0)\bar{\rho}^{\dagger} = -[(\mathbf{c} \odot \mathbf{c})\rho + \gamma \rho^*].$$

We can also exploit the properties of $\mathbf{E}_N$ (see Magnus (1988)) to show that under the null

$$\sqrt{T} vecd(\mathbf{y}_t \mathbf{y}'_{t-1}) \to N(0, \boldsymbol{\Sigma} \odot \boldsymbol{\Sigma}).$$

Finally, to obtain the non-centrality parameter for the serial correlation test of $\mathbf{w}'\mathbf{y}_t$, we simply have to exploit the fact that the relevant influence functions are

$$\mathbf{w}'\mathbf{y}_t \mathbf{y}'_{t-1}\mathbf{w} - \mathbf{w}'\mathbf{G}_y(1)\mathbf{w} = (\mathbf{w}' \otimes \mathbf{w}') vec[\mathbf{y}_t \mathbf{y}'_{t-1} - \mathbf{G}_y(1)],$$

so that the appropriate Jacobian will be $(\mathbf{w}' \otimes \mathbf{w}')\mathbf{M}_l(\boldsymbol{\theta}_s, \mathbf{0})$, whence

$$(\mathbf{w}' \otimes \mathbf{w}')\mathbf{M}_l(\theta_s, \mathbf{0})\bar{\rho}^\dagger = -[(\mathbf{w}'\mathbf{c})^2 \rho + (\mathbf{w}'\boldsymbol{\Gamma} w)\rho^*].$$

Similarly, it is straightforward to show that

$$\sqrt{T}(\mathbf{w}'\mathbf{y}_t\mathbf{y}_{t-1}'\mathbf{w}) \to N[0, (\mathbf{w}'\boldsymbol{\Sigma} w)^2].$$

In the case of the LM test of $H_0 : \boldsymbol{\rho}^\dagger = \mathbf{0}$, the information matrix equality implies that the Jacobian of the scores of $\boldsymbol{\rho}^\dagger$ with respect to $\boldsymbol{\rho}^\dagger$ will be given by (minus) the information matrix, which also gives us the covariance matrix of the scores under the null. By suitably selecting the relevant elements of $\mathcal{I}_{\rho\rho}(\boldsymbol{\phi})$, we can also compute the non-centrality parameters for the tests of the null hypotheses $H_0 : \rho = 0$ and $H_0 : \boldsymbol{\rho}^* = \mathbf{0}$. Analogous comments apply to the Gaussian-based LM tests if we replace the elements of the information matrix by the appropriate elements of $\mathcal{A}_{\rho\rho}(\boldsymbol{\phi})$ or $\mathcal{B}_{\rho\rho}(\boldsymbol{\phi})$.

8.10.2 ARCH TESTS

To keep the algebra simple, we assume once again that $\pi = \mathbf{0}$, that the conditional variances of common and specific factors have been generated according to (8.30), and that the conditional distribution is elliptically symmetric. Hosking's test applied to all the squares and cross-products of $\mathbf{y}_t$ is effectively based on the influence functions that correspond to the first-order autocovariance matrix of $vec(\mathbf{y}_t\mathbf{y}_t')$, $\mathcal{S}_{\mathbf{yy}}(1)$ say, evaluated at $\boldsymbol{\alpha}^\dagger = \mathbf{0}$. More specifically,

$$\mathbf{m}_{st}(\boldsymbol{\theta}_s, \boldsymbol{\alpha}^\dagger) = vec\{[vec(\mathbf{y}_t\mathbf{y}_t' - \boldsymbol{\Sigma})vec'(\mathbf{y}_{t-1}\mathbf{y}_{t-1}' - \boldsymbol{\Sigma})] - \mathcal{S}_{\mathbf{y}}(1)\}.$$

But since

$$E(\mathbf{y}_t\mathbf{y}_t'|I_{t-1}; \boldsymbol{\theta}) = \mathbf{cc}'\lambda_t + \boldsymbol{\Gamma}_t$$

so that

$$vec[E(\mathbf{y}_t\mathbf{y}_t' - \boldsymbol{\Sigma}|I_{t-1}; \boldsymbol{\theta})] = (\mathbf{c} \otimes \mathbf{c})(\lambda_t - 1) + \mathbf{E}_N(\boldsymbol{\gamma}_t - \gamma),$$

and

$$\begin{aligned} vec(\mathbf{y}_{t-1}\mathbf{y}_{t-1}' - \boldsymbol{\Sigma}) = (\mathbf{c} \otimes \mathbf{c})(f_{t-1}^2 - 1) + vec(\mathbf{v}_{t-1}\mathbf{v}_{t-1}' - \boldsymbol{\Gamma}) \\ + (\mathbf{I}_{N_2} + \mathbf{K}_{NN})(\mathbf{c} \otimes \mathbf{I}_N)f_{t-1}\mathbf{v}_{t-1}, \end{aligned}$$

then it follows that

$$\begin{aligned} \mathcal{S}_{\mathbf{y}}(1) &= E[vec(\mathbf{y}_t\mathbf{y}_t' - \boldsymbol{\Sigma})vec'(\mathbf{y}_{t-1}\mathbf{y}_{t-1}' - \boldsymbol{\Sigma})] \\ &= E\{E[vec(\mathbf{y}_t\mathbf{y}_t' - \boldsymbol{\Sigma})|I_{t-1}; \boldsymbol{\phi}]vec'(\mathbf{y}_{t-1}\mathbf{y}_{t-1}' - \boldsymbol{\Sigma})]\} \\ &= E\{[(\mathbf{c} \otimes \mathbf{c})(\lambda_t - 1) + \mathbf{E}_N(\gamma_t - \gamma)][(\mathbf{c}' \otimes \mathbf{c}')(f_{t-1}^2 - 1) + \\ &\quad vec'(\mathbf{v}_{t-1}\mathbf{v}_{t-1}' - \boldsymbol{\Gamma}) + f_{t-1}\mathbf{v}_{t-1}'(\mathbf{c}' \otimes \mathbf{I}_N)(\mathbf{I}_{N_2} + \mathbf{K}_{NN})\} \\ &= (\mathbf{cc}' \otimes \mathbf{cc}')E[(\lambda_t - 1)(f_{t-1}^2 - 1)] + (\mathbf{c} \otimes \mathbf{c})E[(\lambda_t - 1)(\mathbf{v}_{t-1}' \odot \mathbf{v}_{t-1}' - \boldsymbol{\gamma}')]\mathbf{E}_N' \\ &\quad \mathbf{E}_N E[(\boldsymbol{\gamma}_t - \gamma)(f_{t-1}^2 - 1)](\mathbf{c}' \otimes \mathbf{c}') + \mathbf{E}_N E[(\gamma_t - \gamma)(\mathbf{v}_{t-1}' \odot \mathbf{v}_{t-1}' - \boldsymbol{\gamma}')]\mathbf{E}_N' \end{aligned}$$

because of the assumed elliptical symmetry and lack of cross-sectional correlation between f_t and the $v_{it}'s$, and the fact that we are assuming univariate ARCH(1) processes for them. This last assumption also implies that

$$E[(\lambda_t - 1)(f_{t-1}^2 - 1)] = \alpha V(f_{t-1}^2) = \alpha[E(f_{t-1}^4) - 1]$$
$$= \alpha\left[\frac{3(\kappa+1)(1-\alpha^2)}{1-3(\kappa+1)\alpha^2} - 1\right] = \alpha\frac{(3\kappa+2)}{1-3(\kappa+1)\alpha^2},$$

where κ is the multivariate excess kurtosis coefficient. Similarly

$$E[(\gamma_{it} - \gamma_i)(v_{it-1}^2 - \gamma_i)] = \alpha_i V(v_{it-1}^2) = \alpha_i\frac{(3\kappa+2)}{1-3(\kappa+1)\alpha_i^2}\gamma_i^2.$$

In addition, we can show that

$$E[(\gamma_{it} - \gamma_i)(v_{jt-1}^2 - \gamma_j)] = \alpha_i cov(v_{it-1}^2, v_{jt-1}^2) = \alpha_i[E(v_{it-1}^2 v_{jt-1}^2) - \gamma_i\gamma_j]$$
$$= \alpha_i\gamma_i\gamma_j\frac{\kappa}{1-(\kappa+1)\alpha_i\alpha_j},$$
$$E[(\lambda_t - 1)(v_{it-1}^2 - \gamma_i)] = \alpha cov(f_{t-1}^2, v_{it-1}^2) = \alpha\gamma_i\frac{\kappa}{1-(\kappa+1)\alpha\alpha_i},$$
$$E[(\gamma_{it} - \gamma_i)(f_{t-1}^2 - 1)] = \alpha_i cov(f_{t-1}^2, v_{it-1}^2) = \alpha_i\gamma_i\frac{\kappa}{1-(\kappa+1)\alpha\alpha_i}.$$

From here, it is straightforward to see that under the null of conditional homoskedasticity in common and idiosyncratic factors the only nonzero derivatives will be

$$\partial E[(\lambda_t - 1)(f_{t-1}^2 - 1)/\partial\alpha = (3\kappa+2)$$
$$\partial E[(\gamma_{it} - \gamma_i)(v_{it-1}^2 - \gamma_i)]/\partial\alpha_i = (3\kappa+2)\gamma_i^2$$
$$\partial E[(\gamma_{it} - \gamma_i)(v_{jt-1}^2 - \gamma_j)]/\partial\alpha_i = \kappa\gamma_i\gamma_j$$
$$\partial E[(\lambda_t - 1)(v_{it-1}^2 - \gamma_i)]/\partial\alpha = \kappa\gamma_i$$
$$\partial E[(\gamma_{it} - \gamma_i)(f_{t-1}^2 - 1)]/\partial\alpha_i = \kappa\gamma_i$$

whence we can obtain the appropriate Jacobian matrix

$$\mathbf{M}_s(\boldsymbol{\theta}_s, \mathbf{0}) = \partial E[\mathbf{m}_t(\boldsymbol{\theta}_s, \mathbf{0})]/\partial\boldsymbol{\alpha}^{\dagger\prime}.$$

Finally, we will have that

$$\mathbf{M}_s(\boldsymbol{\theta}_s, \mathbf{0})\bar{\alpha}^\dagger = -vec\{(\mathbf{cc}' \otimes \mathbf{cc}')(3\kappa+2)\alpha + (\mathbf{c}\otimes\mathbf{c})\boldsymbol{\gamma}'\mathbf{E}_N'\kappa\alpha$$
$$+ \mathbf{E}_N\boldsymbol{\gamma}(\mathbf{c}'\otimes\mathbf{c}')\kappa\alpha^* + \mathbf{E}_N[2(\kappa+1)(\boldsymbol{\Gamma}\odot\boldsymbol{\Gamma}) + \kappa\gamma\gamma']\mathbf{E}_N'\alpha^*\} \quad (8.58)$$

when

$$\bar{\alpha}^{\dagger\prime} = (\alpha\ \alpha^*\boldsymbol{\iota}_N').$$

As for the asymptotic covariance matrix, the proof of Proposition 8 in Fiorentini and Sentana (2012) implies that if $\boldsymbol{\rho}^\dagger = \mathbf{0}$, then

$$\sqrt{T}\mathbf{m}_{st}(\boldsymbol{\theta}_s, 0) = \sqrt{T} vec[vec(\mathbf{y}_t\mathbf{y}_t' - \boldsymbol{\Sigma})vec'(\mathbf{y}_{t-1}\mathbf{y}_{t-1}' - \boldsymbol{\Sigma})] \to N\{0, [\mathbf{H}(\kappa) \otimes \mathbf{H}(\kappa)]\},$$

when the conditional distribution of $\mathbf{y}_t$ is elliptically symmetric, where

$$\mathbf{H}(\kappa) = (\kappa + 1)(\mathbf{I}_{N^2} + \mathbf{K}_{NN})(\boldsymbol{\Sigma} \otimes \boldsymbol{\Sigma}) + \kappa vec(\boldsymbol{\Sigma})vec'(\boldsymbol{\Sigma}) = \mathbf{H}(\kappa).$$

But given that the autocovariance matrix of $vech(\mathbf{y}_t\mathbf{y}_t')$ will be

$$\mathbf{D}_N^+ E[vec(\mathbf{y}_t\mathbf{y}_t' - \boldsymbol{\Sigma})vec'(\mathbf{y}_{t-1}\mathbf{y}_{t-1}' - \boldsymbol{\Sigma})]\mathbf{D}_N^{+\prime} = \mathbf{D}_N^+ \mathcal{S}_{yy}(1)\mathbf{D}_N^{+\prime},$$

it is straightforward to obtain the relevant limiting mean vector as

$$(\mathbf{D}_N^+ \otimes \mathbf{D}_N^+)\mathbf{M}_s(\boldsymbol{\theta}_s, 0)\bar{\alpha}^\dagger.$$

Similarly, the proof of Proposition 8 in Fiorentini and Sentana (2012) also implies that

$$\frac{\sqrt{T}}{T}\sum_{t=1}^{T} vec[vech(\mathbf{y}_t\mathbf{y}_t' - \boldsymbol{\Sigma})vech'(\mathbf{y}_{t-1}\mathbf{y}_{t-1}' - \boldsymbol{\Sigma})] \to N[0, (\mathbf{D}_N^+\mathbf{H}(\kappa)\mathbf{D}_N^{+\prime} \otimes \mathbf{D}_N^+\mathbf{H}(\kappa)\mathbf{D}_N^{+\prime})],$$

where

$$\mathbf{D}_N^+\mathbf{H}(\kappa)\mathbf{D}_N^{+\prime} = 2(\kappa + 1)\mathbf{D}_N^+(\boldsymbol{\Sigma} \otimes \boldsymbol{\Sigma})\mathbf{D}_N^{+\prime} + \kappa vech(\boldsymbol{\Sigma})vech'(\boldsymbol{\Sigma})].$$

From here, we can obtain the non-centrality parameter for the test that only looks at the marginal autocovariances of $vech(\mathbf{y}_t\mathbf{y}_t')$ by premutiplying by $\mathbf{E}_{N(N+1)/2}'$.

In turn, the diagonalization matrix $\mathbf{E}_N$ allows us to obtain the autocovariance matrix of $vecd(\mathbf{y}_t\mathbf{y}_t' - \boldsymbol{\Sigma})$ as

$$\mathbf{E}_N' E[vec(\mathbf{y}_t\mathbf{y}_t' - \boldsymbol{\Sigma})vec'(\mathbf{y}_{t-1}\mathbf{y}_{t-1}' - \boldsymbol{\Sigma})]\mathbf{E}_N = \mathbf{E}_N'\mathcal{S}_y(1)\mathbf{E}_N,$$

whence we can obtain the non-centrality parameter for the test that only looks at the marginal autocovariances of $vecd(\mathbf{y}_t\mathbf{y}_t')$ by premutiplying $\mathbf{M}_s(\boldsymbol{\theta}_s, 0)\bar{\alpha}^\dagger$ by $(\mathbf{E}_N' \otimes \mathbf{E}_N')$. An analogous manipulation yields the asymptotic covariance matrix of the relevant influence functions.

Finally, it is straightforward to obtain the autocovariance structure of the squares of any linear combination of $\mathbf{y}_t$, $\mathbf{w}'\mathbf{y}_t$ say, by exploiting the fact that

$$E[(\mathbf{w}'\mathbf{y}_t)^2(\mathbf{w}'\mathbf{y}_{t-1})^2] = vec'(\mathbf{w}\mathbf{w}')E[vec(\mathbf{y}_t\mathbf{y}_t')vec'(\mathbf{y}_{t-1}\mathbf{y}_{t-1}')]vec(\mathbf{w}\mathbf{w}').$$

Similarly, it is easy to prove that

$$\frac{\sqrt{T}}{T}\sum_{t=1}^{T}(\mathbf{w}'\mathbf{y}_t)^2(\mathbf{w}'\mathbf{y}_{t-1})^2 \to N[0, (3\kappa + 2)(\mathbf{w}'\boldsymbol{\Sigma}\mathbf{w})^2]$$

under the null.

In the case of the LM test of $H_0 : \boldsymbol{\alpha}^\dagger = \mathbf{0}$, the information matrix equality implies that the Jacobian of the scores of $\boldsymbol{\alpha}^\dagger$ with respect to $\boldsymbol{\alpha}^\dagger$ will be given by (minus) the information matrix, which also gives us the covariance matrix of the scores under the null. By suitably selecting the relevant elements of $\mathcal{I}_{\alpha\alpha}(\boldsymbol{\phi})$, we can also compute the non-centrality parameters for the tests of the null hypotheses $H_0 : \alpha = 0$ and $H_0 :$

$\boldsymbol{\alpha}^* = \mathbf{0}$. Analogous comments apply to the Gaussian-based LM tests if we replace the elements of the information matrix by the appropriate elements of $\mathcal{A}_{\alpha\alpha}(\boldsymbol{\phi})$ or $\mathcal{B}_{\alpha\alpha}(\boldsymbol{\phi})$.

ACKNOWLEDGMENTS

This is an abridged version of Fiorentini and Sentana (2012), which in turn was partly based on Fiorentini and Sentana (2009). We are grateful to Dante Amengual, Manuel Arellano, Javier Mencía, Javier Perote, and Albert Satorra as well as to seminar audiences at Bologna, Cass Business School, CEMFI, Florence, GREQAM, Pablo de Olavide, Toulouse, Finance Forum 2008 (Madrid), ESEM 2009 (Barcelona), SAEe 2009 (Valencia), RCEF 2012 (Toronto), and the Oxford-Man Institute Time Series Econometrics Conference in honor of Andrew Harvey for helpful comments, discussions, and suggestions. Detailed comments by Neil Shephard and two anonymous referees have also substantially improved the presentation. Of course, the usual caveat applies. Sentana gratefully acknowledges financial support from the Spanish Ministry of Science and Innovation through grants ECO 2008-00280 and 2011-26342, while Fiorentini acknowledges funding from MIUR PRIN MISURA–Multivariate models for risk assessment.

9 Inference for models with asymmetric α-stable noise processes

Tatjana Lemke and Simon J. Godsill

9.1 Introduction

A wide range of real-world phenomena exhibit outliers, jumps, and asymmetric characteristics, which cannot be accommodated within the classical Gaussianity assumption. For this reason α-stable distributions have attracted growing interest. Stable distributions include the Gaussian distributions as a special case, and arise from a generalized version of the central limit theorem (CLT) as shown in Gnedenko and Kolmogorov (1968) and Feller (1966). In contrast to the classical CLT the generalized version forgoes the condition of a finite variance and assigns a much less restrictive requirement on tail behavior. Compared to the Gaussian distribution the general stable distribution depends upon two additional parameters, which represent the asymmetry and heavy tailedness, and are therefore more suitable to model phenomena showing these empirical features. Application areas are diverse, including, among others, radar processing, telecommunication, acoustics, and econometrics (see Schoutens (2003); Nolan (2005); Godsill and Rayner (1998)). However, most previously presented works concentrate on symmetric α-stable law and are not flexible enough to address asymmetric behavior.

In the presence of symmetric stable noise, Godsill and Kuruoğlu (see Godsill and Kuruoğlu (1999), Godsill (1999)), introduced Monte Carlo expectation-maximization (MCEM) and Markov chain Monte Carlo methods, which are based on the scale mixtures of normals (SMiN) representation of stable distributions. A related method for inference in models with symmetric Paretian disturbances was proposed by Tsionas (1999). An online Bayesian filtering and smoothing method, also based on the SMiN, was proposed by Lombardi and Godsill (2006). Kuruoğlu (2003) addressed positive α-stable probability distributions, providing an analytical approximation based on a decomposition into a product of a Pearson and another positive stable random variable. Inference for AR processes with possibly asymmetric α-stable innovations have been presented by Gençağa et al. (2008) using a sequential

Bayesian approach. Bayesian inference for stable distribution parameters by exploiting a particular representation involving a bivariate density function was introduced by Buckle (1995), and extended to time series problems by Quiou and Ravishanker (1998). Rosiński (2007) formally introduced the class of tempered stable distribution, which combines both α-stable and Gaussian properties. Alternative classes of distributions, the so-called modified stable laws and normal modified stable laws provide additional flexibility in modeling the dynamics of financial time series and are discussed by Barndorff-Nielsen and Shephard (2001).

In this paper we make use of Poisson series representations (PSR) for asymmetric α-stable distributed random variables and stochastic processes, and use this to provide a conditionally Gaussian framework for inference. By so doing we allow for Bayesian state and parameter estimation using Rao-Blackwellized particle filters and simple Gibbs sampling-based Markov chain Monte Carlo (MCMC) approaches. We have presented parts of this work in conference publications (see Lemke and Godsill (2011, 2012)).

The original contributions of this work include a modified Poisson series representation for stochastic integrals, which makes it into a conditionally Gaussian form, and a novel approach which allows an exact characterization of the mean and covariance of the residual terms when our infinite series is truncated to a finite limit. Our approach can be related for continuous-time processes to the Gaussian approximations for Lévy processes introduced by Rosiński (2001). The approach for the residual approximation is shown detailed for α-stable random variables. A corresponding residual approximation is used for our modified PSR. Furthermore we present the use of the PSRs to perform Bayesian Monte Carlo inference for autoregressive model parameters and demonstrate a Rao-Blackwellized particle filter for state estimation for the continuous-time autoregressive model. Finally, we propose methods using our framework for α-stable distribution parameter estimation in addition to states and model parameters, which will be detailed in future publications.

The remainder of the paper is organized as follows. Section 9.2 gives basic definitions concerning α-stable random variables. In Section 9.3, the Poisson series representation for α-stable random variables is stated and extended to practical application by our two residual approximation approaches. In Section 9.4, the discrete-time state-space models and the discrete-time autoregressive model including the PSR for the innovations are presented, followed by model parameter estimation with a corresponding Markov chain Monte Carlo implementation and evaluation. In Section 9.5, the continuous-time state-space models driven by α-stable Lévy processes are considered and our modified PSR for stochastic integrals is introduced. A Rao-Blackwellized particle filter is presented for state estimation in such models, with some example simulations. In Section 9.6 our conclusions are given.

9.2 The α-stable distribution

The α-stable family of distributions includes the Gaussian as a special case. However, this family of distributions allows in addition for asymmetry and heavy tails. In general, closed-form density functions of $S_\alpha(\sigma, \beta, \mu)$ are not known. The exceptions are for $\alpha = 2$, corresponding to the normal distribution, $\alpha = 1$ and $\beta = 0$, yielding the Cauchy distribution, and $\alpha = 1/2$, $\beta = \pm 1$ for the Lévy distribution. Nevertheless, the α-stable distribution follows a generalized central limit theorem, which states that the sum of a number of random variables with power-law tail distributions decreasing as $1/|x|^{\alpha+1}$ where $0 < \alpha < 2$ will converge to an α-stable distribution as the number of random variables approaches infinity. The α-stable family of distributions $S_\alpha(\sigma, \beta, \mu)$ is defined in terms of its characteristic function $\phi_X(.)$ (see Feller (1966), Applebaum (2004), Samorodnitsky and Taqqu (1994)).

Theorem 9.1 *A real-valued random variable X is stable if and only if there exists* $\sigma > 0$, $-1 \leq \beta \leq 1$, $\mu \in \mathbb{R}$ *such that for all* $t \in \mathbb{R}$

$$\phi_X(t) = \mathbb{E}\left[e^{itX}\right] = \begin{cases} \exp(-\sigma^\alpha |t|^\alpha [1 - i\beta \operatorname{sign}(t) \tan(\frac{\alpha\pi}{2})] + i\mu t) & \text{if } \alpha \neq 1, \\ \exp(-\sigma |t| [1 - i\beta \frac{2}{\pi} \operatorname{sign}(t) \ln |t|] + i\mu t) & \text{if } \alpha = 1. \end{cases} \tag{9.1}$$

The four parameters and their domains are as follows:

- $\alpha \in (0, 2]$ is referred to as the *characteristic exponent* or *index of stability*, which measures the tail thickness.
- $\beta \in [-1, 1]$ is termed the *skewness parameter*. For $\beta = 1$ ($\beta = -1$), the family of distributions $S_\alpha(\sigma, \beta, \mu)$ specializes to the positive (negative) stable family, whereas $\beta = 0$ indicates the symmetric α-stable (SαS) distribution.
- $\sigma > 0$ denotes the *scale parameter*.
- $\mu \in \mathbb{R}$ denotes the *location parameter*.

9.3 General scheme based upon conditional Gaussians

Our aim in this work is to simplify inference in α-stable models by making use of powerful auxiliary variables representations of α-stable random variables and processes. Such formulations are able to represent the α-stable models exactly, in terms of distributions, which may either be evaluated point-wise or sampled using standard techniques. In particular, we will seek conditionally Gaussian representations of these models, where both the mean and precision parameters of the models are considered as random variables. In the symmetric stable case our previous work has demonstrated such a framework, using just scale mixtures of normals (see Godsill and Kuruoğlu (1999), Godsill (1999), Tsionas (1999)). This is in contrast with the MCMC approach of Buckle (1995), in which an exact auxiliary variables approach is proposed, but computations are difficult because no conditionally Gaussian structure arises. Our current work provides a new framework which includes Godsill and Kuruoğlu (1999) and Godsill (1999) as a special case (see Section 9.3.1), and also allows evaluation of stochastic integrals involving α-stable Lévy processes.

In order to motivate the subsequent developments, suppose we have the following marginal-conditional representation for the marginal stable law, $X \sim S_\alpha(\sigma, \beta, \mu)$:

$$(X, \mu_X, \sigma_X) \sim \mathcal{N}(X|\mu_X, \sigma_X^2) p_{\alpha,\beta,\sigma}(\mu_X, \sigma_X),$$

where $p_{\alpha,\beta,\sigma}(\mu_X, \sigma_X)$ denotes the distribution of the possibly dependent latent variables μ_X, σ_X, which in turn depends on the distribution parameters α, β, σ. This is an auxiliary variables representation of the distribution and would facilitate inference procedures. For example, a conditional distribution for the latent variables μ_X, σ_X can be formed as follows:

$$p(\mu_X, \sigma_X|X) \propto \mathcal{N}(X|\mu_X, \sigma_X^2) p_{\alpha,\beta,\sigma}(\mu_X, \sigma_X)$$

and this may be sampled in a Gibbs sampler style of implementation, or used to compute expectations in an EM or variational Bayes framework. Moreover, if observations are conditionally linear and Gaussian distributed, $Y \sim \mathcal{N}(G(\theta)X, H(\theta))$, where θ are parameters of the model, then marginal likelihoods can be calculated using standard Gaussian computations such as the Kalman filter, or a whole sequence of states may be sampled using Kalman filter/simulation smoothers in a Gibbs sampler setting.

The form of $p_{\alpha,\beta,\sigma}(\mu_X, \sigma_X)$, the mixing density, determines the form of the marginal $p(X)$, and in subsequent sections we will show how this can be specified exactly, sampled, and characterized in the α-stable case. We will also show how an equivalent form can be constructed for α-stable Lévy processes. More generally, of course the same framework can be applied with different forms for $p(\mu_X, \sigma_X)$, hence taking us to more general classes than the α-stable class, and one might even consider non-parametric density estimators such as Dirichlet processes for this task. This latter approach will be reported in future work.

9.3.1 POISSON SERIES REPRESENTATION

We now present the form of the series representation for α-stable random variables that forms the basis of our approach. A skewed α-stable random variable X can be represented in terms of a convergent sum based on arrival times of a Poisson process (Theorem 1.4.5 in Samorodnitsky and Taqqu (1994)):

$$\sum_{m=1}^{\infty}\left(\Gamma_m^{-1/\alpha}W_m - k_m^{(\alpha)}\right) \xrightarrow{a.s.} X \sim S_\alpha(\sigma,\beta,0), \tag{9.2}$$

which of course includes the general skewed case ($\beta \neq 0$). Here

$$k_m^{(\alpha)} = \begin{cases} 0 & \text{if } 0<\alpha<1, \\ \mathbb{E}[W_1 \displaystyle\int_{|W_1|/m}^{|W_1|/(m-1)} x^{-2}\sin(x)dx] & \text{if } \alpha = 1, \\ \frac{\alpha}{\alpha-1}\left(m^{\frac{\alpha-1}{\alpha}} - (m-1)^{\frac{\alpha-1}{\alpha}}\right)\mathbb{E}[W_1] & \text{if } 1<\alpha<2 \end{cases} \tag{9.3}$$

and

$$\sigma^\alpha = \frac{\mathbb{E}|W_1|^\alpha}{C_\alpha}, \quad \beta = \frac{\mathbb{E}[|W_1|^\alpha \operatorname{sign} W_1]}{\mathbb{E}|W_1|^\alpha}, \tag{9.4}$$

where $\{W_1, W_2, \ldots\}$ and $\{\Gamma_1, \Gamma_2, \ldots\}$ are two independent sequences of random variables. The W_m are i.i.d. with

$$\mathbb{E}[|W_1|^\alpha] < \infty \qquad \text{if } \alpha \neq 1, \tag{9.5}$$

$$\mathbb{E}[|W_1 \ln(|W_1|)|] < \infty \qquad \text{if } \alpha = 1, \tag{9.6}$$

and Γ_m are arrival times of a unit rate Poisson process, such that $(\Gamma_m - \Gamma_{m-1}) \sim$ Exponential(1). The constant C_α is defined as

$$C_\alpha = \left(\int_0^\infty x^{-\alpha}\sin(x)dx\right)^{-1} = \begin{cases} \frac{1-\alpha}{\Gamma(2-\alpha)\cos(\pi\alpha/2)} & \text{if } \alpha \neq 1, \\ 2/\pi & \text{if } \alpha = 1. \end{cases} \tag{9.7}$$

We further define $\tilde{k}_m^{(\alpha)} := k_m^{(\alpha)}/\mathbb{E}[W_1]$. The constant $k_m^{(\alpha)}$ gives the compensation for the otherwise divergent sum $\sum_{m=1}^{\infty}\Gamma_m^{-1/\alpha}W_m$ when $\alpha > 1$ and the W_m are asymmetric. Asymptotically, as $m \to \infty$, $k_m^{(\alpha)}$ equals to $\mathbb{E}[\Gamma_m^{-1/\alpha}W_m]$. For the sake of simplicity, we do not consider the special case $\alpha = 1$ within this paper.

The convergence in (9.2) demonstrates that skewed α-stable random variables can be simulated using independent and identically distributed random variables $\{W_m\}_{m=1}^{\infty}$ and the arrival times of a unit rate Poisson process $\{\Gamma_m\}_{m=1}^{\infty}$. Note especially that we are free to choose any convenient distribution for the i.i.d. variables W_m, subject to the finite moment conditions. Making use of the above Poisson series representation then permits a Gaussian framework conditional upon $\{\Gamma_m\}_{m=1}^{\infty}$ by choosing the i.i.d. random variables W_m in (9.2) to be Gaussian, $W_m \sim \mathcal{N}(\mu_W, \sigma_W^2)$, which from (9.4) above implies particular values of the α-stable distribution parameters, $\beta = \frac{\mathbb{E}[|W_1|^\alpha \operatorname{sign} W_1]}{\mathbb{E}|W_1|^\alpha}$ and $\sigma^\alpha = \frac{\mathbb{E}|W_1|^\alpha}{C_\alpha}$, and $\mathbb{E}[W_1] = \mu_W$ in (9.3). The conditionally Gaussian structure for $X \sim S_\alpha(\sigma,\beta,0)$ can now be specified as:

$$X|\{\Gamma_m\}_{m=1}^{\infty} \sim \mathcal{N}\left(\mu_W \sum_{m=1}^{\infty}\left(\Gamma_m^{-1/\alpha} - \tilde{k}_m^{(\alpha)}\right), \sigma_W^2 \sum_{m=1}^{\infty} \Gamma_m^{-2/\alpha}\right) := \mathcal{N}\left(\mu_X, \sigma_X^2\right). \tag{9.8}$$

Relationship with the symmetric case

We note here that the symmetric case $\beta = 0$ as studied in Godsill and Kuruoğlu (1999), and Godsill (1999) can be obtained as a special case of the above form, since with $\mu_W = 0$ we have:

$$X|\{\Gamma_m\}_{m=1}^{\infty} \sim \mathcal{N}\left(0, \sigma_W^2 \sum_{m=1}^{\infty} \Gamma_m^{-2/\alpha}\right) := \mathcal{N}\left(\mu_X = 0, \sigma_X^2\right) \tag{9.9}$$

and we have from (9.2) that

$$\sum_{m=1}^{\infty} \Gamma_m^{-2/\alpha} \xrightarrow{a.s.} \lambda \sim S_{\alpha/2}((C_{\alpha/2})^{-2/\alpha}, 1, 0). \tag{9.10}$$

This latter is the fully skewed mixing distribution used in the scale mixtures of normals case (see Godsill and Kuruoğlu (1999) and Godsill (1999)). In that earlier work, the problem was substantially simplified by the full characterization of the mixing density as a skewed stable random variable. Thus, only one auxiliary variable λ needed to be sampled. In the current (asymmetric) setting, however, the problem is more challenging since we have to sample/characterize a bivariate, dependent, random variable μ_X, σ_X^2, for which we do not have any convenient form except for the full infinite summations as given in (9.8).

Residual approximation

In practice the infinite series in (9.8) must be truncated at some finite limit, and we note that this might be a good approximation since the Γ_m terms are strictly increasing, though convergence will be slow for $\alpha > 1$. This truncation, while not necessary in the symmetric $\beta = 0$ case since we have a full characterization of the mixing density as in (9.10), will be required in the full asymmetric case in a practical inference scheme. This is the only source of approximation in our models apart from the errors due to standard Monte Carlo inference procedures. We do, however, substantially mitigate the approximation error by computing very accurate moment-based approximations to the residual error terms and including these in our inference procedures.

Consider first that the summation terminates once Γ_{M+1} exceeds some fixed value c (see Figure 9.1).

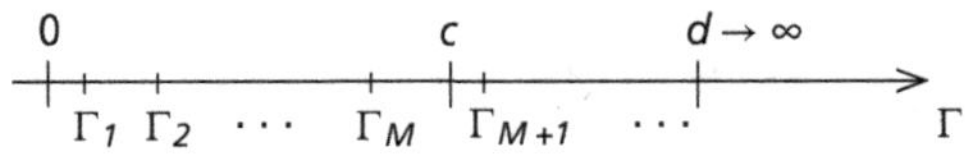

Figure 9.1. Setup of the residual approximation approach, showing the increasing sequence of Poisson arrival times $\{\Gamma_m\}$, the truncation limit c, and the upper limit $d \to \infty$

We first approach the approximation through computing the first two moments of the residual terms in the summations directly:

$$R_1 := \lim_{d\to\infty} \sum_{m:\Gamma_m\in[c,d]} \Gamma_m^{-1/\alpha} - \sum_{l=1}^{d} \tilde{k}_l^{(\alpha)} \tag{9.11}$$

and

$$R_2 := \lim_{d\to\infty} \sum_{m:\Gamma_m\in[c,d]} \Gamma_m^{-2/\alpha}, \tag{9.12}$$

and then approximating those residuals using a bivariate Gaussian distribution, $\mathcal{N}(\mu_R^{(\alpha)}, \Sigma_R^{(\alpha)})$, which takes account of the correlation between R_1 and R_2. We refer to this residual approach as the Gaussian approximation of moments approach (GAMA).

Lemma 9.1 (Gaussian approximation of moments approach) *The moments for a bivariate Gaussian approximation for the residuals (R_1, R_2) of a Poisson series representation as given in (9.8) can be obtained exactly as (see Appendix 9.7):*

$$\mu_R^{(\alpha)} = \begin{bmatrix} \frac{\alpha}{1-\alpha} c^{\frac{\alpha-1}{\alpha}} \\ \frac{\alpha}{2-\alpha} c^{\frac{\alpha-2}{\alpha}} \end{bmatrix} \tag{9.13}$$

and

$$\Sigma_R^{(\alpha)} = \begin{bmatrix} \frac{\alpha}{2-\alpha} c^{\frac{\alpha-2}{\alpha}} & \frac{\alpha}{3-\alpha} c^{\frac{\alpha-3}{\alpha}} \\ \frac{\alpha}{3-\alpha} c^{\frac{\alpha-3}{\alpha}} & \frac{\alpha}{4-\alpha} c^{\frac{\alpha-4}{\alpha}} \end{bmatrix}. \tag{9.14}$$

Hence, the conditionally Gaussian framework for $X \sim S_\alpha(\sigma, \beta, 0)$ can be written as

$$X|\{\Gamma_m\}_{m=1}^{M} \sim \mathcal{N}\left(\mu_X, \sigma_X^2\right), \tag{9.15}$$

where

$$\mu_X = \mu_W \left(\sum_{m=1}^{M} \Gamma_m^{-1/\alpha} + R_1 \right) \tag{9.16}$$

and

$$\sigma_X^2 = \sigma_W^2 \left(\sum_{m=1}^{M} \Gamma_m^{-2/\alpha} + R_2 \right), \tag{9.17}$$

with

$$\begin{pmatrix} R_1 \\ R_2 \end{pmatrix} \sim \mathcal{N}\left(\mu_R^{(\alpha)}, \Sigma_R^{(\alpha)}\right) \tag{9.18}$$

with $\mu_R^{(\alpha)}$ and $\Sigma_R^{(\alpha)}$ as in (9.13) and (9.14).

The GAMA form of approximation is very convenient in that the mean and variance expressions scale directly in terms of the parameters μ_W and σ_W^2. This will mean that in Bayesian inference frameworks these two parameters can be marginalized by direct integration. The downside is that we will have to simulate the residual terms R_1 and R_2 as additional random variables in order to perform inference. An alternative formulation which eliminates these random variables altogether computes the first two moments of the remaining terms of the series, $\lim_{d\to\infty} \sum_{m:\Gamma_m\in[c,d]}(W_m\Gamma_m^{-1/\alpha}) - \sum_{n:\Gamma_n\in[0,d]} k_n^{(\alpha)}$ directly, then approximating the residual with a Gaussian whose mean and variance are matched to those moments. This approach, referred to as the Gaussian approximation approach (GAA), has been presented in our previous work Lemke and Godsill (2012). Then, the approximated conditionally Gaussian framework for $X \sim S_\alpha(\sigma, \beta, 0)$ becomes

$$X|\{\Gamma_m\}_{m=1}^M \sim \mathcal{N}(\mu_X, \sigma_X{}^2), \tag{9.19}$$

where

$$\mu_X = \mu_W \left(\sum_{m=1}^{M} \Gamma_m^{-1/\alpha} + \frac{\alpha}{1-\alpha} c^{\frac{\alpha-1}{\alpha}} \right), \tag{9.20}$$

$$\sigma_X^2 = \sigma_W^2 \sum_{m=1}^{M} \Gamma_m^{-2/\alpha} + (\sigma_W^2 + \mu_W^2)\left(\frac{\alpha}{2-\alpha} c^{\frac{\alpha-2}{\alpha}} \right). \tag{9.21}$$

Note, though, that the GAA terms depend nonlinearly on μ_W and σ_W^2, which will not so easily facilitate marginalization or sampling of these distribution parameters.

Evaluation of the residual approximation

Here we give an example of the accuracy of the Poisson series representation when using our residual GAMA approximation. The distribution parameters are set to $\alpha = 1.5, \beta = 0.8978, \sigma = 2.3967$, and $\mu = 0$, which, according to (9.4), corresponds to $\mu_W = 1$ and $\sigma_W = 1$. Random variables obtained

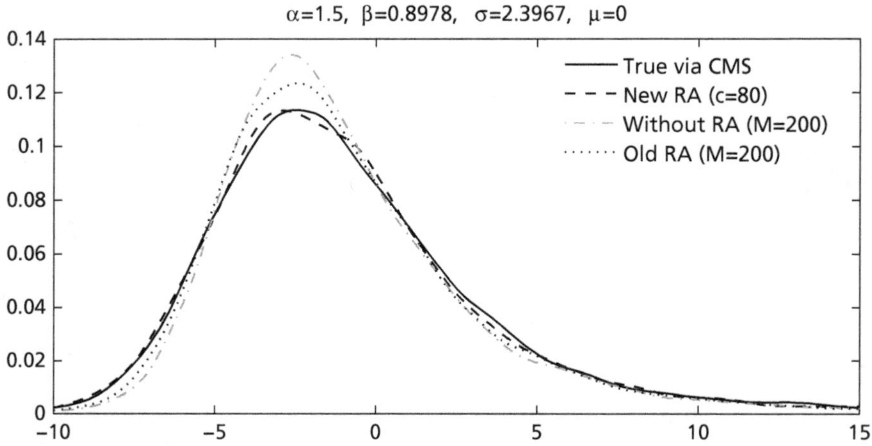

Figure 9.2. Comparing the approximated PSR to the real distribution with $\alpha = 1.5$, $\beta = 0.8978, \sigma = 2.3967, \mu = 0$

from the asymmetric stable law, applying the Chambers-Mallows-Stuck (CMS) method (see Chambers et al. (1976)), served as a benchmark for our comparison of the representations shown in Figure 9.2. The new residual approximation (new RA) with an average number of summation terms of $c = 80$ shows a clear improvement to the previous residual approximation with a simple truncation after a fixed number of summation terms $M = 200$ (old RA) and achieves results almost indistinguishable from the benchmark, as can be seen in Figure 9.2. Similar improvements were obtained for a wide range of different α-stable parameter settings.

9.4 **Discrete-time state-space models**

The above approach to representation of α-stable random variables can be incorporated directly into linear and nonlinear state-space modeling. Take for now a mixed linear/nonlinear state-space model in which we distinguish between linear and nonlinear states, x^L and x^N, respectively. The linear components of the model are assumed to follow a linear Gaussian state-space model with state-space matrices that may depend upon the nonlinear states:

$$x_t^L = A(x_{0:t}^N)x_{t-1}^L + C(x_{0:t}^N) + u_t^L \tag{9.22}$$

$$y_t = B(x_{0:t}^N)x_t^L + v_t^L, \tag{9.23}$$

where $x_{0:t}^N := \{x_0^N, x_1^N, \ldots, x_t^N\}$; u_t^L and v_t^L are independent, zero mean, Gaussian disturbances with covariances $C_u(x_{0:t}^N)$ and $C_v(x_{0:t}^N)$, respectively. Here, the hidden states x_t^L are assumed to propagate according to the state equation (9.22), while the observed data y_t is assumed to be generated via the observation equation (9.23), and the x_t^N are nonlinear elements of the model obeying their own dynamics.

The approach here will be to replace one or both of the Gaussian noise terms with an α-stable distribution, using the conditionally Gaussian form and auxiliary variables. We will consider a simple example of such a framework here, in which parameters of a linear autoregression are estimated in the presence of α-stable driving noise. The extension to more general cases of the model is fairly straightforward, involving auxiliary variables extensions of the standard Monte Carlo or EM-based inference methods for such models.

9.4.1 EXAMPLE: DISCRETE-TIME AR MODEL

Here we consider the discrete-time autoregressive (AR) process of order P with α-stable innovations $\{e_t\}_{t\in\mathbb{N}_0} \sim S_\alpha(\sigma,\beta,0)$. The AR($P$) process expresses a time series as a linear function of its past values. The number of lagged past values is determined by the order P. Formally the definition reads as follows: If $\{e_t\}_{t\in\mathbb{N}_0}$ is a discrete-time α-stable noise process, and $\theta_1,\ldots,\theta_p$ are real numbers, then the discrete-time stochastic process $\{y_t\}_{t\in\mathbb{N}_0}$ is called a AR(P) process if it satisfies the difference equation

$$y_t = \sum_{k=1}^{P} \theta_k y_{t-k} + e_t, \quad e_t \sim S_\alpha(\sigma,\beta,0), \tag{9.24}$$

which we rewrite as

$$\mathbf{y} = G\boldsymbol{\theta} + \mathbf{e}, \tag{9.25}$$

where G is set to be the standard design matrix for a AR(P) process,

$$G = \begin{pmatrix} y_0 & \cdots & y_{1-P} \\ \vdots & & \vdots \\ y_{t-1} & \cdots & y_{t-P} \end{pmatrix}. \tag{9.26}$$

We now transform the α-stable noise process e_t into a conditionally Gaussian distributed noise, $e_t \sim \mathcal{N}(\mu_{e_t},\sigma_{e_t})$. Thus, we use the Poisson series representation presented in (9.2) for the α-stable noise $\{e_t\}_{t\in[0,\infty)}$ to obtain a conditionally Gaussian framework. Then, given $\{\Gamma_m\}_{m=1}^{\infty}$, the model can be written as

$$y_t = \sum_{k=1}^{P} \theta_k y_{t-k} + e_t, \quad e_t|\{\Gamma_m\}_{m=1}^{\infty} \sim \mathcal{N}(\mu_{e_t},\sigma_{e_t}), \tag{9.27}$$

where μ_{e_t} and σ_{e_t} are as μ_X and σ_X in (9.20) and (9.21).

9.4.2 GIBBS SAMPLER FOR MODEL PARAMETER ESTIMATION

For the moment we assume P to be some known model order. Out of many possible Markov chain Monte Carlo strategies, we focus on the approach with a Gibbs sampler here for simplicity. Thus, at the jth iteration, for $j = 1, \ldots I$, a random draw is made with replacement from each of the unknowns:

$$\boldsymbol{\theta}^j \sim p\left(\boldsymbol{\theta}|\boldsymbol{\Gamma}^{j-1}, \boldsymbol{y}\right), \tag{9.28}$$

$$\boldsymbol{\Gamma}^j \sim p\left(\{\boldsymbol{\Gamma}_m\}_{m=1}^M|\boldsymbol{\theta}^j, \boldsymbol{y}\right), \tag{9.29}$$

where $\boldsymbol{\Gamma}^j = \{\{\Gamma^j_{m,t}\}_{m=1}^{M_t}\}_{t=1}^N$ and $\mathbf{y} = \{y_t\}_{t=1}^N$.

We assume conjugate prior framework for the unknowns $\boldsymbol{\theta} = (\theta_1, \ldots, \theta_P)'$:

$$p(\boldsymbol{\theta}) = \mathcal{N}(\boldsymbol{\theta}|\boldsymbol{\mu}, \mathrm{C}). \tag{9.30}$$

The full conditional posterior for $\boldsymbol{\theta}$ can be obtained as Lindley and Smith (1972):

$$p\left(\boldsymbol{\theta}|\boldsymbol{\Gamma}, \mathrm{y}\right) = \mathcal{N}(\boldsymbol{\mu}', \mathrm{C}') \tag{9.31}$$

$$\mathrm{C}' = \left((\mathrm{G}^T\Sigma\mathrm{G} + (\mathrm{C})^{-1}\right)^{-1}, \quad \boldsymbol{\mu}' = \mathrm{C}'\left(\mathrm{G}\Sigma(\mathrm{y} - \boldsymbol{\mu_e}) + \mathrm{C}^{-1}\boldsymbol{\mu}\right), \tag{9.32}$$

where

$$\Sigma = \operatorname{diag}\left[\left(\sigma_{e_1}\right)^{-2}, \ldots, \left(\sigma_{e_N}\right)^{-2}\right]. \tag{9.33}$$

The full conditional for $\boldsymbol{\Gamma}$ will also be required for the chosen Gibbs sampler. Since the $\{\Gamma_{m,t}\}_{m=1}^{M_t}$ are conditionally independent of one another, given $e_t = y_t - \boldsymbol{g_t\theta}$, we obtain

$$p\left(\boldsymbol{\Gamma}|\boldsymbol{\theta}, \mathrm{y}\right) = \prod_{t=1}^{N} p\left(\left\{\Gamma_{m,t}\right\}_{m=1}^{M_t}|e_t\right), \tag{9.34}$$

where

$$p\left(\left\{\Gamma_{m,t}\right\}_{m=1}^{M_t}|e_t\right) \propto \mathcal{N}(e_t|\mu_{e_t}, \sigma^2_{e_t}) \times p\left(\boldsymbol{\Gamma}_t\right) \tag{9.35}$$

with $\boldsymbol{\Gamma}_t = \left\{\Gamma_{m,t}\right\}_{m=1}^{M_t}$. The first step, drawing $\boldsymbol{\theta}^j$ from its posterior, can be performed straightforward, given (9.31) for the linear time series model. The second step consists of obtaining $\boldsymbol{\Gamma}^j = \{\{\Gamma^j_{m,t}\}_{m=1}^{M_t}\}_{t=1}^N$ under the use of (9.34), which involves the product of a normal likelihood and the gamma prior distribution. If we choose the prior as the proposal to a Metropolis-Hastings step then the acceptance probability is obtained as

$$\mathrm{acc}(\boldsymbol{\Gamma}_t;\boldsymbol{\Gamma}'_t) = \min\left(1, \frac{\mathcal{N}(e_t|\mu'_{e_t},\sigma'^2_{e_t})}{\mathcal{N}(e_t|\mu_{e_t},\sigma^2_{e_t})}\right). \tag{9.36}$$

9.4.3 REVERSIBLE JUMP SAMPLER FOR MODEL ORDER ESTIMATION

When fitting an autoregressive model to some real-world data, often the correct model order is unknown as well. That leads us to the model order selection problem, which we address applying a reversible jump sampler (see Troughton and Godsill (1998) and Godsill (2001)).

The model move from the model order P to P' is determined by $q(P,P')$. Then, derived from the M.-H. procedure, the acceptance probability is given as

$$\mathrm{acc}(P;P') = \min\left(1, \frac{\pi(P'|\boldsymbol{\Gamma},\mathrm{y})q(P;P')}{\pi(P|\boldsymbol{\Gamma},\mathrm{y})q(P';P)}\right), \tag{9.37}$$

where

$$\begin{aligned}\pi\left(P|\boldsymbol{\Gamma},\mathrm{y}\right) &= \int_{\boldsymbol{\theta}^{(P)}} \pi\left(P,\boldsymbol{\theta}^{(P)}|\boldsymbol{\Gamma},\mathrm{y}\right) d\boldsymbol{\theta}^{(P)} \\ &\propto \pi(p) \int_{\boldsymbol{\theta}^{(P)}} \pi\left(\mathrm{y}|P,\boldsymbol{\theta}^{(P)},\boldsymbol{\Gamma}\right) \pi\left(\boldsymbol{\theta}^{(P)}\right) d\boldsymbol{\theta}^{(P)}\end{aligned} \tag{9.38}$$

is obtained analytically. $\pi(P)$ denotes a discrete uniform prior on the integers $1,\ldots,P_{\max}$. Two normals, $\pi(\mathrm{y}|P,\boldsymbol{\theta}^{(P)},\boldsymbol{\Gamma})$ arising from the conditionally Gaussian innovations and $\pi(\boldsymbol{\theta}^{(P)})$, the conjugate prior, form the integrand.

9.4.4 RESULTS

To demonstrate the effectiveness of the algorithm for model parameter estimation we provide here just one single exemplary simulation. For the model choice we refer to the AR(5) model used for numerical results in Godsill and Kuruoğlu (1999) for time series with symmetric α-stable noise processes. The innovations $\mathbf{e}$ are obtained applying the CMS method using the α-stable distribution $S_{1.5}(2.3967, 0.8978, 0)$. We perform parameter estimation on a set of one hundred data points synthetically generated from an AR(5) model with parameters $\boldsymbol{\theta} = \{3.54, -5.38, 4.38, -1.93, 0.36\}$. Fast convergence of the model order and model parameters can be observed using the representation, which includes our novel residual approximation (Figure 9.3 solid line, Figure 9.4), while a model that simply truncates the series (9.2) (Figure 9.3 dashed line) reveals a deviation from the true parameter values. These results demonstrate successful parameter estimation for AR(P) models with heavy-tailed and skewed noise processes.

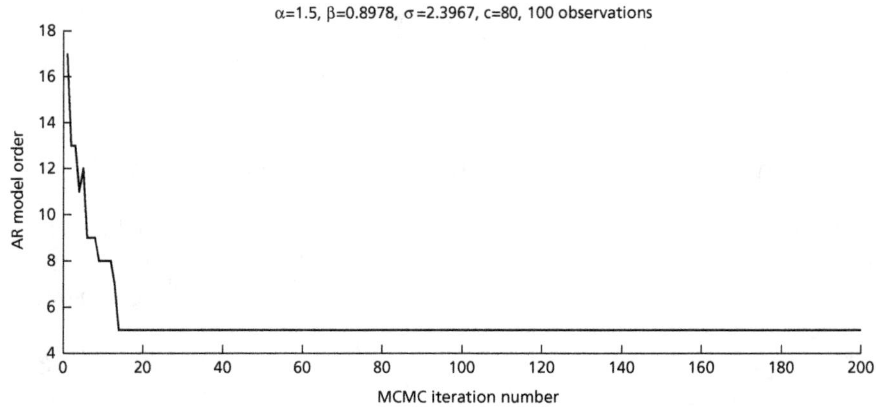

Figure 9.3. RJMCMC sampled AR model order P values

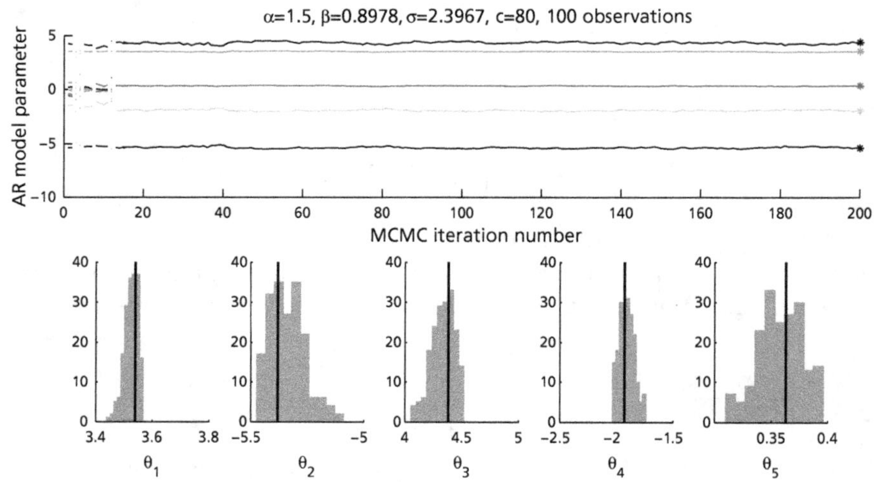

Figure 9.4. Top: MCMC sampled AR model parameter values $\theta_1,\ldots,\theta_5$ using the method, which includes our GAA. The true parameters are marked by "*". Bottom: Histograms from the MCMC output for each parameter. The true parameter values are given by the vertical lines

9.4.5 FURTHER DEVELOPMENTS

In this simple example we have illustrated parameter inference for a standard state-space model driven by α-stable noise. Extension to more elaborate state-space models is a fairly straightforward exercise since our approach to the stable noise is generic. We have shown how to incorporate the Poisson series representation within a Gibbs sampling/reversible jump sampler for both model parameters and model order. We have not in this work attempted to infer the parameters of the α-stable distribution, σ, α, and β. These can in principle be sampled as additional parameters within our framework, and this will be a focus of our future work on the topic. Particular care needs to be taken in the sampling of the auxiliary variables when the stable parameters are unknown, since biases in the estimation can easily occur if these have not properly converged. To overcome this we are developing special sampling schemes involving a combination of rejection sampling and MCMC that will ensure rapid convergence of these parameters over all required ranges of the stable law parameters, and will report these developments in future publications.

9.5 **Continuous-time state-space models**

Real-world data is often of high frequency and arrives at arbitrary times $t_0, t_1, \ldots$. In such cases, a continuous-time formulation is a natural progression. However, we are not aware of any full inference schemes for such processes when the driving noise processes are non-Gaussian and α-stable. One of the main motivations for our work is in fact to obtain suitable representations of the stochastic integrals involved in these models for performing parameter and state inference. It turns out that very elegant extensions of the basic Poisson series representations presented earlier for discrete-time models can be applied in continuous-time cases, enabling fully Bayesian inference about the states and parameters. The basic principles are exactly as for the discrete-time case, with the introduction of additional latent variables which can be thought of as the time of arrival of jumps in the non-Gaussian process.

The class of models we will consider is of the linear stochastic differential equation (SDE) form

$$d\mathbf{X}(t) = A\mathbf{X}(t)dt + \mathbf{h}dL(t) \tag{9.39}$$

where $L(t)$ denotes the α-stable Lévy process; in particular we consider α-stable Lévy motions defined below.

Definition 9.1 *α-stable Lévy motions are stochastic processes; $\{L(t)\}$, where $L(0) = 0$ almost surely and the increments are independent with $L(t) - L(s) \sim S_\alpha((t-s)^{1/\alpha}, \beta, 0)$ for any $0 \le s < t < \infty$ and some $0 < \alpha \le 2, -1 \le \beta \le 1$.*

The P-variate stochastic process has state vector

$$\mathbf{X}(t) = [X^{(0)}(t), \ldots, X^{(P-1)}(t)]^T \tag{9.40}$$

with $X^{(i)}(t)$ being the i-th derivative of $X(t)$.

An equivalent formulation of (9.39) is given by its solution

$$\mathbf{X}(t) = e^{At}\mathbf{X}(0) + \int_0^t e^{A(t-u)}\mathbf{h}dL(u). \tag{9.41}$$

The process also satisfies the recursive relation

$$\mathbf{X}(t) = e^{A(t-s)}\mathbf{X}(s) + \int_s^t e^{A(t-u)}\mathbf{h}dL(u), \quad \text{for all} \quad t > s \geq 0. \tag{9.42}$$

Observations are also assumed linear. Then the continuous-time stochastic process $\{Y_t\}_{t\in[0,\infty)}$ is defined by

$$Y(t) = \mathbf{b}'\mathbf{X}(t) + v(t) \;\; \text{for} \;\; t \in [0, \infty) \tag{9.43}$$

where $\{v(t)\}$ is a Gaussian noise process, and we will assume that discrete observations are available at times t_j.

We can also contemplate nonlinear versions of our models within the same Poisson summation representation, but these would require additional (most likely approximate) discretizations of the system equations, and hence we do not consider them here.

To give a concrete example of such a state-space model, consider the continuous-time autoregression (CAR) model of Brockwell (2001) and Brockwell and Lindner (2009). Let $a_1, a_2, \ldots a_P$ and $b_1, b_1, \ldots b_q$ be real numbers with P and q as given above. Suppose that A is a $P \times P$ matrix defined as

$$A = \begin{cases} -a_1 & \text{if } P = 1 \\ \begin{pmatrix} 0 & 1 & 0 & \cdots & 0 \\ 0 & 0 & 1 & \cdots & 0 \\ \vdots & \vdots & \vdots & & \vdots \\ 0 & 0 & 0 & \cdots & 1 \\ -a_P & -a_{P-1} & -a_{P-2} & \cdots & -a_1 \end{pmatrix} & \text{otherwise,} \end{cases} \tag{9.44}$$

and let $\mathbf{b}$ and $\mathbf{h}$ be P-dimensional vectors

$$\mathbf{b} = \begin{pmatrix} b_0 \\ b_1 \\ \vdots \\ b_{P-1} \end{pmatrix} \quad \text{and} \quad \mathbf{h} = \begin{pmatrix} 0 \\ 0 \\ \vdots \\ 0 \\ 1 \end{pmatrix} \tag{9.45}$$

respectively. For $P = 1$, stationarity is guaranteed if and only if a_1 is positive, but also for $P > 1$ the positivity of the coefficients is a sufficient condition for stationarity. Thus, a_i are assumed to be positive. Such a model has been used in the past to model engineering time series with data arrivals occurring at irregular time instants.

For notational simplicity we will focus here on the first order case, $P = 1$, which leads to the CAR(1), non-Gaussian Ornstein-Uhlenbeck (OU) process. All of the Poisson summation representations presented here can easily be extended, however, to the higher order cases $P > 1$. In the $P = 1$ case, the stochastic integrals required between two observation times t_{j-1} and t_j are

$$X(t_j) = e^{-a_1(t_j - t_{j-1})} X(t_{j-1}) + \int_{t_{j-1}}^{t_j} e^{-a_1(t_j - u)} dL(u), \; j = 2, \ldots, N, \quad (9.46)$$

with not necessarily equidistant time steps $t_j \in [t_{j-1}, t_N)$, where $t_{j-1} > 0$. The driving process is once again an α-stable Lévy process L

$$L(t) - L(s) \sim S_\alpha((t-s)^{1/\alpha}, \beta, 0), \quad t > s.$$

9.5.1 REPRESENTATIONS FOR STOCHASTIC INTEGRALS

We now seek a convenient representation of the stochastic integral required in (9.46). We have modified a basic result given in Samorodnitsky and Taqqu (1994) in order to obtain the conditionally Gaussian form for this integral, the modified Poisson series representation (MPSR).

9.5.2 MODIFIED POISSON SERIES REPRESENTATION

As in the representation for random variables (9.8), we aim for an alternative expression in terms of mean and scale shifted normals for stochastic integrals of the form (9.46). This lemma has previously been presented (without proof) in Lemke and Godsill (2011).

Lemma 9.2 (Modified PSR, Lemke and Godsill (2011)) *Let* $(E, \epsilon, \tilde{m})$ *be a finite measure space and* M *and* α*-stable random measure with* $0 < \alpha < 2$, $\tilde{m}$ *a finite control measure, and* β *the skewness.* $\{\Gamma_1, \Gamma_2, \ldots\}$ *are arrival times of a Poisson process with unit arrival rate,* $\{V_1, V_2, \ldots\}$ *are i.i.d. random variables such that* $V_m \sim \hat{m} = \frac{\tilde{m}}{\tilde{m}(E)}$ *on* E*, and* $\{W_1, W_2, \ldots\}$ *are i.i.d. random variables satisfying*

$$\mathbb{E}[|W_m|^\alpha | V_m] = \mathbb{E}[|W_m|^\alpha] = 1, \quad (9.47)$$

$$\beta(V_m) = \frac{\mathbb{E}[|W_m|^\alpha \operatorname{sign}(W_m) | V_m]}{\mathbb{E}[|W_1|^\alpha]}. \quad (9.48)$$

Then, for any function $f \in F$ *(*f *as defined by (3.4.1), (3.4.2) in Samorodnitsky and Taqqu (1994)) with* $I(f) := \int_E f(u) M(du)$*, the following holds:*

$$S(f) \stackrel{d}{=} I(f), \quad (9.49)$$

where $S(f)$ *is an a.s. converging random series defined as*

$$S(f) = (C_\alpha m(E))^{1/\alpha} \sum_{m=1}^{\infty} [W_m \Gamma_m^{-1/\alpha} f(V_m) - b_m^{(\alpha)} \mu_W \mathbb{E}[f(V_m)]], \quad (9.50)$$

$$b_m^{(\alpha)} = \begin{cases} 0 & if 0 < \alpha < 1, \\ \frac{\alpha-1}{\alpha}(m^{\frac{\alpha-1}{\alpha}} - (m-1)^{\frac{\alpha-1}{\alpha}}) & if 1 < \alpha < 2. \end{cases} \tag{9.51}$$

See Appendix 9.8 for the proof.

The stable integral we address here is

$$I_{t_j}(f) = \int_{E_{t_j}} e^{-a_1(t_j-u)} dL(u), \tag{9.52}$$

where $E_{t_j} = [t_{j-1}, t_j]$, L the Lévy motion, and $\tilde{m}$ is chosen to be the Lebesgue measure. We take the $\{W_m\}_{m=1}^{\infty}$ in (9.49), satisfying $\mathbb{E}[|W_1|^\alpha] = 1$ for the moment, as normally distributed,

$$W_m \sim \mathcal{N}(\mu_W, \sigma_W^2). \tag{9.53}$$

Then, we obtain the stochastic process

$$\{I_{t_j}, t_j \in [0, T]\} \stackrel{d}{=} \tag{9.54}$$

$$\{e^{-a_1 t_j}(C_\alpha m(E_{t_j}))^{1/\alpha} \sum_{m=1}^{\infty} [W_{m,j}\Gamma_{m,j}^{-1/\alpha} e^{a_1 V_{m,j}} - b_m^{(\alpha)} \mu_W \mathbb{E}[e^{a_1 V_1}]]\}, \tag{9.55}$$

where $V_{m,j} \sim \mathrm{U}([t_{j-1}, t_j)]$. The corresponding α-stable distribution $S_\alpha(\sigma_{I_{t_j}}, \beta_{I_{t_j}}, 0)$ is described by the parameters α, $\beta_{\tilde{X}_{t_j}} = \beta$ and

$$\sigma_{I_{t_j}} = \left(\int_{E_{t_j}} e^{-a_1(t_j-x)} dx\right)^{1/\alpha}.$$

In order to remove the restrictive constraint $\mathbb{E}[|W_1|^\alpha] = 1$, we introduce a scaling factor parameter directly into the α-stable distribution of the increments of the model (9.46),

$$L(t) - L(s) \sim S_\alpha((\tilde{\sigma}(t-s))^{1/\alpha}, \beta, 0). \tag{9.56}$$

Free of any additional constraint on μ_W and σ_W^2 the conditionally, Gaussian framework (9.58) remains much as before, but the scaling parameter includes the scaling factor $\tilde{\sigma} = \frac{\mathbb{E}[|W_1|^\alpha]}{C_\alpha}$,

$$\sigma_{\tilde{X}_{t_j}} = \left(\frac{\mathbb{E}[|W_1|^\alpha]}{C_\alpha} \int_{E_{t_j}} e^{-a_1(t_j-x)} dx\right)^{1/\alpha},$$

while the skewness parameter in not affected by scaling. Note that $\left(\frac{\mathbb{E}[|W_1|^\alpha]}{C_\alpha}\right)^{1/\alpha}$ also occurs as the scale parameter in the representation of α-stable random variables.

Therefore the integral (9.52) has a Gaussian distribution conditioned upon $\{\Gamma_{m,j}\}_{m=1}^{\infty}, \{V_{m,j}\}_{m=1}^{\infty}$:

$$I_{t_j}|\{\Gamma_{m,j}\}_{m=1}^{\infty}, \{V_{m,j}\}_{m=1}^{\infty} \tag{9.57}$$

$$\sim \mathcal{N}\left(e^{-a_1 t_j}(t_j - t_{j-1})^{1/\alpha}\mu_W m_{t_j}, e^{-2a_1 t_j}(t_j - t_{j-1})^{2/\alpha}\sigma_W^2 s_{t_j}\right), \tag{9.58}$$

where

$$m_{t_j} = \sum_{m=1}^{\infty}[\Gamma_{m,j}^{-1/\alpha} e^{a_1 V_{m,j}} - b_m^{(\alpha)}\mathbb{E}[e^{a_1 V_1}]], \tag{9.59}$$

$$s_{t_j} = \sum_{m=1}^{\infty} \Gamma_{m,j}^{-2/\alpha} e^{2a_1 V_{m,j}}. \tag{9.60}$$

Closed form result for $P = 1$

We note that, for $P = 1$, we also have a direct expression for the parameters of the α-stable distribution. For the iterative process representation (9.46) we obtain the exact α-stable distribution for the stable integral $I(e^{-a_1(t_j-u)})$ by discretizing the process over some short time intervals of duration $\delta_{t_j} = \frac{t_j - t_{j-1}}{D}$. Hence, using result (1.2.1) from Samorodnitsky and Taqqu (1994), we have

$$I\left(e^{-a_1(t_j-u)}\right) = \int_{u=t_{j-1}}^{t_j} e^{-a_1(t_j-u)}dL(u) \approx \sum_{q=0}^{D-1} e^{-a_1(t_j-q\delta_{t_j})}\delta_{t_j}L_q$$

$$\sim S_\alpha\left(\left(\sum_{q=0}^{D-1}(e^{-a_1(t_j-q\delta_{t_j})})^\alpha \delta_{t_j}\right)^{1/\alpha}, \frac{\sum_{q=0}^{D-1}\beta(e^{-a_1(t_j-q\delta_{t_j})})^\alpha\delta_{t_j}}{\sum_{q=0}^{D-1}(e^{-a_1(t_j-q\delta_{t_j})})^\alpha\delta_{t_j}}, 0\right)$$

$$\overset{D\to\infty}{\sim} S_\alpha\left(\left(\frac{1 - e^{-a_1\alpha(t_j-t_{j-1})}}{a_1\alpha}\right)^{1/\alpha}, \beta, 0\right). \tag{9.61}$$

(The same result can also be obtained directly from results on stochastic integration of α-stable processes (Samorodnitsky and Taqqu (1994), Chapter 3).)

Given this direct characterization of the state transition density, the scale mixture of normals representation can be used for inference in the symmetric $\beta = 0$ version. Otherwise for the asymmetric case the α-stable distribution (9.61) can be represented by the PSR for random variables. However, for the $P > 1$ cases we are not aware of a comparable explicit characterization. Hence, instead of working with this result, we work in our examples with the more general form of (9.58).

Truncation of the series and approximation of the residual terms

Expressing the $P = 1$ model as a state-space model, incorporating our conditionally Gaussian framework for the stochastic integral leads to

$$X_{t_j} = e^{-a_1(t_j - t_{j-1})} X_{t_{j-1}} + I_{t_j}, \tag{9.62}$$

where

$$I_{t_j} | \{\Gamma_{m,j}\}_{m=1}^{\infty}, \{V_{m,j}\}_{m=1}^{\infty} \tag{9.63}$$

$$\sim \mathcal{N}\left(e^{-a_1 t_j}(t_j - t_{j-1})^{1/\alpha} \mu_W m_{t_j}, e^{-2a_1 t_j}(t_j - t_{j-1})^{2/\alpha} \sigma_W^2 s_{t_j}\right). \tag{9.64}$$

As in the PSR for random variables, the infinite series needs to be truncated at some finite $m = M_j$ in practice, such that $m_{t_j} = \tilde{m}_{t_j} + R_{1,j}$ and $s_{t_j} = \tilde{s}_{t_j} + R_{2,j}$, where

$$\tilde{m}_{t_j} = \sum_{m=1}^{M_j} \Gamma_{m,j}^{-1/\alpha} e^{a_1 V m,j}, \quad \tilde{s}_{t_j} = \sum_{m=1}^{M_j} \Gamma_{m,j}^{-2/\alpha} e^{2a_1 V_{m,j}}. \tag{9.65}$$

We approximate the remaining summation terms of the mean and variance of the conditionally Gaussian framework,

$$R_{1,j} := \lim_{d \to \infty} \sum_{m:\Gamma_m > c} \Gamma_{m,j}^{-1/\alpha} e^{a_1 V_{m,j}} - \mathbb{E}[e^{a_1 V_1}] \sum_{m=1}^{d} b_m^{(\alpha)},$$

$$R_{2,j} := \lim_{d \to \infty} \sum_{m:\Gamma_m > c} \Gamma_{m,j}^{-2/\alpha} e^{2a_1 V_{m,j}},$$

assuming some two-dimensional Gaussian distribution $\mathcal{N}(\mu_{R_j}^{(\alpha)}, \Sigma_{R_j}^{(\alpha)})$. We use the GAMA for the residual approximation as presented in Section 9.3.1. Then,

$$\mu_{R_j}^{(\alpha)} = \begin{bmatrix} \frac{\alpha}{1-\alpha} c^{\frac{\alpha-1}{\alpha}} \frac{1}{\Delta t_j a_1} \left(e^{a_1 t_j} - e^{a_1 t_{j-1}}\right) \\ \frac{\alpha}{2-\alpha} c^{\frac{\alpha-2}{\alpha}} \frac{1}{2\Delta t_j a_1} \left(e^{2a_1 t_j} - e^{2a_1 t_{j-1}}\right) \end{bmatrix} \tag{9.66}$$

and

$$\Sigma_{R_j}^{(\alpha)} = \begin{bmatrix} \frac{\alpha}{2-\alpha} c^{\frac{\alpha-2}{\alpha}} \frac{1}{2\Delta t_j a_1} \left(e^{2a_1 t_j} - e^{2a_1 t_{j-1}}\right) & \frac{\alpha}{3-\alpha} c^{\frac{\alpha-3}{\alpha}} \frac{1}{3\Delta t_j a_1} \left(e^{3a_1 t_j} - e^{3a_1 t_{j-1}}\right) \\ \frac{\alpha}{3-\alpha} c^{\frac{\alpha-3}{\alpha}} \frac{1}{3\Delta t_j a_1} \left(e^{3a_1 t_j} - e^{3a_1 t_{j-1}}\right) & \frac{\alpha}{4-\alpha} c^{\frac{\alpha-4}{\alpha}} \frac{1}{4\Delta t_j a_1} \left(e^{4a_1 t_j} - e^{4a_1 t_{j-1}}\right) \end{bmatrix}. \tag{9.67}$$

Thus, the conditionally Gaussian framework for $X(t_j)$ results in

$$X(t_j) | \{\Gamma_{m,j}\}_{m=1}^{M_j}, \{V_{m,j}\}_{m=1}^{M_j} \sim \mathcal{N}\left(\mu_{X_{t_j}}, \sigma_{X_{t_j}}^2\right), \tag{9.68}$$

where

$$\mu_{X_{t_j}} = e^{-a_1 t_j}(t_j - t_{j-1})^{1/\alpha}\mu_W \left(\sum_{m:\Gamma_{m,j}<c} \Gamma_{m,j}^{-1/\alpha} e^{a_1 V_m^j} + R_{1,j} \right) \tag{9.69}$$

$$\sigma^2_{X_{t_j}} = e^{-2a_1 t_j}(t_j - t_{j-1})^{2/\alpha}\sigma^2_W \left(\sum_{m:\Gamma_{m,j}<c} \Gamma_{m,j}^{-2/\alpha} e^{2a_1 V_m^j} + R_{2,j} \right), \tag{9.70}$$

and

$$\begin{pmatrix} R_{1,j} \\ R_{2,j} \end{pmatrix} \sim \mathcal{N}\left(\mu_{R_j}^{(\alpha)}, \Sigma_{R_j}^{(\alpha)}\right) \tag{9.71}$$

with $\mu_{R_j}^{(\alpha)}$ and $\Sigma_{R_j}^{(\alpha)}$ as in (9.66) and (9.67).

---

for $n = 1, \ldots, N$ **do**
  sample $V_{m,0}^{(n)} \sim \mathrm{U}([0, t_0])$, $(\Gamma_{m,0}^{(n)} - \Gamma_{m-1,0}^{(n)}) \sim \mathrm{Exponential}(1)$ and $\left(R_{1,0}^{(n)}, R_{2,0}^{(n)}\right) \sim \mathcal{N}\left(\mu_{R_0}^{(\alpha)}, \Sigma_{R_0}^{(\alpha)}\right)$ (9.66,9.67)
  compute $\tilde{m}_{t_0}^{(n)}$, $\tilde{s}_{t_0}^{(n)}$ (9.65),
  assign $\tilde{w}_0^{(n)}$ and $X_0^{(n)}$
end
for $j = 1, \ldots, T$ **do**
  resample according to weights
  for $n = 1, \ldots, N$ **do**
    sample $V_{m,j}^{(n)} \sim \mathrm{U}([t_{j-1}, t_j])$, $(\Gamma_{m,j}^{(n)} - \Gamma_{m-1,j}^{(n)}) \sim \mathrm{Exponential}(1)$
    and $\left(R_{1,j}^{(n)}, R_{2,j}^{(n)}\right) \sim \mathcal{N}\left(\mu_{R_j}^{(\alpha)}, \Sigma_{R_j}^{(\alpha)}\right)$ (9.66,9.67)
    compute $\tilde{m}_{t_j}^{(n)}$, $\tilde{s}_{t_j}^{(n)}$ (9.65)
    compute weights $w_{t_j}^{(n)} \propto p(y_{t_j}|y_{1:t_{j-1}}, \{\tilde{m}_{1:t_j}^{(n)}\}, \{\tilde{s}_{1:t_j}^{(n)}\})$ using the Kalman Filter
  end
  normalize weights
end

Figure 9.5: CAR(1) state estimation—pseudo code

9.5.3 EXAMPLE: SEQUENTIAL STATE INFERENCE FOR THE $P = 1$ MODEL

In this section we provide example simulations for sequential state inference in a $P = 1$ model, using the representation of the stochastic integral given at the end of the Section 9.5.2. Hence, using the approximated conditionally Gaussian distribution (9.68) we can write

$$X_{t_j} = e^{-a_1(t_j - t_{j-1})} X_{t_{j-1}} + e^{-a_1 t_j}(t_j - t_{j-1})^{1/\alpha} \mu_W (m_{t_j} + R_{1,j}) + u_{t_j} \quad (9.72)$$

$$Y_{t_j} = B_{t_j} X_{t_j} + w_{t_j}, \quad (9.73)$$

where

$$u_{t_j} | \{\Gamma_{m,j}\}_{m=1}^{M_j}, \{V_{m,j}\}_{m=1}^{M_j} \sim \mathcal{N}(0, e^{-2a_1 t_j}(t_j - t_{j-1})^{2/\alpha} \sigma_W^2 (s_{t_j} + R_{2,j})), \quad (9.74)$$

$$w_{t_j} \sim \mathcal{N}(0, \sigma_w^2). \quad (9.75)$$

We note that the MPSR gives us a very similar framework to the PSR for random variables with the main addition that it is conditioned on two sets of auxiliary variables $\{\Gamma_{m,j}\}_{m=1}^{M_j}$ and $\{V_{m,j}\}_{m=1}^{M_j}$. The residual approximations $(R_{1,j}, R_{2,j})$ can be done according to the same scheme as for the PSR for random variables, computing the means and the variance-covariance matrix of the remaining sums in the conditionally Gaussian framework. A brief pseudo-code summarizing the main steps in our Rao-Blackwellized particle filter (RBPF) implementation is now given. For a detailed version of the general RBPF, see Appendix 9.9.

Results

The CAR(1) process is simulated for $t_N = 100$ time steps with the parameters $\alpha = 1.2, \beta = 0.88, \sigma = 1.9$ for the α-stable distribution of the stochastic integral, which corresponds to the parameters $\alpha = 1.2, \mu_W = 1, \sigma_W = 1$ using the MPSR. The CAR(1) model parameter is set to $a_1 = 0.1$ and the observation noise follows the Gaussian distribution $\mathcal{N}(0, 30)$. Thus, the stochastic integral and therefore the dynamic noise follows a heavy-tailed and skewed distribution, which is evidenced by the mainly positive jumps and their size. Figure 9.6 shows the true states, the observed data, and the estimated states in comparison.

9.5.4 FURTHER DEVELOPMENTS

Our simulations here are for state estimation only, in order to provide an example of inference with our approach in a concrete setting. Current work is focusing on joint state and parameter estimation for the α-stable process and also for inference in $P > 1$ cases, for which no direct characterization of the transition density is available, to our knowledge.

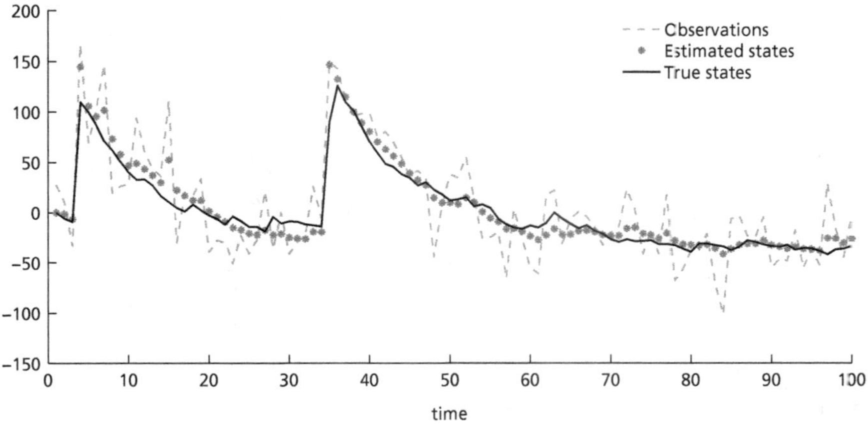

Figure 9.6. RBPF sampled state estimates $X_{0:t_{100}}$

9.6 Conclusions

In this paper we have shown that the conditionally Gaussian framework for α-stable random variables, obtained by the Poisson series representation in combination with our residual approximations, the Gaussian approximation approach (GAA), and the Gaussian approximation of moments approach (GAMA) provide representations which can be used in a straightforward manner to perform Bayesian Monte Carlo inference for AR(P) model parameters. Furthermore, we extended this conditionally Gaussian framework to continuous-time state-space models by introducing a modified Poisson series representation (MPSR) for stochastic integrals. Using our novel residual approximation GAMA also for the MPSR we demonstrated a Rao-Blackwellized particle filter for state estimation for the CAR(1) model.

APPENDICES

9.7 Gaussian approximation of moments

The remaining summation terms are studied by reverting to the Poisson process representation of the Γ_ms on a finite interval $[c, d]$. Specifically, since $\{\Gamma_m\}$ is a unit rate Poisson process, the number of the Γs in the interval follows a Poisson distribution,

$$|\{\Gamma_m : \Gamma_m \in [c, d]\}| \sim \text{Poisson}(d - c) \quad \text{for} \quad d > c, \tag{9.76}$$

and each Γ_m is uniformly and independent distributed on $[c, d]$,

$$\Gamma_m \sim \text{U}([c, d]). \tag{9.77}$$

Taking the limit as $d \to \infty$ accounts for all residual terms. Taking (9.11) and (9.12) with $d \to \infty$ we compute the limits for $\mu_R^{(\alpha)}$ and $\Sigma_R^{(\alpha)}$ as

$$\mu_R^{(\alpha)} = \begin{bmatrix} \mathbb{E}[R_1] \\ \mathbb{E}[R_2] \end{bmatrix} = \begin{bmatrix} \begin{cases} \lim_{d\to\infty} \frac{\alpha}{\alpha-1}\left(d^{\frac{\alpha-1}{\alpha}} - c^{\frac{\alpha-1}{\alpha}}\right) & \text{if } \alpha < 1, \\ \lim_{d\to\infty} \frac{\alpha}{\alpha-1}\left(d^{\frac{\alpha-1}{\alpha}} - c^{\frac{\alpha-1}{\alpha}}\right) - d^{\frac{\alpha-1}{\alpha}} & \text{if } \alpha > 1 \end{cases} \\ \lim_{d\to\infty} \frac{\alpha}{\alpha-2}\left(d^{\frac{\alpha-2}{\alpha}} - c^{\frac{\alpha-2}{\alpha}}\right) \end{bmatrix}$$

$$= \begin{bmatrix} \begin{cases} \lim_{d\to\infty} \frac{\alpha}{\alpha-1}\left(d^{\frac{\alpha-1}{\alpha}} - c^{\frac{\alpha-1}{\alpha}}\right) & \text{if } \alpha < 1, \\ \frac{\alpha}{\alpha-1}\left(-c^{\frac{\alpha-1}{\alpha}}\right) & \text{if } \alpha > 1 \end{cases} \\ \lim_{d\to\infty} \frac{\alpha}{\alpha-2}\left(d^{\frac{\alpha-2}{\alpha}} - c^{\frac{\alpha-2}{\alpha}}\right) \end{bmatrix} = \begin{bmatrix} \frac{\alpha}{1-\alpha} c^{\frac{\alpha-1}{\alpha}} \\ \frac{\alpha}{2-\alpha} c^{\frac{\alpha-2}{\alpha}} \end{bmatrix}. \quad (9.78)$$

In order to obtain the variance-covariance matrix

$$\Sigma_R^{(\alpha)} = \begin{bmatrix} \mathrm{Var}[R_1] & \mathrm{Cov}[R_1, R_2] \\ \mathrm{Cov}[R_2, R_1] & \mathrm{Var}[R_2] \end{bmatrix} \quad (9.79)$$

on the limit $d \to \infty$, we work out each of the components as

$$\mathrm{Var}[R_1] = \lim_{d\to\infty} \frac{\alpha}{\alpha-2}\left(d^{\frac{\alpha-2}{\alpha}} - c^{\frac{\alpha-2}{\alpha}}\right) - \frac{1}{d-c}\frac{\alpha^2}{(\alpha-1)^2}\left(d^{\frac{\alpha-1}{\alpha}} - c^{\frac{\alpha-1}{\alpha}}\right)^2 \quad (9.80)$$

$$= \frac{\alpha}{2-\alpha} c^{\frac{\alpha-2}{\alpha}}, \quad (9.81)$$

$$\mathrm{Var}[R_2] = \lim_{d\to\infty} \frac{\alpha}{\alpha-4}\left(d^{\frac{\alpha-4}{\alpha}} - c^{\frac{\alpha-4}{\alpha}}\right) - \frac{1}{d-c}\frac{\alpha^2}{(\alpha-2)^2}\left(d^{\frac{\alpha-2}{\alpha}} - c^{\frac{\alpha-2}{\alpha}}\right)^2 \quad (9.82)$$

$$= \frac{\alpha}{4-\alpha} c^{\frac{\alpha-4}{\alpha}}, \quad (9.83)$$

and

$$\mathrm{Cov}[R_1, R_2] = \mathrm{Cov}[R_2, R_1] = \lim_{d\to\infty} \sum_{m:\Gamma_m\in[c,d]} \sum_{n:\Gamma_n\in[c,d]} Cov\left(\Gamma_m^{1/\alpha}, \Gamma_n^{-2/\alpha}\right)$$

$$= \frac{\alpha}{\alpha-3}\left(d^{\frac{\alpha-3}{\alpha}} - c^{\frac{\alpha-3}{\alpha}}\right)$$

$$- \frac{1}{d-c}\frac{\alpha^2}{(\alpha-1)(\alpha-2)}\left(d^{\frac{\alpha-1}{\alpha}} - c^{\frac{\alpha-1}{\alpha}}\right)\left(d^{\frac{\alpha-2}{\alpha}} - c^{\frac{\alpha-2}{\alpha}}\right) \xrightarrow{d\to\infty} \frac{\alpha}{3-\alpha} c^{\frac{\alpha-2}{\alpha}}. \quad (9.84)$$

9.8 Modified Poisson series representation

Including, also, the $\alpha = 1$ case for the proof, $S(f)$ in (9.5.1) then reads

$$S(f) = (C_\alpha m(E))^{1/\alpha} \sum_{i=1}^{\infty} [W_i \Gamma_i^{-1/\alpha} f(V_i) - b_i^{(\alpha)} \mu_W \mathbb{E}[f(V_i)]] + \eta_f, \quad (9.85)$$

where

$$\eta_f = \begin{cases} 0 & \text{if } \alpha \neq 1 \\ \frac{2}{\pi} \int_E \mathbb{E}[W_1 \ln(|W_1|) f(V_1)] f(V_1) m(dV_1) + & \\ \qquad + \frac{2}{\pi} \ln(\frac{2}{\pi} m(E)) \int_E \beta(X) |f(x)| \operatorname{sign}(f(x)) m(dx) & \text{if } \alpha = 1. \end{cases} \tag{9.86}$$

Proof. Let $\tilde{W}_i = W_i f(V_i)$, $i = 1, 2, \ldots$. To verify that $S(f)$ converges in distribution to the α-stable integral $I(f)$ we apply Theorem 1.4.5 from Samorodnitsky and Taqqu (1994) and conclude that

$$\sum_{i=1}^{\infty} [W_i \Gamma_i^{-1/\alpha} f(V_i) - b_i^{(\alpha)} \mu_W \mathbb{E}[f(V_i)]] \xrightarrow{a.s.} X \sim S_\alpha(\sigma, \beta, \mu) \tag{9.87}$$

with

$$\sigma^\alpha = \frac{\mathbb{E}[|\tilde{W}_1|^\alpha]}{C_\alpha}, \quad \beta = \frac{\mathbb{E}[|\tilde{W}_1|^\alpha \operatorname{sign}(\tilde{W}_1)]}{\mathbb{E}[|\tilde{W}_1|^\alpha]} \tag{9.88}$$

and

$$\mu = \begin{cases} 0 & \text{if } \alpha \neq 1 \\ -\mathbb{E}[\tilde{W}_1 \ln |\tilde{W}_1|] & \text{if } \alpha = 1. \end{cases} \tag{9.89}$$

Rewriting $\mathbb{E}[|\tilde{W}_1|^\alpha]$ and $\mathbb{E}[|\tilde{W}_1|^\alpha \operatorname{sign}(\tilde{W}_1)]$ in (9.88), leads to

$$\begin{aligned} \mathbb{E}[|\tilde{W}_1|^\alpha] &= \mathbb{E}[|W_1 f(V_1)|^\alpha] \\ &= \int \int_E |W_1|^\alpha |f(V_1)|^\alpha g(W_1|V_1) dW_1 \hat{m}(dV_1) \\ &= \int_E \mathbb{E}[|W_1|^\alpha | V_1] |f(V_1)|^\alpha \hat{m}(dV_1) \\ &= \int_E \mathbb{E}[|W_1|^\alpha] |f(V_1)|^\alpha \hat{m}(dV_1) \\ &= \int_E |f(x)|^\alpha \hat{m}(dx), \end{aligned} \tag{9.90}$$

$$\begin{aligned} \mathbb{E}[|\tilde{W}_1|^\alpha \operatorname{sign}(\tilde{W}_1)] &= \mathbb{E}[|W_1 f(V_1)|^\alpha \operatorname{sign}(W_1 f(V_1))] \\ &= \int \int_E |W_1|^\alpha |f(V_1)|^\alpha \operatorname{sign}(W_1) \operatorname{sign} f(V_1) g(W_1|V_1) \\ &\quad \times dW_1 \hat{m}(dV_1) \\ &= \int_E \mathbb{E}[|W_1|^\alpha \operatorname{sign}(W_1) | V_1] |f(V_1)|^\alpha \operatorname{sign}(f(V_1)) \hat{m}(dV_1) \\ &= \int_E \mathbb{E}[|W_1|^\alpha] \beta(V_1) |f(V_1)|^\alpha \operatorname{sign}(V_1) \hat{m}(dV_1) \\ &= \int_E \beta(x) |f(x)|^\alpha \operatorname{sign}(f(x)) \hat{m}(dx). \end{aligned} \tag{9.91}$$

$$
\begin{aligned}
\mathbb{E}[\tilde{W}_1 \ln(|\tilde{W}_1|)] &= \mathbb{E}[W_1 f(V_1) \operatorname{sign}(|W_1 f(V_1)|)] \\
&= \int \int_E W_1 f(V_1) \ln(|W_1||f(V_1)|) g(W_1|V_1) dW_1 \hat{m}(dV_1) \\
&= \int_E \mathbb{E}[W_1|V_1] f(V_1) \ln(|f(V_1)|) \hat{m}(dV_1) \\
&\quad + \int_E \mathbb{E}[W_1 \ln(|W_1|)|f(V_1)] f(V_1) \hat{m}(dV_1) \\
&= \int_E \beta(x) f(x) \ln(|f(x)|) \hat{m}(dx) \\
&\quad + \int_E \mathbb{E}[W_1 \ln(|W_1|)|f(V_1)] f(V_1) \hat{m}(dV_1).
\end{aligned} \tag{9.92}
$$

Thus, by Property 1.2.3 in Samorodnitsky and Taqqu (1994), p. 11, the random variable $S(f)$ has a $S_\alpha(\tilde{\sigma}, \tilde{\beta}, \tilde{\mu})$ distribution with

$$
\begin{aligned}
\tilde{\sigma}^\alpha &= (C_\alpha m(E))\sigma^\alpha \\
&= \int_E |f(x)|^\alpha m(dx),
\end{aligned} \tag{9.93}
$$

$$
\begin{aligned}
\tilde{\beta} = \beta &= \frac{\int_E \beta(x)|f(x)|^\alpha \operatorname{sign}(f(x)) \hat{m}(dx)}{\int_E |f(x)|^\alpha \hat{m}(dx)} \\
&= \frac{\int_E \beta(x)|f(x)|^\alpha \operatorname{sign}(f(x)) m(dx)}{\int_E |f(x)|^\alpha m(dx)}
\end{aligned} \tag{9.94}
$$

and a shift parameter, which is zero for $\alpha \neq 1$. In the specific case of $\alpha = 1$ where $C_1 = \frac{2}{\pi}$, the shift parameter can be computed as follows:

$$
\begin{aligned}
\tilde{\mu} &= (C_1 m(E))\mu + \eta_f \\
&= -\frac{2}{\pi} m(E) \int_E \beta(x) f(x) \ln(|f(x)|) \hat{m}(dx) \\
&\qquad - \frac{2}{\pi} m(E) \int_E \mathbb{E}[W_1 \ln(|W_1|)|f(V_1)] f(V_1) \hat{m}(dV_1) \\
&\qquad - (\frac{2}{\pi})^2 m(E) \ln(\frac{2}{\pi} m(E)) \sigma\beta + \eta_f \\
&= -\frac{2}{\pi} \int_E \beta(x) f(x) \ln(|f(x)|) m(dx) \\
&\qquad - \frac{2}{\pi} \int_E \mathbb{E}[W_1 \ln(|W_1|)|f(V_1)] f(V_1) m(dV_1) \\
&\qquad - (\frac{2}{\pi})^2 m(E) \ln(\frac{2}{\pi} m(E)) \frac{\pi}{2} \int_E \beta(X)|f(x)| \operatorname{sign}(f(x)) \hat{m}(dx) + \eta_f
\end{aligned}
$$

$$= -\frac{2}{\pi}\int_E \beta(x)f(x)\ln(|f(x)|)m(dx)$$

$$- \frac{2}{\pi}\int_E \mathbb{E}[W_1 \ln(|W_1|)|f(V_1)]f(V_1)m(dV_1)$$

$$- \frac{2}{\pi}\ln(\frac{2}{\pi}m(E))\int_E \beta(X)|f(x)|\,\mathrm{sign}(f(x))m(dx) + \eta_f$$

$$= -\frac{2}{\pi}\int_E \beta(x)f(x)\ln(|f(x)|)m(dx), \tag{9.95}$$

which gives

$$\tilde{\mu} = \begin{cases} 0 & \text{if } \alpha \neq 1, \\ -\frac{2}{\pi}\int_E \beta(x)f(x)\ln(|f(x)|)m(dx) & \text{if } \alpha = 1. \end{cases} \tag{9.96}$$

By Property 3.4.1 in Samorodnitsky and Taqqu (1994), p. 124, the α-stable integral $I(f)$ is a random variable

$$I(f) \sim S_\alpha(\sigma_f, \beta_f, \mu_f), \tag{9.97}$$

where

$$\sigma_f = \left(\int_E |f(x)|^\alpha m(dx)\right)^{1/\alpha}, \tag{9.98}$$

$$\beta_f = \frac{\int_E |f(x)|^\alpha \,\mathrm{sign}(f(x))\beta(x)m(dx)}{\int_E |f(x)|^\alpha m(dx)} \tag{9.99}$$

and

$$\mu_f = \begin{cases} 0 & \text{if } \alpha \neq 1, \\ -\frac{2}{\pi}\int_E \beta(x)f(x)\ln(|f(x)|)m(dx) & \text{if } \alpha = 1. \end{cases} \tag{9.100}$$

Hence $S(f)$ has the same distribution as the α-stable integral.

9.9 Rao-Blackwellized particle filter for state estimation

Assuming an underlying state-space model (9.22, 9.23), we aim to extract the unknown states $x^L_{0:t}$ given the observations $y_{0:t}$. By the chain rule property, the filtering distribution $p(X_{0:t}|y_{0:t})$ can be partitioned into a conditional distribution for the linear part $x^L_{0:t}$ and the nonlinear part $x^N_{0:t}$, resulting in

$$p(X_{0:t}|y_{0:t}) = p(x^L_{0:t}, x^N_{0:t}|y_{0:t}) = p(x^N_{0:t}|y_{0:t})p(x^L_{0:t}|x^N_{0:t}, y_{0:t}). \tag{9.101}$$

The possibility of marginalizing out the linear part suggests the use of a Rao-Blackwellized particle filter (RBPF). Expressing the AR(P) process as a conditionally Gaussian SSM gives us $x^L_{0:t} := x_{0:t}$ and $x^N_{0:t} = \{\{\Gamma_{m,0:t}\}^M_{m=1}, \{V_{m,0:t}\}^M_{m=1}, (R_{1,0:t}, R_{2,0:t})\}$. Then, the Rao-Blackwellized particle filter runs sequential Monte Carlo on the auxiliary variables $x^N_{0:t}$ with target distribution $p(x^N_{0:t}|y_{0:t})$, while $p(x^L_{0:t}|x^N_{0:t}, y_{0:t})$ is obtained analytically using the Kalman filter (see Cappé et al. (2007)).

A particle filter approximates the true posterior distribution by importance weights of particles,

$$\hat{p}(x^N_{0:t}|y_{0:t}) = \sum_{i=1}^{N} w_t^{(i)} \delta_{x^{N,(i)}_{0:t}}(x^N_{0:t}), \tag{9.102}$$

where $\{w_t^{(i)}\}_{i=1,\dots,N}$ denotes the normalized importance weights with N particles. We shall recall the main steps of the sequential Monte Carlo procedure. First, new particles from a proposal density $\pi(x^N_t|x^N_{0:t}, y_{0:t})$ are generated:

$$\tilde{x}_t^{N,(i)} \sim \pi(x^N_t|x^N_{0:t-1}, y_{0:t}). \tag{9.103}$$

Second, the weight $\tilde{w}_t^{(i)}$ of each particle is updated:

$$\tilde{w}_t^{(i)} \propto w_{t-1}^{(i)} \frac{p(y_t|\tilde{x}_t^{N,(i)})p(\tilde{x}_t^{N,(i)}|x^{N,(i)}_{0:t-1})}{\pi(\tilde{x}_t^{N,(i)}|x^{N,(i)}_{0:t-1}, y_{0:t})}. \tag{9.104}$$

To avoid an increasing number of particles with weights close to zero and hence with low impact on the estimated discrete distribution a third resampling step is added. The resampling step adapts the number of the particles according to their weights, such that

$$p(x^{N,(k)}_{0:t} = \tilde{x}^{N,(i)}_{0:t}) = w_t^{(i)}, \quad i = 1, \dots, N. \tag{9.105}$$

For the Rao-Blackwellized particle filter the weight update equation (9.104) is modified due to the non-Markovian marginalized system,

$$p(y_t|y_{0:t}, x^N_{0:t}) \neq p(y_t|x^N_t), \tag{9.106}$$

resulting in recursive weight equation given by

$$\tilde{w}_t^{(i)} \propto w_{t-1}^{(i)} \frac{p(y_t|y_{0:t}, \tilde{x}^{N,(i)}_{0:t})p(\tilde{x}_t^{N,(i)}|x^{N,(i)}_{0:t-1})}{\pi(\tilde{x}_t^{N,(i)}|x^{N,(i)}_{0:t-1}, y_{0:t})}. \tag{9.107}$$

The posterior for the linear states is given as the random Gaussian mixture approximation

$$p(x^L_t|y_{0:t}) \approx \sum_{i=1}^{N} w_t^{(i)} p(x^L_t|x^{N,(i)}_{0:t}, y_{0:t}). \tag{9.108}$$

The conditional densities $p(x^L_t|x^{N,(i)}_{0:t}, y_{0:t})$ are computed using the Kalman filter. At this point we review the basic steps of the Kalman filter. The predictive distribution for the x^L_t based on the previous data $y_{0:t-1}$ and $x^N_{0:t}$ can be obtained by

$$p(x^L_t|x^N_{0:t}, y_{0:t-1}) = \mathcal{N}(x^L_t|\mu^-_t, C^-_t), \tag{9.109}$$

where

$$\mu_t^- = A(x_{0:t}^N)\mu_{t-1} + C(x_{0:t}^N), \tag{9.110}$$

$$C_t^- = A(x_{0:t}^N)C_{t-1}A(x_{0:t}^N)^T + C_u(x_{0:t}^N). \tag{9.111}$$

In the first recursion some initializing values for μ_0 and C_0 are chosen. In a further step the distribution (9.109) is updated through Bayes' theorem including the new observation point y_t, leading to

$$p(x_t^L|x_{0:t}^N, y_{0:t}) = \mathcal{N}(x_t^L|\mu_t, C_t), \tag{9.112}$$

where

$$\mu_t = \mu_t^- + K_t(y_t - B(x_{0:t}^N)\mu_t^-), \tag{9.113}$$

$$C_t^- = (I - K_t B(x_{0:t}^N))C_t^-, \tag{9.114}$$

$$K_t = C_t^- B^T(x_{0:t}^N)(B(x_{0:t}^N)C_t^- B^T(x_{0:t}^N) + C_v)^{-1}. \tag{9.115}$$

K_t is known as the Kalman gain and corrects the prediction μ_t^- due to the new observation to get the optimal estimate μ_t.

To compute the weighting step (9.107), the term $p(y_t|x_{0:t}^N, y_{0:t-1})$ is required. Taking the predictive distribution (9.109) and the known structure of the observation equation (9.23) we can easily obtain

$$p(y_t|x_{0:t}^N, y_{0:t-1}) = \mathcal{N}(y_t|\mu_{y_t}, C_{y_t}), \tag{9.116}$$

where

$$\mu_{y_t} = B(x_{0:t}^N)\mu_t^-, \tag{9.117}$$

$$C_{y_t} = B(x_{0:t}^N)C_t^- B(x_{0:t}^N)^T + C_v. \tag{9.118}$$

10 Martingale unobserved component models

Neil Shephard

10.1 Introduction

When I went to the LSE in 1986 as a new graduate student I wanted to study time series. My first supervisor was Jim Durbin, who was excited about his new paper Harvey and Durbin (1986) which used time series unobserved component models to estimate the size of a public policy intervention. Jim was about to retire and so when Andrew Harvey returned from sabbatical I asked if I could work with him as a Ph.D. student.

The central unobserved component model is the Gaussian "random walk plus noise model" or "local level model." This is a profoundly important model for it: (1) rationalizes exponentially weighted moving average (EWMA) forecasting; (2) is the simplest state-space model that can be handled through the Kalman filter; and (3) through its analysis led to the rational expectations school of forward looking expectations in macroeconomics (for good and for bad). I learnt time series modeling from Andrew through thinking about this model, its extensions, and what it implies for the degree of "discounting the data." Such models should be flexible but sensible and importantly should fit the data. Much of the flavor of this approach can be seen in my still favorite time series book by Harvey (1981), in the exhaustive Harvey (1989), and it was elegantly broadcast in Durbin and Koopman (2001, Ch. 2). The question answered here is "how should one allow this rate of discounting to change through time?"

I advocate the following solution. Replace the random walk with a martingale and the independent and identically distribution (i.i.d.) noise with a martingale difference sequence. I call this class of models "martingale unobserved component models." These martingales are parameterized through stochastic volatility (SV) processes.

I started working on stochastic volatility before I left the LSE in 1991. Charles Goodhart asked me for thoughts on how to remove the time varying diurnality seen in the volatility in exchange rate markets, which had been clearly revealed by the work of Richard Olsen and his colleagues in Zurich. Andrew and Esther Ruiz were working on seasonality, and so we discussed this challenge. To put it into their framework we came up with a "stochastic variance model," where the returns could be transformed into a linear state-space form and so handled using their methods. Once we had that we went back to the simplest model as being interesting in its own right (and forgot about the diurnality). This model is now most accurately called a log-normal stochastic volatility model. At the time we thought "stochastic variance models" were new but we found from Stephen Taylor the existing work on the topic. Our initial multivariate work was published in Harvey et al. (1994). A discussion of the history of SV models is given in Shephard (2005b, Ch. 1). The linkage of SV with both realized volatility (e.g., Barndorff-Nielsen and Shephard (2002) and Barndorff-Nielsen et al. (2008)) and simulation based inference has meant that SV models are now extremely popular in econometrics and have been a common theme to much of my research in the last twenty years.

Martingale unobserved component models parameterized through stochastic volatility innovations are related to Harvey et al. (1992) and Fiorentini et al. (2004), but my direct past connections to it include Shephard (1994) and, for example, Bos and Shephard (2006). The latter paper has an extensive discussion of the literature on this topic. I thought about writing this paper after reading Stock and Watson (2007). Intellectually, one can think of the contribution of this paper as arguing for a different parameterization from that used by Stock and Watson (2007), as well as employing a different computational device. I think there are also some attractions in thinking about the models as martingale unobserved component models, rather than starting with the default Gaussian model associated with the Kalman filter.

Computationally I handle the model using an auxiliary particle filter, implementing it by running thousands of Kalman filters in parallel, allowing the data to select which ones blossom as time evolves. I will use the particular structure of the model to do this statistically efficiently, the most related work to this is Chen and Liu (2000) as well as Fearnhead and Clifford (2003). I should also note the work of Koopman and Bos (2004) and Creal (2012) around this topic, while Stock and Watson (2007) use the Kim et al. (1998) approach to SV. The related work in macroeconomics includes Cogley et al. (2010), D'Agostino et al. (2013), Fernandez-Villaverde et al. (2010), and Caldara et al. (2012).

The rest of this paper has the following form. In Section 10.2 a martingale unobserved component model is defined and various special cases are discussed. A key feature of this model is that it has a simple conditional probabilistic structure. This is discussed in Section 10.3, where the relations to the Kalman filter are brought out. Section 10.4 focuses on how the model can be handled using a particular type of particle filter, which allows both state and parameter estimation. In Section 10.5 the model is used to analyze a time series of quarterly inflation from the United States. In Section 10.6 some conclusions are made.

10.2 Martingale unobserved component models

10.2.1 A FIRST EXAMPLE

I start by considering a univariate local level version of the martingale unobserved component model

$$\begin{aligned} y_t &= \mu_t + \varepsilon_t^*, \\ \mu_{t+1} &= \mu_t + \eta_t^*, \quad t = 1, 2, \ldots, n, \end{aligned}$$

where y_t is observed and the components are the level μ_t and the noise ε_t^*. Here

$$\mathrm{E}\begin{pmatrix} \varepsilon_t^* \\ \eta_t^* \end{pmatrix} |\mathcal{F}_{t-1}^{\varepsilon^*,\eta^*,\mu_0} = \begin{pmatrix} 0 \\ 0 \end{pmatrix},$$

where $\mathcal{F}_t^x$ generically denotes the past and current information of an arbitrary x process, that is $\mathcal{F}_0^x, x_1, \ldots, x_t$, where $\mathcal{F}_0^x$ is some prior. So here ε_t^* and η_t^* are martingale difference sequences with respect to their joint natural filtration $\mathcal{F}^{\varepsilon^*,\eta^*,\mu_0}$, while μ_t is a $\mathcal{F}^{\varepsilon^*,\eta^*,\mu_0}$-martingale. The key idea in martingale unobserved component models is that the filtration is not with respect to the observables but with respect to the components.

10.2.2 PARAMETERIZING THE MODEL

An elegant and mainstream way of parameterizing martingales is through stochastic volatility, e.g., Harvey et al. (1994), Ghysels et al. (1996), and Shephard (2005b). Although at first sight the use of SV looks ad hoc, it is well known that large classes of martingales (i.e., basically continuous sample path martingales with absolutely continuous quadratic variation) can be represented in this way, e.g., Shephard (2005b, Ch. 1).

Here, two "discrete time stochastic volatility" models are used, with

$$\begin{aligned} y_t &= \mu_t + \sigma_{\varepsilon,t}\varepsilon_t, \quad (\varepsilon_t, \eta_t)' \overset{iid}{\sim} N(0, I_2), \\ \mu_{t+1} &= \mu_t + \sigma_{\eta,t}\eta_t, \end{aligned} \tag{10.1}$$

where $t = 1, 2, \ldots, n$. I will call $\sigma_{\varepsilon,t}$ and $\sigma_{\eta,t}$ transitory and permanent component volatilities. Throughout I will assume that the non-negative $(\sigma_{\varepsilon,t}, \sigma_{\eta,t})$ as a process is probabilistically independent from (ε_t, η_t) as a process. More sophisticated dependence is possible, allowing features such as statistical leverage, but that will not be attempted here.

The model is completed by a prior on the initial location. Here I decided to put this on μ_2 and use a simple prior. This is built by making inference conditional on an initial datapoint y_1 and so takes the prior

$$\mu_2|\mathcal{F}_1^{y,\sigma_\varepsilon,\sigma_\eta} \sim N(y_1, \sigma_{\varepsilon,1}^2 + \sigma_{\eta,1}^2). \tag{10.2}$$

This prior can be motivated by noting that $\mu_2 = y_1 - \sigma_{\varepsilon,1}\varepsilon_1 + \sigma_{\eta,1}\eta_1$. Of course more rigorous treatment of initial conditions can be provided, see for example, de Jong (1991a) and Durbin and Koopman (2001, Ch. 2).

The following are important special cases:

- *Random walk plus strict white noise model.* This is where $\sigma_{\varepsilon,t}$ and $\sigma_{\eta,t}$ are both constant, meaning the model is Gaussian. This is the traditional model which can be computationally efficiently handled by the Kalman filter. It is sometimes called the local level model.
- *Martingale plus strict white noise model.* This is where $\sigma_{\varepsilon,t} = \sigma_\varepsilon$ but $\sigma_{\eta,t}$ can vary through time. This means that the location is a martingale, rather than a random walk.
- *Random walk plus SV scale.* This is where $\sigma_{\eta,t} = q^{1/2}\sigma_{\varepsilon,t}$, but $\sigma_{\varepsilon,t}$ can vary through time allowing the scale of the volatilities of change through time but the model has a fixed signal to noise ratio q. This model is studied in, for example, Koopman and Bos (2004) who develop special computational methods to handle it.

Various local models for the volatilities will be defined and analyzed in a moment.

This martingale unobserved component model class extends to cover all Gaussian state-space models replacing all i.i.d. noise terms with martingale difference sequences. Here, we detail two classic examples.

10.2.3 OTHER MARTINGALE UNOBSERVED COMPONENT MODELS

Local linear trend plus cycle

To be explicit, the univariate local linear trend plus cycle version of a martingale unobserved component model would have

$$\begin{aligned} y_t &= \mu_t + \sqrt{1-\rho^2}\psi_t + \varepsilon_t^*, \\ \mu_{t+1} &= \mu_t + \beta_t + \eta_t^*, \\ \beta_{t+1} &= \beta_t + \zeta_t^*, \\ \begin{pmatrix} \psi_{t+1} \\ \dot{\psi}_{t+1} \end{pmatrix} &= \rho \begin{pmatrix} \cos(\lambda_c) & \sin(\lambda_c) \\ -\sin(\lambda_c) & \cos(\lambda_c) \end{pmatrix} \begin{pmatrix} \psi_t \\ \dot{\psi}_t \end{pmatrix} + \begin{pmatrix} \kappa_t^* \\ \dot{\kappa}_t^* \end{pmatrix}. \end{aligned} \tag{10.3}$$

In this model μ_t is interpreted as the trend, β_t as a time varying slope, and $\sqrt{1-\rho^2}\psi_t$ as the cycle. In some applications researchers have preferred to use a "smooth trend" component, which imposes a priori $\eta_t^* = 0$ (which connects to the literature on cubic splines). In the martingale unobserved component model the vector $\left(\varepsilon_t^*, \eta_t^*, \zeta_t^*, \kappa_t^*, \dot{\kappa}_t^*\right)'$ is a martingale difference sequence with respect to its own natural filtration conditional on $\beta_0, \mu_0, \psi_0, \dot{\psi}_0$. Here $\rho \in (0,1)$, which means that if $\left(\kappa_t^*, \dot{\kappa}_t^*\right)'$ is weak white noise then ψ_t has a linear

ARMA(2,1) representation with the complex autoregressive roots. Further λ_c is a frequency in radians. I have parameterized it so that $\text{Var}(\sqrt{1-\rho^2}\psi_t) = \text{Var}(\kappa_t^*) = \text{Var}(\dot{\kappa}_t^*)$, and thus does not vary with ρ or λ_c. The martingale difference sequences can be parameterized as SV processes:

$$\varepsilon_t^* = \sigma_{\varepsilon,t}\varepsilon_t, \quad \eta_t^* = \sigma_{\eta,t}\eta_t, \quad \zeta_t^* = \sigma_{\zeta,t}\zeta_t, \quad \kappa_t^* = \sigma_{\kappa,t}\kappa_t, \quad \dot{\kappa}_t^* = \sigma_{\kappa,t}\dot{\kappa}_t, \tag{10.4}$$

where $\left(\varepsilon_t, \eta_t, \zeta_t, \kappa_t, \dot{\kappa}_t\right)' \overset{iid}{\sim} N(0, I_5)$ and stochastically independent from $\left(\sigma_{\varepsilon,t}, \sigma_{\eta,t}, \sigma_{\zeta,t}, \sigma_{\kappa,t}\right)'$.

Of course a related simpler model writes

$$\begin{aligned} y_t &= \mu_t + \psi_t, \\ \psi_{t+1} &= \rho\psi_t + \sqrt{1-\rho^2}\kappa_t^*, \quad \rho \in (-1, 1), \end{aligned} \tag{10.5}$$

which has an autoregressive measurement error (whose marginal distribution does not depend upon ρ), while a moving average measurement error version can be achieved by writing

$$\begin{aligned} y_t &= \mu_t + (1 \quad 0)\,\psi_t, \\ \psi_{t+1} &= \begin{pmatrix} 0 & 1 \\ 0 & 0 \end{pmatrix} \psi_t + \begin{pmatrix} 1 \\ \lambda \end{pmatrix} \frac{1}{\sqrt{1+\lambda^2}}\kappa_t^*, \quad \lambda \in (-1, 1). \end{aligned} \tag{10.6}$$

Again $(1 \quad 0)\,\psi_t$ is setup so that the implied marginal distribution does not depend upon λ.

Multivariate martingale unobserved component local level model

Likewise, a d-dimensional vector martingale unobserved component version of the local level model would have the form

$$\begin{aligned} y_t &= \mu_t + \varepsilon_t^*, \\ \mu_{t+1} &= \mu_t + \eta_t^*, \end{aligned} \tag{10.7}$$

which is parameterized as $\varepsilon_t^* = \sigma_{\varepsilon,t}\varepsilon_t$ and $\eta_t^* = \sigma_{\eta,t}\eta_t$, with $\left(\varepsilon_t', \eta_t'\right)' \overset{iid}{\sim} N(0, I_{2d})$ which are stochastically independent from $\left(\sigma_{\varepsilon,t}, \sigma_{\eta,t}\right)'$. Here $\sigma_{\varepsilon,t}$ and $\sigma_{\eta,t}$ are $d \times d$ dimensional time-varying volatility matrices.

10.3 **Conditional properties**

10.3.1 GENERAL CASE

This stochastic volatility parameterization of the martingale unobserved component model is a special case of the partially or conditionally Gaussian time series, a class of state-space models defined in Shephard (1994). The core feature of this model is that it can be placed into a Gaussian state-space form by conditioning on, in this special case, the volatilities (note there is a significant and influential literature on the case where we condition on variables with discrete support to deliver a Gaussian state-space form, e.g., Ackerson and Fu (1970), Akashi and Kumamoto (1977), and Carter and Kohn (1994)). To start the discussion, focus on the local level martingale unobserved component model (10.1) and assume that $\sigma_{\varepsilon,t}, \sigma_{\eta,t} > 0$ for all t.

A key characteristic of this model is the "signal/noise ratio process,"

$$q_t = \sigma_{\eta,t}^2 / \sigma_{\varepsilon,t}^2. \tag{10.8}$$

Throughout $q_{1:t} = (q_1, \ldots, q_t)'$.

Remark 10.3.1 *In the univariate martingale local linear trend plus cycle model (10.3) there are three signal/noise processes* $q_{\eta,t} = \sigma_{\eta,t}^2/\sigma_{\varepsilon,t}^2$, $q_{\zeta,t} = \sigma_{\zeta,t}^2/\sigma_{\varepsilon,t}^2$, $q_{\kappa,t} = \sigma_{\kappa,t}^2/\sigma_{\varepsilon,t}^2$. *This has the virtue that the signal/noise processes are invariant to rescaling the data. In the smooth trend plus cycle model there are only two signal/noise processes,* $q_{\zeta,t}$ *and* $q_{\kappa,t}$.

Remark 10.3.2 *In the multivariate martingale local level unobserved component model (10.7) there is a matrix signal/noise process* $Q_t = \Sigma_{\varepsilon,t}^{-1/2} \Sigma_{\eta,t} \Sigma_{\varepsilon,t}^{-1/2\prime}$, *where* $\Sigma_{\varepsilon,t} = \sigma_{\varepsilon,t}\sigma_{\varepsilon,t}'$ *and* $\Sigma_{\eta,t} = \sigma_{\eta,t}\sigma_{\eta,t}' = \Sigma_{\varepsilon,t}^{1/2} Q_t \Sigma_{\varepsilon,t}^{1/2\prime}$. *This signal/noise process is invariant to rescaling and rotation of the data.*

10.3.2 GAUSSIAN STATE-SPACE MODELS

Now take a step backwards and recall the general d-dimensional Gaussian state-space form (e.g., Durbin and Koopman (2001, p. 67)), where μ_t is the general state. This takes on the form of observing

$$y_t = Z_t \mu_t + \varepsilon_t, \quad \varepsilon_t \sim N(0, H_t), \tag{10.9}$$

$$\mu_{t+1} = T_t \mu_t + R_t \eta_t, \quad \eta_t \sim N(0, Q_t), \tag{10.10}$$

where the sequences Z_t, H_t, Q_t, T_t, and R_t are assumed to be non-stochastic and the $\varepsilon_1, \ldots, \varepsilon_n, \eta_1, \ldots, \eta_n$ are stochastically independent. The conditional model is completed by the prior $\mu_1 | \mathcal{F}_0^y \sim N(m_{1|0}, P_{1|0})$.

Then $\mu_{t+1} | \mathcal{F}_t^y \sim N(m_{t+1|t}, P_{t+1|t})$, where the mean and variance updates according to the Kalman filter recursions

$$m_{t+1|t} = T_t m_{t|t-1} + T_t P_{t|t-1} Z_t' F_t^{-1} (y_t - Z_t m_{t|t-1}) \tag{10.11}$$

$$= T_t m_{t|t-1} + K_t (y_t - Z_t m_{t|t-1}), \quad K_t = T_t P_{t|t-1} Z_t' F_t^{-1} \tag{10.12}$$

$$= T_t \left\{ W_t y_t + (I - W_t Z_t) m_{t|t-1} \right\}, \quad W_t = P_{t|t-1} Z_t' F_t^{-1}, \tag{10.13}$$

$$P_{t+1|t} = T_t P_{t|t-1} T_t' - T_t P_{t|t-1} Z_t' F_t^{-1} Z_t P_{t|t-1} T_t' + R_t Q_t R_t' \tag{10.14}$$

$$= T_t P_{t|t-1} (I - W_t Z_t)' T_t' + R_t Q_t R_t', \tag{10.15}$$

$$F_t = Z_t P_{t|t-1} Z_t' + H_t, \tag{10.16}$$

and as a side product,

$$y_t | \mathcal{F}_{t-1}^y \sim N(Z_t m_{t|t-1}, F_t). \tag{10.17}$$

Remark 10.3.3 *In the important special case were $H_t = \sigma_{\varepsilon,t}^2 I$, $R_t = \sigma_{\varepsilon,t}$ and I write $P_{t|t-1} = \sigma_{\varepsilon,t}^2 P_{t|t-1}^*$, where $\sigma_{\varepsilon,t}$ is a non-stochastic sequence, then*

$$F_t = \sigma_{\varepsilon,t}^2 \left(Z_t P_{t|t-1}^* Z_t' + I \right), \quad W_t = P_{t|t-1}^* Z_t' \left(Z_t P_{t|t-1}^* Z_t' + I \right)^{-1} \tag{10.18}$$

$$P_{t+1|t} = \sigma_{\varepsilon,t}^2 \left\{ T_t P_{t|t-1}^* (I - W_t Z_t)' T_t' + Q_t \right\}, \tag{10.19}$$

and so $P_{t+1|t}^* = \frac{\sigma_{\varepsilon,t}^2}{\sigma_{\varepsilon,t+1}^2} \left\{ T_t P_{t|t-1}^* (I - W_t Z_t)' T_t' + Q_t \right\}$. Hence the $P_{t+1|t}^*$ process, and so W_t and $m_{t|t-1}$, depends not on the level of volatility but on the change in the level of volatility. In the special univariate case with $Z_t = T_t = I$, this can be more compactly written as

$$m_{t+1|t} = m_{t|t-1} + \frac{p_{t|t-1}^*}{p_{t|t-1}^* + 1} (y_t - m_{t|t-1}),$$

$$p_{t+1|t}^* = \frac{\sigma_{\varepsilon,t}^2}{\sigma_{\varepsilon,t-1}^2} \left(\frac{p_{t|t-1}^*}{p_{t|t-1}^* + 1} + q_t \right). \tag{10.20}$$

10.3.3 CONDITIONING ON THE VOLATILITIES

For the local level version of the martingale unobserved component model it is useful to condition on the stochastic volatility processes $\left(\sigma_{\varepsilon,t}, \sigma_{\eta,t}\right)$. This conditional model can be handled computationally efficiently using the Kalman filter, which becomes

$$m_{t+1|t} = (1 - \omega_t) m_{t|t-1} + \omega_t y_t, \tag{10.21}$$

where $\mathrm{E}\left(\mu_{t+1} | \mathcal{F}_t^{y,\sigma_\varepsilon,\sigma_\eta}\right) = m_{t+1|t}$. This is driven off the simple recursions

$$\omega_t = \frac{p_{t|t-1}}{p_{t|t-1} + \sigma_{\varepsilon,t}^2}, \quad p_{t+1|t} = \sigma_{\varepsilon,t}^2 \left(\omega_t + q_t\right). \tag{10.22}$$

Of course, here $p_{t+1|t} = \mathrm{E}\left\{ \left(\mu_{t+1} - m_{t+1|t}\right)^2 | \mathcal{F}_t^{y,\sigma_\varepsilon,\sigma_\eta} \right\}$. The conditional likelihood is, via the prediction decomposition,

$$\log f(y_2,\ldots,y_n|\mathcal{F}_1^y,\mathcal{F}_n^{\sigma_\varepsilon,\sigma_\eta}) \tag{10.23}$$

$$= -\frac{1}{2}\sum_{t=2}^{n}\log\left(p_{t|t-1}+\sigma_{\varepsilon,t}^2\right) - \frac{1}{2}\sum_{t=2}^{n}\frac{\left(y_t - m_{t|t-1}\right)^2}{p_{t|t-1}+\sigma_{\varepsilon,t}^2} \tag{10.24}$$

$$= -\frac{1}{2}\sum_{t=2}^{n}\log\sigma_{\varepsilon,t}^2 - \frac{1}{2}\sum_{t=2}^{n}\log\left(p_{t|t-1}^*+1\right) - \frac{1}{2}\sum_{t=2}^{n}\frac{\left(y_t - m_{t|t-1}\right)^2}{\sigma_{\varepsilon,t}^2\left(p_{t|t-1}^*+1\right)}. \tag{10.25}$$

Remark 10.3.4 *For the local linear trend plus cycle model in Section 10.2.3 and Remark 10.3.1 the conditional state-space form becomes*

$$Z_t = (1,0,\sqrt{1-\rho^2},0),\quad h_t = \sigma_{\varepsilon,t}^2,\quad r_t = \sigma_{\varepsilon,t},$$

$$Q_t = diag(q_{\eta,t}, q_{\zeta,t}, q_{\kappa,t}, q_{\kappa,t}),$$

$$T_t = \begin{pmatrix} 1\ 1 & 0 \\ 0\ 1 & 0 \\ 0\ 0\ \rho & \begin{pmatrix} \cos(\lambda_c) & \sin(\lambda_c) \\ -\sin(\lambda_c) & \cos(\lambda_c) \end{pmatrix} \end{pmatrix},$$

and hence (10.18) and (10.19) applies.

10.3.4 GAUSSIAN CASE, NO VOLATILITY CLUSTERING

Suppose there is no volatility clustering, so $\sigma_{\varepsilon,t}^2 = \sigma_\varepsilon^2$ and $\sigma_{\eta,t}^2 = \sigma_\eta^2$. This reproduces the celebrated Gaussian random walk plus noise model which can be handled efficiently by the Kalman filter. All the following is well known.

If t is large and then $\sigma_\varepsilon^2, \sigma_\eta^2 > 0$, then the updating recursion converges to the solution to the Riccati equation

$$\lim_{t\to\infty} p_{t|t-1}^* = p^* = q + \frac{p^*}{p^*+1} = \frac{q+\sqrt{q^2+4q}}{2} \in [0,\infty), \tag{10.26}$$

where, again, $p_{t|t-1} = \sigma_{t,\varepsilon}^2 p_{t|t-1}^*$, which means that the updating equation for the conditional mean has the form of

$$m_{t+1|t} = (1-\omega)\,m_{t|t-1} + \omega y_t,\quad \omega = \frac{p^*}{p^*+1} = \frac{q+\sqrt{q^2+4q}}{2+q+\sqrt{q^2+4q}} \in [0,1], \tag{10.27}$$

a simple discount of the past data. Hence the local level model rationalizes the EWMA updating recursion (Muth (1960))

$$m_{t+1|t} = \omega\sum_{j=0}^{\infty}(1-\omega)^j\, y_{t-j}. \tag{10.28}$$

Of course the first difference of the local level model can be written as a first order moving average model $\Delta y_t = \varsigma_t + \psi \varsigma_{t-1}$, where ς_t is strict white noise and

$$\psi = \omega - 1 = \frac{-2}{2 + q + \sqrt{q^2 + 4q}} \in [-1, 0]. \tag{10.29}$$

Breaking away from the time-invariant model the

$$\psi_t = \frac{-2}{2 + q_t + \sqrt{q_t^2 + 4q_t}} \in [-1, 0], \tag{10.30}$$

is defined here as a "psi process." One way of directly measuring the relative weight of a data point at time t compared to time $t - j$ is through $(1 - \omega)^j$ in (10.28). I determine when j is large enough for $(1 - \omega)^j$ to be around 0.1, which means this is long enough for the data not to have much impact on the forecast. I will call this the "life span process," which may vary through time as

$$\begin{aligned} s_t &= \log 0.1 / \log(1 - \omega_t) \\ &= \log 0.1 / \log(-\psi_t) \in [0, \infty). \end{aligned} \tag{10.31}$$

When s_t is close to zero the forecast is approximately a martingale, while when it is, for example twenty, it roughly averages the data over the last twenty periods.

When there is no transitory component volatility clustering, that is $\sigma_{\varepsilon,t}^2 = \sigma_\varepsilon^2$, then this model is called the martingale plus strict white noise model. This model has some flexibility for forecasting and is still relatively easy to handle. Now, recalling $q_t = \sigma_{\eta,t}^2 / \sigma_\varepsilon^2$,

$$m_{t+1|t} = (1 - \omega_t)\, m_{t|t-1} + \omega_t y_t, \quad p_{t+1|t}^* = q_t + \omega_t, \quad \omega_t = \frac{p_{t|t-1}^*}{p_{t|t-1}^* + 1}, \tag{10.32}$$

where $m_{t+1|t}$ and $p_{t+1|t}^*$ are invariant to σ_ε^2. Clearly $\mu_{t+1}|\mathcal{F}_t^{y,q}, \sigma_\varepsilon^2 \sim N(m_{t+1|t}, \sigma_\varepsilon^2 p_{t+1|t}^*)$.

10.3.5 VOLATILITY MODELS

For the martingale unobserved component models the signal/noise (q_t) and volatility processes ($\sigma_t = \sigma_{\varepsilon,t}$) can evolve over time. I focus on the local level type model and initially make these two processes stochastically independent, with each following log-Gaussian processes

$$\log q_{t+1} = \log q_t + \theta_q \xi_{q,t}, \quad \begin{pmatrix} \xi_{q,t} \\ \xi_{\sigma,t} \end{pmatrix} \overset{iid}{\sim} N(0, I_2), \tag{10.33}$$

$$\log \sigma_{t+1}^2 = \log \sigma_t^2 + \theta_\sigma \xi_{\sigma,t}. \tag{10.34}$$

This means that $\frac{q_{t+1}}{q_t} = \exp(\theta_q \xi_{q,t})$. Hence for moderate θ_q the proportional change is roughly $(q_{t+1} - q_t)/q_t \simeq \theta_q \xi_{q,t}$, which makes sense whatever the current, value, of the volatility due to its attractive scale invariance. In particular then the expected percentage change at any time point is roughly $\mathrm{E}\left|(q_{t+1} - q_t)/q_t\right| \simeq \theta_q \sqrt{\frac{\pi}{2}}$.

This model implies that the level

$$\mu_{t+1} = \mu_t + \sigma_t q_t \eta_t = \mu_t + \sigma_{\mu,t} \eta_t, \quad \sigma_{\mu,t} = \sigma_t q_t, \tag{10.35}$$

is like an integrated stochastic volatility model from financial econometrics, although these models are typically setup with a small degree of mean reversion in the volatility of the level $\sigma_{\mu,t}$—e.g., Harvey et al. (1994), Ghysels et al. (1996), and Shephard (2005b).

Now this model implies that

$$\log \sigma_{\mu,t+1}^2 = \log \sigma_{\mu,t}^2 + \theta_\mu \xi_{\mu,t}, \quad \xi_{\mu,t} \sim N(0,1), \tag{10.36}$$

and $\theta_\mu \xi_{\mu,t} = \theta_\sigma \xi_{\sigma,t} + \theta_q \xi_{q,t}$. This means that $\theta_\mu = \sqrt{\theta_\sigma^2 + \theta_q^2}$. Of course, this model implies $\mathrm{Cov}\left(\xi_{\mu,t}, \xi_{\sigma,t}\right) = \theta_\sigma/\theta_\mu$, $\mathrm{Cov}\left(\xi_{\mu,t}, \xi_{q,t}\right) = \theta_q/\theta_\mu$.

10.3.6 ENCOMPASSES

Stock and Watson (2007) (denoted SW below) parameterize their changing volatilities using independent random walks

$$\log \sigma_{\mu,t+1}^2 = \log \sigma_{\mu,t}^2 + \theta_\mu \xi_{\mu,t}, \quad \log \sigma_{t+1}^2 = \log \sigma_t^2 + \theta_\sigma \xi_{\sigma,t},$$
$$\left(\xi_{\mu,t}, \xi_{\sigma,t}\right)' \overset{iid}{\sim} N(0, I_2). \tag{10.37}$$

This means that

$$\log q_{t+1} = \log \sigma_{\mu,t+1}^2 - \log \sigma_{t+1}^2 = \log q_t + \theta_\mu \xi_{\mu,t} - \theta_\sigma \xi_{\sigma,t} = \log q_t + \theta_q \xi_{q,t}, \tag{10.38}$$

where $\xi_{q,t} \overset{iid}{\sim} N(0,1)$, while

$$\mathrm{Var}\begin{pmatrix} \theta_q \xi_{q,t} \\ \theta_\sigma \xi_{\sigma,t} \end{pmatrix} = \begin{pmatrix} \theta_\mu^2 + \theta_\sigma^2 & -\theta_\sigma^2 \\ -\theta_\sigma^2 & \theta_\sigma^2 \end{pmatrix}. \tag{10.39}$$

This means that a priori $\theta_q^2 \geq \theta_\sigma^2$, while the shocks are negatively correlated. This is a little concerning for if there is a great deal of volatility clustering then this model implies the signal/noise process must also move around a great deal.

To reconcile the two parameterizations, an extended SW parameterization can be introduced that writes

$$\begin{pmatrix} \xi_{\mu,t} \\ \xi_{\sigma,t} \end{pmatrix} = \begin{pmatrix} 1 & 0 \\ \rho & \sqrt{1-\rho^2} \end{pmatrix} \begin{pmatrix} u_{1,t} \\ u_{2,t} \end{pmatrix} = \begin{pmatrix} u_{1,t} \\ \rho u_{1,t} + \sqrt{1-\rho^2} u_{2,t} \end{pmatrix}, \qquad (10.40)$$

for this then encompasses our alternative. We will see later that allowing a correlation between these two innovations will be supported by the data. Then

$$\theta_q \xi_{q,t} = \left(\theta_\mu - \rho\theta_\sigma\right) u_{1,t} - \theta_\sigma \sqrt{1-\rho^2} u_{2,t},$$

$$\theta_q = \sqrt{\left(\theta_\mu - \rho\theta_\sigma\right)^2 + \theta_\sigma^2 \left(1-\rho^2\right)} = \sqrt{\theta_\mu^2 + \theta_\sigma^2 - 2\rho\theta_\mu\theta_\sigma},$$

$$\mathrm{Cov}(\theta_q \xi_{q,t}, \theta_\sigma \xi_{\sigma,t}) = \theta_\sigma \rho \left(\theta_\mu - \rho\theta_\sigma\right) - \theta_\sigma^2 \left(1-\rho^2\right) = \theta_\sigma \left(\rho\theta_\mu - \theta_\sigma\right)$$

$$\mathrm{Cor}(\theta_q \xi_{q,t}, \theta_\sigma \xi_{\sigma,t}) = \frac{\rho\theta_\mu - \theta_\sigma}{\theta_q} = \frac{\rho\theta_\mu - \theta_\sigma}{\sqrt{\theta_\mu^2 + \theta_\sigma^2 - 2\rho\theta_\mu\theta_\sigma}}.$$

Hence if ρ is incorrectly imposed to be zero then the implied θ_q will be too high if the true $\rho > 0$. This is crucial as this parameter determines the degree of discounting used in forecasting. We will see that this is exactly what happens empirically.

Remark 10.3.5 *Suppose the Gaussian local level model is parameterized as*

$$y_t = \mu_t + \sigma\varepsilon_t, \quad \mu_{t+1} = \mu_t + q^{1/2}\sigma\eta_t, \quad \mu_0 \sim N(m_0, P_0),$$

$$(\varepsilon_t, \eta_t)' \overset{iid}{\sim} N(0, I_2). \qquad (10.41)$$

It is helpful to write this time series as the vector using the notation $y \sim LLM(\sigma, q, m_0, P_0)$. *Then if* a *and* b *are scalar constants* $a + by \sim LLM(b\sigma, q, a + bm_0, b^2 P_0)$, *the model is closed under location shifts and rescaling, and the impact of rescaling is just to scale up and down* σ *(as well as some impact on the prior). It has no impact on the key time series properties of the model, which are governed through* q. *If the model is set up as*

$$y_t = \mu_t + \sigma_\varepsilon\varepsilon_t, \;\; \mu_{t+1} = \mu_t + \sigma_\eta\eta_t, \;\; \mu_0 \sim N(m_0, P_0), \;\; (\varepsilon_t, \eta_t)' \overset{iid}{\sim} N(0, I_2), \qquad (10.42)$$

then $a + by \sim LLM(b\sigma_\varepsilon, b\sigma_\eta, a + bm_0, b^2 P_0)$. *This shows that under a scale shift the two volatilities* σ_ε *and* σ_η *must move together, for the implied* q *will otherwise change. This suggests that it is not so attractive to place independent priors on* σ_ε *and* σ_η *(i.e., these two parameters should move together to reflect the scaling of the data), while it may make some sense to make* q *and* σ *independent. This observation is also key in the dynamic case, where scaling of economic data can vary through time. This persuades me to prefer the parameterization (10.41) to (10.42). Of course, in practice it is an empirical question as to which model is empirically more successful in terms of fit and parsimony.*

10.3.7 EWMA AND ADAPTIVE DISCOUNTING

The Gaussian local level model provides a statistical rational for EWMA forecasting

$$m_{t+1|t} = (1-\omega)\, m_{t|t-1} + \omega y_t. \tag{10.43}$$

The above models allow the discount parameter to change through time as an unobservable process which can be efficiently learnt using data and Bayes theory.

There is another literature which has directly allowed the discount factor to move in response to past data. The most well known approach is Trigg and Leach (1967) who specify

$$\omega_t = \frac{|A_t|}{|M_t|}, \quad A_t = \phi e_{t-1} + (1-\phi) A_{t-1}, \quad M_t = \phi\, |e_{t-1}| + (1-\phi) M_{t-1}, \tag{10.44}$$

where $e_t = y_t - m_{t|t-1}$. Typically $\phi = 0.2$, but this does not enforce $\omega_t \in (0,1)$. Discussions of alternatives are given in, for example, Taylor (2004).

Related to this work is the so-called generalized autoregressive score based models of Creal et al. (2013, Section 4.4) which builds an observation driven model based upon a Stock and Watson (2007) type parameterization.

10.4 **Particle filter based analysis**

10.4.1 BASICS OF FILTERING

In this paper I have decided to use a particle filter as the basis of sequential Bayesian inference on μ_t, σ_t, and q_t. Particle filters are now the established extension of the Kalman filter to deal with non-Gaussian and non-linear state-space models. They use simulation to provide filtered estimates of the states and an unbiased estimator of the likelihood. Particle filters can be implemented in various ways, which effect their Monte Carlo statistical efficiency. Early contributions to this include, for example, Gordon et al. (1993), Liu and Chen (1998), Pitt and Shephard (1999), and Doucet et al. (2001). Modern surveys include, for example, Doucet and Johansen (2011) and Creal (2012).

In this section I assume a Markov model $q_{t+1}, \sigma_{t+1}^2 | q_t, \sigma_t^2, \theta$, from which I can simulate. The simplest case is (10.33)–(10.34). For now, assume that the value of θ is known and I will return to that issue in Subsection 10.4.2.

The conditional Gaussian structure of the model can be used to improve upon the bootstrap particle filter of Gordon et al. (1993), integrating out the conditionally Gaussian μ_t for the complete model

$$\begin{aligned} y_t &= \mu_t + \sigma_t \varepsilon_t, \\ \mu_{t+1} &= \mu_t + q_t^{1/2} \sigma_t \eta_t, \quad \left(\varepsilon_t, \eta_t, \xi_{q,t}, \xi_{\sigma,t}\right)' \sim N\left(0, I_4\right), \\ \log q_{t+1} &= \log q_t + \theta_q \xi_{q,t}, \\ \log \sigma_{t+1}^2 &= \log \sigma_t^2 + \theta_\sigma \xi_{\sigma,t}. \end{aligned}$$

In this approach the only "particle state variable" is

$$\alpha_t = \left(\sigma_t^2, q_t\right)', \tag{10.45}$$

for the remaining state, μ_t, can be integrated out analytically. The genesis of this appears in the auxiliary particle filter of Pitt and Shephard (1999), but it is dealt with more extensively by Doucet et al. (2000) and in particular Chen and Liu (2000) and Fearnhead and Clifford (2003). It has also been used by Creal (2012) in this context and by Creal et al. (2010). This approach is sometimes called a mixture Kalman filter and is also related to Rao-Blackwellization.

In effect this auxiliary particle filter simply runs M Kalman filters in parallel, each with an individual signal/noise ratio and volatility process which nudges the signal/noise rate and volatility in random directions at each time increment. Some of these Kalman filters will run up pretty high likelihoods through time, others will perform poorly—as they have uncompetitive signal/noise ratios or volatilities. When there is sufficient imbalance in the likelihoods between the different filters, the Kalman filters are resampled. Resampling samples with probability in proportion to the likelihood to produce a new set of M Kalman filters from their parents. Thus Kalman filters that generate large likelihoods survive and replicate; those that have poor fit die. The particles $\alpha_t^{(j)}$ are indexed by $j = 1, 2, \ldots, M$. Like the Kalman filter, the particle filter is run conditioned on $\theta = \left(\theta_q, \theta_\sigma\right)'$.

Algorithmically it takes on the following form.

Auxiliary particle filter

1. Set $t = 1$, draw $\alpha_1^{(j)}$ from a prior $f(\alpha_1)$. Here I have taken

$$q_1^{(j)} \stackrel{L}{=} 0.3\chi_1^2, \quad \sigma_1^{2(j)} \stackrel{L}{=} 0.25\chi_1^{-2}, \tag{10.46}$$

where $\stackrel{L}{=}$ denotes being equal in law (i.e., distribution). Set

$$m_{2|1}^{(j)} = y_1, \quad p_{2|1}^{(j)} = \sigma_1^{2(j)}\left(1 + q_1^{(j)}\right), \quad w_{2|1}^{(j)} = 0 \tag{10.47}$$

and sample $\alpha_2^{(j)} \sim \alpha_2 | \alpha_1^{(j)}, \theta$.

2. Set $t = t + 1$. Compute for each $j = 1, 2, \ldots, M$ in parallel the conditional Kalman filter (10.21) and (10.22), recording

$$l_t^{(j)} = -\frac{1}{2}\log\left(p_{t|t-1}^{(j)} + \sigma_t^{2(j)}\right) - \frac{1}{2}\frac{\left(y_t - m_{t|t-1}^{(j)}\right)^2}{\left(p_{t|t-1}^{(j)} + \sigma_t^{2(j)}\right)}, \tag{10.48}$$

updating $w_{t+1|t}^{(j)} = w_{t|t-1}^{(j)} + l_t^{(j)}$, and computing

$$\omega_t^{(j)} = \frac{p_{t|t-1}^{(j)}}{p_{t|t-1}^{(j)} + \sigma_t^{2(j)}}, \tag{10.49}$$

then

$$m_{t+1|t}^{(j)} = \left(1 - \omega_t^{(j)}\right) m_{t|t-1}^{(j)} + \omega_t^{(j)} y_t, \quad p_{t+1|t}^{(j)} = \sigma_t^{2(j)}\left(q_t^{(j)} + \omega_t^{(j)}\right). \tag{10.50}$$

Then simulate forward the particle state

$$\alpha_{t+1}^{(j)} \sim \alpha_{t+1}|\alpha_t^{(j)}, \theta. \tag{10.51}$$

3. Record some summary results if desired. This will be discussed in Subsection 10.4.2.
4. Resample every three (an ad hoc choice) increments in time, by resampling with replacement from the

$$\left\{\alpha_t^{(1)}, m_{t+1|t}^{(1)}, p_{t+1|t}^{(1)}\right\}, \dots, \left\{\alpha_t^{(M)}, m_{t+1|t}^{(M)}, p_{t+1|t}^{(M)}\right\} \tag{10.52}$$

   with probability proportional to $w_{t+1|t}^{*(1)}, \dots, w_{t+1|t}^{*(M)}$, where $w_{t|t-1}^{*(j)}$ is given in (10.53) below. Having done this set $w_t^{(j)} = 0$ for $j = 1, 2, \dots, M$.
5. Go to 2.

The fact that the key step 2 is entirely parallel in j means it can be implemented using matrix computations. This is important as M may be quite large.

Finally, we note that for the Stock-Watson parameterization discussed in Section 10.3.6, the only change would be that $\alpha_t = (\sigma_{\varepsilon,t}^2, \sigma_{\eta,t}^2)'$, $\sigma_{\eta,1}^{2(j)} = \sigma_1^{2(j)} q_1^{(j)}$, $p_{t+1|t}^{(j)} = \sigma_{\eta,t}^{2(j)} + \sigma_{\varepsilon,t}^{2(j)} \omega_t^{(j)}$, and replacing $\sigma_t^{2(j)}$ by $\sigma_{\varepsilon,t}^{2(j)}$ everywhere.

10.4.2 RECORDING SOME OUTPUT

As the particle filter steps through time the output needs to be recorded. Here I highlight some important quantities that can be saved; the extension to many others is straightforward.

3. Record some output.
   (a) Compute

$$w_{t|t-1}^{*(j)} = \frac{\exp(w_{t|t-1}^{(j)})}{\sum_{i=1}^{M} \exp(w_{t|t-1}^{(i)})}. \tag{10.53}$$

(b) Then a variety of things can be estimated, such as
  i. Particle estimates of the log-likelihood contribution

$$\log \widehat{f}(y_t|\mathcal{F}_{t-1}^y, \theta) = \log \left[\sum_{j=1}^{M} w_{t|t-1}^{*(j)} \exp\left\{ l_t^{(j)} \right\} \right], \tag{10.54}$$

  ii. Particle estimates of the level $\widehat{\mathrm{E}}(\mu_t|\mathcal{F}_{t-1}^y, \theta) = \sum_{j=1}^{M} w_{t|t-1}^{*(j)} m_{t|t-1}^{(j)}$.
  iii. Particle estimates of the state process $\widehat{\mathrm{E}}(\alpha_t|\mathcal{F}_{t-1}^y, \theta) = \sum_{j=1}^{M} w_{t|t-1}^{*(j)} \alpha_t^{(j)}$.

In practice it is often better to use quantiles as summaries of $\alpha_t|\mathcal{F}_{t-1}^y, \theta$ and $\mu_t|\mathcal{F}_{t-1}^y, \theta$. To do this recall that the distribution of the posterior $\mu_t|\mathcal{F}_{t-1}^y, \theta$ can be estimated as $\widehat{F}(\mu_t|\mathcal{F}_{t-1}^y, \theta) = \sum_{j=1}^{M} w_{t|t-1}^{*(j)} F_N(\mu_t; m_{t|t-1}^{(j)}, P_{t|t-1}^{(j)})$, where F_N is the normal distribution function. The composition method can be used to simulate from this and then the corresponding empirical quantiles can be calculated. Alternatively this distribution can be first analytically marginalized (i.e., using the properties of the normal distribution) and then numerically inverted to compute the quantiles. To estimate the quantiles of $\alpha_t|\mathcal{F}_{t-1}^y, \theta$ the weighted particles $w_{t|t-1}^{*(j)}, \alpha_t^{(j)}$ can be resampled and the corresponding empirical quantiles computed.

10.4.3 PARTICLE MCMC

The particle filter delivers an unbiased estimator of the likelihood. This property is important and so it is worthwhile reflecting upon it for a moment. Here I draw upon Andrieu et al. (2010) and the exposition in Flury and Shephard (2011). See also Pitt et al. (2012).

Think of all the uniform random numbers behind a run of a particle filter as u, then the particle filter based estimated likelihood can be thought of as

$$\widehat{L}_u(\theta|y) = f(y|\theta, u), \tag{10.55}$$

an artificial conditional density, that has the property

$$\mathrm{E}_U \left\{ \widehat{L}_U(\theta|y) \right\} = \int f(y|\theta, u) f(u) du, \tag{10.56}$$

where $f(u) \propto 1$ is the density of the uniforms.

This means that an artificial joint posterior can be constructed as

$$f(u, \theta|y) \propto \widehat{L}_u(\theta|y) f(u) f(\theta), \tag{10.57}$$

by thinking of u as a set of auxiliary variables. Hence inference can be carried out by simulating from $u, \theta|y$ and discarding the draws from u. This approach is called particle MCMC and is due to Andrieu et al. (2010), while outside the particle context it can be traced back to Beaumont (2003). It simply makes a proposal to move θ by drawing a proposed move to θ' but where the likelihood is estimated using a particle filter based upon some draws u'. If the proposal is accepted both θ', u' are accepted. The resulting chain is a correct draw from $\theta|y$ even though an estimated likelihood is being used. This contrasts markedly with the maximum simulated likelihood literature discussed by, for example, Gourieroux and Monfort (1996).

10.5 **Illustration using inflation data**

10.5.1 THE COMPLETE MODEL

I follow an example given in Harvey (1981) on a time series of inflation, that he modeled by a Gaussian random walk plus noise model. Here the martingale unobserved component model

$$\begin{aligned} y_t &= \mu_t + \sigma_t \varepsilon_t, \\ \mu_{t+1} &= \mu_t + \sigma_t q_t^{1/2} \eta_t, \\ \log \sigma_{t+1}^2 &= \log \sigma_t^2 + \theta_\sigma \zeta_{\sigma,t}, \\ \log q_{t+1} &= \log q_t + \theta_q \zeta_{q,t}, \end{aligned} \tag{10.58}$$

will be used where σ_t is called "SV scaling," as it effects all aspects of the scale of movements in the series, and $\left(\varepsilon_t, \eta_t, \zeta_{\sigma,t}, \zeta_{q,t}\right)' \overset{iid}{\sim} N(0, I_4)$ is assured. The virtue of the martingale unobserved component model here is that it produces rather simple forecasts of inflation a few steps out, say one year, just extrapolating the current estimate of the level μ_t. At that time horizon it is likely to be somewhat robust to any moderate unmodeled serially correlation in ε_t.

Here the addition of the time varying q allows the EWMA type forecasts implied by the local level model to adapt the rate of discounting to the recent data.

This model is indexed by three "parameters" with priors that are integrated directly over: $\mu_2|y_1, q_1, \sigma_1^2 \sim N(m_1, p_{2|1})$, where $m_1 = y_1$ and $p_{2|1} = \sigma_1^2\left(1 + q_1\right)$, and the independent initial conditions

$$q_1 \overset{L}{=} 0.3\chi_1^2, \quad \sigma_1^2 \overset{L}{=} 0.25\chi_1^{-2}. \tag{10.59}$$

The focus here will be on the last parameters θ_σ and θ_q and they will be assumed to be a priori independent and that

$$\theta_\sigma \overset{L}{=} 0.15\left(\frac{1}{3}\chi_3^2\right), \quad \theta_q \overset{L}{=} 0.3\left(\frac{1}{3}\chi_3^2\right). \tag{10.60}$$

These are ad hoc choices, but I selected them to ensure the mode is away from zero, centered on roughly plausible values, and the tails of these prior are quite thin to crush implausible values.

Stock and Watson (2007) also focused on this problem but they used a different parameterization (the results for their model will be given at the end of Section 10.5.3). However, I will use the same data they used. The series is the main U.S. inflation series CPI-U and starts in January 1947. The Stock and Watson version of the data finishes December 2004. I follow them in computing quarterly price levels as the average monthly price level in each quarter. In our graphs of the raw quarterly data, estimated volatilities and diagnostics will be extended to August 2012, keeping the estimated parameters θ_σ and θ_q fixed at the December 2004 estimates.

Throughout the series analyzed will be one hundred times the first difference of the log of the series, which is roughly the percentage quarterly inflation series. This raw series is given in Figure 10.1, where the updating of the series to reflect new data is indicated by the vertical line.

10.5.2 ESTIMATED LIKELIHOOD

The first step to understanding the empirical content of the model will be to graph the particle filter based estimates of the log-likelihood as a function of

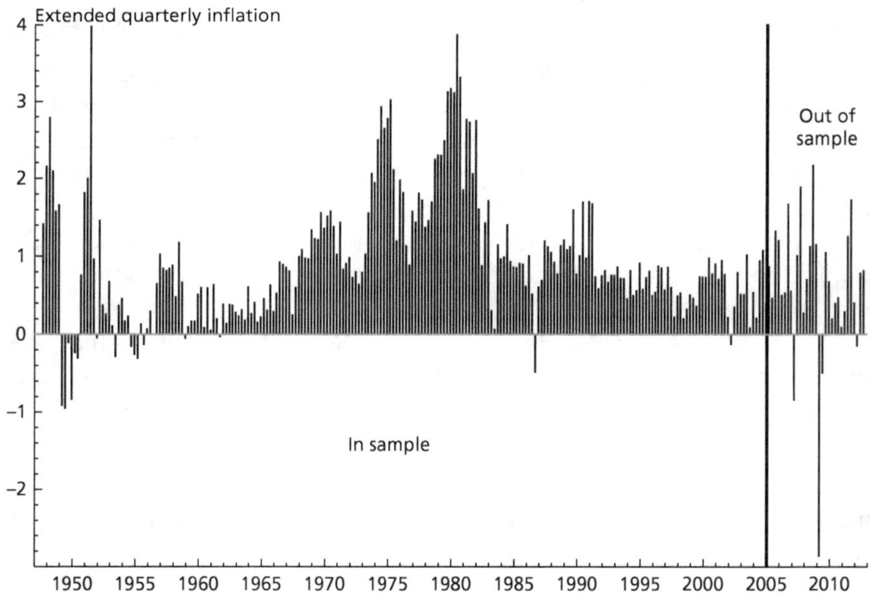

Figure 10.1. Computed quarterly U.S. inflation series, measured through CPI-U. This is constructed as one hundred times the first difference of the log of the average of the price index during the quarter. The vertical line indicates when the out of sample new data starts, which is the first quarter of 2005

the two parameters in θ. For the plot as a function of θ_q I fixed $\theta_\sigma = 0.25$ and when the function in θ_σ is drawn I took $\theta_q = 0.1$. The particle filter is run using M as 250, 500, 1000, and 5000. Figures 10.2 and 10.3 shows the results and indicate, as expected, that the filter becomes more precise as M increases. The top row of graphs shows the particle filter estimates of the log-likelihood, throughout using a log10 scale for θ_q and θ_σ. It indicates quite a flat likelihood for θ_q but with some support for a value of away from $\theta_q = 0$, which is the random walk plus SV scale model special case. It shows quite a heavily peaked likelihood for θ_σ away from $\theta_\sigma = 0$. Taken together this suggests a full martingale unobserved component model is needed, but volatility clustering in the scale is the main feature to be added to the Gaussian model.

10.5.3 PARTICLE MCMC OUTPUT

The pictures in Figures 10.2 and 10.3 shows that the likelihood is quite tightly estimated and so the mixing properties of the particle MCMC should not change very much as M varies. The output from an MCMC chain is placed in Figure 10.4, which is based on the results from running a MCMC chain

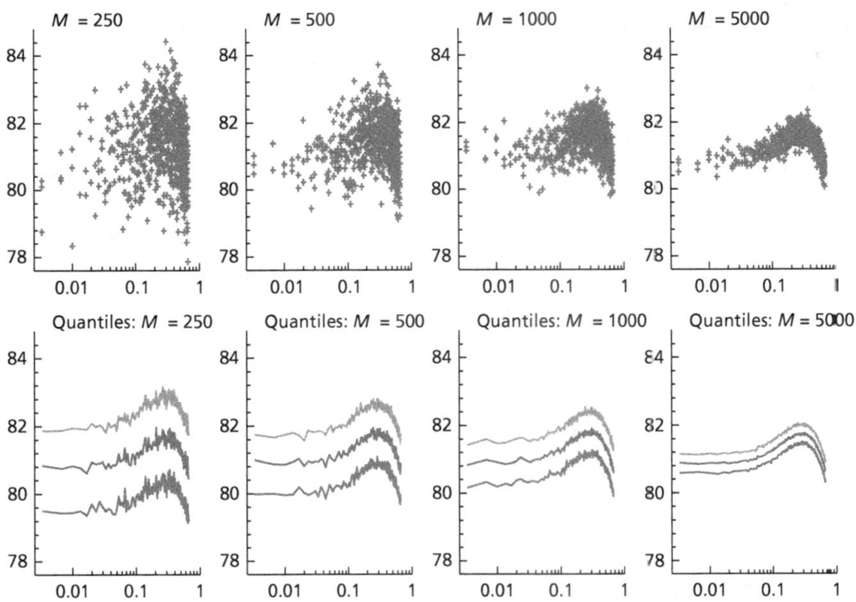

Figure 10.2. Estimated log-likelihood function for the inflation series for θ_q fixing throughout $\theta_\sigma = 0.25$. Top row of graphs shows the particle filter estimator of the log-likelihood for various values of M using three replications evaluating θ_q at 200 different values. Bottom graphs are the pointwise 0.1, 0.5, and 0.9 quantiles of the estimates. Four hundred replications are used for each parameter point in estimating the quantiles. The likelihood is quite flat for this parameter

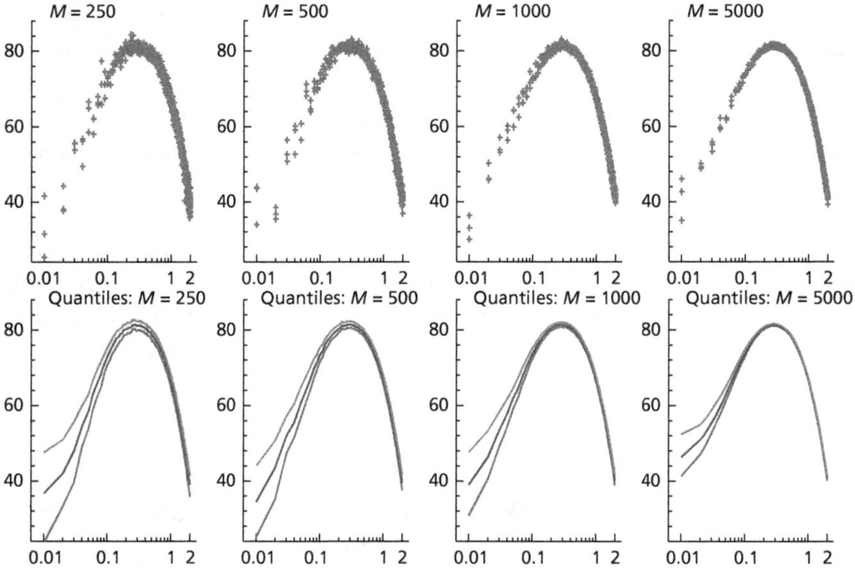

Figure 10.3. Estimated log-likelihood function for the inflation series for θ_σ fixing throughout $\theta_q = 0.1$. Top row of graphs shows the particle filter estimator of the log-likelihood for various values of M using three replications evaluating θ_σ at 200 different values. Bottom graphs are the pointwise 0.1, 0.5, and 0.9 quantiles of the estimates. Four hundred replications are used for each parameter point. The parameter is shown on a log10 scale. This shows a highly informative likelihood

for 20 000 iterations. The MCMC estimation results are reported in Table 10.2. This uses a random walk proposals for $\log\theta_q$ and $\log\theta_\sigma$, with independent proposals with standard deviations of 0.25—which are large moves. Writing the proposals as $\theta_q,\theta_\sigma,\widehat{f}$ (recalling $\widehat{f}(y|\theta_q,\theta_\sigma)$ is the particle filter estimate of $f(y|\theta_q,\theta_\sigma)$) and the existing values in the MCMC chain as $\theta_q',\theta_\sigma',\widehat{f}'$, then the probability of accepting these values is

$$\min\left\{1, \frac{\widehat{f}(y|\theta_q,\theta_\sigma)f(\theta_q,\theta_\sigma)}{\widehat{f}'(y|\theta_q',\theta_\sigma')f(\theta_q',\theta_\sigma')} \frac{\theta_q e^{-\frac{1}{2\times 0.25^2}\left(\log\theta_q-\log\theta_q'\right)^2}\theta_\sigma e^{-\frac{1}{2\times 0.25^2}\left(\log\theta_\sigma-\log\theta_\sigma'\right)^2}}{\theta_q' e^{-\frac{1}{2\times 0.25^2}\left(\log\theta_q-\log\theta_q'\right)^2}\theta_\sigma' e^{-\frac{1}{2\times 0.25^2}\left(\log\theta_\sigma-\log\theta_\sigma'\right)^2}}\right\}$$

$$= \min\left\{1, \frac{\widehat{f}(y|\theta_q,\theta_\sigma)f(\theta_q,\theta_\sigma)}{\widehat{f}'(y|\theta_q',\theta_\sigma')f(\theta_q',\theta_\sigma')}\frac{\theta_q\theta_\sigma}{\theta_q'\theta_\sigma'}\right\}.$$

The key feature here is that the value of the estimated likelihood is accepted or rejected, as well as the proposed parameter values. Further, $f(\theta_q,\theta_\sigma)$ is the joint prior.

In all eight, such chains are run independently in parallel, using a multicore processor. The output is then thought of as a cross-section of long independent

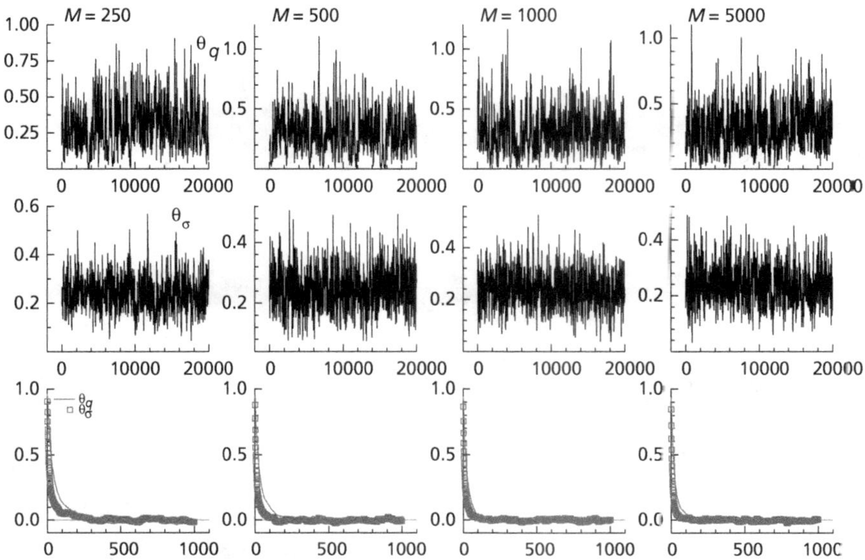

Figure 10.4. Particle MCMC inference for the θ_q and θ_σ parameters for the inflation series. Resulting MCMC chain for a variety of values of M being 250, 500, 1000, and 5000. Suggests the mixing of the chain is relatively fast

time series with the same marginal distribution. Figure 10.4 shows the path from a single chain, while the autocorrelation function is the average autocorrelation function from the eight independent chains.

The posteriors are typically summarized using quantiles. This is carried out by computing the quantile for each chain separately and then cross-sectionally averaging the resulting quantiles. The uncertainty of this estimate can be measured by using the standard error of this arithmetic mean, but when I did this the errors were so small there was little utility in reporting them here.

Table 10.1 shows quantile summaries of $\theta|y$ for different values of M, showing that the results are entirely comparable. There is an improvement in precision as M increases, but the computational cost of running the algorithm is proportional to M. The results suggest that using a small value of M may be the most computationally effective for this problem.

The posterior $\theta_q|y$ shows quite a high degree of spread and non-symmetry, with 80% of the mass roughly between 0.09 and 0.55. The posterior $\theta_\sigma|y$ is much tighter and symmetric, with 80% of the mass roughly between 0.14 and 0.33. The posterior means are $\theta_q = 0.37$ and $\theta_\sigma = 0.23$, respectively, while the standard deviation measures reflect the results from the quantiles. The posterior correlation between the parameters is around -0.4.

Table 10.1 also shows summaries for the fitted Stock-Watson parameterization (note the corresponding likelihood profiles are given in Figures 10.5 and 10.6). In their empirical work they impose a priori that $\theta_{\sigma_\varepsilon} = \theta_{\sigma_\eta} =$

Table 10.1 Results from a random walk based particle MCMC algorithm analysis of the posterior $\theta|y$ for the inflation series. Generically $Q_q(X)$ denotes the q-th quantile of the random variable X, computed by cross-sectionally averaging the individual quantiles from each chain. IAC_X denotes the integrated autocorrelation $1+2\sum_{j=1}^{1500}\widehat{\rho}_j$, where $\widehat{\rho}_j$ is the cross-sectional average of the autocorrelation of the MCMC chain for X using this level of M

M	$Q_{.1}(\theta_q\|y)$	$Q_{.5}(\theta_q\|y)$	$Q_{.9}(\theta_q\|y)$	$Q_{.1}(\theta_\sigma\|y)$	$Q_{.5}(\theta_\sigma\|y)$	$Q_{.9}(\theta_\sigma\|y)$	
250	0.097	0.279	0.558	0.141	0.230	0.333	
500	0.093	0.281	0.551	0.146	0.232	0.333	
1,000	0.103	0.285	0.560	0.143	0.230	0.334	
5,000	0.096	0.281	0.548	0.145	0.232	0.333	
M	$E(\theta_q\|y)$	$sd(\theta_q\|y)$	IAC_{θ_q}	$E(\theta_\sigma\|y)$	$sd(\theta_\sigma\|y)$	$\text{IAC}_{\theta_\sigma}$	$Cor(\theta_q,\theta_\sigma\|y)$
250	0.307	0.171	90	0.234	0.074	54	-0.382
500	0.306	0.173	81	0.237	0.073	26	-0.339
1,000	0.313	0.187	54	0.235	0.075	25	-0.416
5,000	0.306	0.174	51	0.236	0.074	21	-0.384
			Stock-Watson parameterization				
M	$Q_{.1}(\theta_{\sigma\eta}\|y)$	$Q_{.5}(\theta_{\sigma\eta}\|y)$	$Q_{.9}(\theta_{\sigma\eta}\|y)$	$Q_{.1}(\theta_{\sigma\varepsilon}\|y)$	$Q_{.5}(\theta_{\sigma\varepsilon}\|y)$	$Q_{.9}(\theta_{\sigma\varepsilon}\|y)$	
250	0.294	0.478	0.711	0.112	0.202	0.335	
500	0.299	0.480	0.713	0.110	0.200	0.332	
1,000	0.301	0.482	0.718	0.109	0.199	0.330	
5,000	0.299	0.482	0.713	0.111	0.199	0.332	
M	$E(\theta_{\sigma\eta}\|y)$	$sd(\theta_{\sigma\eta}\|y)$	$\text{IAC}_{\theta_{\sigma\eta}}$	$E(\theta_{\sigma\varepsilon}\|y)$	$sd(\theta_{\sigma\varepsilon}\|y)$	$\text{IAC}_{\theta_{\sigma\varepsilon}}$	$Cor(\theta_{\sigma\eta},\theta_{\sigma\varepsilon}\|y)$
250	0.494	0.165	35	0.215	0.089	52	-0.344
500	0.496	0.157	23	0.213	0.092	37	-0.321
1,000	0.499	0.162	19	0.211	0.087	28	-0.316
5,000	0.497	0.164	17	0.213	0.090	30	-0.372

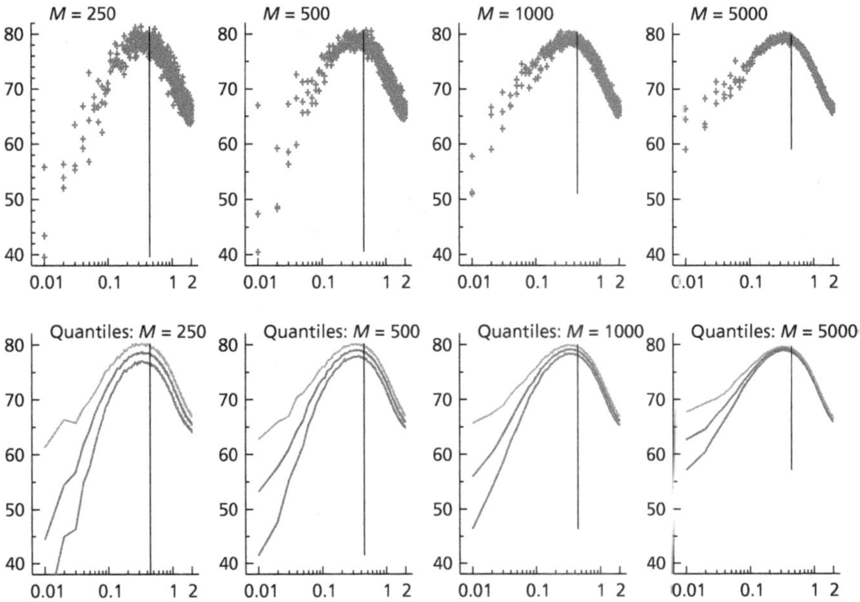

Figure 10.5. Estimated log-likelihood function for the inflation series using the SW model for θ_{σ_η} fixing $\theta_{\sigma_\varepsilon} = 0.25$. Top row: Particle filter estimator of the log-likelihood for various values of M using three replications evaluating θ_{σ_η} at 200 different values. Bottom: Pointwise 0.1, 0.5, and 0.9 quantiles of the estimates. Four hundred replications are used for each parameter point in estimating the quantiles

$\sqrt{0.2} \simeq 0.447$, which is not a million miles away from the values supported by the data, although the data would prefer a lower value for $\theta_{\sigma_\varepsilon}$. What is most striking about these estimated values is that under (10.37) the implied $\theta_q \simeq \sqrt{0.47^2 + 0.18^2} \simeq 0.50$, which is high compared to the estimated directly parameterized model.

Results for the extended model

Figure 10.7 reconciles the results from the two different models, by allowing a correlation between the two SV innovations in (10.40), and showing the resulting estimated likelihood profile as a function of the correlation. This shows the log-likelihood is substantially more sensitive to ρ for the SW parameterization, while for both parameterizations the graphs support a correlation which is very close to one, say around 0.8 (when we subsequently look at the Bayesian posteriors the median and expected values are substantially lower than this as I have used a flat prior over the interval $-1, 1$ and the likelihood is not tightly peaked, but the posterior summaries are higher for the extended SW as the corresponding score is larger). For the extended model the change in log-likelihood

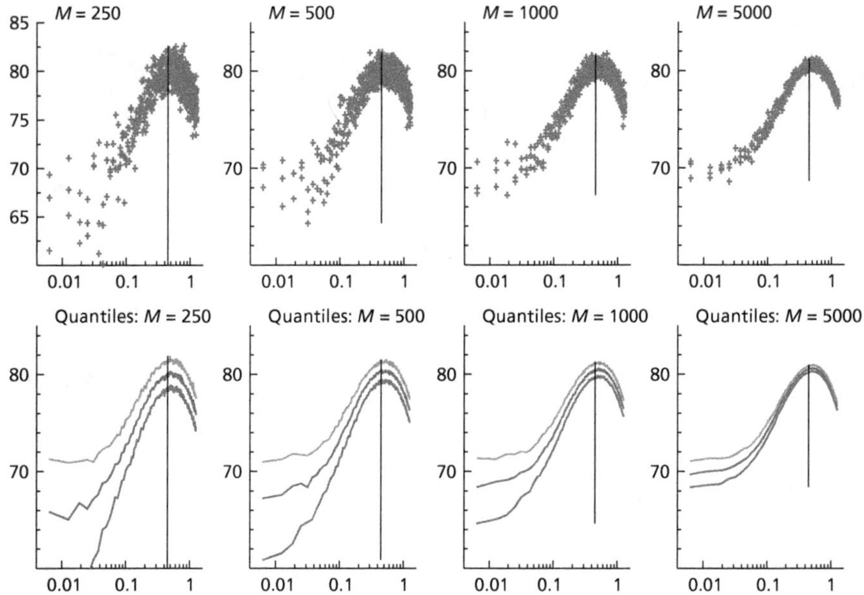

Figure 10.6. Estimated log-likelihood function for the inflation series using the SW model for $\theta_{\sigma_\varepsilon}$ fixing throughout $\theta_{\sigma_\eta} = 0.25$. Top row: Particle filter estimator of the log-likelihood for various values of M using three replications evaluating $\theta_{\sigma_\varepsilon}$ at 200 different values. Bottom: Pointwise 0.1, 0.5, and 0.9 quantiles of the estimates. Four hundred replications are used for each parameter point in estimating the quantiles

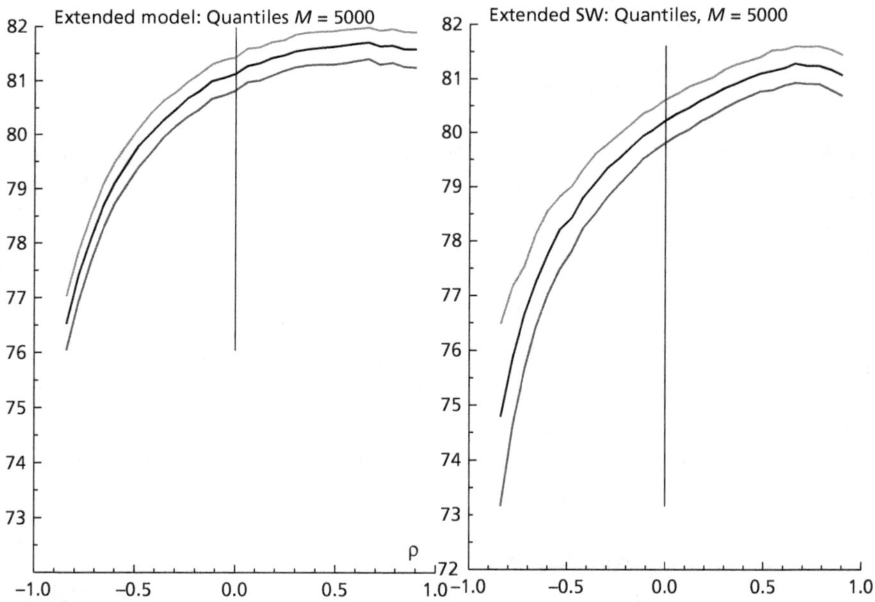

Figure 10.7. Estimated log-likelihood function for the inflation series for $\rho \in (-1, 1)$ fixing the other parameters for the extended model (fixing $\theta_q = 0.18$, $\theta_\sigma = 0.22$) and the extended SW parameterization (fixing $\theta_{\sigma_\eta} = 0.46$, $\theta_{\sigma_\varepsilon} = 0.16$). Shows pointwise 0.1, 0.5, and 0.9 quantiles of the estimates based on 400 replications

by setting $\rho = 0$ looks to be around 0.5, while for the extended SW model it is around 1.5. When we focus on the SW parameterization and take this $\rho = 0.8$ value then the implied $\theta_q \simeq \sqrt{(0.48 - 0.8 \times 0.20)^2 + 0.36 \times 0.20^2} \simeq 0.34$, which is not a great deal from the estimate value for the original model.

10.5.4 HEADLINE FEATURES OF THE FITTED MARTINGALE UNOBSERVED COMPONENT MODEL

Figure 10.8 shows the results from the fit of the martingale unobserved component model for the quarterly inflation series. This was computed using the parameter values $\theta_q = 0.27$ and $\theta_\sigma = 0.23$.

The first graph is the time series of the raw quarterly inflation data (denoted by a cross) together with the one-step ahead forecast from the model (shown by a time series line).

The second graph shows the time series of the 0.1, 0.5, and 0.9 quantiles of

$$\sigma_t \sqrt{q_t} | \mathcal{F}_t, \theta \tag{10.61}$$

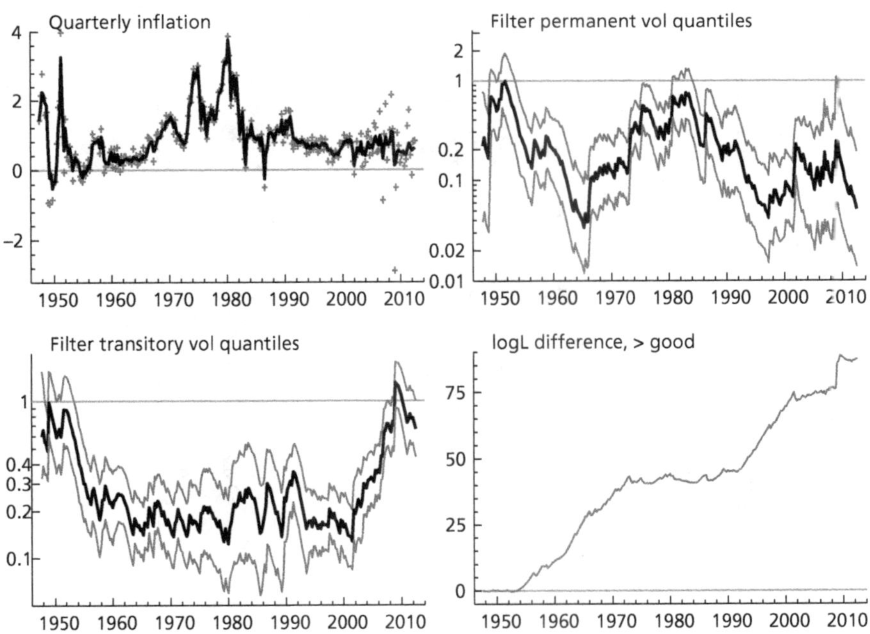

Figure 10.8. Main fit from the inflation series. Top left shows the raw quarterly inflation with the one-step ahead forecasts. Top right shows the 0.1, 0.5, and 0.9 quantiles of the filtered volatility of the permanent component, and the bottom left shows the volatility of the transitory component. The filtering looks at the distributions $\sigma_t|\mathcal{F}_t$ and $\sigma_t q_t^{1/2}|\mathcal{F}_t$. They are plotted on the same log10 scale. Bottom right shows the cumulative difference in the log-likelihood compared to the linear model

Table 10.2 Results from a random walk based particle MCMC algorithm analysis of the posterior $\theta|y$ for the inflation series. Generically $Q_q(X)$ denotes the q-th quantile of the random variable X, computed by cross-sectionally averaging the individual quantiles from each chain. IAC_X denotes the integrated autocorrelation $1+2\sum_{j=1}^{1500}\widehat{\rho}_j$, where $\widehat{\rho}_j$ is the cross-sectional average of the autocorrelation of the MCMC chain for X using this level of M

M	$Q_p(\theta_q\|y)$				$Q_p(\theta_\sigma\|y)$				$Q_p(\rho\|y)$			
	$p=.1$	$p=.5$	$p=.9$	IAC	$p=.1$	$p=.5$	$p=.9$	IAC	$p=.1$	$p=.5$	$p=.9$	IAC
250	0.080	0.260	0.532	106	0.141	0.228	0.332	47	-0.368	0.278	0.775	41
500	0.087	0.266	0.528	74	0.142	0.228	0.334	31	-0.364	0.273	0.776	29
1,000	0.082	0.261	0.526	63	0.142	0.229	0.334	26	-0.359	0.276	0.780	23
5,000	0.081	0.254	0.528	45	0.144	0.230	0.335	18	-0.366	0.276	0.785	22

Extended SW parameterization

M	$Q_p\left(\theta_{\sigma_\eta}\|y\right)$				$Q_p(\theta_{\sigma_\varepsilon}\|y)$				$Q_p(\rho\|y)$			
	$p=.1$	$p=.5$	$p=.9$	IAC	$p=.1$	$p=.5$	$p=.9$	IAC	$p=.1$	$p=.5$	$p=.9$	IAC
250	0.272	0.443	0.674	43	0.121	0.206	0.323	39	-0.117	0.487	0.861	48
500	0.270	0.437	0.665	26	0.119	0.206	0.327	31	-0.113	0.487	0.860	29
1,000	0.271	0.435	0.665	17	0.121	0.207	0.324	26	-0.104	0.499	0.862	24
5,000	0.270	0.436	0.668	16	0.119	0.205	0.321	22	-0.125	0.486	0.855	24

which represents the standard deviation of the quarterly change in the level of underlying inflation. The results show that the volatility changes by an order of magnitude through time, sometimes being around 0.05 while reaching a peak which is close to 0.9. The quantiles show the precision with which this volatility is estimated changes through time.

The third graph shows the time series of the 0.1, 0.5, and 0.9 quantiles of

$$\sigma_t|\mathcal{F}_t, \theta \tag{10.62}$$

that represents the time series of the volatility of the temporary component of inflation. This has a minimum of around 0.2 and maximizes at around one in the early- to mid-1950s and again around 2010. The volatility of the temporary component did not rise during the 1970s and 1980s, which significantly differentiates it from the volatility of the permanent component. Stock and Watson (2007) report that σ_t is relatively constant between the 1960s and 1990s, which is true but it is importantly not the case outside this period and enforcing homogeneity would be highly dangerous to forecasting and understanding. A further note is that during periods of high transitory volatility it is hard to precisely estimate the permanent component and so the permanent component's volatility. Thus high periods of transitory volatility, such as the recent times, are associated with wide confidence intervals for the permanent volatility.

The fourth plot shows the cumulative difference in the log-likelihood between the martingale unobserved component model and the Gaussian random walk plus noise model. This shows a consistent improvement from the martingale model, except during the late-1970s to around 1990, where the models fitted roughly similarly.

10.5.5 ANALYSIS

Figure 10.9 looks underneath some of the time series properties of the fitted model. The spot moving average process ψ_t is shown on the top left, plotted as $-\psi_t$. Recall that as $\psi_t \to 0$ the model becomes a martingale, while when it is close to one it is using a lot of recent data to average away the noise. The movements during the sample are considerable and important, largely replicating the observation of Stock and Watson (2007). At the start of the 1980s the process went close to zero, which is suggesting a martingale forecast. At the end of the sample the process is close to one.

The top right shows the estimate of the signal-noise ratio

$$\mathrm{E}\left(q_t|\mathcal{F}_t, \theta\right) \tag{10.63}$$

which shows very large moves, with changes of roughly two orders of magnitude. The current value of the signal-noise ratio is very low by historical standards.

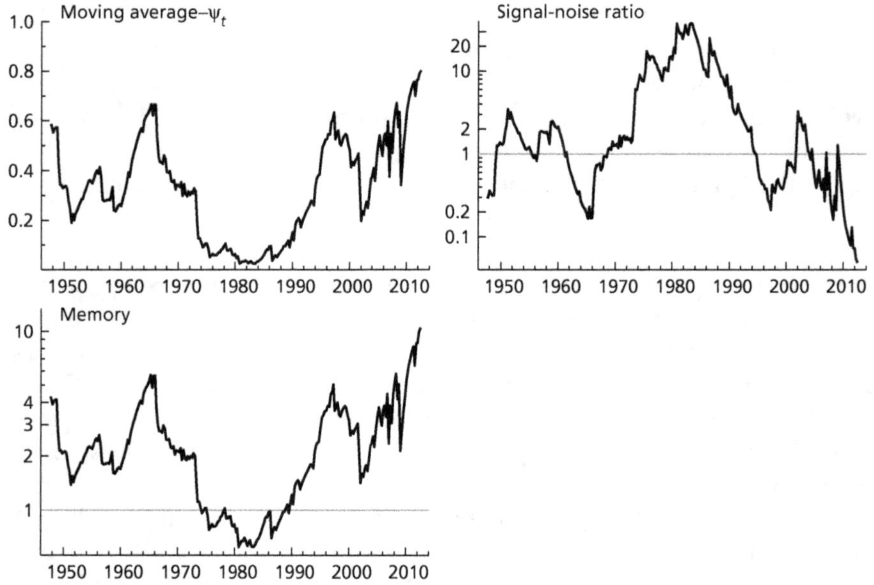

Figure 10.9. Analysis of a martingale unobserved component model applied to quarterly inflation. Top left: Filtered estimate of the moving average coefficient. Top right: Filtered estimate of the signal/noise ratio. Bottom left: Filtered estimate of the memory of the time series

The bottom left picture is the time series of the memory measure s_t through time. Recall this works out how long it takes before the weight on a past data point is under 10% and is computed via (10.31). This picture is drawn on the log10 y-scale but has little extra information in it compared to the plot of the moving average process.

10.5.6 DIAGNOSTIC STATISTICS

Figure 10.10 shows some plots to diagnose the empirical effectiveness of the models. The first column of figures looks at the innovations from the fitted model

$$v_t = \mathrm{E}\left(\frac{y_t - m_{t|t-1}}{\sigma_t \sqrt{p^*_{t|t-1} + 1}} | \mathcal{F}_{t-1} \right), \tag{10.64}$$

where the posterior mean is averaged over $q_{1:t}$ and $\sigma_{1:t}$. This should be roughly i.i.d. standard normal if the model fits well. In the top plot v_t is drawn against t and there is little apparent structure visible to the eye. The bottom plot shows the correlogram for v_t, drawn using an index plot, together with the

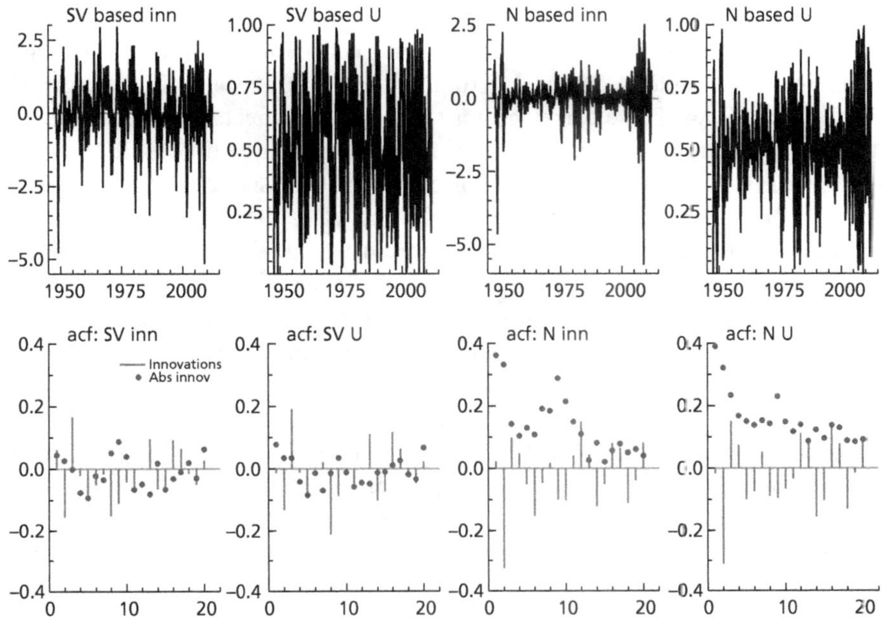

Figure 10.10. Diagnostics from the model for the extended inflation series. Top are raw innovations and probability integral transforms (U), for two models: Martingale plus noise and random walk plus noise. Bottom denotes correlograms either of the raw series or the absolute value or reflected version $2|U-1/2|$. Estimation is based on data up until 2004; log-likelihood is computed at these estimated parameter points and includes all data up to 2012

correlogram for $|v_t|$, shown by a series of dots. The latter is designed to pick up any missing volatility clustering in the innovations, the former looks for linear dependence in the innovations. The results are somewhat encouraging; there seems very little volatility clustering. There are some quite large autocorrelations appearing for v_t in lags 2 and 3, while the results for $|v_t|$ look very strong. This indicates the model could be improved for very short-term forecasting by allowing some very short memory into ε_t.

The second column looks at the plot of the probability integral transforms

$$U_t = \mathrm{E}\left[\Phi\left(\frac{y_t - m_{t|t-1}}{\sigma_t\sqrt{p^*_{t|t-1}+1}}\right)|\mathcal{F}_{t-1}\right] \tag{10.65}$$

against time, where Φ is the standard normal distribution function. The U_t should be an i.i.d. $U(0,1)$ sequence if the model is correctly specified and so the plot should appear like the scatter of a two dimensional homogenous Poisson process on $[0,1]\times[1,n]$. These types of transforms have been used by many researchers, e.g., Rosenblatt (1952), Shephard (1994), and Diebold et al. (1998).

The correlogram of the U_t and the so-called reflected uniforms $2\,|U_t - 1/2|$ (see Shephard (1994) who introduced them to check for volatility clustering, realizing they should also be i.i.d. $U(0,1)$ if the model is well fitting) are given in the plot below. These correlograms should be less affected by outliers than the corresponding ones for v_t, but in this case there is little difference.

The same diagnostics are also reported for the Gaussian random walk plus noise model in the third and fourth columns. Here there are some obvious failings, most dramatically due to volatility clustering. The correlogram in the third column also shows a very high correlation at lag 2 for the v_t (much higher than for the martingale plus noise model, and this feature also appears in the corresponding results for U_t). Overall the fit of the linear model is quite poor.

All of these results are summarized in Table 10.3, which shows the fitted martingale unobserved component model's parameters, the corresponding log-likelihood, and Box-Pierce summaries from the above correlograms. These were computed using twenty lags. Also given is the result from the linear model (where $\theta_q = \theta_\sigma = 0$), which has a poorer likelihood and worse dependence measures. Most of the gain is made by dealing with the heteroskedasticity in the data. Table 10.3 also reports results from the random walk plus SV noise model ($\theta_q = 0$) and the martingale plus strict white noise model ($\theta_\sigma = 0$). These models have parameters that need to be estimated and this is carried out using particle MCMC in the usual way. Note that these two constrained models deliver parameter estimates that are considerably higher than for the martingale unobserved component model, as they try to use their flexibility to deal with the heteroskedasticity in the data. That the parameters jump upwards is not surprising given the posterior for the general model is negatively correlated and these model simplifications wrongly impose one of the parameters as zero. Overall the martingale unobserved component model is slightly better than the random walk plus SV noise model. This is in turn better than the martingale plus strict white noise model, but the degree of difference is surprisingly small. This latter model boosts up θ_q to such a large degree that it can deal with some of the heteroskedasticity in the data.

10.5.7 MULTISTEP OUT OF SAMPLE FORECASTING

I finish by looking at how the models perform out of sample, forecasting a year ahead, that is—four steps ahead. The results are given in Figure 10.11 for the martingale unobserved component model and the Gaussian random walk plus noise model. This out of sample period includes the start of the great recession and so considerably stresses these forecasting devices. Table 10.3 also shows the out of sample likelihood for the fitted models.

Throughout, the parameters θ_σ and θ_q are fixed at their December 2004 values. Then running through the data, the estimated level μ_t is extrapolated

Table 10.3 Analysis based upon M = 25,000. logL is the estimated log-likelihood evaluated at that parameter point, while Δ logL is the change in the log-likelihood caused by the extended sample size—so is the out of sample log-likelihood (parameter values are unchanged with the extended data). Box-Pierce (BP) with 20 degrees of freedom. Here, v denotes the innovations from the model and U denotes the probability integral transforms form the model. First model is the martingale unobserved component model ($\mathrm{E}(\theta_q, \theta_\sigma | y)$). Second is the SV scaled random walk plus noise ($\mathrm{E}(\theta_\sigma | y, \theta_q = 0)$). Third is the martingale plus strict white noise ($\mathrm{E}(\theta_q | y, \theta_\sigma = 0)$). Fourth is the Stock-Watson model ($\mathrm{E}(\theta_{\sigma_\varepsilon}, \theta_{\sigma_\eta} | y)$). Fifth is the martingale plus the white noise model ($\mathrm{E}(\theta_{\sigma_\varepsilon} | y, \theta_{\sigma_\eta} = 0)$). Sixth is the random walk plus SV noise ($\mathrm{E}(\theta_{\sigma_\eta} | y, \theta_{\sigma_\varepsilon} = 0)$). Seventh is the Gaussian random walk plus noise model ($\theta_q = 0, \theta_\sigma = 0$). Last two results are from the nested model ($\mathrm{E}(\theta_{\sigma_\eta} | y, \theta_{\sigma_\varepsilon} = 0)$)

	θ_q	θ_σ	ρ	$\theta_{\sigma\eta}$	$\theta_{\sigma\varepsilon}$	$\log L$	$\Delta \log L$	BP_v	$BP_{\lvert v \rvert}$	BP_U	$BP_{2\lvert U-1/2 \rvert}$
			Insample								
$\mathrm{E}(\theta_q, \theta_\sigma \lvert y)$	0.31	0.23				81.6		44.7	14.9	43.5	10.3
$\mathrm{E}(\theta_\sigma \lvert y, \theta_q = 0)$	0.00	0.27				80.7		45.9	14.3	42.2	10.4
$\mathrm{E}(\theta_q \lvert y, \theta_\sigma = 0)$	0.61	0.00				76.1		40.9	11.6	43.1	9.4
$\mathrm{E}(\theta_{\sigma\varepsilon}, \theta_{\sigma\eta} \lvert y)$				0.49	0.21	80.6		43.6	13.9	43.9	10.0
$\mathrm{E}(\theta_{\sigma\varepsilon} \lvert y, \theta_{\sigma\eta} = 0)$				0.00	0.39	72.4		40.8	13.9	42.1	12.4
$\mathrm{E}(\theta_{\sigma\eta} \lvert y, \theta_{\sigma\varepsilon} = 0)$				0.61	0.00	75.9		42.1	11.7	43.9	10.0
$\theta_q = 0, \theta_\sigma = 0$	0.00	0.00				14.5		55.1	57.7	52.2	62.4
$\mathrm{E}(\theta_q, \theta_\sigma, \rho \lvert y)$	0.26	0.23	0.28			81.8		45.1	14.9	43.2	9.6
$\mathrm{E}(\theta_{\sigma\varepsilon}, \theta_{\sigma\eta}, \rho \lvert y)$			0.49	0.44	0.21	81.4		44.2	14.4	43.1	9.8
			Extended dataset–out of sample								
$\mathrm{E}(\theta_q, \theta_\sigma \lvert y)$	0.31	0.23				66.0	-15.6	36.0	14.0	43.7	9.7
$\mathrm{E}(\theta_\sigma \lvert y, \theta_q = 0)$	0.00	0.27				62.2	-18.5	37.9	14.8	41.6	10.9
$\mathrm{E}(\theta_q \lvert y, \theta_\sigma = 0)$	0.61	0.00				54.4	-21.7	39.9	24.8	41.6	14.4
$\mathrm{E}(\theta_{\sigma\varepsilon}, \theta_{\sigma\eta} \lvert y)$				0.49	0.21	65.8	-14.8	35.6	12.9	44.3	9.3
$\mathrm{E}(\theta_{\sigma\varepsilon} \lvert y, \theta_{\sigma\eta} = 0)$				0.00	0.39	57.4	-15.0	34.9	13.3	43.3	12.2
$\mathrm{E}(\theta_{\sigma\eta} \lvert y, \theta_{\sigma\varepsilon} = 0)$				0.61	0.00	54.1	-21.8	40.1	25.2	41.7	14.6
$\theta_q = 0, \theta_\sigma = 0$	0.00	0.00				-14.8	-29.3	59.2	115	64.5	144
$\mathrm{E}(\theta_q, \theta_\sigma, \rho \lvert y)$	0.26	0.23	0.28			65.6	-16.2	35.4	13.8	43.1	9.1
$\mathrm{E}(\theta_{\sigma\varepsilon}, \theta_{\sigma\eta}, \rho \lvert y)$			0.49	0.44	0.21	65.4	-16.0	35.7	12.9	43.5	8.7

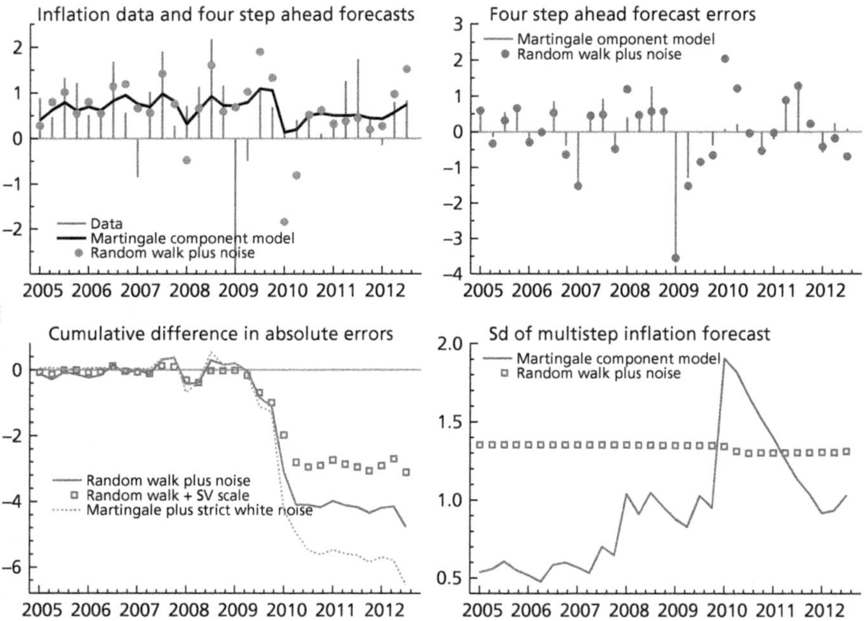

Figure 10.11. Out of sample multistep ahead (year ahead) forecasts for the inflation series. Top left are the raw quarterly data and the multistep ahead forecasts of those series using the martingale unobserved component model and random walk plus noise model. Top right are the forecast errors from the models. Bottom left is the cumulate sum of the difference of the absolute forecast errors from the three models compared to the martingale unobserved component model. If the cumulative sum is below zero this is some evidence in favor of the forecasts from the martingale unobserved component model. Bottom right is the standard deviation of the four steps ahead forecast of the inflation series, shown for the two models. The results here are very different

four periods into the future as well as computing the forecast's standard error under the various models.

Top left shows the data together with the two sets of forecasts. The results are dominated by the large negative inflation shock at the start of the recession. Of course the forecasts take four periods to react to this innovation. The Gaussian random walk plus noise model overreacts and this causes massive forecast errors later in the period. The martingale unobserved component model reacts more appropriately. The difference in the cumulative absolute forecast errors compared to the same quantity for other models is given in the bottom left. Numbers below zero suggest some support for the martingale unobserved component model. This suggests the random walk plus noise with SV scaling is the most competitive alternative to the martingale unobserved component model, followed by the random walk plus noise model. The martingale plus strict white noise model is poor in comparison.

The associated forecast standard errors are drawn in the bottom right, with the results for the Gaussian random walk plus noise obviously being time invariant. The martingale unobserved component model ramps up the standard error before the start of the crisis and ever more so at the onset. The standard errors have tailed off considerably in recent times, down to roughly their long run average values.

10.6 Conclusion

This paper looks at the Gaussian local level model and inflation forecasting, generalizes it by allowing for stochastic volatility scaling and allowing the signal/noise ratio to change through time. I call this type of model a martingale unobserved component model. The model is relatively easy to handle and extends to a wide class of important time series models. The diagnostics of the model are calculated and compared to the results for various special cases.

The particle filter is used to handle the model, which extends the Kalman filter to allow for non-Gaussianity and nonlinearity. The particle filter is used to generate an estimate of the log-likelihood which is used inside a MCMC algorithm, in order to make inferences on the parameters of the model. The MCMC chains are quite well behaved and are simple to parallelize to exploit multicore computers or indeed GPUs.

The martingale unobserved component model generalizes in various ways, to allow for trends, cycles, and for seasonal components. The methods developed here can be extended in the same way. The martingale unobserved component model can also be set in continuous-time and used to look at high frequency financial data, where stochastic volatility scaling is clearly very important. Likewise, non-parametric regression can be analysed using this kind of model, as noted by for example Wecker and Ansley (1983), Kohn et al. (1992), and Harvey and Koopman (2000). By allowing for the signal/noise ratio to change through time, the approach discussed here allows the smoothing to be carried out with, in effect, a local bandwidth that might be important for some applications.

ACKNOWLEDGMENTS

This paper was written in honor of Andrew Harvey's sixty-fifth birthday. I am grateful to James Taylor for introducing me to the non-model based EWMA literature which allows the discount parameter to change through time as a function of past data. The detailed comments of Siem Jan Koopman and two referees were also very helpful, as well as suggestions from Mike Pitt.

11 More is not always better: Kalman filtering in dynamic factor models

Pilar Poncela and Esther Ruiz

11.1 Introduction

Dynamic factor models (DFM), originally introduced by Geweke (1977) and Sargent and Sims (1977), assume that there is a small number of underlying states common to the variables in a system see Stock and Watson (2011) for a recent survey. The procedures available for factor extraction are based either on using the Kalman filter and smoothing (KFS) algorithms or on principal components (PC) see Harvey (1989, pp. 100–66) for a detailed description of the KFS algorithms, and Stock and Watson (2002a) and Forni et al. (2005) for PC procedures.

Both PC and KFS factor estimators are consistent as the cross-sectional and time dimensions increase see Stock and Watson (2002a) for PC and Doz et al. (2011, 2012) for KFS estimators. Therefore, many authors argue that including all available variables in the system seems to be a natural choice when extracting factors see, among many others, Stock and Watson (2002b) who consider 215 variables, Forni et al. (2003) who incorporate 447 variables, Amengual and Watson (2007) 124 variables, Eickmeier (2009) 173 variables, Altissimo et al. (2010) 145 variables, Gupta and Kabundy (2011) 267 variables, or Proietti (2011) 149 variables. In contrast, several authors argue that, in real life problems, introducing many variables is not always a good strategy. For example, Watson (2003) shows that the predictive precision of the common factors extracted using PC does not increase when increasing the cross-sectional dimension beyond forty or fifty variables. Later, Boivin and Ng (2006) point out that if by adding an extra variable we are not adding information about the factor but rather simply extra cross-sectional correlation among the idiosyncratic disturbances, then the estimated factors deteriorate, while Bai and Ng (2008a) conclude that by selecting the variables, the predictive performance of the estimated factors can increase with respect to that obtained when all available variables are included. More recently, Caggiano et al. (2011) conclude that between twelve and twenty variables are enough to obtain the best performance when predicting euro area gross domestic product (GDP) using extracted factors see also, Banbura and Runstler (2011).

The objective of this paper is to contribute to the debate about whether it is better to estimate the underlying factors including all available variables or by selecting an appropriate subset. We do so by analyzing the uncertainty associated with the estimated common factors when the cross-sectional and temporal dimensions are finite. We focus on KFS extraction which requires the assumption of a particular specification of the DFM and estimation of the model parameters. In spite of the computational burden involved in this estimation, KFS procedures have several advantages. First, data irregularities, as mixed frequencies or missing data, can be handled. Second, they can be implemented in real time as individual data are released. Third, they provide a framework for incorporating restrictions derived from economic theory. Fourth, KFS procedures are more efficient than PC for a flexible range of specifications that include non-stationary DFM and idiosyncratic noises with strong cross-correlations. Finally, they allow one to obtain uncertainty measures associated with the estimated factors when the cross-sectional dimension is finite, while only asymptotic mean square errors (MSEs) are available in the context of PC.

When, as usual, the Kalman filter is run with the unknown parameters substituted by consistent estimates, the MSEs of the estimated factors have two sources of uncertainty, one stemming from the filtering process itself and the other from the estimated parameters. Our first contribution is to analyze separately the effect of the number of variables on the filter and parameter uncertainties. Furthermore, by focusing on the Kalman filter, we also contribute to the literature by considering a wide range of specifications of the factors and idiosyncratic noises and by analyzing how different characteristics of the DFM affect the finite sample MSE. In particular, we consider stationary and non-stationary specifications together with idiosyncratic noises with weak and strong contemporaneous correlations. Finally, we derive expressions not only of the MSE of the smoothed but also of the real time one-step-ahead and filtered factor estimates. We show that, as far as the idiosyncratic noises are serially uncorrelated and regardless of whether their contemporaneous correlations are weak or strong, the filter uncertainty is a non-increasing function of the cross-sectional dimension. Furthermore, in situations of empirical interest, if the cross-sectional dimension is beyond a relatively small number, the filter uncertainty only decreases marginally. However, the limiting behavior of the MSE depends on the properties of the covariances of the idiosyncratic noises and the weights of the factors. Weak cross-correlations together with pervasive factors is a sufficient condition for the uncertainty of the filtered and smoothed factor estimates to converge to zero with the cross-sectional dimension. To measure the contribution of the parameter uncertainty to the total uncertainty when both the cross-sectional and temporal dimensions are finite, we carry out Monte Carlo experiments incorporating the parameter uncertainty into the Kalman filter MSE using the proposal of Delle Monache and Harvey (2011). We show that, in cases of empirical interest, the parameter uncertainty could represent a large percentage of the total uncertainty associated with the estimation of the underlying factors. Furthermore, even for relatively large sample sizes, the parameter and total uncertainties could increase with respect to the cross-sectional dimension.

The rest of the paper is organized as follows: Section 11.2 describes the DFM; Section 3 analyzes the filtering uncertainty; and Section 11.4 measures the uncertainty added when the parameters are estimated. Finally, Section 11.5 concludes.

11.2 Dynamic factor model

In this section, we consider a DFM with a single factor; see Engle and Watson (1981), Stock and Watson (1991), Proietti and Moauro (2006), Aruoba et al. (2009), Camacho and Perez-Quiros (2010), and Frale et al. (2011) for empirical applications with just one factor. The underlying factor, F_t, is given by the following autoregressive model of order 1, the AR(1) model, as given by

$$F_t = \phi F_{t-1} + \eta_t, \tag{11.1}$$

where ϕ is the autoregressive parameter such that $-1 < \phi \leq 1$. The disturbance, η_t, is a Gaussian white noise with variance σ_η^2.[1] Note that when $\phi = 1$, the common factor is non-stationary and represents a stochastic level; see, for example, Peña and Poncela (2004, 2006), Eickmeier (2009), and Lam et al. (2011) for non-stationary factors.

The latent factor, F_t, affects the evolution of an unknown number of variables. Consider that the first N of these variables, contained in the vector $Y_t^{(N)} = (y_{1t}, \ldots, y_{Nt})'$, are observed by the econometrician and that the following DFM is used for factor extraction

$$Y_t^{(N)} = \mu^{(N)} + P^{(N)} F_t + \varepsilon_t^{(N)} \tag{11.2}$$

where $\varepsilon_t^{(N)}$ is an $N \times 1$ vector of idiosyncratic noises which follows the following VAR(1) process

$$\varepsilon_t^{(N)} = \Gamma^{(N)} \varepsilon_{t-1}^{(N)} + a_t^{(N)} \tag{11.3}$$

where $a_t^{(N)}$ is a Gaussian white noise vector with finite and positive definite covariance matrix $\Sigma_a^{(N)}$. The idiosyncratic noises, $\varepsilon_t^{(N)}$, are distributed

[1] Alternatively, as proposed, for example by Harvey and Streibel (1998) and Onatski (2012), the factor can be assumed to have a fixed finite variance. The main conclusions are similar to those obtained in this chapter; see Poncela and Ruiz (2012) for details.

independently of $\eta_{t-\tau}$ for all leads and lags. The vector of constants is $\mu^{(N)} = (\mu_1, \ldots, \mu_N)'$ and $P^{(N)} = (p_1, \ldots, p_N)'$ is the factor loading vector. When the factor is stationary, the VAR(1) process in (11.3) can be either stationary or not; see, for example, Bai and Ng (2004) for DFM with unit roots both in the factors and the idiosyncratic components. However, when the factor is non-stationary, we assume that the idiosyncratic noises are either stationary or they have at most $N-1$ unit roots. In this case, the DFM is observable and the Kalman filter reaches the steady-state; see Harvey (1989, pp. 118–19).

The DFM in Equations (11.2) and (11.3) with F_t defined as in (11.1) is conditionally Gaussian. Consequently, when the idiosyncratic noises are serially uncorrelated, the Kalman filter and smoothing algorithms provide minimum MSE estimates of the underlying factors. On the other hand, when the idiosyncratic noises are serially correlated, the DFM can be reformulated as follows:

$$Y_t^{(N)} = \mu^{(N)*} + \Gamma^{(N)} Y_{t-1}^{(N)} + \left[P^{(N)} \; -\Gamma^{(N)} P^{(N)} \right] \begin{bmatrix} F_t \\ F_{t-1} \end{bmatrix} + a_t^{(N)}$$

$$\begin{bmatrix} F_t \\ F_{t-1} \end{bmatrix} = \begin{bmatrix} \phi & 0 \\ 1 & 0 \end{bmatrix} \begin{bmatrix} F_{t-1} \\ F_{t-2} \end{bmatrix} + \begin{bmatrix} \eta_t \\ 0 \end{bmatrix}, \qquad (11.4)$$

where $\mu^{(N)*} = \mu^{(N)}(I^{(N)} - \Gamma^{(N)})$ with $I^{(N)}$ being the $N \times N$ identity matrix; see Reis and Watson (2010) and Jungbacker et al. (2011) for implementations of the model in (11.4). Alternatively, one can deal with the autocorrelation of the idiosyncratic noises by augmenting the state vector by $\varepsilon_t^{(N)}$; see, for example, Jungbacker et al. (2011). Although both formulations lead to the same results when the initialization issues are properly accounted for, the latter enlarges the dimension of the state vector so that the Kalman filter becomes computationally cumbersome when N is large. Consequently, when dealing with DFM with known parameters, we consider the reformulation in (11.4).

11.3 Known parameters: filtering uncertainty

It is well known that, in conditionally Gaussian models, as the DFM considered in this paper, the estimates of the underlying factor provided by the KFS algorithms are given by the corresponding conditional means. Denoting by $f_{t|\tau}$ the estimate of F_t obtained with the information available up to time τ, if the model parameters are known, the filter delivers

$$f_{t|\tau} = E\left[F_t | Y_1^{(N)}, \ldots, Y_\tau^{(N)}\right]$$

where $\tau = t-1$, for one-step-ahead predictions, $\tau = t$ for filtered estimates and $\tau = T$ for smoothed factor estimates. Therefore, by construction, regardless of the cross-sectional dimension, the filter delivers unbiased estimates of the factor. Including more variables implies more information to estimate the factor and, consequently, under mild conditions, the MSEs of the factor estimates are expected to be non-increasing functions of the cross-sectional dimension. Next, we derive the MSE of $f_{t|\tau}$ as a function of the cross-sectional dimension. Because the filter is run in two different state-space models depending on whether the idiosyncratic noises are or not serially correlated, we consider both cases separately.

11.3.1 SERIALLY UNCORRELATED IDIOSYNCRATIC NOISES

In this subsection, we analyze how the MSEs of $f_{t|\tau}$ depend on N in the DFM with serially uncorrelated idiosyncratic noises. Given that the system matrices are time-invariant, the Kalman filter reaches the steady-state in which the MSE of $f_{t|t-1}$ and $f_{t|t}$ are constant; see Harvey (1989, pp. 118–19). The MSE of the smoothed estimates are also constant in the middle of the sample. From now on, we focus on steady-state MSEs.

Consider first the MSE of the one-step-ahead factor predictions, denoted by $V(N)$, which is obtained after solving the following Riccati equation

$$V(N) = \phi^2 \left[V(N) - V(N)P^{(N)\prime} \left(P^{(N)} V(N) P^{(N)\prime} + \Sigma_\varepsilon^{(N)} \right)^{-1} P^{(N)} V(N) \right] + \sigma_\eta^2 \tag{11.5}$$

where $\Sigma_\varepsilon^{(N)} = \Sigma_a^{(N)}$. As in any linear Gaussian model, the MSE of the one-step-ahead predicted factors is independent of the particular data available; see Harvey (1989, p. 111). The following lemma characterizes the solution of the Riccati equation.

Lemma 1 Given the DFM in (11.2) with the factor defined as in (11.1), and $\varepsilon_t^{(N)}$ being a serially uncorrelated vector process with contemporaneous covariance matrix given by $\Sigma_\varepsilon^{(N)}$, not necessarily diagonal, the one-step-ahead MSE is given by the solution of the Riccati equation in (11.5) which is

$$V(N) = \frac{\sigma_\eta^2 Q(N) - 1 + \phi^2 + \sqrt{\left(\sigma_\eta^2 Q(N) - 1 + \phi^2\right)^2 + 4\sigma_\eta^2 Q(N)}}{2Q(N)} \tag{11.6}$$

where $Q(N) = P^{(N)\prime}\Omega^{(N)}P^{(N)}$ with $\Omega^{(N)} = \left(\Sigma_\varepsilon^{(N)}\right)^{-1}$. Furthermore, $V(N+1) = V(N)$, if i) $\phi = 0$, or ii) $p_{N+1} = 0$ and ε_{N+1}, is not correlated with any of the other N variables previously included in the system. Otherwise, $V(N+1) < V(N)$.

Proof. See: Appendix 11.6.

Lemma 1 establishes that, when the factor is white noise, the MSE of $f_{t|t-1}$ is σ_η^2 and does not depend on the cross-sectional dimension. When $\phi \neq 0$, adding a new variable to the system never decreases the precision in the estimation of the underlying factor. It is important to remark that this result is satisfied regardless of whether the cross-correlations between the idiosyncratic noises are weak or strong or whether the factor is stationary or not. Furthermore, when adding an additional variable with zero loading, if this variable is correlated with the variables already included in the model, the steady-state MSE also decreases. This result somehow contradicts the conclusion of Boivin and Ng (2006) about the deterioration of factor estimates when adding an extra variable which is not adding information about the factor but rather simply extra cross-sectional correlation among the idiosyncratic disturbances. We guess that their conclusion could be attributed to the estimation method.

For a given cross-sectional dimension, N, the one-step-ahead MSE also depends on the dynamics of the underlying factor. First, it is obvious that for fixed ϕ, $V(N)$ always increases with σ_η^2. On the other hand, when σ_η^2 is fixed, the precision of $f_{t|t-1}$ decreases with ϕ. This could be expected as the variability of the factor is larger as it approaches a random walk. Therefore, as ϕ is larger, more variables are needed to estimate the underlying factor with a given precision.

Consider now the limiting behavior of $V(N)$ when N increases. There are situations in which, by definition, there is a finite number of variables, N^*, that depend on the factor; see, for example, the factor model in Chamberlain and Rothschild (1983). In these cases, according to (11.6), the minimum MSE is given by

$$V(N^*) = 0.5\left(\sigma_\eta^2 - 1 + \phi^2\right)\left(1 - \sqrt{1 + \frac{4\sigma_\eta^2\overline{Q}}{\left(\sigma_\eta^2\overline{Q} - 1 + \phi^2\right)^2}}\right), \qquad (11.7)$$

where $\overline{Q} = Q(N^*)$. The MSE in (11.7) is always larger than σ_η^2 for finite $\overline{Q}$.

Alternatively, there are applications in which there are potentially infinitely many variables that depend on the factor. In this case, the limiting behavior of $V(N)$ depends on whether the contemporaneous correlations between the idiosyncratic noises are weak or strong. Note that the MSE of the one-step-ahead factor predictions in expression (11.6) depends on the cross-sectional dimension, N, through the term $Q(N)$. It is straightforward to show that, if $Q(N)$ converges to infinity with N, then $V(N)$ tends to σ_η^2. Notice that this result could be expected as the one-step-ahead factor predictions always involve the uncertainty associated with η_t. However, when $Q(N)$ converges to a constant, $\overline{Q}$, then the limit of the MSE is given by expression (11.7).

The following lemma establishes the conditions for $Q(N)$ to diverge and, consequently, the MSE of $f_{t|t-1}$ to converge to its minimum σ_η^2.

Lemma 2 Let $g_N^2 = \sigma_{N|N-1}^{-2}\left(p_N - P^{(N-1)'}\Omega^{(N-1)}\Sigma_{N,N-1}\right)^2$ where $\sigma_{N|N-1}^2 = \sigma_N^2 - \Sigma_{N,N-1}'\Omega^{(N)}\Sigma_{N,N-1}$ is the variance of ε_N conditional on ε_i for $i = 1, \ldots, N-1$ and $\Sigma_{N,N-1}$ is the $N-1$ vector containing the covariances between ε_N and ε_i for $i = 1, \ldots, N-1$ and let $Q(N)$ be defined as in expression (11.6). If $\lim_{N\to\infty} \frac{g_{N+1}^2}{g_N^2} = l$ exists and $l > 1$, then

$$\lim_{N\to\infty} Q(N) = \infty.$$

Proof. In the proof of Lemma 1, we show that $Q(N+1) = Q(N) + g_{N+1}^2$. Therefore, $Q(N)$ is a series of positive terms and the result of the lemma is obtained as a direct consequence of the D'Alembert criterion; see, for example, Piskunov (1969, p. 829).

Lemma 2 has several important implications for the empirical implementation of DFM. First, it is important to note that the usual assumption about the idiosyncratic noises being weakly correlated and the factors being pervasive is a sufficient condition for $Q(N)$ to go to infinity with N. In this case, when $N \to \infty$, $\frac{1}{N}\sum_{i=1}^{N} p_i^2$ is asymptotically larger than the maximum eigenvalue of $\Sigma_\varepsilon^{(N)}$ and the explanatory content of the factor, defined as in Onatski (2012), strongly dominates the explanatory content of the idiosyncratic noises. Second, the condition in Lemma 2 for $Q(N)$ to go to infinity is sufficient but not necessary. If $\lim_{N\to\infty} \frac{g_{N+1}^2}{g_N^2}$ does not exist or is equal to 1, the series $Q(N)$ can either converge or diverge when the number of variables increase, and it is necessary to use alternative criteria to solve the problem. Third, from an empirical point of view, we expect the variables being introduced in the model according to a criterion that implies some kind of ordering. If, for example, the variables are introduced in the model according to their explanatory content with respect to the factor, so that the less pervasive variables are those introduced later, we expect the conditions in Lemma 2 for $Q(N)$ to go to infinity with N not to be satisfied and the MSE could never be as small as σ_η^2.

Next, we illustrate the results of Lemmas 1 and 2, focusing on the issue of how many variables should be used to extract the factors and the role of the contemporaneous correlations on the MSE of $f_{t|t-1}$. For this purpose, we first consider a DFM with contemporaneously uncorrelated idiosyncratic noises (strict DFM) and the factor specified as in equation (11.1) with $\sigma_\eta^2 = 1$ and $\phi = 0.4, 0.8$, and 1 and $q_i = \frac{p_i^2}{\sigma_i^2} = 1, i = 1, \ldots, N$, where σ_i^2 is the variance of the ε_{it}; see Camacho et al. (2015) for the empirical adequacy of these values. In this case, the minimum MSE of the one-step-ahead factor predictions is 1. The top left panel of Figure 11.1 displays the MSE in (11.6) as a function of the cross-sectional dimension. Observe that the larger is ϕ, the larger the MSEs of $f_{t|t-1}$ and a larger number of variables is required to converge. However,

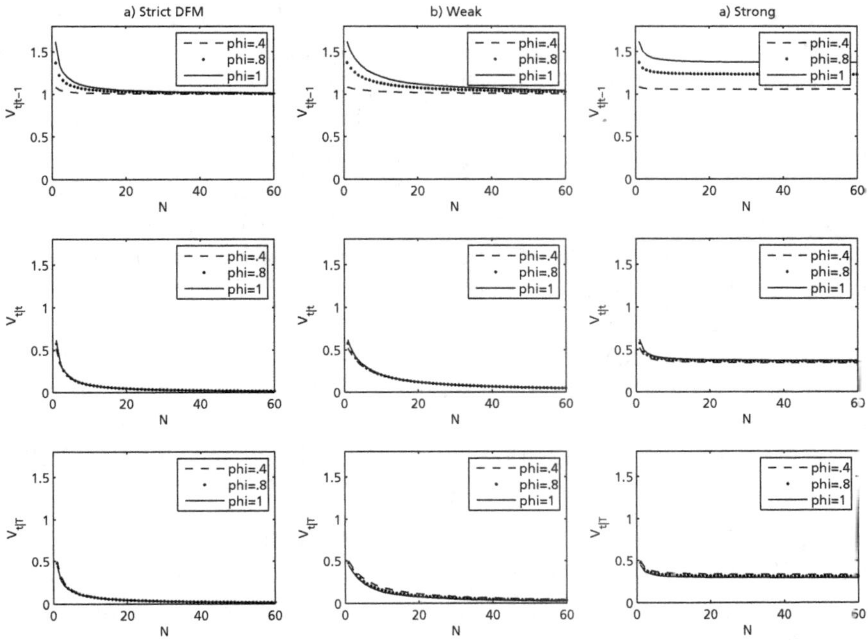

Figure 11.1. Steady-state MSE of one-step-ahead (first row), filtered (second row), and smoothed (third row) estimates of the underlying factor in a strict DFM (first column) and contemporaneously correlated idisyncratic noises with weak (second column) and strong (third column) correlations

even when $\phi = 1$, if the number of variables is around thirty, adding additional variables only decreases the MSE marginally. It is surprising to observe that this number of variables is similar to those found by Watson (2003), Banbura and Runstler (2011), and Caggiano et al. (2011) when the factor is estimated by PC. On the other hand, when the autoregressive dependence of the underlying factor is small, $\phi = 0.4$, the MSEs are very close to $\sigma_\eta^2 = 1$ regardless of the number of variables included in the system. In this case, using the information of approximately five variables, the MSE is already quite close to its minimum.

Next, we analyze the effects of the presence of contemporaneous correlations on the previous conclusions. For this purpose, we consider a DFM with contemporaneously correlated idiosyncratic noises. The factor is defined with the same relative weights, q_i, and parameters ϕ and σ_η^2 as above. The idiosyncratic noises have weak cross-correlations with covariance matrix, $\Sigma_\varepsilon^{(N)}$, given by a Toeplitz matrix with parameter $b = 0.5$ so that the ij-th element of the correlation matrix of the idiosyncratic noises is given by $0.5^{(j-i)}$. The second column of the first row of Figure 11.1 plots the corresponding MSE. Comparing the first and second columns of Figure 11.1, we can observe that the MSE are approximately the same, although, when there is weak cross-correlation, they decay slightly slower than in the strict DFM. Therefore, adding weak cross-correlations only increases the uncertainty marginally.

Finally, we consider a DFM with strong cross-correlations among the idiosyncratic noises. The same DFM is again considered but now all off-diagonal elements of $\Sigma_\varepsilon^{(N)}$ are equal to 0.5 so that the eigenvalues of $\Sigma_\varepsilon^{(N)}$ are not bounded when $N \to \infty$. The third column of the first row of Figure 11.1 plots the corresponding MSE in equation (11.6). We can observe that, as established in Lemma 2, the MSEs do not converge to σ_η^2; see, also, Onatski (2012) for the same result in the context of PC factor extraction. Finally, it is important to note that the presence of strong autocorrelations among the idiosyncratic noises can be interpreted as the existence of a second factor that is not explicitly specified. In this case, the estimator of the included factor is not consistent.

Now, we consider the properties of real time filtered and smoothed MSE which are established in the following lemma.

Lemma 3 Given the DFM in (11.2) with the factor defined as in (11.1), and $\varepsilon_t^{(N)}$ being a serially uncorrelated vector process with contemporaneous covariance matrix given by $\Sigma_\varepsilon^{(N)}$, non-necessarily diagonal, the filtered and smoothed MSE are given by

$$W(N) = \frac{V(N)}{1 + V(N)Q(N)} \tag{11.8}$$

$$S(N) = \frac{V(N)\left(1 + V(N)Q(N) - \phi^2\right)}{(1 + V(N)Q(N))^2 - \phi^2}, \tag{11.9}$$

respectively. Furthermore, $W(N+1) = W(N)$ and $S(N+1) = S(N)$ if i) $\phi = 0$, or ii) $p_{N+1} = 0$ and ε_{N+1}, is not correlated with any of the other N variables previously included in the system. Otherwise, $W(N+1) < W(N)$ and $S(N+1) < S(N)$.

Proof. See: Appendix 11.6.

The filtered and smoothed uncertainties never increase as more variables are used to estimate the underlying factor. Note that the asymptotic behavior of $W(N)$ and $S(N)$ depends on the convergence of $Q(N)$. If $Q(N)$ tends to ∞ with N, $W(N)$ and $S(N)$ converge to zero. Therefore, the filtered and smoothed factor estimates are consistent when the cross-sectional dimension tends to infinity, being the conditions for consistency the same as those established in Lemma 2 for $Q(N)$ to diverge. This result can be compared with Doz et al. (2011, 2012) who prove the consistency of Kalman filter smoothed estimates assuming a more restrictive stationary DFM.

As an illustration of the performance of the filtered and smoothed MSE, we consider, once more, the same DFM considered above. The second and third rows of Figure 11.1 plot the MSE of $f_{t|t}$ and $f_{t|T}$, respectively. In the first column, we consider the strict DFM while the second and third columns plot the MSE when the idiosyncratic noises have weak and strong correlations, respectively. Figure 11.1 shows that the MSEs are very similar for all values of ϕ considered, and for filtered and smoothed estimates. The MSEs decrease very slowly for N larger than twenty. Furthermore, when N is larger than thirty, the MSEs are approximately equal to zero unless the idiosyncratic noises are strongly correlated. For instance, for the strict DFM, if $\phi = 1$, the reduction in the MSE is only about 3% when going from $N = 20$ to 30 variables. We can see that by adding weak cross-correlation, the filtered MSEs increase with respect to those in the strict DFM. The MSEs still converge to zero with the cross-sectional dimension but the rate is slower. However, if the cross-correlations are strong, then the filtered MSEs do not converge to zero. It is important to note that, in this latter case, having around ten variables in the system already generates filtered MSEs which are very close to the minimum. One could conclude that smoothing is irrelevant when the cross-sectional dimension goes to infinity.

11.3.2 SERIALLY CORRELATED IDIOSYNCRATIC NOISES

As mentioned above, if the idiosyncratic noises are serially correlated, the Kalman filter is optimal when implemented in the state-space model in (11.4) in which the state vector is not scalar, as it contains both F_t and F_{t-1}. In this case, the corresponding Riccati equation does not necessarily have a closed-form solution as a function of the parameters of the model; see Lancaster and Rodman (1995) and Rojas (2011) for solutions of the Riccati equation. In this subsection, we analyze the effects of the serial correlation of the idiosyncratic noises on the MSE by running the prediction equations of the Kalman filter until the steady-state is reached for several particular DFM. Note that the Kalman filter prediction equations deliver both one-step-ahead and filtered MSE without requiring the simulation of the series. We also obtain the value of the smoothed MSE in the middle of the sample where they are constant.

The particular DFM considered include both stationary and non-stationary factors. The factor is specified as in equation (11.1) with $\sigma_\eta^2 = 1$ and $\phi = 0.8$ and 1. The idiosyncratic noises are assumed to be contemporaneously uncorrelated, i.e., $\Sigma_a^{(N)} = I^{(N)}$. The autoregressive matrix in equation (11.4) is given by $\Gamma^{(N)} = \rho I^{(N)}$, with $\rho = 0, 0.5$, and 0.9. Note that when $\rho = 0$, the same strict DFM considered above is obtained for comparative purposes. As mentioned above, the existence of the steady-state requires the DFM to be observable. Consequently, the case when both the underlying factor and the idiosyncratic noises are random walks has been ruled out. Furthermore, we should note that, when both ρ and ϕ are close to one, the steady-state is only reached after a very large number of steps of the Kalman filter.

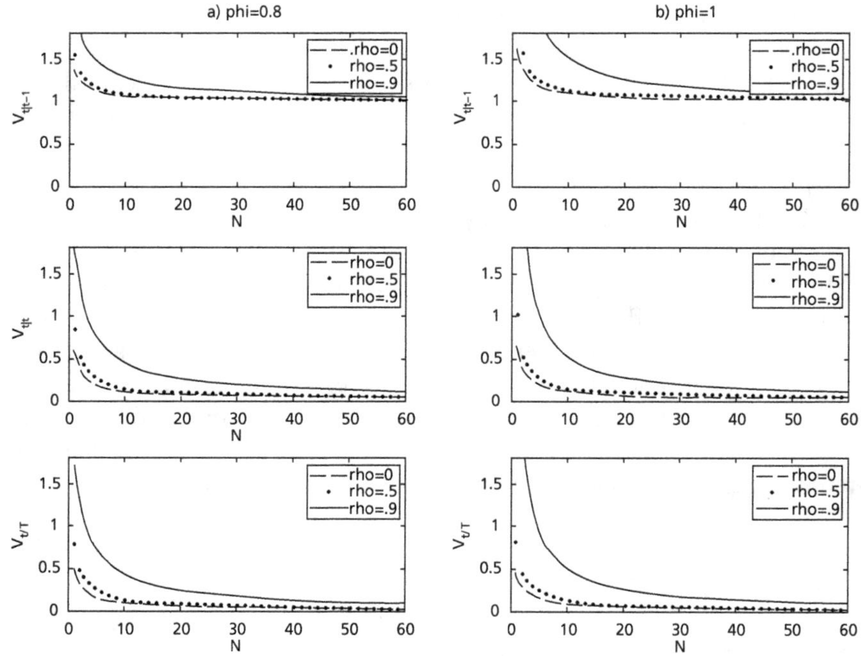

Figure 11.2. Steady-state MSE of one-step-ahead (first row), filtered (second row), and smoothed (third row) estimates of the underlying factor in a DFM with serially correlated idiosyncratic noises with parameter ρ, for stationary (left column) and non-stationary (right column) factors

The left column of Figure 11.2 plots the MSE as a function of the cross-sectional dimension for one-step-ahead, filtered, and smoothed factor estimates for the DFM with a stationary factor, i.e., when $\phi = 0.8$, while the right column plots the same quantities when the factor is non-stationary, i.e., $\phi = 1$. We can observe that, regardless of whether the factor is stationary or not, the MSE of $f_{t|\tau}$, $\tau = t-1$, t, and T, are very similar when the idiosyncratic noises are serially uncorrelated and when they have moderate temporal dependences, i.e., when $\rho = 0.5$. The conclusions about the number of variables needed in order to have estimates of the underlying factor with a precision close to the maximum are very similar to those obtained in the Section 11.3.1. However, when the idiosyncratic noises are very persistent, i.e., $\rho = 0.9$, we observe much larger MSE for each cross-sectional dimension. Furthermore, the convergence of the MSE toward their minimum is slower than when the idiosyncratic errors are moderately autocorrelated. Therefore, to have the same precision in the estimation of the factor, it is necessary to introduce in the model a larger number of variables. This result could be expected given that, if both the factor and the idiosyncratic noises are highly persistent, it could be difficult for the filter to distinguish between them.

Summarizing, it seems that unless the idiosyncratic noises are highly persistent, the previous conclusions about the number of variables to be included in the system to estimate the underlying factors are maintained in the presence of serially correlated idiosyncratic noises.

11.3.3 GENERALIZATION TO MORE THAN ONE FACTOR

The results above have been obtained assuming that there is a unique common factor in the system. However, in practice, a larger number of common underlying factors could be expected in large systems. The natural question to ask is whether the conclusions are still the same when we need to estimate more than one factor. As mentioned above, we do not consider the model uncertainty and, therefore, we assume that the number of factors is known and given by r. The vector of factors is then given by

$$F_t = \Phi F_{t-1} + \eta_t \tag{11.10}$$

where Φ is an $r \times r$ diagonal matrix containing the autoregressive parameters and η_t is an $r \times 1$ Gaussian vector with diagonal covariance matrix Σ_η. The variables in the system are related with the underlying factors through equation (11.2) where $P^{(N)}$ is now an $N \times r$ matrix of factor loadings. To simplify the discussion, we consider a DFM with serially uncorrelated idiosyncratic noises with contemporaneous covariance matrix $\Sigma_\varepsilon^{(N)}$. Given that the state vector is not scalar, the corresponding Riccati equation does not need to have a closed-form solution in terms of the parameters of the model. Therefore, once more, in this subsection, we obtain the MSE associated with the estimates of the vector of underlying factors by running the prediction equations of the Kalman filter until the steady-state is reached. The particular DFM considered is defined as in (11.10) with $diag(\Phi) = (0.8, 0.4)$ and $\Sigma_\eta = I^{(r)}$. The weights are given by $p_{i1} = 1$, $i = 1, \ldots, N$, while $p_{12} = 0$ and the remaining weights of the second factor, p_{i2}, $i = 2, \ldots, N$, have been randomly generated from a uniform $[0, 1]$ distribution. Therefore $\Sigma_\varepsilon^{(N)} = I^{(N)}$. Figure 11.3 plots the filtered MSE for the first and second factors in the first and second rows, respectively. Finally, the third row of Figure 11.3 plots the corresponding covariances delivered by the filter. We can observe that the absolute covariances between the estimated factors decrease with the cross-sectional dimension, N. It is also clear that the MSE of the second factor decrease very quickly with the first variables added to the system but then, after having around thirty variables, the decrease is rather slow.

Therefore, it seems that the presence of more than one factor, requires, in general, more variables to be estimated with a given precision. This could be due to the correlation between the estimated factors.

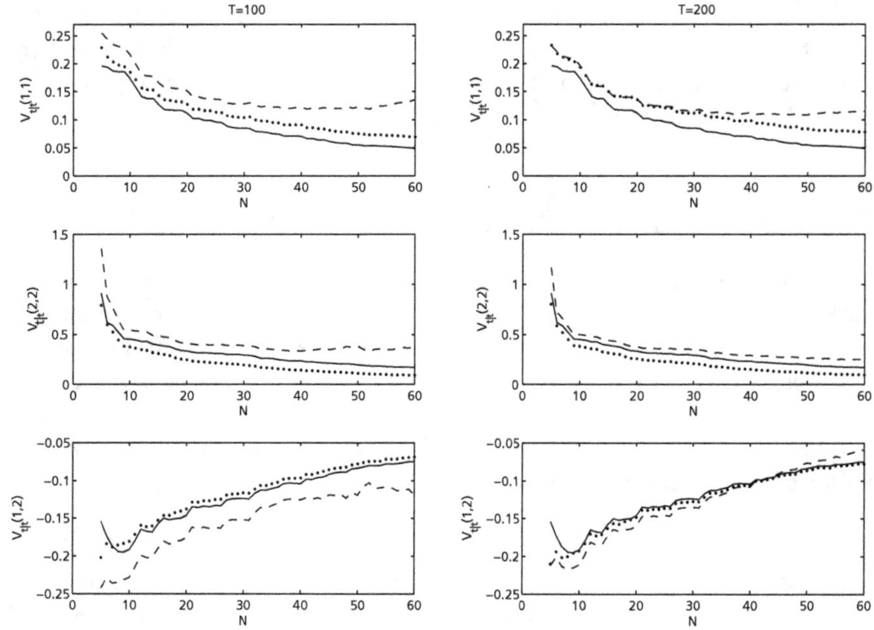

Figure 11.3. Steady-state "true" MSE (dashed lines), filter (known parameters) MSE (continuous lines), and estimated parameters MSE (dotted lines) matrices of filtered estimates in a strict DFM with two factors and parameters estimated by ML with $T = 100$ (left column) and $T = 200$ (second column). The first two rows represent the MSE of each of the factors while the third row represents the covariances

11.4 **Estimated parameters**

In Section 11.3, we assume known parameters and analyze how the filter uncertainty depends on the cross-sectional dimension. However, in practice, the parameters are unknown and are usually substituted by consistent estimates. In this case, the total uncertainty associated with the estimation of the underlying factors has two components, one related to the stochastic error uncertainty, considered in Section 11.3, and another to the parameter uncertainty. Note that the MSEs delivered by the filter run with estimated parameters, which are usually reported in practice by many authors, underestimate the true uncertainty as they do not incorporate the additional uncertainty due to the parameter estimation; see, for example, Jazwinski (1970, pp. 266–329), Hamilton (1986), Quenneville and Singh (2000a), and Rodríguez and Ruiz (2012). In this section, we measure the additional uncertainty attributable to parameter estimation and its relation with the cross-sectional and time dimensions. Note that the parameter uncertainty is expected to increase with the cross-sectional dimension as more parameters need to be estimated when adding additional variables to the system. On the other hand, if the parameter estimator is consistent, increasing the temporal dimension decreases the parameter uncertainty which disappears in the limit. Because of its popularity and given that the model considered in this paper is assumed to be conditionally Gaussian, we focus on the ML estimator of the parameters with the log-likelihood maximized using the EM algorithm. This algorithm has the attractiveness of being derivative free and only requires one pass of the smoother in each iteration. Therefore, it is computationally convenient when estimating the parameters in large DFM ignoring the dependence of the idiosyncratic noises; see Doz et al. (2012).

In order to measure the total MSE of one-step-ahead factor predictions when the Kalman filter is implemented with estimated parameters, we treat the estimated model as if it were misspecified and use the results in Delle Monache and Harvey (2011), who establish a general framework to compute the MSE in misspecified linear state-space models.[2] Consider the strict DFM with r factors given by equations (11.2) and (11.10). Denoted by $\widehat{K}_t$ the filter gain and by $\widehat{V}_{t+1|t}$ the one-step-ahead MSE matrices delivered by the filter run with ML estimated parameters. Note that although both $\widehat{K}_t$ and $\widehat{V}_{t+1|t}$ depend on the cross-sectional and temporal dimensions, we do not make this dependence explicit to simplify the notation. Then, according to the formulae derived by Delle Monache and Harvey (2011), the true MSE of $\widehat{f}_{t|t-1}$, the one-step-ahead factor predictions delivered by the Kalman filter with the true parameters substituted by estimated parameters, denoted by $\widehat{\Phi}$ and $\widehat{P}^{(N)}$, is given by

$$\begin{aligned} V_{t+1|t} = & \left[\widehat{\Phi} - \widehat{K}_t\widehat{P}^{(N)}\right] V_{t|t-1} \left[\widehat{\Phi} - \widehat{K}_t\widehat{P}^{(N)}\right]' + \\ & \left[(\Phi - \widehat{\Phi}) - \widehat{K}_t \left(P^{(N)} - \widehat{P}^{(N)}\right)\right] X_t \left[(\Phi - \widehat{\Phi}) - \widehat{K}_t \left(P^{(N)} - \widehat{P}^{(N)}\right)\right]' + \\ & \widehat{K}_t \Sigma_\varepsilon^{(N)} \widehat{K}_t' + \Sigma_\eta + \left[(\Phi - \widehat{\Phi}) - \widehat{K}_t \left(P^{(N)} - \widehat{P}^{(N)}\right)\right] C_{t|t-1} \left[\widehat{\Phi} - \widehat{K}_t\widehat{P}^{(N)}\right]' \\ & + \left[\widehat{\Phi} - \widehat{K}_t\widehat{P}^{(N)}\right] C'_{t|t-1} \left[(\Phi - \widehat{\Phi}) - \widehat{K}_t \left(P^{(N)} - \widehat{P}^{(N)}\right)\right]' \end{aligned} \tag{11.11}$$

where $X_{t+1} = \Phi X_t \Phi' + \Sigma_\eta$ and $C_{t+1|t} = \Phi C_{t|t-1}\left[\widehat{\Phi} - \widehat{K}_t\widehat{P}^{(N)}\right]' + \Phi X_t \left[(\Phi - \widehat{\Phi}) - \widehat{K}_t \left(P^{(N)} - \widehat{P}^{(N)}\right)\right]' + \Sigma_\eta$ with $X_0 = C_0 = V_{1/0} = 0$. Delle Monache and Harvey (2011) show that the true MSE in expression (11.11) has a steady-state.

The MSE of the filtered estimates of the underlying factors obtained when the filter is run with estimated parameters, $\widehat{f}_{t|t}$, can also be derived from the

[2] One could also use Jazwinski (1970, pp. 266–329).

results in Delle Monache and Harvey (2011) as follows:

$$W_{t|t} = V_{t|t-1} + \widehat{M}_t G_t \widehat{M}_t' - V_{t|t-1}\widehat{P}^{(N)\prime}\widehat{M}_t' - \widehat{M}_t P^{(N)} V_{t|t-1}' - C_{t|t-1}'\left(P^{(N)} - \widehat{P}^{(N)}\right)'\widehat{M}_t' - \widehat{M}_t\left(P^{(N)} - \widehat{P}^{(N)}\right)C_{t|t-1}, \quad (11.12)$$

where $\widehat{M}_t = \widehat{V}_{t|t-1}\widehat{P}^{(N)\prime}\left[\widehat{P}^{(N)}\widehat{V}_{t|t-1}\widehat{P}^{(N)\prime} + \widehat{\Sigma}_\varepsilon^{(N)}\right]^{-1}$ and

$$G_t = \widehat{P}^{(N)} V_{t|t-1}\widehat{P}^{(N)\prime} + \Sigma_\varepsilon^{(N)} + \left(P^{(N)} - \widehat{P}^{(N)}\right) X_t \left(P^{(N)} - \widehat{P}^{(N)}\right)' + \widehat{P}^{(N)} C_{t|t-1}'\left(P^{(N)} - \widehat{P}^{(N)}\right)' + \left(P^{(N)} - \widehat{P}^{(N)}\right) C_{t|t.-1}\widehat{P}^{(N)\prime}.$$

Delle Monache and Harvey (2011) do not provide results for the true MSE of the fixed interval smoothed estimates considered in this paper. However, we have seen before that the MSE of the filtered and smoothed estimates are nearly indistinguishable. Therefore, we expect the MSE of the smoothed factor estimates obtained when the filter is run with estimated parameters to be similar to that of the filtered estimates and only consider the latter in this section.

The "true" MSEs in equations (11.11) and (11.12) depend on the parameter estimates obtained in a particular data set. Consequently, we carry out Monte Carlo experiments to measure the uncertainty associated with $\widehat{f}_{t|t-1}$ and $\widehat{f}_{t|t}$ in the context of a strict DFM. For a given specification, we generate $R = 500$ replicates of sizes $T = 100$ and 200 and for each replicate, we obtain the ML estimates of the parameters using the EM algorithm, and compute the true MSE of the one-step-ahead and filtered factor estimates, using expressions (11.11) and (11.12) until they reach the steady-state. Then we average the MSE through all replicates. The resulting averages are denoted as $\overline{V}(N, T)$ and $\overline{W}(N, T)$, respectively. As an illustration, we consider the same DFM with two factors defined in Section 3.3. The usual identifying restrictions are imposed before estimation. First, as the model is stationary, $Y_t^{(N)}$ is assumed to be zero mean and, consequently, all series in $Y_t^{(N)}$ are centered prior to their analysis. Finally, we restrict $\Sigma_\eta = I^{(2)}, p_{11} > 0$, and $p_{12} = 0$; see Harvey (1989, p. 451). We also compute the MSE associated with the stochastic error. In order to save space, we only report the results related the filtered estimates of the factors. The first row of the left column of Figure 11.3 plots the total MSE of the filtered estimates of the first factor together with the corresponding stochastic error MSE when $T = 100$. The second row plots the same quantities for the second factor. Finally, the third row plots the corresponding covariances. We can observe that, obviously, the total MSEs are larger than the MSEs obtained when the filter is run with known parameters. Furthermore, while the filter MSE approaches zero for relatively small cross-sectional dimensions, the total uncertainty has a U shape. As more variables are introduced in the system, more parameters need to be estimated and, consequently, the total uncertainty could even increase with N. The difference between the total and the filter uncertainty is relatively small when the number of variables in the system is small but increases with the cross-sectional dimension. When dealing with a system with $N = 60$, we can observe that most of the uncertainty can be attributed to the parameter estimation. The right column of Figure 11.3 plots the same quantities when $T = 200$. In this case, we can observe that the total MSEs of $\widehat{f}_{t|t}$ are obviously smaller for each N. This is expected given that the ML estimator of Θ is consistent and, consequently, (11.12) tends to the MSE with known parameters when $T \to \infty$. However, the parameter uncertainty is still important when $N = 60$.

Figure 11.3 also illustrates the biases incurred when using the MSE delivered by the Kalman filter implemented with estimated parameters. In this case, the "true" MSEs are underestimated.

To assess the importance of the parameter uncertainty when extracting the factors by running the Kalman filter with estimated parameters, Figure 11.4 plots the relative difference between the average total MSE, $\overline{V}(N, T)$, and the MSE attributable to the filter uncertainty, $V(N)$, i.e., the proportion of the total uncertainty that can be attributed to the parameter uncertainty for one-step-

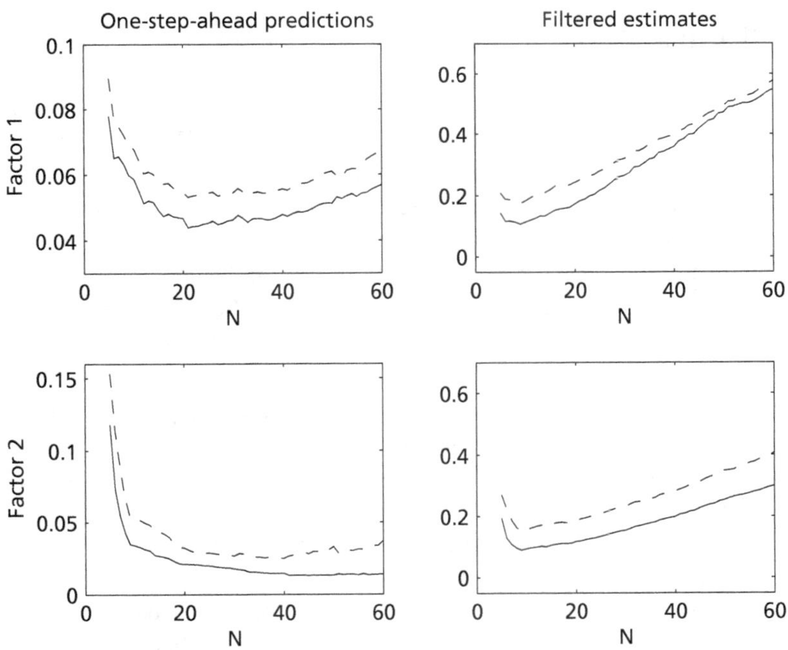

Figure 11.4. Proportion of the total MSE represented by the parameter uncertainty in a strict DFM with two factors for one-step-ahead (left column) and filtered (right column) estimates when the parameters are estimated with $T = 100$ (dashed lines) and $T = 200$ (continuous lines). The first row represents the results for the first factor while the second row corresponds to the second factor

ahead (left column) and filtered (right column) factor estimates, when $T = 100$ and 200. We can observe that, this percentage is a quadratic function of the cross-sectional dimension for one-step-ahead predictions of the first factor. It is minimum when the number of variables in the system is between twenty and thirty variables. In this case, if $T = 200$, the parameter uncertainty represents around a 4.5% of the total uncertainty. However, if there is a small number of variables in the system, it can represent around 8% and when $N = 60$, it represents around a 6%. When looking at the results for the second factor, we can see that the percentage decreases with N, being around 12% when there are few variables in the system and being as small as 2% when the number of variables is sixty. Nevertheless, the percentages of the filtered estimates are completely different. Regardless of whether we look at the results for the first or second factor, we can observe that the percentage of the parameter uncertainty over the total uncertainty is minimum when the number of variables is around ten. In this case, the percentage is around ten. However, as the number of variables increases, this percentage also increases and could be as large as 55% in the first factor or 30% in the second.

11.5 **Conclusions**

In this paper, we contribute to the issue about the finite sample uncertainty associated with the extraction of unobserved factors in DFM in the context of KFS procedures. We also extend available consistency results to real-time filtered estimates of the factors. Assuming that the model specification is known, if the Kalman filter is implemented, as usual, substituting the unknown parameters by consistent estimates the total MSE can be decomposed into the part attributable to the filter uncertainty and that attributable to the parameter uncertainty. When looking at the former component of the MSE we show that, regardless of whether the idiosyncratic noises are weakly or strongly correlated, a relatively small number of variables, typically around thirty, is enough to estimate the factors with an uncertainty close to its potential minimum. However, a larger number of variables could be needed if the idiosyncratic noises are highly persistent. Furthermore, when looking at the parameter uncertainty, we show that it can represent a large percentage of the total uncertainty. For a given temporal dimension, the parameter uncertainty and the total uncertainty can even be an increasing function of the cross-sectional dimension. Therefore, it seems that there is no point in including a huge number of variables for factor extraction. This result suggests that it could be worth going back to the Kalman filter, improving the efficiency in the estimation of unobserved factors with respect to PC whose main advantage is being able to deal with very large DFM.

In this paper, the results about the parameter uncertainty have been obtained using the ML estimator based on the EM algorithm. However, other alternative estimators, as those proposed by Jungbacker and Koopman (2015) or Doz et al. (2011, 2012), could be considered. The results could then be compared with those of PC. This issue is left for further research. Comparing the performance of Kalman filter and PC estimators in the context of misspecified models is also in our research agenda.

APPENDIX

11.6 **Proof of Lemmas**

PROOF OF LEMMA 1

Using the Woodbury formula for the inverse of the sum of two matrices given, for instance, by Press et al. (2007, p. 80), and after some straightforward algebra, the following equation is obtained:

$$Q(N)V^2(N) - \left(\sigma_\eta^2 Q(N) - 1 + \phi^2\right) V(N) - \sigma_\eta^2 = 0. \qquad (11.13)$$

Taking the positive solution of the equation (11.13), we obtain expression (11.6) for the steady-state MSE.

To prove that $V(N)$ is a non-increasing function of N, subtract expression (11.13) evaluated at $N+1$ from the same expression evaluated at N,

$$V(N)^2 Q(N) - V(N+1)^2 Q(N+1) + V(N)\left(-\sigma_\eta^2 Q(N) + 1 - \phi^2\right) - V(N+1)\left(-\sigma_\eta^2 Q(N+1) + 1 - \phi^2\right) = 0. \qquad (11.14)$$

First, we need to prove that $Q(N+1) \geq Q(N)$. Consider the following partition of the covariance matrix of the idiosyncratic noises when the cross-sectional dimension is $N+1$

$$\Sigma_\varepsilon^{(N+1)} = \begin{bmatrix} \Sigma_\varepsilon^{(N)} & \Sigma_{N,N+1} \\ \Sigma'_{N,N+1} & \sigma_{N+1}^2 \end{bmatrix},$$

where $\Sigma_{N,N+1}$ is an $N \times 1$ vector that collects the covariances between ε_i, $i = 1,\ldots,N$ and ε_{N+1}. Using the formula for the inverse of the partitioned matrix, the following expression for $\Omega^{(N+1)} = (\Sigma_\varepsilon^{(N+1)})^{-1}$ is obtained:

$$\Omega^{(N+1)} = \begin{bmatrix} B & -\frac{1}{\sigma_{N+1}^2} B\Sigma_{N,N+1} \\ -\frac{1}{\sigma_{N+1}^2}\Sigma'_{N,N+1}B & \frac{1}{\sigma_{N+1}^2} + \frac{1}{\sigma_{N+1}^4}\Sigma'_{N,N+1}B\Sigma_{N,N+1} \end{bmatrix}, \qquad (11.15)$$

where $B = \left(\Sigma_\varepsilon^{(N)} - \frac{\Sigma_{N,N+1}\Sigma'_{N,N+1}}{\sigma_{N+1}^2}\right)^{-1}$. Applying the formula of the inverse of the sum of two matrices, B can be rewritten as

$$B = \Omega^{(N)} + \left(1 - \frac{1}{\sigma^2_{N+1}} \Sigma'_{N,N+1} \Omega^{(N)} \Sigma_{N,N+1}\right)^{-1} \frac{1}{\sigma^2_{N+1}} \Omega^{(N)} \Sigma_{N,N+1} \Sigma'_{N,N+1} \Omega^{(N)}$$

$$= \Omega^{(N)} + \frac{1}{\sigma^2_{N+1|N}} \Omega^{(N)} \Sigma_{N,N+1} \Sigma'_{N,N+1} \Omega^{(N)}, \qquad (11.16)$$

where $\sigma^2_{N+1|N} = \sigma^2_{N+1} - \Sigma'_{N,N+1} \Omega^{(N)} \Sigma_{N,N+1}$ is the variance of ε_{N+1} conditional on ε_i, $i = 1, ..., N$. Finally, let the factor loading vector be partitioned as $P^{(N+1)'} = \left[P^{(N)'} \;\; p_{N+1} \right]$. Considering the previous partitions and using the results in (11.15) and (11.16), it is possible to write $Q(N+1)$ as

$$\begin{aligned} Q(N+1) &= P^{(N+1)'} \Omega^{(N+1)} P^{(N+1)} = \\ &= P^{(N)'} \Omega^{(N)} P^{(N)} + \tfrac{1}{\sigma^2_{N+1|N}} \left(P^{(N)'} \Omega^{(N)} \Sigma_{N,N+1}\right)^2 + \tfrac{p^2_{N+1}}{\sigma^2_{N+1}} \\ &+ \tfrac{p^2_{N+1}}{\sigma^4_{N+1}} \Sigma'_{N,N+1} \Omega^{(N)} \Sigma_{N,N+1} \left(1 + \tfrac{1}{\sigma^2_{N+1|N}} \Sigma'_{N,N+1} \Omega^{(N)} \Sigma_{N,N+1}\right) \\ &- 2 \tfrac{p_{N+1}}{\sigma^2_{N+1}} P^{(N)'} \Omega^{(N)} \Sigma_{N,N+1} \left(1 + \tfrac{1}{\sigma^2_{N+1|N}} \Sigma'_{N,N+1} \Omega^{(N)} \Sigma_{N,N+1}\right). \end{aligned} \qquad (11.17)$$

Using $1 + \frac{1}{\sigma^2_{N+1|N}} \Sigma'_{N,N+1} \Omega^{(N)} \Sigma_{N,N+1} = \frac{\sigma^2_{N+1}}{\sigma^2_{N+1|N}}$, in (11.17), it follows that

$$\begin{aligned} Q(N+1) = Q(N) &+ \frac{1}{\sigma^2_{N+1|N}} \left(P^{(N)'} \Omega^{(N)} \Sigma_{N,N+1}\right)^2 \\ &+ \frac{p^2_{N+1}}{\sigma^2_{N+1}} + \frac{p^2_{N+1}}{\sigma^2_{N+1}} \frac{1}{\sigma^2_{N+1|N}} \Sigma'_{N,N+1} \Omega^{(N)} \Sigma_{N,N+1} \\ &- 2 \frac{p_{N+1}}{\sigma^2_{N+1|N}} P^{(N)'} \Omega^{(N)} \Sigma_{N,N+1} = Q(N) + g^2_{N+1} \end{aligned} \qquad (11.18)$$

where $g^2_{N+1} = \frac{1}{\sigma^2_{N+1|N}} \left(p_{N+1} - P^{(N)'} \Omega^{(N)} \Sigma_{N,N+1}\right)^2$. Consequently, $Q(N+1) \geq Q(N)$, regardless of the properties of the covariances between the idiosyncratic noises. Note that if ε_{N+1} is uncorrelated with ε_i, $i = 1, \ldots, N$, then $g^2_{N+1} = q_{N+1} = \frac{p^2_{N+1}}{\sigma^2_{N+1}}$. If further, the variable y_{N+1} is non-informative so that $p^2_{N+1} = 0$, or $\sigma^2_{N+1} = \infty$, then $g^2_{N+1} = 0$ and, consequently, $Q(N+1) = Q(N)$. However, if only $p^2_{N+1} = 0$, then $g^2_{N+1} = \frac{1}{\sigma^2_{N+1|N}} \left(P^{(N)'} \Omega^{(N)} \Sigma_{N,N+1}\right)^2 > 0$ and $Q(N+1) > Q(N)$.

To prove that $V(N)$ is non-increasing, consider again equation (11.14) and take into account (11.18) to obtain the following expression:

$$\begin{aligned} &\left(V(N)^2 - V(N+1)^2\right) Q(N) - V(N+1)^2 g^2_{N+1} + \\ &V(N) \left(-\sigma^2_\eta Q(N) + 1 - \phi^2\right) - V(N+1) \left(-\sigma^2_\eta Q(N) + 1 - \phi^2\right) + \\ &V(N+1) \sigma^2_\eta g^2_{N+1} \end{aligned}$$

$$= (V(N) - V(N+1))\,(V(N) + V(N+1))\,Q(N) +$$
$$(V(N) - V(N+1))\left(-\sigma_\eta^2 Q(N) + 1 - \phi^2\right) +$$
$$V(N+1)\left(\sigma_\eta^2 - V(N+1)\right) g_{N+1}^2 = 0.$$

Rearranging terms,

$$V(N+1)\left(V(N+1) - \sigma_\eta^2\right) g_{N+1}^2 \tag{11.19}$$
$$= (V(N) - V(N+1))\left[\left(1-\phi^2\right) + \left(V(N) + V(N+1) - \sigma_\eta^2\right) Q(N)\right].$$

Next, we show that if $\phi > 0$, then $V(N+1) > \sigma_\eta^2$. Consider the expression of $V(N)$ in (11.6). Then,

$$V(N+1) > \frac{\sigma_\eta^2 Q(N+1) - 1 + \sqrt{\left(\sigma_\eta^2 Q(N+1) - 1\right)^2 + 4\sigma_\eta^2 Q(N+1)}}{2Q(N+1)}$$
$$= \frac{\sigma_\eta^2 Q(N+1) - 1 + \sigma_\eta^2 Q(N+1) + 1}{2Q(N+1)} = \sigma_\eta^2.$$

Therefore, the left hand side of expression (11.19) is always positive if $g_{N+1}^2 > 0$. On the other hand, $\left[\left(1-\phi^2\right) + \left(V(N) + V(N+1) - \sigma_\eta^2\right) Q(N)\right] > 0$ and consequently, $V(N) > V(N+1)$. Finally, if $g_{N+1}^2 = 0$, then $V(N) = V(N+1)$.

PROOF OF LEMMA 3

The steady-state MSE of the filtered estimates of the underlying factors are given by

$$W(N) = V(N) - V(N)P^{(N)\prime}\left[P^{(N)}V(N)P^{(N)\prime} + \Sigma_\varepsilon^{(N)}\right]^{-1} P^{(N)}V(N).$$

Using the formula for the inverse of the sum of matrices given by Rao (1973), expression (11.8) of the MSE of $f_{t|t}$ is directly obtained. In order to prove that it is non-decreasing, we have to show that

$$\frac{V(N)}{1 + V(N)Q(N)} - \frac{V(N+1)}{1 + V(N+1)Q(N+1)} =$$
$$\frac{V(N)\left[1 + V(N+1)Q(N+1)\right] - V(N+1)\left[1 + V(N)Q(N)\right]}{\left[1 + V(N)Q(N)\right]\left[1 + V(N+1)Q(N+1)\right]} \geq 0.$$

Since the denominator is positive, the proof is reduced to show that the numerator is positive. After some straightforward algebra

$$V(N)\left[1 + V(N+1)Q(N+1)\right] - V(N+1)\left[1 + V(N)Q(N)\right] =$$
$$V(N) - V(N+1) + V(N)V(N+1)\left[Q(N+1) - Q(N)\right] \geq 0.$$

Lemma 1 establishes that if $g_{N+1}^2 = 0$, then $Q(N+1) = Q(N)$ and $V(N) = V(N+1)$. Otherwise, $Q(N+1) > Q(N)$ and $V(N) > V(N+1)$, so the inequality is already proved.

As regards the variance of the smoothed factor, $S(N)$, we can show that under the same hypothesis as Lemma 1, it is also a decreasing function of the number of series N. We have to prove that $S(N) \geq S(N+1)$, where $S(N)$ is given in (11.9):

$$\frac{V(N)\left(1+V(N)Q(N)-\phi^2\right)}{(1+V(N)Q(N))^2-\phi^2} \geq \frac{V(N+1)\left(1+V(N+1)Q(N+1)-\phi^2\right)}{(1+V(N+1)Q(N+1))^2-\phi^2}.$$

Notice that since $V(N) \geq V(N+1)$, then

$$\frac{V(N)\left(1+V(N+1)Q(N+1)-\phi^2\right)}{(1+V(N+1)Q(N+1))^2-\phi^2} \geq \frac{V(N+1)\left(1+V(N+1)Q(N+1)-\phi^2\right)}{(1+V(N+1)Q(N+1))^2-\phi^2} = $$
$$= S(N+1).$$

Therefore, if we can prove that

$$S(N) = \frac{V(N)\left(1+V(N)Q(N)-\phi^2\right)}{(1+V(N)Q(N))^2-\phi^2} \geq \frac{V(N)\left(1+V(N+1)Q(N+1)-\phi^2\right)}{(1+V(N+1)Q(N+1))^2-\phi^2}, \tag{11.20}$$

then it will be already proved that $S(N) \geq S(N+1)$. To prove (11.20), it suffices to prove that

$$\begin{aligned}&\left(1+V(N)Q(N)-\phi^2\right)\left((1+V(N+1)Q(N+1))^2-\phi^2\right)\\ &-\left(1+V(N+1)Q(N+1)-\phi^2\right)\left((1+V(N)Q(N))^2-\phi^2\right) \geq 0,\end{aligned}$$

since the denominators are positive and $V(N)$ can be simplified from both sides of the inequality (11.20).

After some straightforward algebra,

$$\begin{aligned}&\left[(1-\phi^2)+V(N)Q(N)\right]\left[1-\phi^2+V^2(N+1)Q^2(N+1)+2V(N+1)Q(N+1)\right]\\
&-\left[(1-\phi^2)+V(N+1)Q(N+1)\right]\left[1-\phi^2+V^2(N)Q^2(N)+2V(N)Q(N)\right]\\
=&\ [V(N)Q(N)-V(N+1)Q(N+1)]\,(1-\phi^2)+\\
&\left[V^2(N+1)Q^2(N+1)-V^2(N)Q^2(N)\right](1-\phi^2)\\
&+V(N)V(N+1)Q(N)Q(N+1)\,[V(N+1)Q(N+1)-V(N)Q(N)]\\
&+2(1-\phi^2)\,[V(N+1)Q(N+1)-V(N)Q(N)]\\
=&\left[(1-\phi^2)\,(1+(V(N+1)Q(N+1)+V(N)Q(N)))+V(N)V(N{+}1)Q(N)Q(N{+}1)\right]\\
&\times[V(N+1)Q(N+1)-V(N)Q(N)].\end{aligned}$$

Since $1-\phi^2 \geq 0$ and both $V(N)$ and $Q(N)$ are positive for all N, it suffices to prove that $V(N+1)Q(N+1) \geq V(N)Q(N)$. By (11.8),

$$V(N)Q(N) = \frac{\sigma_\eta^2 Q(N) - 1 + \phi^2 + \sqrt{\left(\sigma_\eta^2 Q(N) - 1 + \phi^2\right)^2 + 4\sigma_\eta^2 Q(N)}}{2}.$$

Then, also taking into account (11.18),

$$V(N+1)Q(N+1) - V(N)Q(N)$$

$$= \frac{\sigma_\eta^2 g_{N+1}^2}{2} + \frac{\sqrt{\left(\sigma_\eta^2 Q(N+1) - 1 + \phi^2\right)^2 + 4\sigma_\eta^2 Q(N+1)}}{2} - \frac{\sqrt{\left(\sigma_\eta^2 Q(N) - 1 + \phi^2\right)^2 + 4\sigma_\eta^2 Q(N)}}{2}.$$

Since $\frac{\sigma_\eta^2 g_{N+1}^2}{2}$ is positive, we have to prove that

$$\sqrt{\left(\sigma_\eta^2 Q(N+1) - 1 + \phi^2\right)^2 + 4\sigma_\eta^2 Q(N+1)} \geq \sqrt{\left(\sigma_\eta^2 Q(N) - 1 + \phi^2\right)^2 + 4\sigma_\eta^2 Q(N)}.$$

After taking squares and taking into account (11.18),

$$\left(\sigma_\eta^2 \left(Q(N) + g_{N+1}^2\right) - \left(1 - \phi^2\right)\right)^2 + 4\sigma_\eta^2 \left(Q(N) + g_{N+1}^2\right) -$$
$$\left(\sigma_\eta^2 Q(N) - \left(1 - \phi^2\right)\right)^2 + 4\sigma_\eta^2 Q(N)$$
$$= \sigma_\eta^4 g_{N+1}^4 + 2\sigma_\eta^2 Q(N) g_{N+1}^2 - 2\sigma_\eta^2 g_{N+1}^2 \left(1 - \phi^2\right) + 4\sigma_\eta^2 g_{N+1}^2$$
$$= \sigma_\eta^4 g_{N+1}^4 + 2\sigma_\eta^2 Q(N) g_{N+1}^2 + 2(1 + \phi^2)\sigma_\eta^2 g_{N+1}^2 \geq 0,$$

which is positive since it is the sum of three positive terms. Therefore, the result is proved and $S(N) \geq S(N+1)$.

ACKNOWLEDGMENTS

This paper was presented at the workshop in honour of the sixty-fifth birthday of Andrew Harvey held in the Oxford Man Institute on June 2012. We are very grateful to all participants. Financial support from the Spanish Government project ECO2012-32854 is acknowledged by the first author while the second author acknowledges support from project ECO2012-32401. We also appreciate very useful comments from the participants in the Computational and Financial Econometrics Conference held in London in December 2011, the Nonlinear Dynamics and Econometrics Conference held in Istanbul in April 2012 and the II_t Workshop on Time Series Econometrics held in Zaragoza in May 2012. We especially appreciate comments from Davide Delle Monache and Gabriel Perez-Quiros. Finally, the comments of the editors and two anonymous referees have been very useful to make a more readable paper. Obviously, any remaining errors are our responsibility.

12 On detecting end-of-sample instabilities

Fabio Busetti

12.1 Introduction

The detection of changepoints and parameter instabilities has attracted considerable attention in the statistics and econometrics literature. If the location of a potential changepoint is known in advance, the Wald test of Chow (1960) has optimal properties against the hypothesis of a one-time shift in the parameters in a standard regression framework. If no prior information is available on the location, the solution proposed by Quandt (1960) is to take the maximum of the Wald statistics computed over the set of possible changepoints. The asymptotic representation and the critical values of this "sup test" (supremum test) have been derived by Andrews (1993). Andrews and Ploberger (1994) have shown that better properties can be obtained by taking some averages of the Wald statistics instead of the maximum. Similar, and asymptotically equivalent, tests can be obtained using the LM and LR statistics.

A different class of parameter instabilities tests has been derived against the alternative hypothesis of random walk coefficients. In a linear model with potentially time-varying parameters, locally most powerful (LMP) tests have been proposed by Nyblom and Makelainen (1983), King and Hillier (1985), and Nyblom (1989); the latter paper is in fact concerned with a more general hypothesis of martingale time-variation in the parameter which nests the cases of random walk coefficients and "discrete" parameter shifts randomly occurring in the sample. Elliott and Muller (2006) follow a similar approach, but they focus on tests that maximize the power against a chosen, fixed alternative hypothesis, not necessarily close to the null as for the LMP statistics. Their paper provides a generalization of previous results of Franzini and Harvey (1983) and Shively (1988).

A further alternative way to detect parameter instabilities is to examine the sequence of regression coefficients estimated with an increasingly large data set, as in the "fluctuation tests" of Ploberger et al. (1989). Similarly, the CUSUM tests (cumulative sum tests) of Brown et al. (1975) look at the behavior of the partial sum process of (squared) recursive OLS residuals; more generally, partial sums of Kalman filter residuals were considered in Harvey (1989, pp. 256–8) for unobserved components models. Unlike the Wald and LMP statistics, these tests have been proposed without reference to any specific alternative hypothesis.[1] A detailed survey of the testing methods described above is given in Stock (1994).[2]

The ability of the tests to reject the null hypothesis of stability clearly depends on the number of 'post-changepoint' observations relative to the sample size. The tests are then likely to display low power when a changepoint occurs in the late part of the sample. Failing to detect end-of-sample parameter changes has particularly pernicious implications if the models are used for making predictions, since these instabilities will largely affect the model's forecasts.

This paper therefore considers various modifications of existing tests and introduces new statistics designed to have high power in such circumstances. The properties of both Wald-type tests of a one-time shift in the parameters and locally most powerful (LMP) tests against the hypothesis of random walk coefficients are examined. It is proposed to take functionals of the Wald and LMP statistics such that either the set of possible changepoints is restricted to the last part of the sample or the occurrence of changepoints is given increasing weight throughout the sample. Asymptotic critical values of the tests are provided and their properties are evaluated in finite samples.

The Wald-type tests examined here extend the range of applications of the results of Andrews and Ploberger (1994). The LMP-type tests are derived against the alternative hypothesis of a switch from stable to random walk coefficients at some point in the sample, thus generalizing Nyblom (1989), who considered time varying coefficients throughout the whole sample. A similar hypothesis was considered by Busetti and Taylor (2004), where the purpose was to detect persistence changes of time series, and by Andrews and Kim (2006) for the residuals of a cointegrating regression.

Our results show that, for the case of an unknown end-of-sample changepoint, the LMP-type tests appear to have, in general, better properties than the Wald-type tests, even against the hypothesis of a one-time shift in the parameters. However, for dynamic models, the Wald-type tests turn out to be more (less) powerful in the case of a decrease (increase) in the degree of persistence of the data. When the end-of-sample changepoint is known, the subsampling tests in the spirit of Andrews (2003) have the best properties, as expected, but they display low power if the instability is not correctly located.

In summary, this paper proceeds as follows. Section 12.2 describes the testing framework and provides the new and modified test statistics aimed at

[1] In discussing the article by Brown et al. (1975), Harvey notes that the power of these tests may however be low "...in the presence of many types of structural change likely to occur in practice" (p. 180).

[2] The related issue of estimating the time of changepoints is considered in Bai (1997) and Bai and Perron (1998) in a linear regression set up. A further strand of the literature has instead been concerned with testing stability of the whole distribution of a time series; see, inter alia, Picard (1985), Inoue (2001), Lee and Nah (2005), and Busetti and Harvey (2010, 2011).

detecting end-of-sample instabilities with the associated critical values. The size and properties of the tests are evaluated by means of Monte Carlo simulations in Section 12.3. Section 12.4 contains two empirical illustrations on the use of the tests. Section 12.5 concludes.

12.2 End-of-sample instability tests in a linear regression model

We consider a linear regression model with $K = k_1 + k_2$ regressors $x_t = (x'_{1t}, x'_{2t})'$ and $T = n + m$ observations, where a breakpoint can occur in the second subsample of size m,

$$y_t = \begin{cases} x'_{1t}\beta_1 + x_2{}'_t\beta_2 + u_t & \text{for } t = 1, \ldots, n \\ x'_{1t}(\beta_1 + \delta_t) + x'_{2t}\beta_2 + u_t & \text{for } t = n+1, \ldots, n+m \end{cases} \tag{12.1}$$

with u_t being an i.i.d. disturbance such that $E(u_t | x_t) = 0$ and $E\left(u_t^2 \middle| x_t\right) = \sigma^2$; the regressors x_t satisfy standard assumptions of stationarity with $p \lim T^{-1} \sum_{t=1}^{T} x_t x'_t$ being a positive definite matrix. The null hypothesis of parameter stability is $H_0 : \delta_t = 0$ for all $t = n + 1, \ldots, n + m$. Under the alternative hypothesis $\delta_t \neq 0$ for some t. The model allows parameter instability to occur only for a subset of the regressors; the term $x'_{2t}\beta_2$ disappears if the presence of instabilities is investigated for all the coefficients.

12.2.1 WALD-TYPE TESTS

A standard F-test has optimal properties against the hypothesis of a one-time structural change in the coefficients that occurs at a known fraction, $\pi = 1 - \frac{m}{T} \in (0, 1)$, of the sample size. Let $x_t(\pi) = \left(x'_{1t}, x'_{2t}, x'_{1t}1(t \geq [\pi T])\right)'$, where the notation $[z]$ indicates the nearest integer to z. Denote by Q and $Q(\pi)$ the sum of squared OLS residuals from regressing y_t on x_t and y_t on $x_t(\pi)$, respectively. The F-statistic is

$$F(\pi) = \frac{(Q - Q(\pi))/k_1}{Q(\pi)/(T - 2k_1 - k_2)}, \tag{12.2}$$

that compares the statistical fit of the restricted and unrestricted models, the latter allowing a structural change in the coefficients of x_{1t}. This is a standard Wald test; the null limiting distribution of $k_1 F(\pi)$ is a χ^2 with k_1 degrees of freedom.

If the location of the parameter change is not known a priori, Quandt (1960) proposes to take the maximum of the F-statistics over the set of possible breakpoints,

$$\text{Sup-F} = \underset{\pi \in \Pi}{Sup}\, F(\pi) \tag{12.3}$$

where Π is a closed subset of (0, 1). Andrews and Ploberger (1994) show that taking averages of the F-statistics yields better properties than those of the Sup-F test. Here we consider their so-called Exp-F test statistic, defined as

$$\text{Exp-F} = \log \int_{\pi \in \Pi} \exp\left(\frac{k_1}{2} F(\pi)\right) dJ(\pi), \tag{12.4}$$

where $J(\pi)$ is a chosen weight function (i.e., probability measure) on the values of $\pi \in \Pi$.

Andrews (1993) shows that, under the null hypothesis of no structural change, for each $\pi \in \Pi$ the limit distribution of $F(\pi)$ can be represented as a quadratic form of a Brownian Bridge process,

$$F(\pi) \xrightarrow{d} F_{\infty}(\pi) \equiv k_1^{-1} \frac{B_{k_1}(\pi)' B_{k_1}(\pi)}{\pi(1-\pi)},$$

where $B_{k_1}(\pi) = W_{k_1}(\pi) - \pi W_{k_1}(\pi)$ and $W_{k_1}(\pi)$ is a k_1-dimensional Brownian motion. Thus, by an application of the continuous mapping theorem, the limiting distributions of the statistics (12.3)–(12.4) are obtained by replacing $F(\pi)$ with $F_{\infty}(\pi)$ in their definitions.

The asymptotic critical values for k_1Sup-F are given in Andrews (1993, p. 840) for different Πs; note that the critical values provided by Andrews need to be divided by k_1 to be used with our statistic (12.3) that is defined in terms of F-statistics instead of the equivalent Wald statistics considered in those papers.

As noted by Andrews and Ploberger (1994), in order to get higher power against end-of-sample instability, the Sup-F and Exp-F tests can be computed restricting Π to the latest part of the sample; here we consider the latest 25% and 10% of the sample.

The asymptotic critical values for Exp-F provided by Andrews and Ploberger (1994) pertain to the case of equal weights for all possible breakpoints (i.e., a uniform measure on Π).[3] A simple modification of the Exp-F test aimed at achieving higher power against end-of-sample instabilities is to abandon the case of uniform weights in favor of giving higher weights against change-points occurring later in the sample. Here we propose a weighting scheme that increases linearly throughout the sample, i.e., define the modified statistic

[3] In practice, for a uniform measure on $\Pi = [\pi_0, \pi_1]$, the test statistic is computed as Exp-F$= \log \sum_{t=[\pi_0 T]}^{[\pi_1 T]} \exp\left(\frac{k_1}{2} F(\pi_t)\right) / T^*$, where $\pi_t = t/T$ and $T^* = [\pi_1 T] - [\pi_0 T] - 1$.

$$\text{Exp-F}_{LIN} = \log \int_{\pi \in \Pi} \exp\left(\frac{k_1}{2} F(\pi)\right) \pi \, d\pi. \tag{12.5}$$

The critical values of Exp-F_{LIN} are provided in Table 12.1 for $\Pi = [.01, .99]$ and for a number of potentially changing parameters k_1 between 1 and 5. Table 12.1 also provides the critical values for the Sup-F and Exp-F where $\Pi = [.05, 95]$, $[.01, .99]$, $[.75, 99]$, and $[.90, .99]$.[4]

12.2.2 LMP-TYPE TESTS

Under Gaussianity, a locally most powerful test of the null hypothesis of parameter stability against the alternative of random walk coefficients has been derived by Nyblom and Makelainen (1983) for a level plus noise model, and then adapted to a regression framework by King and Hillier (1985) and Nyblom (1989); an extension to the multivariate time series model is considered in Nyblom and Harvey (2000). The LMP test statistic is given by

$$L = \widehat{\sigma}^{-2} T^{-2} \sum_{t=1}^{T} S_t' V^{-1} S_t, \tag{12.6}$$

where $\widehat{u}_t = y_t - x_t'\widehat{\beta}$ are the OLS residuals from regressing y_t on x_t, $\widehat{\sigma}^2 = (T-K)^{-1} \sum_{t=1}^{T} \widehat{u}_t^2$, $S_t = \sum_{j=t}^{T} \widehat{u}_j x_{1j}$, and $V = T^{-1} \sum_{t=1}^{T} x_{1t} x_{1t}'$. The test can be made robust to heteroskedasticity if $\widehat{\sigma}^{-2} V^{-1}$ is replaced by V_*^{-1}, with $V_* = (T-K)^{-1} \sum_{t=1}^{T} \widehat{u}_t^2 x_t x_t'$.

This is a locally most powerful invariant test against random walk coefficients throughout the whole sample. Since we are interested in detecting end-of-sample instabilities, we propose a modification of the test that focuses on breaks potentially occurring only in the last fraction of the sample $\pi = 1 - \frac{m}{T}$. The test therefore has optimal properties against the alternative hypothesis that

$$\delta_t = \begin{cases} 0 & \text{for } t = 1, \ldots, T - [\pi T] \\ \delta_{t-1} + \eta_t & \text{for } t = [\pi T] + 1, \ldots, T \end{cases},$$

where η_t is a Gaussian $iid(0, \theta\sigma^2 I_T)$ disturbance independent of x_t and u_t, with $\theta \geq 0$. The LMP statistic is

$$L(\pi) = \widehat{\sigma}^{-2} (T - [\pi T])^{-2} \sum_{t=[\pi T]+1}^{T} S_t' V^{-1} S_t, \tag{12.7}$$

[4] The asymptotic critical values reported in Table 12.1 were obtained by direct simulation of the relevant functionals over 50 000 replications, approximating Brownian bridges with Gaussian processes in a sample of size 10 000.

Table 12.1 Critical values of the Wald-type and LMP-type tests

		k_1=1			k_1=2			k_1=3			k_1=4			k_1=5		
Test	Π	0.10	0.05	0.01	0.10	0.05	0.01	0.10	0.05	0.01	0.10	0.05	0.01	0.10	0.05	0.01
Sup-F	[.05,.95]	8.206	9.789	13.157	5.593	6.436	8.235	4.532	5.127	6.438	3.915	4.405	5.450	3.540	3.948	4.826
	[.01,.99]	9.041	10.639	14.152	6.065	6.907	8.696	4.865	5.463	6.751	4.189	4.662	5.659	3.762	4.158	5.000
	[.75,.99]	7.152	8.682	11.999	4.972	5.829	7.647	4.082	4.699	6.070	3.574	4.052	5.104	3.249	3.653	4.516
	[.90,.99]	6.479	7.978	11.402	4.584	5.432	7.264	3.786	4.402	5.729	3.337	3.816	4.854	3.038	3.446	4.309
Exp-F	[.05,.95]	1.513	2.035	3.356	2.635	3.281	4.734	3.611	4.308	5.939	4.496	5.270	7.050	5.356	6.202	8.155
	[.01,.99]	1.523	2.048	3.341	2.668	3.296	4.739	3.647	4.357	5.953	4.555	5.307	7.056	5.419	6.263	8.176
	[.75,.99]	1.483	2.016	3.346	2.542	3.194	4.701	3.487	4.249	5.958	4.344	5.157	6.967	5.185	6.054	7.977
	[.90,.99]	1.461	2.026	3.394	2.514	3.180	4.722	3.423	4.196	5.923	4.263	5.066	6.885	5.073	5.927	7.854
Exp-F_{LIN}	[.01,.99]	0.748	1.047	1.892	1.421	1.839	2.912	2.086	2.610	3.902	2.731	3.331	4.770	3.398	4.082	5.677
$L(.75)$		0.976	1.319	2.158	1.641	2.099	3.335	2.249	2.748	3.940	2.887	3.497	4.737	3.464	4.119	5.401
$L(.90)$		1.097	1.511	2.539	1.854	2.365	3.486	2.578	3.160	4.347	3.303	3.878	5.131	3.940	4.496	6.001
Sup-L	[.05,.95]	1.623	2.088	3.117	2.533	3.116	4.519	3.335	3.900	5.308	4.115	4.785	6.134	4.777	5.399	6.939
	[.01,.99]	1.949	2.489	3.596	2.971	3.531	4.917	3.798	4.409	5.797	4.600	5.281	6.644	5.352	6.145	7.466
	[.75,.99]	1.929	2.473	3.596	2.940	3.495	4.911	3.781	4.383	5.780	4.571	5.277	6.644	5.342	6.129	7.466
	[.90,.99]	1.809	2.367	3.441	2.842	3.388	4.813	3.662	4.235	5.656	4.434	5.144	6.593	5.229	6.042	7.393
Exp-L	[.05,.95]	0.776	1.028	1.710	1.343	1.668	2.671	1.898	2.333	3.359	2.527	2.980	3.997	3.033	3.558	4.742
	[.01,.99]	0.774	1.029	1.696	1.363	1.693	2.726	1.940	2.367	3.411	2.581	3.063	4.114	3.128	3.633	4.886
	[.75,.99]	1.081	1.448	2.326	1.866	2.341	3.483	2.565	3.100	4.220	3.308	3.886	5.238	3.980	4.535	5.912
	[.90,.99]	1.124	1.521	2.546	2.002	2.522	3.734	2.726	3.279	4.532	3.438	4.065	5.407	4.156	4.830	6.163
Exp-L_{LIN}	[.01,.99]	0.494	0.668	1.172	0.915	1.214	2.005	1.368	1.725	2.561	1.893	2.340	3.396	2.421	2.837	4.011

which corresponds to (12.6) if $\pi = 0$.[5] Under the null hypothesis of constant coefficients, the limiting distribution can be represented as a quadratic form of k_1-dimensional Brownian bridge

$$L(\pi) \xrightarrow{d} (1-\pi)^{-2} \int_{\pi}^{1} B_{k_1}(s)' B_{k_1}(s) ds. \tag{12.8}$$

If the location of a possible parameter instability is unknown, we proceed as for the Wald-type tests using the statistics

$$\text{Sup-}L = \underset{\pi \in \Pi}{Sup}\ L(\pi) \tag{12.9}$$

$$\text{Exp-}L = \log \int_{\pi \in \Pi} \exp(L(\pi))\, dJ(\pi) \tag{12.10}$$

for $\Pi = [.05, 95], [.01, .99], [.75, .99], [.90, .99]$, and

$$\text{Exp-}L_{LIN} = \log \int_{\pi \in \Pi} \exp(L(\pi))\, \pi\, d\pi. \tag{12.11}$$

for $\Pi = [.01, .99]$. The null limiting distributions of (12.9)–(12.11) are immediately obtained from (12.8) by an application of the continuous mapping theorem. Asymptotic critical values are shown in Table 12.1.

12.2.3 FEW END-OF-SAMPLE OBSERVATIONS

When the number of post-changepoint observations is "small," the distribution of the F-statistic (12.2) cannot be approximated by a χ^2 and thus the χ^2 critical values are no longer appropriate.[6] Andrews (2003) has proposed a variant of the F test that can be used even for a very small end-of-sample size, where the critical values are obtained by a simple "parametric subsampling" method. For the case of serially uncorrelated disturbances u_t (and $m \geq K$) the statistic is defined as

$$\mathcal{S} = \mathcal{S}_{n+1}\left(\widehat{\beta}, \widehat{\sigma}^2\right) \tag{12.12}$$

where $\widehat{\beta}$ is the OLS estimate of β (using all the $n+m$ observations), $\widehat{\sigma}^2$ is the usual estimate of the error variance, and for $j = 1, 2, \ldots, n+1$,

$$\mathcal{S}_j\left(\beta, \sigma^2\right) = \sigma^{-2}\left(Y_j(m) - X_j(m)\beta\right)' P_j(m) \left(Y_j(m) - X_j(m)\beta\right), \tag{12.13}$$

[5] A similar statistic was proposed by Busetti and Taylor (2004) for detecting a switch from a I(0) to a I(1) process at the fraction π of the sample.

[6] However, under Gaussian errors and strictly exogenous regressors the statistic follows an F distribution in finite samples under the null hypothesis.

with $P_j(m) = X_j(m)\left[X_j(m)'X_j(m)\right]^{-1}X_j(m)'$ is the usual projection matrix. $X_j(m)$ is the $m \times K$ matrix $\left(x_j', x_{j+1}', \ldots, x_{j+m-1}'\right)'$, $Y_j(m)$ is the $m \times 1$ vector $\left(y_j, y_{j+1}, \ldots, y_{j+m-1}\right)'$.

Andrews (2003) shows that for large T and m fixed the distribution function of $\mathcal{S}$ converges to the empirical distribution function of $\left\{\mathcal{S}_j\left(\beta,\sigma^2\right) : j = 1,\ldots,n-m+1\right\}$ evaluated at consistent estimators of β and σ^2. The critical value of the test is therefore the $1-\alpha$ sample quantile of $\left\{\mathcal{S}_j\left(\beta,\sigma^2\right) : j = 1,\ldots,n-m+1\right\}$. In our simulations we use the estimators $\widehat{\beta}_{(n)}$ and $\widehat{\sigma}^2_{(n)}$ obtained by a single OLS regression over the stability subsample $(t = 1,\ldots,n)$.[7]

When the end-of-sample size is small the distribution of the proposed $L(\pi)$ test (12.7) can also be obtained by the same parametric subsampling approach. It just requires to obtain the empirical distribution function of $\left\{\mathcal{L}_j\left(\widehat{\beta}_{(n)},\widehat{\sigma}^2_{(n)}\right) : j = 1,\ldots,n-m+1\right\}$, where, for $j = 1, 2, \ldots, n+1$,

$$\mathcal{L}_j\left(\widehat{\beta}_{(n)},\widehat{\sigma}^2_{(n)}\right) = \widehat{\sigma}^{-2}_{(n)}m^{-2}\sum_{t=j}^{j+m-1} S_{t,j}\left(\widehat{\beta}_{(n)}\right)' V^{-1} S_{t,j}\left(\widehat{\beta}_{(n)}\right)$$

where $S_{t,j}\left(\widehat{\beta}_{(n)}\right) = \sum_{h=t}^{j+m-1}\left(y_h - x_h'\widehat{\beta}_{(n)}\right)x_{1h}$. We call this test $\mathcal{L}(\pi)$.[8]

12.3 **Size and power properties of the tests**

The size and power properties of the tests described in Section 2 are evaluated by means of Monte Carlo simulations in the context of simple linear regression models. The first subsection considers both cases of a one-time change in the parameters and of random walk coefficients for a static regression model. The second section looks at dynamic models, providing results for the case of a change in the persistence parameter of an autoregression. In all experiments the number of Monte Carlo replications is set to 50 000.

12.3.1 STATIC REGRESSION MODELS

The data generating process in the Monte Carlo simulations corresponds to the model (12.1) with $x_{1t} = \left(1,(-1)^t\right)$, $\sigma^2 = 1$ and where there are no other

[7] For each j Andrews considers a "leave-$[m/2]$-out" rule that uses $n + [m/2]$ observations (i.e., it leaves out $t = j, j+1, \ldots, j+[m/2]-1$); this is obtained by looking at the size/power trade-off in Monte Carlo experiments. Our choice on the other hand reflects our focus for the cases of: (i) relatively large m, which discourages using observations in the second subsample; (ii) dynamic regression models where one cannot leave out central observations in the sample.

[8] A special case of this test occurs by setting $x_{1t} = 1$ in the statistics; this was proposed by Andrews and Kim (2006) as a test of "cointegration breakdown" at the end of the sample.

Table 12.2 Empirical size of the tests for the static regression model

Test	Π	Gaussian errors T=100	Gaussian errors T=200	Gaussian errors T=400	t(3) T=200	χ^2 T=200
Sup-F	[.05,.95]	.04	.04	.05	.08	.07
	[.01,.99]	.04	.04	.04	.12	.11
	[.75,.99]	.03	.03	.04	.09	.08
	[.90,.99]	.02	.03	.04	.08	.08
Exp-F	[.05,.95]	.06	.05	.05	.08	.07
	[.01,.99]	.06	.05	.05	.13	.11
	[.75,.99]	.06	.05	.06	.10	.09
	[.90,.99]	.05	.05	.05	.10	.09
Exp-F_{LIN}	[.01,.99]	.06	.05	.06	.10	.09
$L(0)$		.05	.05	.05	.05	.04
$L(.75)$		.06	.06	.06	.07	.06
$L(.90)$		.11	.09	.09	.11	.10
Sup-L	[.05,.95]	.06	.05	.05	.08	.07
	[.01,.99]	.08	.07	.06	.10	.10
	[.75,.99]	.07	.07	.06	.10	.10
	[.90,.99]	.08	.07	.06	.10	.10
Exp-L	[.05,.95]	.06	.06	.06	.07	.07
	[.01,.99]	.07	.06	.05	.09	.08
	[.75,.99]	.07	.06	.05	.10	.09
	[.90,.99]	.09	.07	.06	.10	.09
Exp-L_{LIN}	[.01,.99]	.07	.06	.05	.09	.09

regressors x_{2t}. For the alternative hypothesis we consider both cases of one-time change and random walk coefficients. The set-up is broadly similar to the one considered in Andrews et al. (1996). We look specifically at the properties of the tests when the breakpoint occurs in the latest part of the sample, affecting the last 25%, 10%, 5%, and 2% observations.

Tables 12.2 and 12.3 presents the empirical size of the tests for sample sizes of $T = 100$, 200, and 400 observations for both Gaussian and non-Gaussian disturbances (the latter only for $T = 200$) and tests run at 5% significance; note that the statistics reported in Table 12.3 are computed for a given changepoint and—in the case of subsampling—with the empirical distribution that depends on π. Consider first the Gaussian case. For $T = 100$ the size is accurate for the standard F test with a known breakpoint and the L test of Nyblom (1989). The asymptotic F-type tests tend to present smaller distortions than the LMP type tests, which tend to be somewhat oversized; however, in all cases but one the empirical rejection frequencies do not exceed 8%. The subsampling $F(\pi_0)$ and $L(\pi_0)$, that require a "small" post-break subsample, have good

Table 12.3 Empirical size of the standard F-test and of the subsampling based tests for the static regression model for different error distributions

		Gaussian errors												*t(3) errors*				*χ^2 errors*			
		T=100				T=200				T=400				T=200				T=200			
Test	π	.75	.90	.95	.98	.75	.90	.95	.98	.75	.90	.95	.98	.75	.90	.95	.98	.75	.90	.95	.98
F (π_0)		.05	.05	.05	.06	.05	.05	.05	.05	.05	.05	.05	.05	.05	.07	.07	.07	.05	.05	.06	.07
F (π_0)	subsampling	.17	.08	.06	.06	.17	.08	.06	.06	.17	.08	.07	.05	.16	.08	.06	.06	.17	.08	.06	.06
$L(\pi_0)$	subsampling	.15	.08	.06	.06	.15	.07	.06	.05	.15	.08	.06	.05	.14	.07	.06	.06	.15	.07	.06	.05

size properties for $\pi_0 \geq 0.95$ but they are significantly oversized otherwise. Increasing the sample to 200 and 400 observations yields better sized tests, except for the case of the subsampling statistics (however their oversizing is still minor if $\pi_0 \geq 0.95$). For $T = 400$ nearly all asymptotic tests present rejection frequencies equal to the nominal size. In the case of non-Gaussian distributions (and $T = 200$), the tests tend to be oversized but in most cases the empirical rejection frequencies do not exceed 10%; the deterioration of the size properties is more evident for the F-type statistics.

We now turn to the power of the tests under a local deviation from the null hypothesis.[9] Consider first the case of one-time change in the coefficients, where the local alternative hypothesis is $\delta_{t,T} = \delta/\sqrt{T}\mathbf{i}$ for all $t = n + 1, n + 2, \ldots, n + m$, and i is a two-dimensional vector of ones. Table 12.4 contains rejection frequencies for $\delta = 4.8,\ 7.2,\ 9.6,\ 12$ that yields the power of the tests (not size-adjusted).

If the changepoint is known, the Wald F-test has optimal properties and it therefore presents the highest rejection frequencies (because of the oversizing, in a few cases other tests have slightly higher rejection frequencies); for $\pi_0 \geq 0.95$ the subsampling version of the F-test has very similar size and power properties as the asymptotic test. It is, however, interesting to observe that as $\pi_0 \to 1$ the Wald test $F(\pi_0)$ behaves very similarly to the LMP test, $L(\pi_0)$.

If the changepoint is unknown, we first confirm the result of Andrews et al. (1996) that the Exp-F test has higher power than the Sup-F test; the same is true for the comparison between Sup-L and Exp-L. However, and interestingly, the rejection frequencies of the Exp-L (Sup–L) tests appear generally higher than those of the Exp-F (Sup-F). For example, for $\delta = 9.6$ and $\pi_0 = 0.95$, the rejection frequency of Exp-L over [.01,.99] is 0.63 against 0.44 of Exp-F; these two figures become 0.70 and 0.60 for tests computed over the interval [.75,.99] of possible breakpoints. The properties of Exp-F_{LIN} are very similar to that of Exp-F when the latter is computed over $\Pi = [.75, .99]$; both tests are however dominated by Exp-L_{LIN} and Exp-L over $\Pi = [.75, .99]$. For late changepoints, when $0.95 \leq \pi_0 \leq 0.98$, the highest rejection frequencies are displayed by Exp-L over $\Pi = [.90, .99]$, that however suffers from non-negligible oversizing when $T = 100$. The Exp-L_{LIN} is on balance a better option in this case.[10]

Figure 12.1 summarizes the main findings in terms of power under the alternative hypothesis of a one-time shift in the coefficients occurring in the last 5% of the sample, $\pi = 0.95$.

Table 12.5 reports the power properties of the tests against the case of random walk coefficients; specifically, the local alternative hypothesis is

[9] In these simulations the sample size is $T = 200$, but the rejection frequencies provided are in principle not affected by the sample size since they represent an approximation of the local asymptotic power of the tests.

[10] As expected, unreported simulations show that—when the changepoint occur earlier in the sample—the power of the end-of-sample stability tests is lower than that of the standard Sup-F, Sup-L, Exp-F, and Exp-L over $\Pi = [.05, .95]$.

fraction π_0 of the sample size (T=200) under Gaussianity; δ represents the distance from the null hypothesis

Test	Π	π_0	δ=4.8				δ=7.2				δ=9.6				δ=12			
			0.75	0.90	0.95	0.98	0.75	0.90	0.95	0.98	0.75	0.90	0.95	0.98	0.75	0.90	0.95	0.98
F (π_0)			.75	.43	.25	.13	.98	.79	.50	.24	1.00	.96	.76	.38	1.00	1.00	.92	.56
F (π_0)	subsampling		.83	.47	.26	.14	.99	.79	.52	.24	1.00	.96	.76	.39	1.00	.99	.92	.56
F (.95)	subsampling		.40	.29	.26	.08	.56	.51	.52	.11	.72	.73	.76	.16	.84	.89	.92	.24
Sup-F	[.05,.95]		.50	.22	.10	.05	.90	.51	.21	.06	1.00	.83	.43	.07	1.00	.97	.68	.09
	[.01,.99]		.45	.19	.09	.05	.88	.47	.21	.07	.99	.79	.42	.12	1.00	.96	.67	.21
	[.75,.99]		.50	.26	.13	.06	.90	.58	.30	.10	1.00	.87	.54	.19	1.00	.98	.77	.31
	[.90,.99]		.15	.24	.13	.06	.34	.57	.32	.12	.60	.86	.57	.21	.83	.98	.80	.34
Exp-F	[.05,.95]		.58	.24	.10	.06	.93	.53	.20	.07	1.00	.83	.38	.08	1.00	.97	.61	.10
	[.01,.99]		.57	.24	.11	.07	.92	.53	.23	.09	1.00	.83	.44	.13	1.00	.97	.68	.20
	[.75,.99]		.60	.35	.18	.08	.93	.69	.36	.13	1.00	.92	.60	.21	1.00	.99	.82	.33
	[.90,.99]		.22	.33	.22	.10	.45	.66	.44	.17	.71	.91	.69	.28	.89	.99	.88	.42
Exp-F-LIN	[.01,.99]		.62	.33	.17	.08	.94	.66	.35	.12	1.00	.91	.59	.21	1.00	.99	.81	.33
L(0)			.50	.13	.07	.05	.88	.25	.09	.05	.99	.43	.13	.06	1.00	.65	.18	.06
L(.75)			.66	.33	.13	.07	.95	.64	.23	.08	1.00	.88	.38	.10	1.00	.98	.56	.13
L(.90)			.36	.49	.29	.13	.63	.80	.53	.19	.84	.96	.76	.25	.95	1.00	.91	.35
L (p0)	subsampling		.74	.42	.26	.15	.96	.73	.49	.26	1.00	.93	.73	.42	1.00	.99	.90	.60
L (.95)	subsampling		.38	.29	.26	.06	.54	.50	.49	.12	.69	.72	.73	.21	.81	.88	.90	.33
Sup-L	[.05,.95]		.51	.36	.23	.09	.90	.70	.46	.14	1.00	.92	.71	.22	1.00	.99	.89	.33
	[.01,.99]		.43	.32	.23	.15	.85	.63	.44	.27	.99	.89	.68	.42	1.00	.98	.87	.60
	[.75,.99]		.39	.31	.23	.15	.81	.63	.44	.27	.98	.89	.68	.42	1.00	.98	.87	.60
	[.90,.99]		.21	.30	.24	.16	.41	.60	.46	.28	.65	.87	.70	.44	.85	.98	.88	.61
Exp-L	[.05,.95]		.64	.36	.17	.08	.95	.69	.35	.10	1.00	.92	.59	.15	1.00	.99	.80	.21
	[.01,.99]		.61	.35	.20	.10	.93	.68	.39	.16	1.00	.91	.63	.27	1.00	.99	.84	.41
	[.75,.99]		.44	.39	.24	.12	.82	.72	.46	.20	.98	.93	.70	.33	1.00	.99	.88	.48
	[.90,.99]		.24	.33	.27	.15	.44	.62	.51	.26	.66	.87	.75	.40	.85	.97	.91	.57
Exp-L-LIN	[.01,.99]		.54	.37	.23	.11	.90	.70	.44	.20	.99	.92	.68	.32	1.00	.99	.87	.47

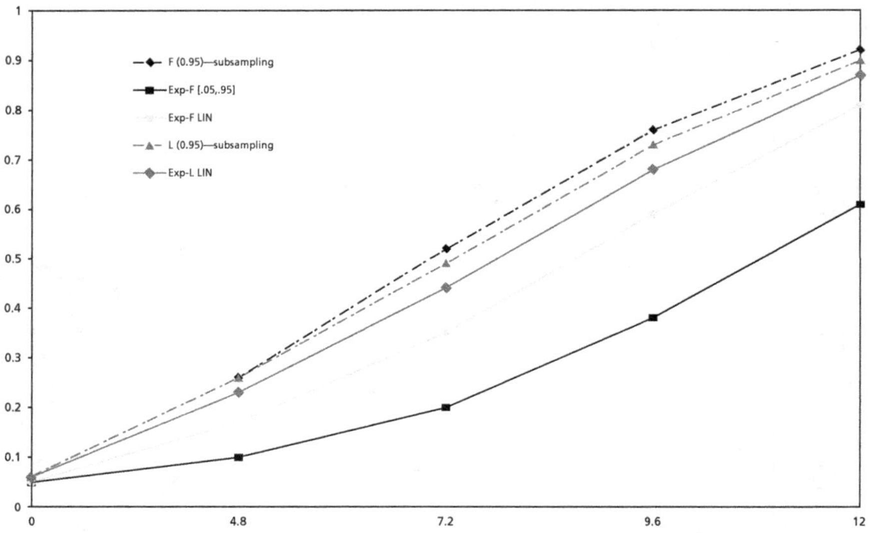

Figure 12.1. Power of selected tests against a one-time shift in the coefficients at $\pi = 0.95$ (on the x-axis selected values of the parameter δ of the local alternative hypothesis)

that $\delta_t = \delta_{t-1} + \eta_t$ for $t = n+1, \ldots, n+m$, where η_t is i.i.d. $N(0, \omega_T^2)$, $\omega_T = q/T$; simulation results are presented for values of $q = 0, .15, .3, .5, 1$. If the changepoint is known, as expected the subsampling version of the LMP test achieves the highest power and it is preferable to the subsampling F-test. For the case of unknown changepoint, the results are qualitatively similar to the one reported in Table 12.4, with the LMP-type tests displaying higher power than the Wald-type tests. The main findings for $\pi = .95$ are summarized graphically in Figure 12.2.

Overall, our simulations suggest that—for the case of unknown changepoint occurring toward the end of the sample—good choices are the Exp-L_{LIN} test and the Exp-L computed over $\Pi = (.75, .99)$, as their power is close to that of the optimal Wald and LMP statistics under both cases of one-time parameter shift and of random walk coefficients respectively; these tests are however slightly oversized when the sample size tends to be small. These end-of-sample stability tests are clearly not as powerful as the standard tests when the changepoint occurs earlier in the sample, e.g., at $\pi_0 = 0.5$; unreported simulations show that for earlier changepoints the Exp-F_{LIN} and Exp-L_{LIN} tests work better than (the end-of-sample) Exp-F and Exp-L over $\Pi = [.75, .99]$.

12.3.2 DYNAMIC REGRESSION MODELS

For dynamic regression models it is of interest to evaluate the properties of the tests when there is a change in the degree of persistence in the data. Here, the sample size (T=200) under Gaussianity; q represents the distance from the null hypothesis

		q=0.15				q=0.3				q=0.5				q=1			
Test	Π \ π_0	0.75	0.90	0.95	0.98	0.75	0.90	0.95	0.98	0.75	0.90	0.95	0.98	0.75	0.90	0.95	0.98
F (π_0)		.52	.18	.09	.06	.82	.46	.20	.08	.92	.71	.41	.14	.97	.91	.74	.36
F (π_0)	subsampling	.63	.22	.10	.06	.86	.49	.21	.09	.94	.72	.42	.14	.98	.91	.75	.37
F (.95)	subsampling	.55	.23	.10	.05	.81	.50	.21	.06	.91	.74	.42	.08	.96	.91	.75	.17
Sup-F	[.05,.95]	.48	.11	.05	.04	.84	.37	.09	.05	.95	.65	.22	.05	.99	.90	.58	.07
	[.01,.99]	.45	.10	.05	.04	.83	.36	.10	.04	.95	.66	.27	.06	.99	.91	.68	.18
	[.75,.99]	.51	.13	.06	.04	.86	.43	.14	.05	.96	.71	.34	.07	1.00	.93	.73	.24
	[.90,.99]	.35	.13	.06	.03	.73	.43	.15	.05	.88	.72	.36	.08	.96	.94	.75	.26
Exp-F	[.05,.95]	.52	.12	.06	.05	.85	.37	.09	.06	.95	.65	.21	.06	.99	.89	.55	.08
	[.01,.99]	.52	.13	.07	.05	.85	.39	.11	.06	.95	.67	.28	.07	1.00	.92	.67	.18
	[.75,.99]	.59	.19	.08	.06	.89	.49	.17	.07	.97	.75	.38	.09	1.00	.94	.75	.26
	[.90,.99]	.43	.20	.09	.06	.77	.51	.21	.07	.91	.77	.43	.11	.97	.95	.79	.31
Exp-F-LIN	[.01,.99]	.57	.18	.08	.06	.89	.47	.17	.07	.97	.74	.37	.09	1.00	.94	.75	.25
L(0)		.41	.07	.05	.05	.76	.17	.06	.05	.89	.38	.09	.05	.97	.71	.23	.06
L(.75)		.60	.16	.07	.06	.88	.41	.11	.06	.96	.67	.22	.07	.99	.89	.55	.11
L(.90)		.53	.28	.13	.09	.82	.58	.25	.11	.93	.80	.46	.14	.97	.95	.78	.26
L (π_0)	subsampling	.68	.23	.10	.06	.91	.53	.23	.09	.97	.77	.47	.15	1.00	.94	.79	.39
L (.95)	subsampling	.53	.23	.10	.03	.78	.50	.23	.04	.90	.74	.47	.07	.96	.92	.79	.22
Sup-L	[.05,.95]	.55	.20	.09	.06	.87	.51	.20	.07	.96	.76	.42	.09	1.00	.94	.76	.23
	[.01,.99]	.51	.20	.11	.07	.85	.50	.23	.10	.96	.76	.45	.16	1.00	.95	.80	.39
	[.75,.99]	.50	.20	.11	.07	.85	.50	.23	.10	.95	.76	.45	.16	1.00	.95	.80	.39
	[.90,.99]	.40	.20	.11	.08	.75	.50	.24	.11	.90	.76	.47	.16	.97	.95	.81	.40
Exp-L	[.05,.95]	.59	.18	.08	.06	.88	.47	.16	.06	.96	.73	.33	.08	1.00	.93	.69	.16
	[.01,.99]	.58	.19	.08	.06	.88	.48	.18	.07	.97	.75	.40	.10	1.00	.94	.76	.28
	[.75,.99]	.55	.22	.10	.07	.87	.52	.22	.08	.96	.77	.44	.13	1.00	.95	.79	.32
	[.90,.99]	.42	.22	.12	.08	.76	.52	.26	.10	.90	.77	.49	.16	.97	.95	.82	.38
Exp-L-LIN	[.01,.99]	.57	.21	.10	.06	.88	.51	.21	.08	.96	.77	.44	.12	1.00	.95	.78	.31

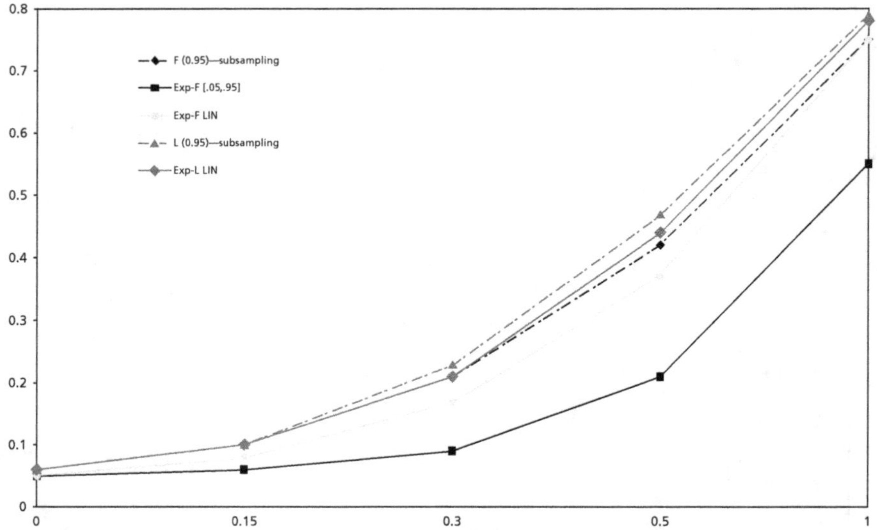

Figure 12.2. Power of selected tests against random walk coefficients starting at $\pi = 0.95$ (on the x-axis selected values of the parameter q of the local alternative hypothesis)

we consider a simple AR(1) process and evaluate the tests against a one-time change in the autoregressive coefficient. The data generating process corresponds to the model (12.1) with $x_{1t} = y_{t-1}$, $x_{2t} = 1$, $\sigma^2 = 1$; we consider a one-time change in both direction of lower and higher persistence, $\delta_t = -0.4$ and 0.4 for $t = n + 1, \ldots, n + m$, starting from $\beta_1 = 0.5$. Table 12.6 provide the rejection frequencies of the tests for a sample size of $T = 400$ observations. The Wald-type tests display once again very good size properties, while the empirical size of the LMP-type tests tend to be slightly higher than the nominal 5% even for $T = 400$. For all tests, the power of the tests tends to be much higher for an increase than a decrease of persistence. More interestingly, the Wald-type tests appear significantly more powerful than the LMP-type of test when there is a decrease in persistence ($\rho = 0.1$), while the opposite is true for $\rho = 0.9$. For example, for $\pi_0 = 0.95$ the rejection frequencies of Exp-F_{LIN} (Exp-L_{LIN}) are 0.22 (0.13) for $\rho = 0.1$ and 0.56 (0.73) for $\rho = 0.9$.

12.4 Empirical illustrations

We use the tests to detect possible instabilities in two simple "nowcasting" regression models for Italian output. Nowcasting quarterly gross domestic product (GDP) is usually based on indicators available more timely and/or with higher frequency. One well known good predictor for GDP is industrial

Table 12.6 Empirical rejection frequencies of the tests for the static regression model against a one-time change in the persistence coefficient at the fraction π_0 of the sample size ($T = 400$)

		ρ=0.5 size of the tests				ρ=0.1 decrease in persistence				ρ=0.9 increase in persistence			
	π_0	0.75	0.90	0.95	0.98	0.75	0.90	0.95	0.98	0.75	0.90	0.95	0.98
Test	Π												
F (π_0)		.05	.05	.05	.05	.91	.57	.32	.14	.99	.86	.62	.35
F (π_0)	subsampling	.16	.08	.06	.05	.95	.61	.33	.14	.99	.84	.61	.34
F (.95)	subsampling	.24	.09	.06	.05	.47	.35	.33	.10	.53	.59	.61	.25
Sup-F	[.05,.95]	.05	.05	.05	.05	.71	.33	.15	.06	.97	.73	.45	.15
	[.01,.99]	.04	.04	.04	.04	.67	.28	.14	.06	.96	.71	.46	.22
	[.75,.99]	.05	.05	.05	.05	.70	.38	.19	.08	.97	.77	.53	.27
	[.90,.99]	.04	.04	.04	.04	.23	.33	.20	.09	.46	.77	.55	.29
Exp-F	[.05,.95]	.05	.05	.05	.05	.76	.32	.13	.06	.98	.75	.46	.16
	[.01,.99]	.05	.05	.05	.05	.75	.32	.15	.07	.98	.76	.48	.21
	[.75,.99]	.05	.05	.05	.05	.78	.46	.22	.09	.98	.83	.57	.27
	[.90,.99]	.05	.05	.05	.05	.30	.41	.26	.11	.50	.81	.61	.31
Exp-F-LIN	[.01,.99]	.05	.05	.05	.05	.80	.43	.22	.09	.98	.82	.56	.27
L(0)		.04	.04	.04	.04	.60	.14	.06	.05	.98	.71	.40	.13
L(.75)		.05	.05	.05	.05	.70	.29	.09	.05	.99	.88	.61	.27
L(.90)		.08	.08	.08	.08	.28	.36	.20	.08	.69	.91	.75	.42
L (π_0)	subsampling	.16	.07	.05	.05	.78	.29	.12	.05	.99	.88	.70	.43
L (.95)	subsampling	.18	.08	.05	.04	.26	.15	.12	.04	.61	.69	.70	.39
Sup-L	[.05,.95]	.07	.07	.07	.07	.54	.25	.13	.06	.98	.89	.73	.41
	[.01,.99]	.08	.08	.08	.08	.44	.20	.12	.08	.98	.89	.74	.49
	[.75,.99]	.08	.08	.08	.08	.36	.21	.12	.08	.96	.89	.74	.49
	[.90,.99]	.09	.09	.09	.09	.13	.18	.12	.08	.63	.87	.74	.49
Exp-L	[.05,.95]	.06	.06	.06	.06	.70	.30	.12	.05	.99	.89	.69	.35
	[.01,.99]	.07	.07	.07	.07	.66	.27	.12	.07	.99	.89	.71	.43
	[.75,.99]	.07	.07	.07	.07	.40	.28	.13	.07	.94	.90	.73	.45
	[.90,.99]	.08	.08	.08	.08	.14	.19	.15	.08	.61	.85	.75	.48
Exp-L-LIN	[.01,.99]	.07	.07	.07	.07	.55	.27	.13	.07	.98	.90	.73	.45

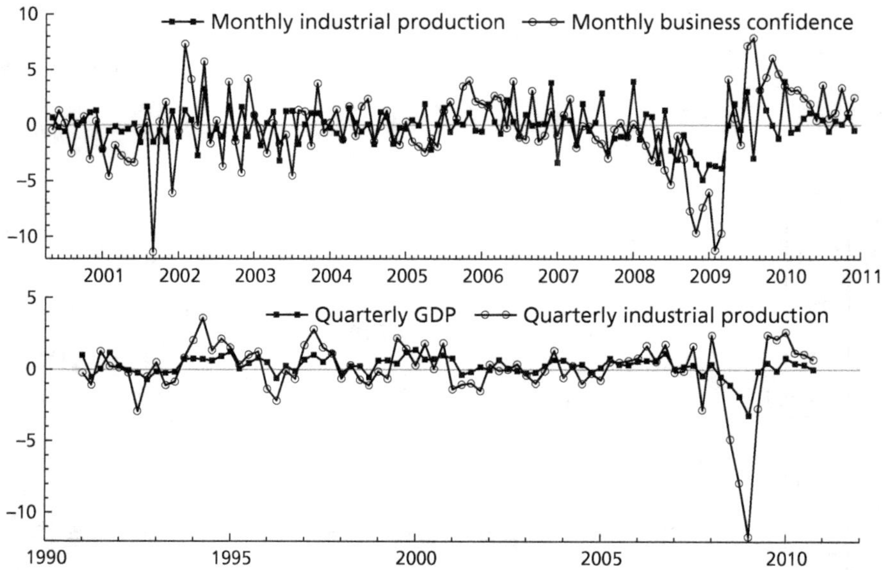

Figure 12.3. Business confidence, industrial production, and GDP in Italy

production. However, the monthly data for industrial production are only available with a substantial lag, which requires a further predictive model to be constructed, based on qualitative indicators coming from surveys that are available before the end of the same month.

Here we consider two simple models, one for quarterly GDP and the other for monthly industrial production, which can be used in conjunction for the purpose of nowcasting Italian GDP growth. Figure 12.3 shows in the top panel the monthly series of Italian industrial production and that of a business confidence indicator over the period 2000–10 (in terms of percentage growth rates). The question is whether the association between a sentiment indicator and the corresponding "hard data" breaks down at time of crisis, as during the profound recession of 2008–09. We capture the association between (the log of) industrial production, y_t, and the (log of the) confidence indicator, x_t, by the simple linear regression

$$\Delta y_t = \alpha_0 + \alpha_1 \Delta y_{t-1} + \alpha_2 \Delta x_t, \tag{12.14}$$

where Δ is the first difference operator. We investigate end-of-sample instability by estimating this model for samples of different lengths. The shortest sample consists of ninety-eight observations, ending in 2008.H1, where H1 indicates the first half of the year; then we add six observations at time and recompute the tests with samples that end at 2008.H2, 2009.H1, and 2009.H2, 2010.H1, 2010.H2. All the end-of-sample tests considered in this paper—computed allowing for time variation in all three coefficients

Table 12.7 End-of-sample LMP-type tests for the linear relation between GDP and industrial production, estimated with different end-dates of the sample; *=reject at 10%, **=reject at 5%, ***=reject at 1%

Test	Π	End sample	2008.H1	2008.H2	2009.H1	2009.H2	2010.H1	2010.H2
L(.95)	subsampling		*	-	***	***	*	-
Sup-L	[.05,.95]		-	-	***	***	*	*
	[.01,.99]		***	-	***	***	**	-
	[.75,.99]		***	-	***	***	**	*
	[.90,.99]		***	-	***	***	**	*
Exp-L	[.05,.95]		-	-	***	*	-	-
	[.01,.99]		**	-	***	***	-	-
	[.75,.99]		**	-	***	***	*	-
	[.90,.99]		**	-	***	***	**	*
Exp-L_{LIN}	[.01,.99]		**	-	***	***	*	-

α_0, α_1, α_2—indicate stability of the model (12.14) if it is estimated using data up to the first half of 2008, before the financial crisis became acute; detailed results are available upon request. Thereafter, using data up to the end of 2008 and later, nearly all tests strongly reject the null hypothesis of stability; the evidence provided by both the Wald and the LMP-type tests is very similar.

The same exercise is then carried out for the association between GDP and industrial production at the quarterly frequency. The bottom panel of Figure 12.3 shows the percentage growth rates of these two series over the period 1991–2010. Table 12.7 reports the results for the LMP-type tests for the linear regression model (12.14) estimated with different end-points, starting from 2008.H1. Instability is generally rejected, particularly when the data of 2009 are included in the sample; indeed the signs of instabilities begin earlier.[11] For brevity we have not included in Table 12.7 the corresponding results for the Wald-type tests, since they never lead to a rejection (consistently with the simulation evidence reported in Section 12.3 section that showed generally higher rejection frequencies for the LMP-type tests, despite some oversizing in small samples). For this specific example, a rejection of the hypothesis of stability seems a plausible outcome.[12]

Overall, the two empirical illustrations convey the message that model based predictions should be interpreted with caution in the presence of unusually large fluctuations of the indicators toward the end of the sample; in these cases adding a dose of forecaster's judgment would be a wise option to choose.

[11] This may be related to the fact that the recessive phase of the Italian industrial sector started at the beginning of 2007 while GDP growth turned negative only in last quarter of that year.

[12] The heteroskedasticity robust version of the LMP-type tests, described in Section 2.2, provides very similar results to the standard version; details are available upon request.

12.5 Conclusions

The paper has investigated the properties of several tests aimed at detecting instabilities that may occur toward the end of the sample. Tests constructed in terms of LMP statistics appear in general more powerful than those based on Wald statistics; the latter however possess better size properties. Overall, a LMP-type test that gives increasing weight to possible changepoints along the sample appears to be a good choice against both alternative hypotheses of one-time parameter shift and of random walk coefficients. Its power is in fact quite close to that of the optimal tests based on knowing the end-of-sample changepoint in advance.

One obvious drawback of the end-of-sample stability tests is that they are not as powerful if changepoints occur earlier in the sample, in which case the standard approach of Andrews et al. (1996) remains preferable.

The practical usefulness of the tests proposed in this paper has been illustrated with nowcasting linear regression models for industrial production and GDP in Italy: model instabilities have been uncovered at the time of the "Great Recession" of 2008–09.

13 Improved frequentist prediction intervals for autoregressive models by simulation

Jouni Helske and Jukka Nyblom

13.1 Introduction

A traditional approach to time series forecasting usually involves a selection of a family of suitable models, e.g., the class of autoregressive integrated moving average (ARIMA) models. Then, using different model selection criteria within the family based of autocorrelation and partial autocorrelation functions together with formal criteria such as Akaike or Bayesian information criterion, the analyst chooses a suitable "best fitting" representative from the model family, estimates the parameters, makes diagnostic checks, and if he is happy with his choice computes predictions and the prediction intervals. The prediction intervals are usually computed as if the chosen model were correct and the parameters completely known, with no reference to the process of model selection. Chatfield (1993, 1996) has strongly criticized the omission of model uncertainty in forecasting. Clements and Hendry (1999, Sections 1.3.8 and 2.2) introduce a detailed taxonomy of forecast errors, and stress the effects of the structural breaks in time series forecasting. As a remedy they propose robustifying forecasts, for example by differencing and intercept corrections.

It is the common view of references given in the previous paragraph that the parameter uncertainty is often a minor source of prediction errors in practical applications when the sample size is large enough. Clements and Hendry (1999, p. 128) remark that although the parameter uncertainty is unlikely to lead to serious forecast failure, it may have larger effect in conjunction with model misspecification. Nevertheless, we believe that it is justified to handle also this part of the model uncertainty. In textbooks it is a common topic, see for example Harvey (1993, pp. 58–9). Here we show how to make corrections in a fairly simple way under autoregressive (AR) models.

Several proposals have been made for improving prediction intervals when parameters are estimated. One group of solutions focus on finding a more accurate prediction mean squared error in the presence of estimation; see, for example Phillips (1979), Fuller and Hasza (1981), Ansley and Kohn (1986), Quenneville and Singh (2000b), and Pfeffermann and Tiller (2005). Both analytic and bootstrap approaches are tried.

Barndorff-Nielsen and Cox (1996) give general results for prediction intervals in the presence of estimated parameters. These results are further developed for time series models by Vidoni (2004, 2009). Unfortunately fairly complicated expressions appear already in rather simple models. Bootstrap solutions are given by several authors; see for example Beran (1990), Masarotto (1990), Grigoletto (1998), Kim (2004), Pascual et al. (2004), Clements and Kim (2007), Kabaila and Syuhada (2008), and Rodríguez and Ruiz (2009).

Our aim is to find solutions for prediction interval problems using a Bayesian viewpoint with a requirement of objectivity for our predictions. Berger (2006) strongly encourages the use of the term "objective Bayes" in such situations. Operationally, "objectivity" in our application means adopting priors which produce prediction intervals that exhibit approximately the same coverage probabilities in both Bayesian and frequentist sense.

There is a vast literature on matching posterior and frequentist inferences to some degree of approximation. Methods based on invariance arguments, on information criteria, or divergence measures as well as on asymptotic expansions are tried. The introduction section of Fraser et al. (2010) gives a short review on these as well as a list of relevant references to their own works and others. The starting point of their approach is to replace p value computations in a sample space with posterior integration over the parameter space. Matching the Bayesian and frequentist coverage probabilities of prediction intervals in regression models with independent cases are treated in Datta and Mukerjee (2003) and Datta and Mukerjee (2004). A common feature in all these approaches is that often the so-called Jeffreys prior or its modification shows up as a solution. As we will see it happens also in this article. In an article by Arellano and Bonhomme (2009), where the main issue is the bias reduction in panel data models, the authors use a prior related to Jeffreys prior. A predecessor in the bias reduction with the help of Jeffreys prior is given by Firth (1993).

Early examples using the Bayesian approach in autoregressive models are given by (Zellner, 1971, p. 188) and Chow (1974). Later, Broemeling and Land (1984) showed, among other things, that using a normal-gamma prior for the parameters the one-step ahead prediction follows a t distribution (understood as a location-scale family). They further deduced that all predictions up to k step ahead have a joint predictive density which consists of the product of k univariate t densities. Thompson and Miller (1986) propose a simulation method for computing prediction intervals based on the Bayesian arguments called "sampling the future." This differs from our approach. We compute the prediction interval by simulating directly the posterior prediction probability which is more accurate and considerably less time consuming. Liu (1994) develops an approximation to their method, and Snyder et al. (2001) make an approximate extension of it to ARIMA models. They also compare several types of prediction intervals in terms of the frequentist coverage probabilities.

After the emergence of Markov chain Monte Carlo simulation the Bayesian paradigm has gained increasing popularity in time series and econometrics; see for example Geweke (2005), Prado and West (2010), and references therein. A thorough exposition of the Bayesian forecasting is given by Geweke and Whiteman (2006). Nevertheless, in forecasting under an AR model either direct simulation or importance sampling is fast and sufficient. Based on independent simulation replicates it also renders rather simple formulas for assessing the Monte Carlo simulation error.

13.2 Motivation

Consider an AR(p) process

$$y_t = \beta_0 + \beta_1 y_{t-1} + \cdots + \beta_p y_{t-p} + \varepsilon_t, \quad t = 1, \ldots, n, \ldots, \tag{13.1}$$

where the errors ε_t are independently drawn from $N(0, \sigma^2)$, and coefficients $\beta_j, j = 0, \ldots, p$, are arbitrary fixed values. Assume that we have observed $y_{-p+1}, \ldots, y_n$. Write $\boldsymbol{x}_t = (1, y_{t-1}, \ldots, y_{t-p})'$, and let $\boldsymbol{X}$ be the matrix with rows $\boldsymbol{x}_t'$, $t = 1, \ldots, n$. Further, let $Y = (y_1, \ldots, y_n)'$ and $\boldsymbol{\beta} = (\beta_0, \beta_1, \ldots, \beta_p)'$. Then the model (13.1) can be written as $Y = X\boldsymbol{\beta} + \boldsymbol{\varepsilon}$. If we condition on the starting values $y_0' = (y_{-p+1}, \ldots, y_0)$ we are led to the conditional likelihood

$$L_c(\boldsymbol{\beta}, \sigma) = (2\pi)^{-\frac{n}{2}} \sigma^{-n} \exp\left(-\frac{1}{2\sigma^2}(Y - X\boldsymbol{\beta})'(Y - X\boldsymbol{\beta})\right). \tag{13.2}$$

The conditional maximum likelihood estimates for $\boldsymbol{\beta}$ coincide with the least squares estimates

$$\widehat{\boldsymbol{\beta}} = (X'X)^{-1}X'Y, \tag{13.3}$$

and the maximum likelihood estimate for σ^2, corrected by the degrees of freedom, is

$$s^2 = \frac{(Y - X\widehat{\boldsymbol{\beta}})'(Y - X\widehat{\boldsymbol{\beta}})}{n - p - 1}. \tag{13.4}$$

The predictive value for y_{n+1} is $\boldsymbol{x}_{n+1}'\widehat{\boldsymbol{\beta}}$, and the standard prediction interval with approximate coverage probability $1 - 2\alpha$ is

$$\boldsymbol{x}_{n+1}'\widehat{\boldsymbol{\beta}} \pm s z_\alpha, \tag{13.5}$$

where z_α is the α quantile of the standard normal distribution. In practice, the true coverage may be considerably below the nominal value $1-2\alpha$.

Let us suppose for a moment that we have an ordinary regression model with some truly exogeneous variables. Then using the same notation, X and $\boldsymbol{x}_{n-1}$, as before the exact coverage probability is obtained by

$$\boldsymbol{x}'_{n+1}\widehat{\boldsymbol{\beta}} \pm s t_{\alpha,n-p-1}\sqrt{1+\boldsymbol{x}'_{n+1}(X'X)^{-1}\boldsymbol{x}_{n+1}}. \tag{13.6}$$

As is well known, the extra factor here compared to (13.5) involving the exogeneous variable takes into account the estimation error in regression coefficients. In addition, a minor correction is done by replacing the normal quantile with that from Student's t with $n-p-1$ degrees of freedom. Although the assumptions of AR models do not satisfy the assumptions leading to (13.6), our simulations show that the very same intervals (13.6) have practically correct coverage probabilities also under the AR models.

Combining the formulas (13.2)–(13.4) with the identity

$$\begin{aligned}(Y-X\boldsymbol{\beta})'(Y-X\boldsymbol{\beta}) &= (Y-X\widehat{\boldsymbol{\beta}})'(Y-X\widehat{\boldsymbol{\beta}})\\ &\quad + (\widehat{\boldsymbol{\beta}}-\boldsymbol{\beta})'(X'X)(\widehat{\boldsymbol{\beta}}-\boldsymbol{\beta}),\end{aligned}$$

the conditional likelihood function (13.2) can be written as

$$\begin{aligned}L_c(\boldsymbol{\beta},\sigma) &= (2\pi)^{-\frac{n}{2}}\sigma^{-n}\exp\left(-\frac{(n-p-1)s^2}{2\sigma^2}\right)\\ &\quad\times\exp\left(-\frac{1}{2\sigma^2}(\widehat{\boldsymbol{\beta}}-\boldsymbol{\beta})'(X'X)(\widehat{\boldsymbol{\beta}}-\boldsymbol{\beta})\right).\end{aligned} \tag{13.7}$$

Until now we have had the frequentist approach. But it is also illuminating to see the prediction interval (13.6) from a Bayesian point of view, where $\boldsymbol{\beta}$ and σ are now treated as a random vector and variable. If we multiply the likelihood $L_c(\boldsymbol{\beta},\sigma)$ by the improper prior $p(\boldsymbol{\beta},\sigma)=1/\sigma$, we find, as is well known in the ordinary regression, that a posteriori

$$\begin{aligned}\frac{(n-p-1)s^2}{\sigma^2}\,\bigg|\, Y &\sim \chi^2(n-p-1),\\ \boldsymbol{\beta}\mid Y,\sigma &\sim N(\widehat{\boldsymbol{\beta}},\sigma^2(X'X)^{-1}).\end{aligned} \tag{13.8}$$

These together lead to the result that the interval (13.6) is a Bayesian predictive interval with exact coverage probability $1-2\alpha$. The corresponding frequentist coverage probability is, however, only approximate

$$P\left(y_{n+1}\in \boldsymbol{x}'_{n+1}\widehat{\boldsymbol{\beta}} \pm s t_{\alpha,n-p-1}\sqrt{1+\boldsymbol{x}'_{n+1}(X'X)^{-1}\boldsymbol{x}_{n+1}}\,\bigg|\,\boldsymbol{\beta},\sigma\right)\approx 1-2\alpha. \tag{13.9}$$

However, in Section 13.5 we will find that the approximation (13.9) is very good when the coefficients are well within the stationarity region.

The approximation seems to be worst in the nearly unit root cases. The explanation is likely to be as follows. If the coefficients are not too close to the boundary of the stationarity region, the sampling distribution of $\widehat{\boldsymbol{\beta}}$ is approximately as in (13.8) when the roles of $\boldsymbol{\beta}$ and $\widehat{\boldsymbol{\beta}}$ are interchanged. On the other hand this not true in the nearly unit root case. So the question arises what is "good" prior in AR models in general. It turns out that using a certain modification of Jeffreys prior seems to be preferred to uniform prior in nearly unit root models. One solution to the nearly unit root problem is provided by Clements and Hendry (1999, p. 92) who recommend in some cases to impose a unit root albeit it were not warranted by the unit root test.

At this point it might be interesting to recall the vigorous debate that broke out in the early 1990s between Peter Phillips and Christopher Sims on the value of unit root econometrics as such and its relation to Bayesian approach in statistics and econometrics; see Sims and Uhlig (1991), Phillips (1991), Sims (1991), and their references. The unit root inference is not an issue here. The Bayesian controversy focused on the choice of an appropriate prior distribution. Nevertheless, we do not want to revive this controversy here simply because we are not so much interested in inferences on autoregressive coefficients themselves but rather in prediction intervals. The role of the prior in this article is to produce a good weighting scheme for the predictive distribution.

13.3 **Predictive distributions**

13.3.1 UNIFORM PRIOR

In this section we develop prediction formulas under the Bayesian paradigm employing the noninformative uniform prior for $(\boldsymbol{\beta}, \log \sigma)$, i.e., the prior takes the form $p(\boldsymbol{\beta}, \sigma) = 1/\sigma$, for $\beta \in \mathbb{R}^{p+1}$ and $\sigma > 0$. We have already seen that when $k = 1$ the predictive distribution is Student's t. But for $k > 1$ the predictive distribution is not known to be any common distribution. Therefore we have to rely on simulation.

We use a generic notation $p(\cdot)$ for a density. The Bayesian predictive density of y_{n+k} is defined as the conditional density of y_{n+k} given Y. It is obtained by integration as

$$p(y_{n+k} \mid Y) = \int p(y_{n+k} \mid \boldsymbol{\beta}, \sigma, Y)\, p(\boldsymbol{\beta}, \sigma \mid Y)\, \mathrm{d}\boldsymbol{\beta}\, \mathrm{d}\sigma, \tag{13.10}$$

where $p(\boldsymbol{\beta}, \sigma \mid Y)$ is the posterior density of $(\boldsymbol{\beta}, \sigma)$. In AR models with normally distributed errors the density $p(y_{n+k} \mid \boldsymbol{\beta}, \sigma, Y)$ can be written explicitly. Recall the prediction formulas given $\boldsymbol{\beta}, \sigma^2, Y$ (see, for example Box et al. (2008, Chapter 5)) and designate $\hat{y}_n(k) = E(y_{n+k} \mid Y, \boldsymbol{\beta})$, $k = 0, \pm 1, \pm 2, \ldots$.

Note that when $k \leq 0$ we have $\hat{y}_n(k) = y_{n+k}$. The equation (13.1) immediately yields recursion for predicted values

$$\hat{y}_n(k) = \beta_0 + \beta_1 \hat{y}_n(k-1) + \cdots + \beta_p \hat{y}_n(k-p), \quad k = 1, 2, \ldots .$$

It is also plain from (13.1) that for some constants $\psi_1, \psi_2, \ldots$ depending on $\boldsymbol{\beta}$ we have

$$y_{n+k} - \hat{y}_n(k) = \varepsilon_{n+k} + \psi_1 \varepsilon_{n+k-1} + \cdots + \psi_{k-1} \varepsilon_{n+1}.$$

Box et al. (2008) gives recursions for the coefficients ψ_j as follows:

$$\psi_j = \beta_1 \psi_{j-1} + \cdots + \beta_p \psi_{j-p}, \quad j \geq 1, \ \psi_0 = 1, \text{ and } \psi_j = 0 \text{ for } j < 0.$$

We find $\psi_1 = \beta_1, \psi_2 = \beta_1^2 + \beta_2$, and so on. The prediction error variance, given $\boldsymbol{\beta}, \sigma^2$, is then $\sigma^2(1 + \psi_1^2 + \cdots + \psi_{k-1}^2) = \sigma^2 v^2(\boldsymbol{\beta})$.

In the further development it is useful to introduce a more detailed notation $\hat{y}_n(k; \boldsymbol{\beta}) = E(y_{n+k} \mid Y, \boldsymbol{\beta})$. Then,

$$y_{n+k} \mid \boldsymbol{\beta}, \sigma^2, Y \sim N(\hat{y}_n(k; \boldsymbol{\beta}), \sigma^2 v_k^2(\boldsymbol{\beta})).$$

Combining this with (13.10) and changing the order of integration we find

$$P(y_{n+k} \leq b \mid Y) = E\left[\Phi\left(\frac{b - \hat{y}_n(k; \boldsymbol{\beta})}{\sigma v_k(\boldsymbol{\beta})}\right) \middle| Y\right], \tag{13.11}$$

where Φ is the cumulative distribution function of the standard normal distribution. When $k > 1$ we cannot do the integration involved analytically. Nevertheless, a Monte Carlo solution is straightforward. We can proceed as follows:

1. Draw independent q_i, $i = 1, \ldots, N$ from $\chi^2(n-p-1)$, and let $\sigma_i^2 = (n-p-1)s^2/q_i$.
2. Draw $\boldsymbol{\beta}_i$ from $N(\widehat{\boldsymbol{\beta}}, \sigma_i^2(X'X)^{-1})$, independently for $i = 1, \ldots, N$.
3. Compute the average:

$$\bar{P}_N(b) = \frac{1}{N} \sum_{i=1}^{N} \Phi\left(\frac{b - \hat{y}_n(k; \boldsymbol{\beta}_i)}{\sigma_i v_k(\boldsymbol{\beta}_i)}\right). \tag{13.12}$$

The prediction interval is then found by solving separately both $\bar{P}_N(b) = \alpha$ and $\bar{P}_N(b) = 1 - \alpha$. Let the solutions be $\hat{b}_\alpha, \hat{b}_{1-\alpha}$, respectively. Then $(\hat{b}_\alpha, \hat{b}_{1-\alpha})$ is the prediction interval with posterior coverage probability $1 - 2\alpha$ when N is large. Broemeling and Land (1984) noted that multi-step predictions can be constructed through a sequence of one-step predictions, each step involving t distribution. However, they do not suggest any computational method as to how to utilize this in practice. Thompson and Miller (1986) continued the work of Broemeling and Land by proposing the Bayesian simulation of the future values $y_{n+1}, \ldots, y_{n+k}$. The prediction limits are then derived from the quantiles of the simulated values. This is more time consuming as well as being less accurate than what we suggest here.

13.3.2 PREDICTION WITH GENERAL PRIOR

Combine the conditional likelihood (13.2) and a general prior $p(\boldsymbol{\beta},\sigma)/\sigma$. Then by (13.7) and (13.8) the joint posterior is

$$
\begin{aligned}
p(\boldsymbol{\beta},\sigma \mid Y) \propto\ & \sigma^{-(n-p)} e^{-(n-p-1)s^2/(2\sigma^2)} \\
& \times \sigma^{-p-1} \exp\left(-\frac{1}{2\sigma^2}(\widehat{\boldsymbol{\beta}}-\boldsymbol{\beta})'(X'X)(\widehat{\boldsymbol{\beta}}-\boldsymbol{\beta})\right) \\
& \times p(\boldsymbol{\beta},\sigma).
\end{aligned}
$$

It can be evaluated using importance sampling. In the simulation algorithm only item three need be changed to:

3' Compute the weighted average:

$$
\bar{P}_{N,w}(b) = \frac{\sum_{i=1}^{N} w_i \Phi\left(\frac{b-\hat{y}_n(k;\boldsymbol{\beta}_i)}{\sigma_i v_k(\boldsymbol{\beta}_i)}\right)}{\sum_{i=1}^{N} w_i}, \tag{13.13}
$$

$$
w_i = p(\boldsymbol{\beta}_i,\sigma_i). \tag{13.14}
$$

The prediction interval is then solved as before.

It is also possible to incorporate the starting values $(y_{-p+1},\ldots,y_0) = y_0'$ into the likelihood. Denote $\text{cov}(y_0 \mid \boldsymbol{\beta},\sigma) = \sigma^2 V = \sigma^2 V(\boldsymbol{\beta})$, and $E(y_i \mid \boldsymbol{\beta}) = \mu = \mu(\boldsymbol{\beta})$, $i = -p+1,\ldots,0$. The matrix V can be obtained from the Yule-Walker equations, see Box et al. (2008, p. 58), and $\mu = \beta_0/(1-\beta_1-\cdots-\beta_p)$. To obtain the full likelihood, the conditional likelihood (13.2) is just multiplied by

$$
p(y_0 \mid \beta,\sigma) = (2\pi)^{-\frac{p}{2}}\sigma^{-p}|V|^{-\frac{1}{2}} \exp\left(-\frac{1}{2\sigma^2}(y_0-\mu\mathbf{1})'V^{-1}(y_0-\mu\mathbf{1})\right).
$$

This leads to changing the weights in step 3'. The new weights are

$$
w_i = I(\boldsymbol{\beta}_i \in \mathbb{R}\times S_p)\, p(\boldsymbol{\beta}_i,\sigma_i) p(y_0 \mid \boldsymbol{\beta}_i,\sigma_i), \tag{13.15}
$$

where $I(\cdot)$ is an indicator and S_p is the stationarity region of AR(p) process. The first coordinate of $\boldsymbol{\beta}_i$ can, of course, take any real value.

13.3.3 STANDARD ERRORS FOR MONTE CARLO PREDICTION LIMITS

Here we give simple formulas for approximate standard errors when computing the prediction limits by Monte Carlo simulation. Let b_α be such that

$P(y_{n+k} \leq b_\alpha \mid Y) = \alpha$ in (13.11), and $\hat{b}_\alpha$ be such that $\bar{P}_N(\hat{b}_\alpha) = \alpha$ in (13.12). The first order Taylor expansion at $\hat{b}_\alpha$ leads to

$$\hat{b}_\alpha - b_\alpha \approx \frac{\alpha - \bar{P}_N(b_\alpha)}{\bar{P}'_N(\hat{b}_\alpha)},$$

where $\bar{P}'_N$ is the derivative of $\bar{P}_N$. The variance of the numerator on the right side can be estimated from the sample values by S^2/N, where

$$S^2 = \frac{1}{N-1} \sum_{i=1}^{N} \left[\Phi\left(\frac{\hat{b}_\alpha - \hat{y}_n(k; \boldsymbol{\beta}_i)}{\sigma_i v_k(\boldsymbol{\beta}_i)} \right) - \alpha \right]^2.$$

Define

$$\text{s.e.}(\hat{b}_\alpha) = \frac{S/\sqrt{N}}{\bar{P}'_N(\hat{b}_\alpha)} = \frac{S}{\sum_{i=1}^{N} \frac{1}{\sigma_i v_k(\boldsymbol{\beta}_i)} \varphi\left(\frac{\hat{b}_\alpha - \hat{y}_n(k; \boldsymbol{\beta}_i)}{\sigma_i v_k(\boldsymbol{\beta}_i)} \right) / \sqrt{N}},$$

where φ is the density of the standard normal distribution. Omitting the details we can show that $(\hat{b}_\alpha - b_\alpha)/\text{s.e.}(\hat{b}_\alpha)$ tends to the standard normal distribution as $N \to \infty$.

In case we use a weighted average as in (13.14) and (13.15), a similar technique leads to the standard error

$$\text{s.e.}(\hat{b}_\alpha) = \frac{S}{\sum_{i=1}^{N} \frac{w_i}{\sigma_i v_k(\boldsymbol{\beta}_i)} \varphi\left(\frac{\hat{b}_\alpha - \hat{y}_n(k; \boldsymbol{\beta}_i)}{\sigma_i v_k(\boldsymbol{\beta}_i)} \right) / \sqrt{N}},$$

$$S^2 = \frac{1}{N-1} \sum_{i=1}^{N} \left[\alpha w_i - w_i \Phi\left(\frac{\hat{b}_\alpha - \hat{y}_n(k; \boldsymbol{\beta}_i)}{\sigma_i v_k(\boldsymbol{\beta}_i)} \right) \right]^2.$$

13.4 **Priors**

Here we give some examples of the priors that are likely to give improvements for frequentist coverage probabilities. A popular principle to generate priors, often improper, is to follow Jeffreys rule which leads to the square root of the determinant of the information matrix. Applying this to the conditional likelihood (13.2), we first find that

$$\frac{\partial^2 L_c}{\partial \boldsymbol{\beta} \partial \boldsymbol{\beta}'} = -\frac{1}{\sigma^2} \sum_{t=1}^{n} x_t x'_t.$$

The information matrix is obtained by taking the expectation given the parameters and changing the sign. Assuming stationarity yields

$$-E\left[\frac{\partial^2 L_c}{\partial \boldsymbol{\beta} \partial \boldsymbol{\beta}'}\right] = \frac{n}{\sigma^2} E\left[x_1 x_1'\right] = \frac{n}{\sigma^2} \begin{pmatrix} 1 & \mu \mathbf{1}' \\ \mu \mathbf{1} & V + \mu^2 \mathbf{1}\mathbf{1}' \end{pmatrix}.$$

The determinant of the matrix involved is easily seen to equal $|V|$ (note that V depends only on β_j, $j = 1, \ldots, p$). Now we take the convention that the parameter groups $\{\sigma\}$, $\{\beta_0\}$, and $\{\beta_1, \ldots, \beta_p\}$ are independent a priori, and that $\log \sigma$, β_0 are uniform. This leads to the prior we here call Jeffreys prior:

$$p_J(\boldsymbol{\beta}, \sigma)/\sigma = I(\sigma > 0) I(\boldsymbol{\beta} \in \mathbb{R} \times S_p) \sigma^{-1} \sqrt{|V(\boldsymbol{\beta})|}. \tag{13.16}$$

The uniform prior over the stationarity region is

$$p_U(\boldsymbol{\beta}, \sigma)/\sigma = I(\sigma > 0) I(\boldsymbol{\beta} \in \mathbb{R} \times S_p) \sigma^{-1}.$$

In our simulations we combine these priors with the full likelihood, i.e., the weights are $p(y_0 \mid \boldsymbol{\beta}_i, \sigma_i) p_J(\boldsymbol{\beta}_i, \sigma_i)$ and $p(y_0 \mid \boldsymbol{\beta}_i, \sigma_i) p_U(\boldsymbol{\beta}_i, \sigma_i)$ for Jeffreys and uniform on the stationarity region priors, respectively. Note that when Jeffreys prior is used the determinant $|V|$ cancels when forming the weights.

For the stationary AR(1) model Jeffreys marginal prior density of β_1 is

$$\begin{aligned} p_J(\beta_1) &= \frac{1}{\pi \sqrt{1 - \beta_1^2}}, & |\beta_1| < 1 \\ &= 0, & \text{otherwise,} \end{aligned}$$

and for AR(2) the marginal prior density of (β_1, β_2) is

$$\begin{aligned} p_J(\beta_1, \beta_2) &= \frac{1}{(1 + \beta_2) \sqrt{(1 - \beta_2)^2 - \beta_1^2}}, & \beta_1 + \beta_2 < 1,\ \beta_2 - \beta_1 < 1, \\ & & |\beta_2| < 1, \\ &= 0, & \text{otherwise.} \end{aligned}$$

Note that the first density is proper whereas the second one is not; the latter property is true for all AR(p) with $p > 1$.

Without a stationarity restriction Berger and Yang (1994) defined an alternative proper marginal prior density for the AR(1) model as

$$\begin{aligned} p_R(\beta_1) &= \frac{1}{2\pi \sqrt{1 - \beta_1^2}}, & |\beta_1| < 1, \\ &= \frac{1}{2\pi |\beta_1| \sqrt{\beta_1^2 - 1}}, & |\beta_1| > 1, \end{aligned} \tag{13.17}$$

which they call a reference prior. It is easily seen that the reference prior is invariant under the transformation $\beta_1 \to 1/\beta_1$. Note that the reference prior should be combined with the conditional likelihood (13.2), and then it produces peaks also in the posterior at $\beta_1 = \pm 1$. There is no use combining it with the full likelihood because then it reduces to Jeffreys prior. There seems to be no generalization available of the reference prior to higher order models.

13.5 Simulation experiments

In this section we compare different priors as regards their ability to match the frequentist coverage probabilities to the Bayesian ones. Thus, we now turn back to the frequentist interpretation of the probability. Suppose that we have a realization from an autoregressive process with parameters $\boldsymbol{\beta}, \sigma$ which are fixed but unknown. Despite this, we wish to compute the prediction interval $(b_\alpha, b_{1-\alpha})$ via a Bayesian route from the probability distribution (13.11). Although the posterior coverage probability is exactly $1 - 2\alpha$, the corresponding frequentist probability usually is not. It refers to an infinite sequence of new realizations from the same model with parameters $\boldsymbol{\beta}, \sigma$, and it is defined by the probability

$$P(b_\alpha \le y_{n+k} \le b_{1-\alpha} \mid \boldsymbol{\beta}, \sigma),$$

where all $y_{n+k}, b_\alpha, b_{1-\alpha}$ are random. In short, the frequentist coverage probability is an average coverage probability over the realizations. The conditional frequentist coverage probability $P(b_\alpha \le y_{n+k} \le b_{1-\alpha} \mid Y, \boldsymbol{\beta}, \sigma)$ is a random variable, and in an actual application we do not know this probability. In our simulation experiment the values $b_\alpha, b_{1-\alpha}$ are replaced by $\hat{b}_\alpha, \hat{b}_{1-\alpha}$ obtained from (13.12) (apart from the case of the uniform prior with $k = 1$).

The chosen priors are uniform, uniform on the stationarity region, Jeffreys prior (13.16), and the reference prior (13.17) of Berger and Yang (1994) for AR(1). In all comparisons we have used 50 000 replicates, where each replicate is a realization from a stationary AR(p) process of length $n + p$. Within each replicate the Monte Carlo sample size is $N = 50$. In this type of simulation experiments N need not be large, because the main variation in coverage probabilities is due to the different replicates. Nevertheless, in an actual application N should be considerably larger as is seen in Section, 13.6. All computations are done in the (R environment see R Development Core Team (2012)).

Start with AR(1). Figure 13.1 shows the coverage probabilities of one-step-ahead prediction intervals (13.6) based on the t distribution for several values of β_1, when $n = 30$. Here, and in all other experiments, we are aiming for the coverage probability of .9. For negative β_1 the coverage probabilities tend to be slightly above the nominal value whereas when β_1 is close to 1, they drop below the nominal value. However, note that even for $\beta_1 = .9$ the coverage probability is .894. The standard errors of coverage probabilities are less than 3×10^{-4} in all cases. Although not shown in the figure, the prediction intervals appear to have approximately equal tail probabilities. The coverage probabilities of the standard prediction intervals stay below .88.

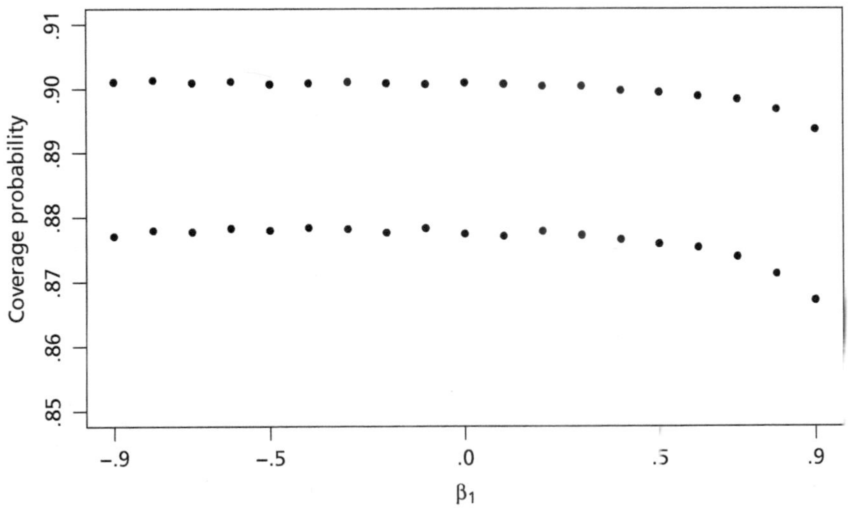

Figure 13.1. Coverage probabilities of one-step-ahead predictions in AR(1) models when $n = 30$. The nominal coverage is .9. Dots are the estimated values, based on 50 000 replicates. The upper dots relate to the Bayesian intervals with uniform prior and the lower ones to standard intervals

Figure 13.2 shows the coverage probabilities of multi-step prediction for AR(1) processes with different prior choices and four different parameter values for β_1. The standard error of coverage probability was less than 7×10^{-4} in all cases. The reference prior seems to be slightly preferred compared to others. Especially when β_1 is near to 1, the reference prior leads to coverage probabilities that are closest to the nominal frequentist coverage probability. Elsewhere the priors give almost the same coverage probabilities.

In order to examine the differences between plug-in method and the Bayesian method with different priors in the case of AR(2) processes, we use nine different parameter combinations for β_1 and β_2. The parameters are defined through the roots of the characteristic equation

$$\beta(r) = 1 - \beta_1 r - \beta_2 r^2 = 0.$$

Let the roots be r_1 and r_2. Then the parameters β_1 and β_2 can be written as

$$\beta_1 = \frac{1}{r_1} + \frac{1}{r_2}, \; \beta_2 = -\frac{1}{r_1 r_2}.$$

The reciprocals of the roots and the corresponding parameters β_1 and β_2 are in Table 13.1.

Figure 13.3 shows the coverage probabilities for each process with the nominal coverage probability of .9 and $n = 30$. The standard error of the coverage probability was less than 5×10^{-4} in all cases. The Bayesian methods perform

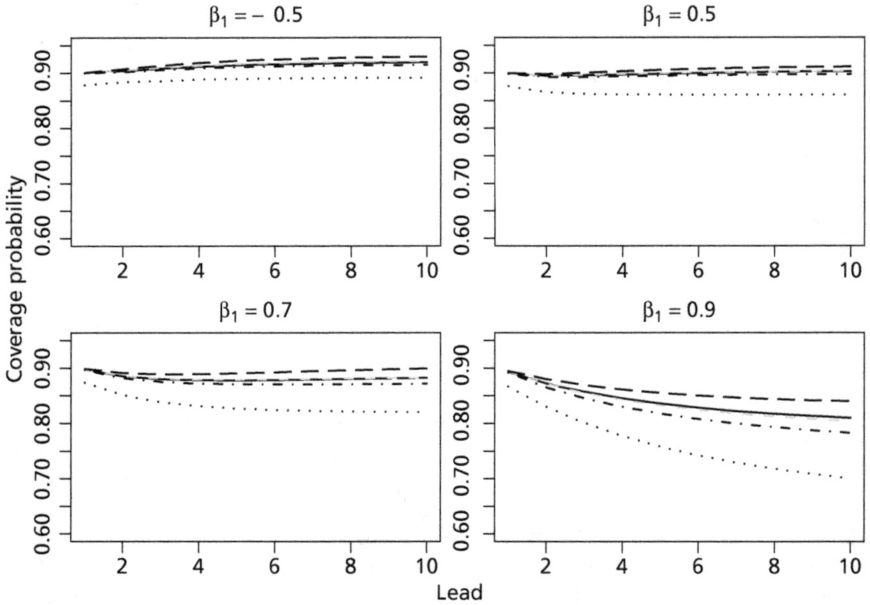

Figure 13.2. Coverage probabilities of multi-step-ahead predictions in AR(1) models with different values of β_1 when $n = 30$. The nominal coverage is .9. The top dashed black line is based on the reference prior and the bottom dotted line represents the standard coverage probabilities. The lines based on the uniform (______), Jeffreys (- - - -), and uniform stationary (. _ . _ . _ .) prior are almost indiscernible apart from the case $\beta_1 = 0.9$

Table 13.1 The AR(2) models used in the simulation experiments

r_1^{-1}	r_2^{-1}	β_1	β_2
0.9	0.5	1.4	−0.45
0.9	−0.5	0.4	0.45
−0.9	0.5	−0.4	0.45
−0.9	−0.5	−1.4	−0.45
0.5	0.5	1.0	−0.25
0.5	−0.5	0	0.25
$0.9\exp(\frac{i}{5})$	$0.9\exp(-\frac{i}{5})$	1.76	−0.81
$0.9\exp(\frac{i\pi}{2})$	$0.9\exp(-\frac{i\pi}{2})$	0	−0.81
$0.9\exp(\frac{i3\pi}{4})$	$0.9\exp(-\frac{i3\pi}{4})$	−1.27	−0.81

much better in all cases. Figures 13.4 and 13.5 show the spectral densities and the associated autocorrelation functions of the corresponding processes. The spectral densities are scaled such that the total density integrates to 1.

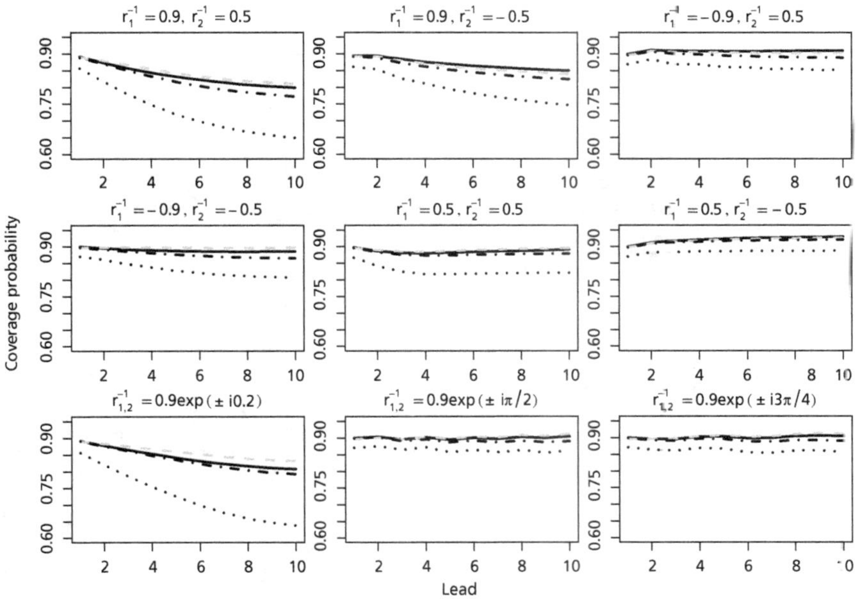

Figure 13.3. The coverage probabilities of the nine AR(2) processes given in Table 13.1, with $n = 30$. The nominal coverage is .9. The black dotted lines (. . . . .) shows the coverage probabilities of the traditional plug-in method. The other lines are related to the Bayesian methods; the solid dark grey line (________) corresponds to the uniform prior, the light grey dash (_ _ _ _) to Jeffreys prior, and the dot-and-dash line (. _ . _ . _ .) to the uniform stationary prior

As in the case of the AR(1) process, the standard plug-in method gives much smaller coverage probabilities than the Bayesian method. We further see that when the mass of the spectral density function in Figure 13.4 is mostly on the smaller frequencies, corresponding to the slowly decaying autocorrelation function in Figure 13.5, the coverage probabilities stay under the nominal coverage probability. Furthermore, Jeffreys prior performs slightly better than the other priors. The plug-in method is very poor, especially for longer forecast horizons.

When the mass of the spectral density function is mostly peaked on the medium or high frequencies and the corresponding autocorrelation functions alternate, the forecast horizon does not seem to affect the coverage probabilities either for the Bayesian or for the plug-in methods. The Bayesian methods are almost exactly equal to the nominal coverage or somewhat exceed it, while the plug-in method is staying clearly below the nominal coverage. In the case where the spectral density function is very flat, the Bayesian methods give coverage probabilities slightly over the nominal level, while the plug-in method stays below. Assuming stationarity with uniform prior seems to provide coverage probabilities little below those of the uniform and Jeffreys priors.

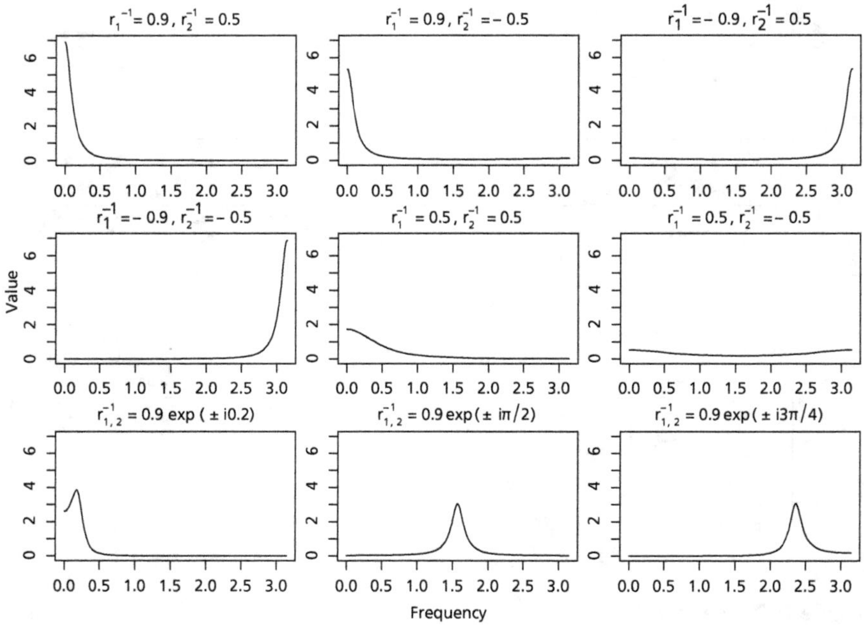

Figure 13.4. The scaled spectral density functions of the nine AR(2) processes from Table 13.1

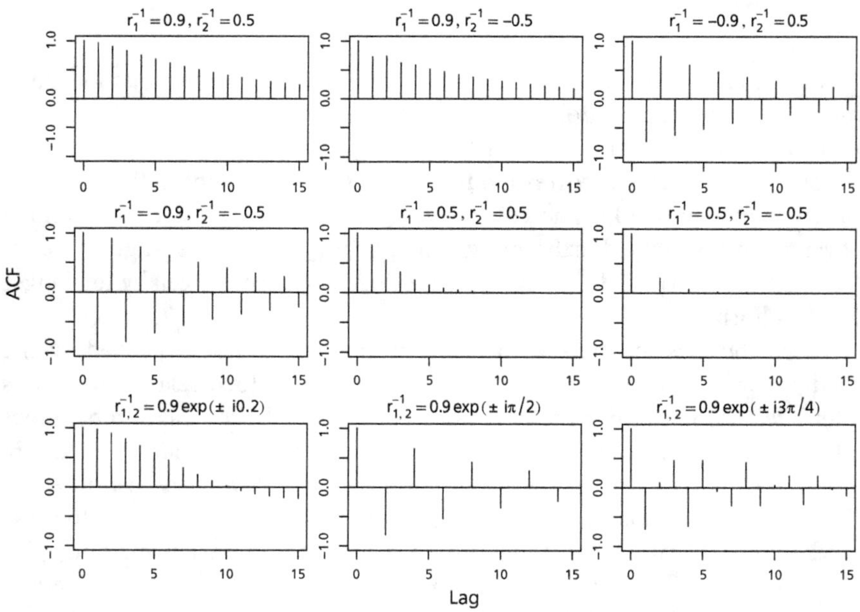

Figure 13.5. The autocorrelation functions of the nine AR(2) processes from Table 13.1

13.6 Annual gross domestic product growth

As an empirical example we applied our method to forecasting the annual gross domestic product (GDP) changes (in percentages) in the United Kingdom (U.K.) and Spain (Figure 13.6). The data is from World Bank's databank (http://databank.worldbank.org). Observations from 1962 to 2001 are used for finding an appropriate model for each series. Then the series is forecast for ten years ahead, 2002–11. In the following we first pretend, as far as possible, that we do not know what has happened in the forecast period. Afterwards when future has been revealed, we try to learn from it.

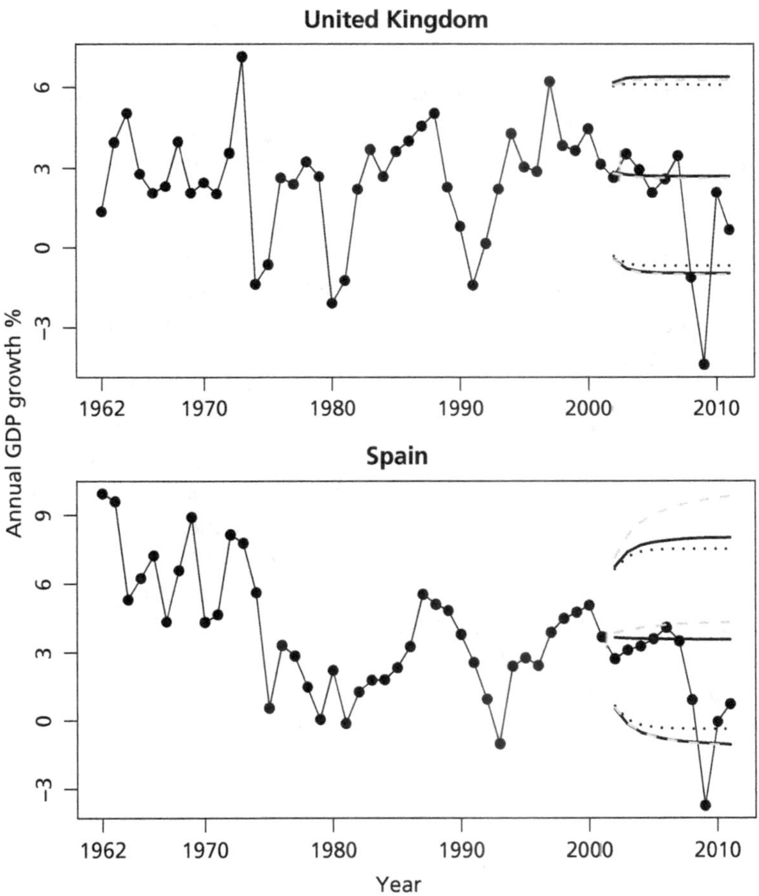

Figure 13.6. The annual GDP 1962–2001 in the U.K. and Spain together with the point predictions, actual values, and 90% prediction intervals for the years 2002–11. Solid grey lines (———) represent intervals and point estimates computed by the Bayesian method with uniform prior, the grey dashed lines (- - - -) correspond to the Bayesian method with Jeffreys prior, and the black dotted lines (........) correspond to the standard plug-in method

Based on the autocorrelation and partial autocorrelation functions, we assume an AR(1) model for both series. The least squares estimates of the autoregressive coefficients are .35 for the U.K. and 0.65 for Spain. The estimates for β_0 are 1.77 (U.K.) and 1.25 (Spain) and for σ are 1.93 (U.K.) and 1.83 (Spain).

The residuals of the Spanish series are well behaved. There is neither apparent autocorrelation nor deviation from normality. As for the residuals from the U.K. series, there appear to be two fairly large negative residuals which stem from the large drops in 1974 and 1980 (i.e., two in forty years). Therefore we should not be surprised to see at least one aberrant value in the forecast period of ten years. No autocorrelation is left in the residuals of the U.K. series. Granted, we should take into account possible outlying values, but it would require models outside autoregressive family, which is beyond the scope of this article. Therefore we also forecast with the estimated model a in this case. In summary, we expect that the Spanish GDP is likely to stay within a 90% interval during the forecast period, with possibly at most one minor violation. But we are more uncertain as to whether this holds for the GDP of the U.K.

Now, of course, we know that the preceding reasoning is too optimistic. In 2001 there was not any indication of the financial crisis starting in 2008. Figure 13.6 shows the actual values, the 90% prediction intervals, and the point predictions for both series. The latter values in the Bayesian forecasting are posterior medians, computed by setting $\alpha = .5$. As expected, the Bayesian method produces wider prediction intervals than the standard plug-in method for both series, and substantially wider for the Spanish series. We expected at most one minor violation in the Spanish series, whereas we see one large drop below the 90% prediction limit. In the U.K. series there is one value at the boundary (in 2008) and one significantly below the boundary (in 2009). The former is plausible, bearing in mind the drops in 1974 and 1980, whereas the latter is exceptionally low. But in both cases the main body of values in 2002–11, especially those in 2010 and 2011, are well within the boundaries. In summary, the values in 2009 are exceptionally low in both series compared to what our model predicts.

There is also one more thing we can learn. Although the differences between the Bayesian intervals are mostly negligible, we see that Jeffreys prior leads to a significantly higher upper limit for the Spanish series. The same is also true for the point forecasts (i.e., medians). A closer examination of the importance weights using Jeffreys prior shows that large values are associated with a large values of the autoregressive coefficient. This seems to be due to the large difference between the first observation and the stationary mean, leading to doubts on the stationarity of the series. Our conclusion is that the uniform prior combined with the conditional likelihood is more preferable to Jeffreys prior in this case.

What should we think of the drops in 2009 occurring in both series? Looking at the series as such and ignoring other information, they seem to be single unlikely events rather than the sign of a structural break. There are at least two arguments for this. First, the values after 2009 (in 2010 and 2011) fit rather well into the main body of the data. Second, we computed prediction bands such that the observed values of the year 2009 are lying exactly on the boundary. The corresponding coverage probabilities are $1 - \alpha$ with $\alpha = 0.0016$ for the U.K. and with $\alpha = .0078$ for the Spanish series. Moreover, the values in 2009 are also extremely aberrant compared with those in the forty year period 1962–2011. In summary, the values in 2009 are highly improbable in the light of the rest of the data.

Nevertheless, the accumulated information on world economy and the crisis concerning the euro countries untill the end of 2012 makes us think that the assumption of a structural break should still be considered seriously. If we are to forecast from 2012 onwards we should carefully explore the methods suggested by Clements and Hendry (1999).

Table 13.2 gives numeric information on the predictions and prediction intervals related to our application under the assumption that our model is correct. First, it contains the actual prediction limits with their Monte Carlo standard errors obtained from the formulas in Section 13.3.3. Second, it shows the coverage probabilities for both models in case they were true. The forecasting horizon is $k = 1, 10$ and the nominal coverage is .9. The two Bayesian

Table 13.2 Coverage probabilities and prediction limits, with standard errors, related to the fitted models for the U.K. and SPA GDP series. The forecast horizon is 2002–2011, and the nominal coverage probability is .9

	Uniform prior		Jeffreys prior		Plug-in	
	$k = 1$	$k = 10$	$k = 1$	$k = 10$	$k = 1$	$k = 10$
United Kingdom						
Coverage	0.900	0.906	0.900	0.907	0.883	0.881
$\hat{b}_{1-\alpha}$	6.164	6.388	6.089	6.269	6.037	6.091
s.e.$(\hat{b}_{1-\alpha})$	0.002	0.003	0.002	0.003	–	–
$\hat{b}_{\alpha}$	−0.448	−0.960	−0.462	−1.003	−0.321	−0.687
s.e.$(\hat{b}_{\alpha})$	0.002	0.003	0.002	0.003	–	–
Spain						
Coverage	0.899	0.892	0.899	0.895	0.881	0.850
$\hat{b}_{1-\alpha}$	6.766	8.014	7.159	9.825	6.647	7.521
s.e.$(\hat{b}_{1-\alpha})$	0.002	0.005	0.003	0.021	–	–
$\hat{b}_{\alpha}$	0.498	−1.066	0.546	−1.097	0.616	−0.395
s.e.$(\hat{b}_{\alpha})$	0.002	0.006	0.003	0.011	–	–

methods give one-step-ahead prediction intervals which are practically correct. When $k = 10$, the coverage probabilities for the U.K. slightly differ from the nominal ones. The standard method gives intervals which have coverage probabilities below the nominal level in both cases. These comparisons concern the chosen models only, not the actual series: the reported coverage probabilities are averages corresponding to AR(1) models with coefficients .35 and .65. This explains why the actual intervals for the Spanish series with uniform and Jeffreys priors are of considerably different widths, although their coverage probabilities are close to each other on average. The deviance may also be related to the aberrant starting value.

We have used the sample size of $N = 100\,000$ in our Monte Carlo simulation, and therefore the standard errors are fairly small. Standard errors should decrease at the rate $1/\sqrt{N}$ that is also confirmed by our experiments. With $N = 1000$ the standard errors are approximately tenfold compared to those in Table 13.2.

We have also compared the probabilities that the future value lies below the lower limit or above the upper limit. They seem to be approximately equal. Thus, our method seems to produce equal tail prediction intervals.

13.7 **Discussion**

We have shown the benefits of the Bayesian approach to prediction interval calculations under autoregressive schemes. Our message to the practitioners is that there are appropriate prior distributions leading to improved prediction intervals compared to those obtained by the common plug-in method. It has turned out that the uniform and Jeffreys priors meet most practical goals for such intervals. Jeffreys prior might have a slight advantage as regards coverage probabilities, although the dependence on the initial observations may have detrimental effects if the starting values are too far from their mean. Our simulation method is straightforward and easy to understand and to implement. An estimate for the Monte Carlo error can also be obtained.

It is plain that when the length of the time series increases, the parameter uncertainty decreases and thus also the coverage probabilities get closer to the nominal level. For example in the AR(1) case, with $n = 100$, $h = 10$, and $\beta_1 = .5$, the coverage probability of the plug-in method roughly achieves the nominal probability. But any guidelines as to when to use the simulation method proposed here, instead of the plug-in method, seems to require massive Monte Carlo experimenting. The reason is that the outcome depends on many variables: the length of the series and forecasting horizon, the order of the model, and the chosen coefficients. We believe that the proposed approach could be taken as a default method, because it is computationally so light and hardly ever gives results worse than the plug-in method.

The prediction intervals are computed under the Gaussian assumption. We have made some simulation experiments under AR(1) and AR(2) models, where the true errors are from Student's t distribution with five degrees of freedom and from Laplace distribution, but we still compute the prediction intervals under Gaussian assumption. In both cases the coverage probabilities are smaller than in the case where the true errors are Gaussian, but usually the difference is about one percentage point or less. Although such a small experiment does not afford any general conclusions, it makes us conjecture that the method is fairly robust against minor deviation from normality.

Although we have handled univariate AR processes only, the method could be extended to general ARIMA and vector autoregressive models, but more careful considerations of prior distributions are then needed. In addition, these complex models can be more sensitive to structural breaks and other issues discussed by Clements and Hendry (1999).

Finally, when dealing with a particular data set, some evidence of the adopted prior is obtained by simulating the estimated model, as we have done in Section 13.6.

ACKNOWLEDGMENTS

The authors thank the editors of the volume and two anonymous referees for their comments and constructive criticism which has led to considerable improvements. We also thank the participants of the Conference in honor of Andrew Harvey's sixty-fifth birthday for the discussion and comments. Jouni Helske would like to thank the Emil Aaltonen Foundation for its financial support.

14 The superiority of the LM test in a class of econometric models where the Wald test performs poorly

Jun Ma and Charles R. Nelson

14.1 Introduction

This paper is concerned with inference in models that can take the form:

$$y_i = \gamma \cdot g(\beta, \vartheta, \mathbf{x}_i) + \varepsilon_i;\ i = 1, \ldots, N. \tag{14.1}$$

Here, the error ε_i is a martingale difference sequence (MDS) with respect to its own history and the history of other variables y_i and $\mathbf{x}_i$, and the function $g(.)$ is continuous and differentiable. Parameters β, of interest for testing, and ϑ, a vector of other parameters, are identified only if $\gamma \neq 0$. This class is broader than it may appear at first, including, in addition to nonlinear regression models, the workhorse ARMA model, the GARCH model, and the unobserved components models. When errors ε_i are i.i.d. $N(0, \sigma^2)$, maximum likelihood estimates are obtained by nonlinear least squares. Standard inference for $\hat{\beta}$ uses the estimated asymptotic standard error to construct a Wald t-test statistic, denoted by $t(\hat{\beta})$, or confidence interval. What models in this class have in common is that standard inference works poorly in finite samples when the identifying parameter γ is small relative to its estimation error. Nelson and Startz (2007)—hereafter NS—show that the estimated standard error for $\hat{\beta}$ is generally smaller than implied by asymptotic theory, essentially because it depends on $\hat{\gamma}$. Further, the distribution of $\hat{\beta}$ will generally be displaced away from the true value. Although these two effects might seem to imply that the size of the test based on $t(\hat{\beta})$ would be excessive, NS show that it may be too large or too small since the numerator and denominator are correlated in finite samples, and the direction depends on specifics of the model and data. Further, while asymptotic theory is valid when the identification condition is met, the limiting distribution of $t(\hat{\beta})$ takes hold slowly as sample size increases in the range relevant for applied work in economics.

In this paper we consider a Lagrange multiplier (LM) test based on linearization of $g(\beta, \vartheta, \mathbf{x}_i)$, in the spirit of Breusch and Pagan (1980) and Harvey (1990). The test may, alternatively, be interpreted as in Fieller's (1954) test for a ratio of regression coefficients in the linearized model, or as a reduced-form test, in the spirit of Anderson and Rubin (1949), both of which have correct size (and are equivalent) if $g(.)$ is linear. While the LM test is asymptotically equivalent to Wald and likelihood ratio (LR) tests, its relation to tests that have correct size in the linear case suggests that the LM test will be close to exact when the linear approximation to $g(\beta, \vartheta, \mathbf{x}_i)$ is close. Section 14.2 presents the "modified LM" test and compares its size and power in small samples with the standard Wald test in the archetypal case of linearity. Section 14.3 studies the small sample performance of the LM test in several nonlinear models that are of considerable interest in practice, namely nonlinear regression, ARMA, the unobserved components model of trend and cycle, and GARCH models. Section 14.4 concludes.

14.2 The modified LM test as an approximation to a test that is exact in finite samples

We begin with the model in (14.1) and linearize $g(\beta, \vartheta, \mathbf{x}_i)$ in β around the null hypothesis, $H_0 : \beta = \beta_0$, obtaining

$$y_i = \gamma \cdot g(\beta_0, \vartheta, x_i) + \lambda \cdot g_\beta(\beta_0, \vartheta, x_i) + e_i;\ \lambda = \gamma \cdot (\beta - \beta_0), \qquad (14.2)$$

where $g_\beta = dg(.)/d\beta$ and e_i includes a remainder. The LM test may be computed in two-steps:

(1) Fit the restricted model $y_i = \gamma \cdot g(\beta_0, \vartheta, \mathbf{x}_i) + e_i$, obtaining the restricted estimates $\widetilde{\gamma}$ and ϑ and restricted residuals: $\widetilde{e}_i = y_i - \widetilde{\gamma} \cdot g(\beta_0, \widetilde{\vartheta}, x_i)$.
(2) Regress $\widetilde{e}_i$ on $g(\beta_0, \widetilde{\vartheta}, x_i)$ and $g_\beta(\beta_0, \widetilde{\vartheta}, x_i)$ to see if the second term adds significant explanatory power.

However, following Harvey (1990), a "modified LM" test can be done in one step by running regression (14.2), noting the significance of the second term in a Wald t-test, using statistic $t(\hat{\lambda})$. He notes that in the case of linear regression the resulting statistic will be essentially the F-statistic for additional variables, suggesting that little is lost in doing one step in the nonlinear case. Indeed, we find this to be the case.

Note too that since the null hypothesis $\beta = \beta_0$ implies $\lambda = 0$, the modified LM test may also be thought of as a reduced-form test. The RF test will have exact size in finite samples when $g(\beta, \vartheta, \mathbf{x}_i)$ is linear, since in that case (14.2) is a classical linear regression. Alternatively, Fieller (1954) noted that an exact test may be obtained for a ratio of regression coefficients from the reduced form, in this case by making use of the relation $\lambda = \gamma \cdot (\beta_0 + \Delta\beta)$ where $\Delta\beta$ is the amount by which β differs from its value under the null hypothesis. Thus the three approaches all use the reduced-form regression obtained from the first order expansion of $g(\beta, \mathbf{x}_i)$ and the resulting Wald t-statistic, $t(\hat{\lambda})$.

Recall that standard inference for $\hat{\beta}$ uses the estimated asymptotic standard error, obtained from the last iteration of nonlinear least squares or by the "delta method," to construct a Wald t-statistic, $t(\hat{\beta})$. Because the modified LM is also based on a Wald t-statistic, $t(\hat{\lambda})$ in the RF, we try to reduce the potential confusion by referring to them as the "standard test" and "LM test", respectively. To demonstrate the problems with the standard test in this setting, and thus the need for an alternative test, it is useful to consider the archetypal model,

$$y_i = \gamma \cdot (x_i + \beta \cdot z_i) + \varepsilon_i. \tag{14.3}$$

An example in practice is the Phillips curve model of Staiger et al. (1997b), where y is the change in inflation, $g = (x_i + \beta)$, x is actual unemployment, and β is the natural rate of unemployment, or NAIRU. The standard test for β may be based on the asymptotic standard error for $\hat{\beta}$ or equivalently the "delta method." Alternatively, to test the hypothesis $\beta = 0$ using the modified LM approach we expand $g(.)$, obtaining the reduced form

$$y_i = \gamma \cdot x_i + \lambda \cdot z_i + \varepsilon_i \tag{14.4}$$

and test the hypothesis $\lambda = \gamma \cdot \beta = 0$.

NS showed that the size of the Wald test using $t(\hat{\beta})$ will depend on the correlation between the reduced-form coefficients as determined by correlation between the regressors. When the regressors are orthogonal the size is too small, in fact for any given sample size $t(\hat{\beta}) < t(\hat{\lambda})$. In contrast, when correlation between the regressors is strong, this inequality reverses; despite the fact that the asymptotic standard error for $\hat{\beta}$ is always downward biased; see Appendix 14.5 for further details. The modified LM test based on $t(\hat{\lambda})$ has exact size since the RF is a classical linear regression.

The identification condition $\gamma \neq 0$ is a maintained hypothesis underlying the asymptotic standard error and t-statistic for $\hat{\beta}$. If it does not hold then the information matrix for the model is singular. However, $t(\hat{\lambda})$ still has an exact t-distribution when identification fails because the reduced form (14.4) is a properly specified classical regression regardless of the value of γ. Of course the data do not contain information about β in that case so the test has no power.

To illustrate the relative performance of the two tests for this archetypal model, we report Monte Carlo experiments where the true $\beta = 0$, the regressors have unit variance and are fixed in repeated samples, and errors are i.i.d. $N(0, 1)$. Estimation is done in EViews™ using the nonlinear regression routine, so the evaluation of $t(\hat{\beta})$ is representative of what would be reported

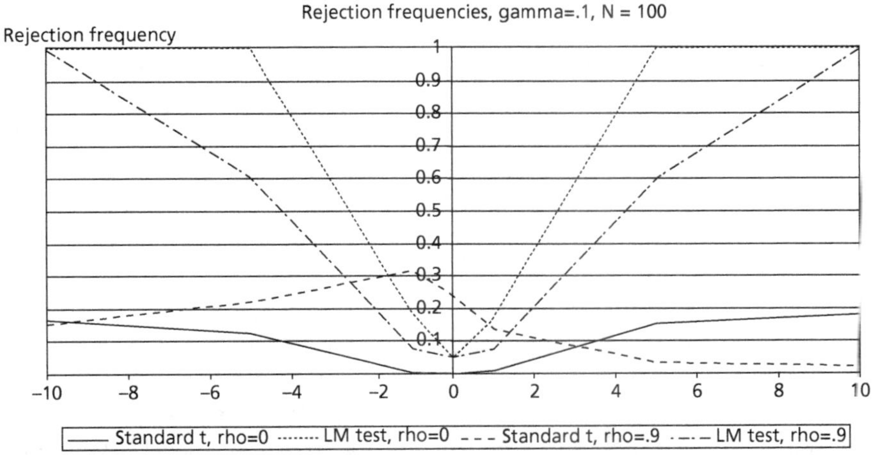

Figure 14.1. Rejection frequencies for tests of $H_0 : \beta = 0$, $N = 100$, $\gamma = .10$

in applied research. The number of replications is 10 000 in all experiments in this paper, and we note that the standard deviation of estimated size is .002 when the true size is .05.

Figure 14.1 explores the response of rejection frequencies for tests based on $t(\hat{\beta})$ and $t(\hat{\lambda})$ to departures of the true value of β from the null hypothesis value of zero. The true value of $\gamma = .10$, and the independent variables are, alternatively, orthogonal and strongly correlated (correlation .9). Since size is not correct for $t(\hat{\beta})$, this response can at best suggest whether the test conveys some information about the null hypothesis. For the case of orthogonal regressors, what we see is that the frequency of rejection using $t(\hat{\beta})$ increases very slowly as a function of the true β. In contrast, the power of the correctly sized test based on $t(\hat{\lambda})$ rises steeply as the true β departs from zero. The corresponding comparison when the independent variables are strongly correlated reveals that rejections by the standard test become *less* frequent as the true value of β departs further from the null of zero, rather than more frequent. In contrast, the power of the correctly sized LM test does increase as expected as the null departs from the true value. Thus we conclude that the Wald test is not only poorly sized but contains little if any information about the null hypothesis. This echoes the findings of Staiger et al. (1997a) who report that in Monte Carlo experiments in the more restrictive NAIRU model "the coverage rate of the delta method confidence interval is poorly controlled over empirically relevant portions of the parameter space." Based on this finding, the confidence intervals for the NAIRU presented in their well-known 1997 paper use Fieller's method, and are wider than intervals using delta method standard errors.

Asymptotic theory does, necessarily, take hold as sample size becomes large, albeit very slowly, as is evident in Table 14.1. Note here that $V_{\hat{\gamma}}$ is the variance

Table 14.1 The effect of sample size N on the distribution of $\hat{\beta}$ and size of $t(\hat{\beta})$ with orthogonal regressors

Sample size N	100	10 000	1 000 000	10 000
True γ	0.01	0.01	0.01	0.1
Asymptotic $\gamma/\sqrt{V_{\hat{\gamma}}}$	0.1	1	10	10
Median $\hat{\beta}$	0.1	0.01	−0.00	0.00
Range (.25, .75)	(−0.95, 1.17)	(−0.64, 0.63)	(−0.07, 0.07)	(−0.07, 0.07)
Size for $t(\hat{\beta})$	0.0001	0.0006	0.043	0.045

of the estimator of γ. The size of the test based on $t(\hat{\beta})$ approaches its nominal level only as the quantity $\gamma/\sqrt{V_{\hat{\gamma}}}$ approaches 10, requiring a sample size as large as 10 000 for $\gamma = .1$, and 1 000 000 for $\gamma = .01$! This provides a metric for "weak identification" and a benchmark for where we expect asymptotic tests to perform satisfactorily. Distribution and testing theory for weakly identified models is an active area of research in econometrics and the focus of a rapidly growing literature. Among others, Andrews and Cheng (2012) study the asymptotic properties of a range of estimators in a class of models in which parameters are unidentified or weakly identified in some parts of the parameter space. Andrews and Mikusheva (2015) investigate robust approaches to testing in the context of a DSGE model.

Since the LM test is not expected to have exact size when $g(.)$ is not linear we use simulation to evaluate its performance relative to the standard t-test in four models of practical interest in Section 14.3. Based on NS we expect that the size of the standard test will depend on the correlation between $g(.)$ and $g_\beta(.)$, which are fixed in the linear case. In the nonlinear case, an estimation routine like Gauss-Newton iterates on β_0 to obtain least squares estimates, the final standard errors and the resulting t-statistic being based on the evaluation of $g(.)$ and $g_\beta(.)$ at $\beta = \hat{\beta}$. Thus the correlation between the "regressors" is not fixed in the general case but rather depends on the provisional value of β at each iteration. As we see below, this co-determination affects the distribution of the point estimate and the size of the standard test, but not the reduced-form test which relies on evaluation of the $g(.)$ and $g_\beta(.)$ under the null hypothesis.

14.3 Small sample performance of the modified LM test in four models

14.3.1 NONLINEAR REGRESSION: A PRODUCTION FUNCTION

Consider the Hicks-neutral Cobb-Douglas production function:

$$y_i = \gamma \cdot x_i^{\beta} + \varepsilon_i;\ \gamma \neq 0, \tag{14.5}$$

Table 14.2 Small sample distribution of $\hat{\beta}$ and test sizes, true $\gamma = .01$, $N = 100$

True β	0	0.1	0.5	0.9
$\rho_{g(\beta),g_\beta(\beta)}$	0.07	0.29	0.77	0.92
Asymptotic $\gamma/\sqrt{V_{\hat{\gamma}}}$	0.10	0.10	0.09	0.11
Median $\hat{\beta}$	−0.04	−0.09	−0.05	0.12
Range (.25, .75)	(−0.53, 0.50)	(−0.59, 0.42)	(−0.56, 0.48)	(−0.41, 0.71)
Size for $t(\hat{\beta})$	0.027	0.037	0.114	0.179
Size for $t(\hat{\lambda})$	0.053	0.054	0.054	0.054

where y_i and x_i are per capita output and capital input respectively, γ is total factor productivity, and β is the share of capital input. The linear reduced-form approximation is

$$y_t = \gamma \cdot x_i^{\beta_0} + \lambda \cdot x_i^{\beta_0} \log(x_i) + e_i, \tag{14.6}$$

where $\lambda = \gamma \cdot (\beta - \beta_0)$. Based on the analysis of the linear model we expect the point estimate $\hat{\beta}$ and the size of the standard t-test to be biased in directions indicted by the correlation between x_i^{β} and $x_i^{\beta} \log(x_i)$, corresponding to $g(\beta, x_i)$ and $g_\beta(\beta, x_i)$. Alternatively, the LM test will be based on the reduced-form coefficient λ, which we expect to have close to correct size. To see if these implications hold, we drew one sample of x_i from the log-normal distribution and paired it with 10 000 paths of standard Normal ε_i, each of sample size 100. Estimation is done in EViews™ using the nonlinear regression routine.

Table 14.2 reports estimation results for values of β in the economically relevant range, zero to .9, with $\gamma = .01$. The second line is the uncentered correlation between the "regressors" in the linear reduced form, x_i^{β} and $x_i^{\beta} \log(x_i)$ which is of interest because analysis of the linear case suggests that bias in $\hat{\beta}$ should vary inversely with it; see Appendix 14.5. Thus when the true value of β is .9, and the regressors are evaluated at .9 the correlation between regressors is .92 and $\hat{\beta}$ should be strongly biased downward. As we see below, $\hat{\beta}$ is biased downward as expected, but the relation of bias to true β is attenuated because Gauss-Newton evaluates the regressors at the estimated value rather than true value of β. Note also that the standard Wald t-test rejects the null too infrequently when the true β is zero but rejects too often when true β is large. While the relation of size distortion to correlation is again in the expected direction based on the linear case, it is attenuated because regressors are evaluated at the downwardly biased point estimates. Finally, we also report the size of the LM test which is close to correct in all cases.

We report in Table 14.3 the corresponding results as γ increases from .01 to 1, fixing the true value of β at .5. As $\gamma/\sqrt{V_{\hat{\gamma}}}$ increases identification becomes better, the asymptotic distribution of $t(\hat{\beta})$ gradually takes hold, and its size

Table 14.3 Small sample distribution of $\hat{\beta}$ and test sizes, $N = 100$, true $\beta = .5$

True γ	0.01	0.1	1
Asymptotic $\gamma/\sqrt{V_{\hat{\gamma}}}$	0.09	0.91	9.10
Median $\hat{\beta}$	−0.05	0.27	0.50
Range (.25, .75)	(−0.56, 0.48)	(−0.26, 0.64)	(0.46, 0.54)
Size for $t(\hat{\beta})$	0.114	0.103	0.052
Size for $t(\hat{\lambda})$	0.054	0.054	0.054

approaches the nominal level of .05. We find that a value of ten for the metric $\gamma/\sqrt{V_{\hat{\gamma}}}$ seems to provide a rough rule of thumb for good identification and correct size. In contrast, the LM test maintains approximately correct size across the range of parameter values.

A special case of interest is $\gamma = 0$ corresponding to failure of the identification condition for β, when the asymptotic theory underling the Wald standard error and t-statistic for $\hat{\beta}$ is not valid. However, the reduced-form test does not depend on that assumption, and we find that its empirical size is .054, close to its nominal size.

Figure 14.2 compares rejection frequencies for the two tests as the true β departs from the fixed null $H_0 : \beta = .5$. The standard test using $t(\hat{\beta})$ starts with a higher level of rejection frequency when the true β is .5, reflecting its size distortion, and rejections decline as true β increases from .5. In contrast,

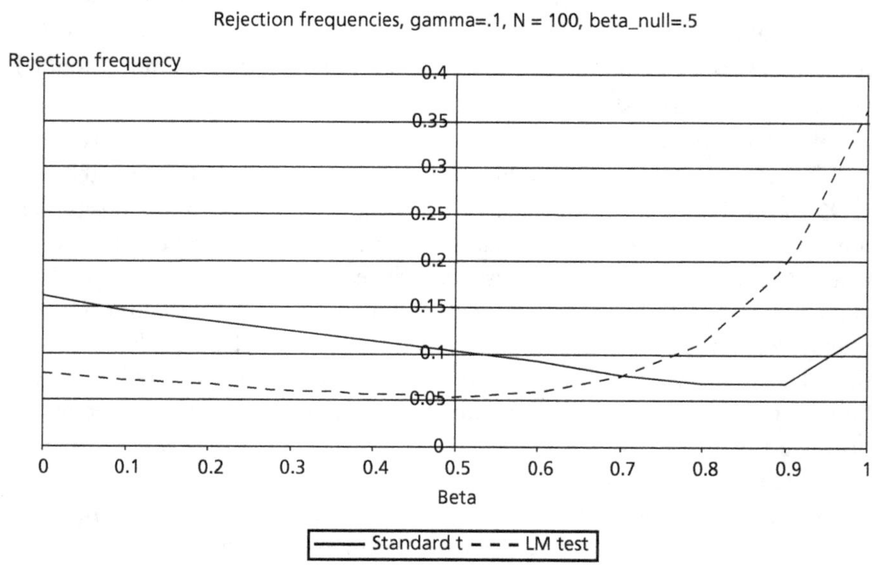

Figure 14.2. Rejection frequencies for tests of $H_0 : \beta = .5, N = 100$, True $\gamma = .1$

the LM test based on $t(\hat{\lambda})$ starts with correct size when the null is true and power increases *monotonically* as true β deviates from the null. Neither test is sensitive to departure from the null in the direction of zero; we surmise that the nonlinearity of the model accounts for this asymmetry.

14.3.2 THE ARMA(1,1) MODEL

Ansley and Newbold (1980), in their study of ARMA model estimation, noted that in the case of near parameter redundancy standard Wald confidence intervals are too narrow. NS also study the failure of Wald inference in that situation. Fortunately, ARMA models fall into the class we are concerned with here. We begin with the workhorse ARMA(1,1) and inference for the moving average coefficient. The results are then extended to the autoregressive coefficient and higher order models. Consider then:

$$\begin{aligned} y_t &= \phi \cdot y_{t-1} + \varepsilon_t - \theta \cdot \varepsilon_{t-1};\ t = 1, \ldots, T \\ \varepsilon_t &\sim \text{i.i.d.}\, N(0, \sigma_\varepsilon^2),\ |\phi| < 1,\ |\theta| < 1. \end{aligned} \tag{14.7}$$

Given invertibility of the moving average, we may express it in the form:

$$y_t = \gamma \cdot g(\theta, \vec{y}_{t-1}) + \varepsilon_t, \tag{14.8}$$

where, $\gamma = (\phi - \theta)$, $g(\theta, \vec{y}_{t-1}) = \sum_{i=1}^{\infty} \theta^{i-1} y_{t-i}$ and $\vec{y}_{t-1} = (y_{t-1}, y_{t-2}, \ldots)$. NS show that when γ is small relative to the sample variation the estimated standard error for either $\hat{\phi}$ or $\hat{\theta}$ is too small and the standard t-test rejects the null too often.

Linearizing the $g(.)$ around the null to obtain the LM test for θ:

$$y_t = \gamma \cdot g(\theta_0, \vec{y}_{t-1}) + \lambda \cdot g_\theta(\theta_0, \vec{y}_{t-1}) + e_t, \tag{14.9}$$

where $g_\theta(\theta, \vec{y}_{t-1}) = \frac{\partial g(\theta, \vec{y}_{t-1})}{\partial \theta} = \sum_{i=2}^{\infty} (i-1) \cdot \theta^{i-2} y_{t-i}$, $\lambda = \gamma \cdot (\theta - \theta_0)$, and e_i incorporates a remainder term. If the null $\theta = \theta_0$ is correct, the second term in regression (14.9) should not be significant. In practice, to evaluate the regressors $[g(\theta_0, \vec{y}_{t-1}), g_\theta(\theta_0, \vec{y}_{t-1})]$, we set y_t at its unconditional mean for all $t \leq 0$. Once again we are interested in comparing the small sample performance of the usual Wald test based on asymptotic theory, $t(\hat{\theta})$, with the LM best based on reduced-form equation (14.9), using $t_{\hat{\lambda}}$; Monte Carlo simulations were done in EViews™.

Table 14.4 explores the effect of γ for true $\theta = 0$ with a sample size $T = 1000$. When γ is small relative to sample variation, as indicated by a small value of the metric $\gamma / \sqrt{V_{\hat{\gamma}}}$, the Wald t-test rejects the null too often. As γ gets larger and the key metric $\gamma / \sqrt{V_{\hat{\gamma}}}$ approaches ten, asymptotic theory gradually

Table 14.4 Effect of γ on inference for ARMA(1,1), true $\theta = 0$, $T = 1000$

True $\gamma(=\phi)$	0.01	0.1	0.2	0.3
Asymptotic $\gamma/\sqrt{V_{\hat{\gamma}}}$	0.32	3.16	6.32	9.49
Median $\hat{\theta}$	−.02	−.01	−.00	−.00
Range (.25, .75)	(−.65, .64)	(−.26, .24)	(−.11, .11)	(−.07, .07)
Size for $t(\hat{\theta})$	0.4585	0.2237	0.1051	0.0734
Size for $t(\hat{\lambda})$	0.0506	0.0518	0.0526	0.0522

takes hold and the size of conventional t-test based on $t(\hat{\theta})$ becomes closer to its nominal level, 0.05. The fact that the sampling distribution of $t(\hat{\theta})$ depends on the nuisance parameter γ implies that the test is not pivotal. Note that the LM test based on $t(\hat{\lambda})$ in this case is equivalent to testing the second lag in an AR(2) regression, which is approximately the Box-Ljung Q-test with one lag for the residuals from an AR(1) regression. The estimated size of the LM test is correct within sampling error.

One may wonder how the LM test performs when true γ is zero, corresponding to the case of identification failure. The empirical size of the LM test in this case is 0.0509, close to correct.

We note that the median and inter-quartile range of $\hat{\theta}$ suggest that the sampling distribution of $\hat{\theta}$ is centered on zero. However, the histogram of $\hat{\theta}$ in Figure 14.3 for the case $\gamma = .01$ shows that the estimates tend to be concentrated close to boundaries of the parameter space, reflecting the well-known "pile-up" effect in ARMA models; see Hannan (1982) and Hauser et al. (1999). Figure 14.4 plots the uncentered correlation ρ between the "regressors" $g(\theta, \vec{y}_{t-1})$ and $g_\theta(\theta, \vec{y}_{t-1})$ as a function of provisional estimate θ. At $\theta = 0$ the correlation is zero but becomes larger in absolute value as θ moves away from zero in either direction and toward the boundaries where $\hat{\theta}$ occurs with greatest frequency. The excessive size of the test based on $t_{\hat{\theta}}$ reflects this strong correlation between the reduced-form regressors when $\hat{\theta}$ falls far from zero and moves toward boundaries, not simply the too-small standard error as surmised by NS. The relative success of the LM test comes from the fact that it evaluates the test statistic under the null hypothesis $\theta_0 = 0$ instead of at $\hat{\theta}$.

Table 14.5 explores the effect of increasing sample size when true $\gamma = .01$. Asymptotic theory does take hold, but the conventional t-test approaches correct size very slowly (requiring a sample size as large as 10 000 for $\gamma = .1$!). In contrast, the reduced-form test consistently has correct size within sampling error.

Often it is the AR root ϕ that is of a great economic interest since it measures persistence. For instance, in an influential work Bansal and Yaron (2004) show that if consumption growth g_t follows an ARMA(1,1) process a large value of ϕ

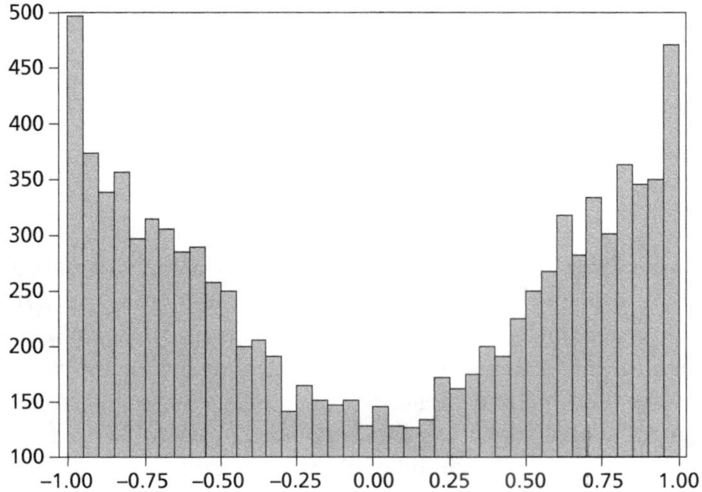

Figure 14.3. Histogram $\hat{\theta}$ in the Monte Carlo experiments. True $\gamma = .01, \theta = 0, T = 1000$

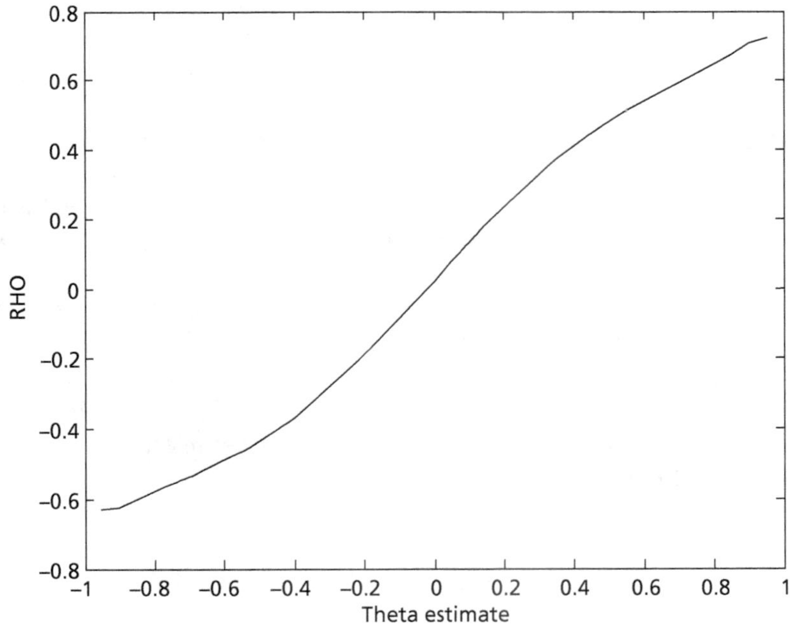

Figure 14.4. Computer uncentered correlation between $g(\theta, y_{t-1})$ and $g_\theta(\theta, y_{t-1})$ based on one sample draw. True $\gamma = .01, \theta = 0, T = 1000$

Table 14.5 Sample size and inference in the ARMA(1, 1), True $\theta = 0$

Sample size	100	1000	10 000	10 000
True $\gamma(=\phi)$	0.01	0.01	0.01	0.1
Asymptotic $\gamma/\sqrt{V_{\hat{\gamma}}}$	0.1	0.32	1	10
Median $\hat{\theta}$	–0.04	–0.02	–0.02	–0.00
Range (.25, .75)	(–0.69, 0.67)	(–0.65, 0.64)	(–0.58, 0.55)	(–0.07, 0.07)
Size for $t(\hat{\theta})$	0.483	0.458	0.399	0.066
Size for $t(\hat{\lambda})$	0.051	0.051	0.049	0.048

implying a very persistent consumption growth expectation, namely the long-run risk, may explain the infamous equity premium puzzle. Ma (2013) finds that the $\hat{\gamma}$ in the estimated ARMA(1,1) is small relative to its sampling variance and explores the implications of possible test size distortion in the conventional test as well as valid inference following the strategy suggested in this paper. In another influential paper, van Binsbergen and Koijen (2010) propose the AR(1) process to model the conditional expected returns as a latent variable resulting in an ARMA(1,1) process for the realized returns. In their estimation results, the persistence parameter is estimated to be very high with a rather small standard error. However, their estimated variance of the shock to the conditional expected returns process is very small, resulting in very close AR and MA roots, which raises a concern of the validity of conventional inferences. Ma and Wohar (2014) extend the LM test in current work to reexamine the issue of van Binsbergen and Koijen and find that the corrected confidence interval for the persistence parameter of the expected returns is much wider.

In Appendix 14.6, we offer details about how to obtain the reduced-form and LM test for AR coefficient ϕ. For the case $\gamma = 0.1, \phi = 0$, and $T = 100$ the rejection frequency of the LM test is 0.046 in contrast to 0.423 for the standard t-test.

The reduced-form test can also be generalized to address any higher order ARMA model, and the extensions are illustrated in Appendix 14.7. For the ARMA(2,2) model with parameter values $\phi_1 = 0.01, \phi_2 = 0.01, \theta_1 = 0, \theta_2 = 0$, and $T = 100$ we find that the standard t-test for $\hat{\theta}_1$ and $\hat{\theta}_2$ has empirical sizes of .571 and .698. In contrast the reduced-form test gives rejection frequencies of .049 and .049 respectively.

14.3.3 THE UNOBSERVED COMPONENT MODEL FOR DECOMPOSING TREND AND CYCLE

The unobserved component model (hereafter UC) of Harvey (1985) and Clark (1987) is widely used to decompose the log of real gross domestic product (GDP) into trend and cycle. Thus

$$y_t = \tau_t + c_t, \tag{14.10}$$

where trend is assumed to be a random walk with drift

$$\tau_t = \tau_{t-1} + \mu + \eta_t,\ \eta_t \sim \text{i.i.d.}\, N(0, \sigma_\eta^2), \tag{14.11}$$

and cycle has a stationary AR(1) representation:

$$c_t = \phi \cdot c_{t-1} + \sqrt{1-\phi^2} \cdot \varepsilon_t,\ \varepsilon_t \sim \text{i.i.d.}\, N(0, \sigma_\varepsilon^2). \tag{14.12}$$

Note that we focus on the AR(1) cycle to simplify the illustration. And with this particular specification, the size of the cycle ($\text{Var}(c_t) = \sigma_\varepsilon^2$) is invariant to changes in ϕ and the latter only controls the persistence of the cycle.

The UC model is estimated by maximizing the likelihood computed using the Kalman filter under the assumption that trend and cycle shocks are uncorrelated. In practice the largest AR root is estimated to be close to unity, implying that the cycle is very persistent, and the trend variance is estimated to be very small, implying that the trend is very smooth. The question we wish to investigate here is whether standard inference about cycle persistence may be spurious and whether the approach in this paper can provide a correctly sized test.

Following Morley et al. (2003), we note that the univariate representation of this particular UC model is ARMA(1,1) with parameters implied by the equality (dropping the constant for illustration purpose)

$$(1-\phi L)\Delta y_t = (1-\phi L)\eta_t + \sqrt{1-\phi^2} \cdot (\varepsilon_t - \varepsilon_{t-1}) = u_t - \theta u_{t-1}, \tag{14.13}$$

where $u_t \sim$ i.i.d. $N(0, \sigma_u^2)$. Thus the AR coefficient of the ARMA(1,1) is simply ϕ, while the MA parameter θ is identified, under the restriction $\sigma_{\eta,\varepsilon} = 0$, by matching the zero and first-order autocovariances of the equivalent MA parts:

$$\psi_0 = (1+\phi^2)\sigma_\eta^2 + 2(1-\phi^2)\sigma_\varepsilon^2 = (1+\theta^2)\sigma_u^2 \tag{14.14}$$

$$\psi_1 = -\phi\sigma_\eta^2 - (1-\phi^2)\sigma_\varepsilon^2 = -\theta\sigma_u^2. \tag{14.15}$$

We may then solve for the unique θ that permits invertibility:

$$\theta = \frac{(1+\phi^2) + 2(1-\phi^2)\left(\sigma_\varepsilon^2/\sigma_\eta^2\right) - \sqrt{\left[(1+\phi)^2 + 4(1-\phi^2)\left(\sigma_\varepsilon^2/\sigma_\eta^2\right)\right] \cdot (1-\phi)^2}}{2\left[\phi + (1-\phi^2)\left(\sigma_\varepsilon^2/\sigma_\eta^2\right)\right]}. \tag{14.16}$$

It is straightforward to show that θ becomes arbitrarily close to ϕ as $\sigma_\varepsilon/\sigma_\eta$ approaches zero. By an analogy to the ARMA(1,1) model, the estimated standard error for $\hat{\phi}$ may be too small when $\sigma_\varepsilon/\sigma_\eta$ is small relative to sampling variation, and a standard t-test may be incorrectly sized.

To visualize spurious inference in this case, we implement a Monte Carlo simulation experiment. Data is generated from the UC model given by (14.10)–(14.12) with true parameter values $\mu = .8$, $\phi = 0$, $\sigma_\eta^2 = .95$, $\sigma_\varepsilon^2 = .05$, corresponding roughly to quarterly U.S. GDP if almost all the variation were due to trend while the cycle is small with no persistence at all. Estimation is done in MATLAB 6.1 and the routine is available on request. Sample size T is 200, approximately what is encountered in practice for postwar data. To avoid local maxima, various starting values are used.

The standard t-test for $\hat{\phi}$ indeed rejects the null much too often; size is .481. This is partly because the standard error for $\hat{\phi}$ is underestimated; the median is .2852 compared with its true value of 1.4815. Furthermore, $\hat{\phi}$ is upward biased as illustrated in Figure 14.5, its median being .58. Many $\hat{\phi}$s occur close to the positive boundary. This is consistent with the finding of Nelson (1988) that a UC model with persistent cycle variation fits better than the true model even when all variation is due to stochastic trend; the case where $\sigma_\varepsilon^2 = 0$.

At the same time, the cycle variance estimate $\hat{\sigma}_\varepsilon^2$ is upward biased, having a median of .24, while the trend innovation variance estimate $\hat{\sigma}_\eta^2$ is instead

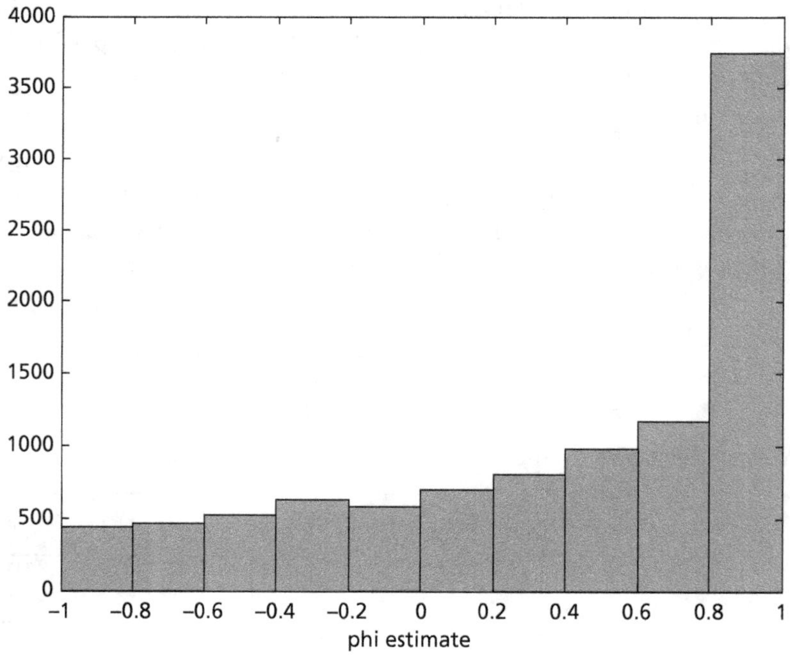

Figure 14.5. Plot of $\hat{\phi}$, the estimated persistence of the cycle component, in the Monte Carlo experiment, with the following true parameters: $\mu = .08$, the drift term in the permanent component; $\phi = 0$, the persistence of the cycle component; $\sigma_g^2 = .05$, the variance of the cycle component and in this case coincides with the variance of the cycle innovation since $\phi = 0$; and $\sigma_n^2 = .95$, the variance of the trend innovation shock

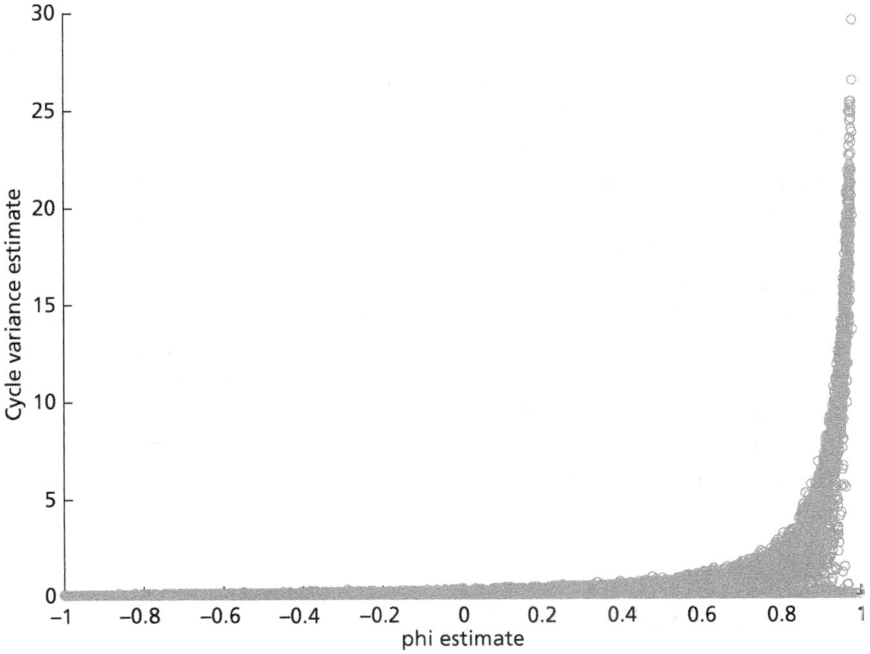

Figure 14.6. Scatter plot of $\hat{\phi}$, the estimated persistence of the cycle component, and $\hat{\sigma}_\varepsilon^2$, the estimated variance of the cycle component in the unobserved components model; in the Monte Carlo experiment the true parameters are $\mu = 0.8$, the drift term in the permanent component; $\phi = 0$, the persistence of the cycle component; $\sigma_\varepsilon^2 = .05$, the variance of the cycle component and in this case coincides with the variance of the cycle innovation since $\phi = 0$; and $\sigma_n^2 = .95$, the variance of the trend innovation shock

downward biased, with a median of .73. What is the underlying driving force for the upward bias of $\hat{\phi}$ and $\hat{\sigma}_\varepsilon^2$ and the downward bias of $\hat{\sigma}_\eta^2$? The scatter plot in Figure 14.6 shows that there is a strong positive comovement between $\hat{\phi}$ and $\hat{\sigma}_\varepsilon^2$, thus persistence in the estimated cycle tends to occur in samples that also show large variance in the cycle. This is driven by the necessity that the model must account for the small amount of serial correlation in the data generating process for Δy_t. Setting the autocovariance at lag one equal to the true value for the sake of illustration, one obtains the restriction $-(1-\phi)^2\sigma_\varepsilon^2 = -.05$. One solution is the combination of true values, $\phi = 0$; $\sigma_\varepsilon^2 = .05$, but another is $\phi = .9$; $\sigma_\varepsilon^2 = 5$. Thus $\hat{\sigma}_\varepsilon^2$ will be far greater than its true value when $\hat{\phi}$ is close to its positive boundary, implying a dominating persistent cycle that tends to mimic the true underlying stochastic trend. Finally, Figures 14.5 and 14.6 show that large negative values of $\hat{\phi}$ are possible but infrequent because positive variances place restrictions on the parameter space.

In light of the connection between UC model and ARMA model we implement the LM test in the following steps: first impose the null $\phi = \phi_0$ and estimate all other parameters in the UC model; secondly, impute from (14.16) the restricted estimate $\widetilde{\theta}$ and $\widetilde{u}_t$ in the reduced-form ARMA(1,1) model; lastly, compute the reduced form t-test statistic by following the strategy in Appendix 14.6. Using the same set of simulated data as above for true parameter values $\mu = 0.8, \phi = 0, \sigma_\eta^2 = 0.95$, and $\sigma_\varepsilon^2 = 0.05$, the rejection frequency for the LM test for ϕ is 0.054.

One may also be interested in the case where all variation is due to stochastic trend, i.e., $\sigma_\varepsilon^2 = 0$. For this case, the identification for ϕ fails and the standard t-test is not well-defined. However, the reduced-form LM test works well and gives estimated size 0.0581 in the Monte Carlo with true parameter values $\mu = 0.8, \phi = 0, \sigma_\eta^2 = 1$, and $\sigma_\varepsilon^2 = 0$.

The reduced-form test can also be generalized to address a UC model with higher AR orders in the cycle by following the strategy discussed in Appendix 14.7.

14.3.4 THE GARCH(1,1) MODEL

The GARCH model developed by Bollerslev (1986) is perhaps one of the most popular approaches in capturing the time-varying volatility for time series data. The archetypal GARCH(1,1) may be written:

$$\varepsilon_t = \sqrt{h_t} \cdot \xi_t,\ \xi_t \sim \text{i.i.d. } N(0,1) \tag{14.17}$$

$$h_t = \omega + \alpha \cdot \varepsilon_{t-1}^2 + \beta \cdot h_{t-1}. \tag{14.18}$$

To see why GARCH is among the models we are concerned with, write out its ARMA representation and make an analogy to the ARMA(1,1) model:

$$\varepsilon_t^2 = \omega + (\alpha + \beta) \cdot \varepsilon_{t-1}^2 + w_t - \beta \cdot w_{t-1}. \tag{14.19}$$

The innovation $w_t = \varepsilon_t^2 - h_t = h_t(\xi_t^2 - 1)$ is a martingale difference sequence (MDS) with time-varying variance.

Define the persistence of volatility by $\rho = \alpha + \beta$. Since ρ and β correspond to the AR and MA roots, respectively, α controls the information about β and ρ. Ma et al. (2007) show that when α is small relative to its sampling variation, the standard error for $\hat{\beta}$ or $\hat{\rho}$ is underestimated and the standard t-test rejects the null too often, implying a significant and persistent GARCH effect even when there is none.

The persistence of volatility is of particular economic interest. For example, Bansal and Yaron (2004) show that a large value of ρ for consumption volatility, interpreted as long-run risk, may help to resolve the equity premium puzzle. Ma (2013) investigates the inference issue of ρ when the model is weakly identified and its implication to the equity premium puzzle in the long-run risk model.

It is also useful to write out the state-space representation of the GARCH model to see the root cause of the weak identification and its likely implications for statistical inference. Rewrite (14.19) to give the state equation that describes the dynamic evolution of the unobserved volatility:

$$h_t = \omega + \rho \cdot h_{t-1} + \alpha \cdot w_{t-1}, \tag{14.20}$$

where $w_{t-1} = \varepsilon_{t-1}^2 - h_{t-1}$. The measurement equation is simply:

$$\varepsilon_t^2 = h_t + w_t. \tag{14.21}$$

This is a particular state-space model since the shock in the state equation $(\alpha \cdot w_{t-1})$ is one period lag of the shock in the measurement equation (w_t). The essence of the state-space model is to filter out the noise (w_t) in order to extract the unobservable signal sequence (h_t), and the parameter α in the GARCH model measures the size of the signal shock relative to that of the noise shock. When the signal is small relative to noise, the uncertainty of the signal dynamics ought to be fairly large and a correct test must produce results consistent with this intuition.

To extend the LM test to test the null $\rho = \rho_0$, rewrite (14.18) to obtain:

$$h_t = \frac{\omega}{1-\rho} + \alpha \cdot g(\rho, \vec{w}_{t-1}), \tag{14.22}$$

where $g(\rho, \vec{w}_t) = \sum_{i=1}^{\infty} \rho^{i-1} w_{t-i}$ and $\vec{w}_{t-1} = (w_{t-1}, w_{t-2}, \ldots)$. Take a linear expansion of $g(.)$ around the null:

$$h_t = \frac{\omega}{1-\rho} + \alpha \cdot g(\rho_0, \vec{w}_{t-1}) + \lambda \cdot g_\rho(\rho_0, \vec{w}_{t-1}), \tag{14.23}$$

where $\lambda = \alpha \cdot (\rho - \rho_0)$ and $g_\rho(\rho_0, \vec{w}_{t-1}) = \sum_{i=2}^{\infty} (i-1) \cdot \rho_0^{i-2} w_{t-i}$. The LM test then is the t-statistic for λ. To make the test feasible one needs to have a consistent estimate for w_t which can be readily obtained through estimation under the restriction of null.

Table 14.6 presents the comparison of the LM test based on $t(\hat{\lambda})$ and the standard t-test based on $t(\hat{\rho})$ for a range of values of α where $\beta = 0$ and sample size $T = 1000$ (the MATLAB 6.1 code is available on request). When the key metric $\gamma/\sqrt{V_{\hat{\gamma}}}$ is small, the standard t-test rejects the null too often. The LM test however has consistently better size.

In the case $\alpha = 0$, identification fails and the standard t-test does not have the usual asymptotic distribution. The LM test for $\hat{\rho}$, however, is still valid and has estimated size of .074 for the case with true $\beta = 0$ and $T = 1000$.

We apply the LM test to monthly S&P 500 index returns from the DRI Database for the period January 1947 to September 1984 that was studied in Bollerslev (1987). The GARCH estimates with Bollerslev and Wooldridge (1992) robust standard errors, after accounting for the "Working (1960) effect," are:

$$\hat{\omega} = 0.16 \cdot 10^{-3} (0.14 \cdot 10^{-3}),\ \hat{\alpha} = 0.077(0.048),\ \hat{\beta} = 0.773(0.169).$$

Table 14.6 LM test and standard t-tests for GARCH(1,1): True $\beta = 0$, $T = 1000$

True $\gamma(=\alpha)$	0.01	0.05	0.1	0.2
True $\rho(=\alpha+\beta)$	0.01	0.05	0.1	0.2
Asymptotic $\gamma/\sqrt{V_{\hat{\gamma}}}$	0.32	1.59	3.19	6.60
Median $\hat{\rho}$	0.34	0.10	0.11	0.20
Range (.25, .75)	(−0.27, 0.75)	(−0.24, 0.54)	(−0.11, 0.33)	(0.09, 0.30)
Size of $t(\hat{\rho})$	0.470	0.338	0.188	0.078
Size for $t(\hat{\lambda})$	0.072	0.069	0.071	0.070

The standard t-test implies a significant GARCH effect as indicated by the 95% confidence interval for β: $(0.44, 1)$. The GARCH effect also appears fairly persistent since $\hat{\rho} = 0.85$ and its standard error computed using the delta method is 0.13. However, the small value of $\hat{\alpha}$ (the upper bound for α at a 95% significance level is 0.173) relative to the sample size $T = 453$, raises concern about the possibility of spurious inference for β. To obtain a confidence interval for β based on the LM test, we numerically invert $t(\hat{\lambda})$ over a grid of positive values for ρ_0, compute the corresponding $t(\hat{\lambda})$, and plot the latter against the former in Figure 14.7. Note that since the volatility is bound to be positive it is the positive region of the volatility persistence that is most relevant and is what we focus on. Based on Figure 14.7, the resulting 95% confidence interval for ρ appears to cover almost the entire admissible parameter space. That this can happen in practice should not surprise us, in light of Dufour's theorem (1997), which says that the probability that a valid confidence interval covers the entire parameter space must be greater than zero if identification is weak enough.

Finally we note that we have assumed a normal distribution for the error term ξ_t in (14.17) throughout our exercises to focus on the issue of weak identification. Previous studies such as Dueker (1997) have investigated the effect of excess kurtosis of the data on the GARCH estimation and found that the leptokurtic characteristic of the data tends to provide another source for the upward bias of the GARCH effect.

14.4 Summary and conclusions

This paper presents a modified LM test as an alternative to the standard Wald t-test in a class of models where the latter is associated with spurious inference when identification is weak. There are models that have a representation of the form $y = \gamma \cdot g(\beta, x) + \varepsilon$, where β is the parameter of interest and the

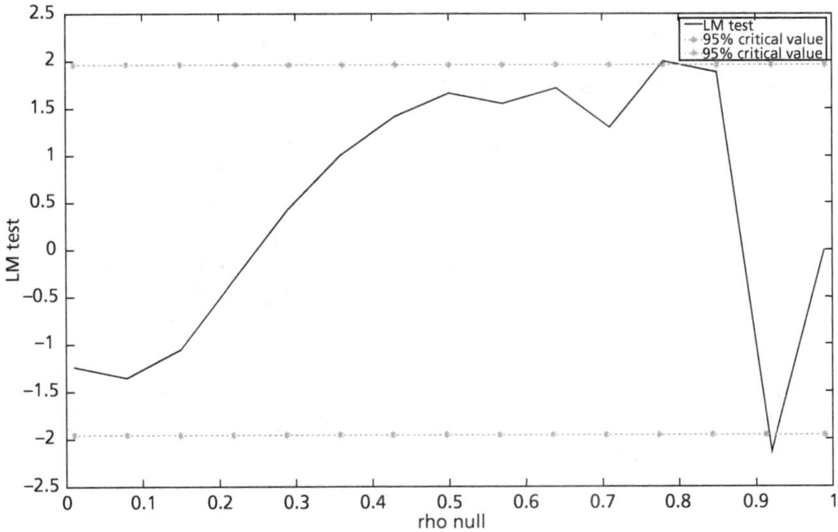

Figure 14.7. The 95% confidence interval for $\hat{\rho}$ based on the LM test for the monthly S&P 500 stock return data

amount of information about that parameter available from the data depends on the unknown identifying parameter γ. This class includes not only the obvious nonlinear regression model but also the workhorse ARMA model and, by extension, GARCH and unobserved components models. NS showed that inference is problematic because the standard error for $\hat{\beta}$ depends on $\hat{\gamma}$. While the estimated standard error is downward biased in finite samples in the class of models that satisfy the zero-information-limit condition, the t-statistic can be either too large or too small depending on the data generating process. In this paper we show that small sample inference in this class is usefully studied by working with the approximation:

$$g(\beta, x) \approx g(\beta_0, x) + (\beta - \beta_0) \cdot g_\beta(\beta_0, x)$$

and the corresponding reduced-form regression:

$$y_i = \gamma \cdot g(\beta_0, x_i) + \lambda \cdot g_\beta(\beta_0, x_i) + e_i; \text{ where } \lambda = \gamma \cdot (\beta - \beta_0).$$

A t-test that exploits the fact that the null hypothesis $\beta = \beta_0$ implies $\lambda = 0$ and thus is exact when $g(.)$ is linear. We show that it has nearly correct size when the reduced-form model is only an approximation. This test may be interpreted as a modified LM test in the spirit of Harvey (1990). The paper illustrates the superior performance of the LM test in the presence of weak identification with examples from nonlinear regression, ARMA, GARCH, and unobserved components models.

APPENDICES

14.5 The linear regression case

Consider the regression model (14.2). The parameter of interest β may be estimated as the ratio of two parameters estimates in the reduced form (14.3) $\hat{\beta} = \hat{\lambda}/\hat{\gamma}$. Although the moments of the ratio of normal random variables do not in general exist, see Fieller (1932) and Hinkley (1969), we can nevertheless draw some conclusions about the sampling distribution of $\hat{\beta}$. Noting that $\hat{\lambda}$ and $\hat{\gamma}$ are jointly normal across samples, the conditional mean of the former given the latter implies:

$$\hat{\lambda} = \alpha + \kappa \cdot \hat{\gamma} + \nu \tag{14.24}$$

where α and κ are parameters and ν is normal and uncorrelated with $\hat{\gamma}$ by construction. To simplify exposition we focus on the case $\beta = 0$ and standardized regressors with sample correlation ρ. It is straightforward to show that $\alpha = \rho \cdot \gamma$, $\kappa = -\rho$, and the variance of ν is σ^2/N. Making these substitutions and dividing by $\hat{\gamma}$ one obtains:

$$\hat{\beta} = -\rho + \rho \cdot \left(\frac{\gamma}{\hat{\gamma}}\right) + \frac{\nu}{\hat{\gamma}}. \tag{14.25}$$

Consider now how the distribution of $\hat{\beta}$ is affected by γ, which controls the amount of information in the data about β, and by correlation between regressors ρ, for given sample size. A larger value of γ means that the ratio $\gamma/\hat{\gamma}$ tends to be closer to unity, since the standard deviation of $\hat{\gamma}$, given by $\sqrt{\sigma^2 \cdot N^{-1}/(1-\rho^2)}$, is not a function of γ. The second term in (14.25) will tend toward ρ, canceling out the first term, and the third term will be relatively small, so the sampling distribution of $\hat{\beta}$ will be located more tightly around its true value, zero. However, a smaller value of γ means that $\gamma/\hat{\gamma}$ will typically be small, thus locating the sampling distribution of $\hat{\beta}$ around $-\rho$ but with greater dispersion since the third term will typically be large. Shifting now to the effect of ρ, stronger correlation will increase sampling variation in $\hat{\gamma}$, so the second and third terms will tend to be small, concentrating the distribution of $\hat{\beta}$ around $-\rho$. (In this paper we refer to these shifts of central tendency away from the true value as "bias" for the sake of brevity.)

Turning now to hypothesis testing, the asymptotic variance of $\hat{\beta}$ derived either from the information matrix for (14.2) under maximum likelihood, or using the "delta method" for indirect least squares, is given by:

$$V_{\hat{\beta}} = \frac{1}{\gamma^2} \cdot \frac{\sigma^2}{N} \cdot \frac{m_{xx} + 2\beta \cdot m_{xz} + \beta^2 \cdot m_{zz}}{m_{xx} \cdot m_{zz} - m_{xz}^2} \tag{14.26}$$

where "m" denotes the raw sample second moment of the subscripted variables. In practice the parameters are unknown and are replaced in standard software packages by the point estimates. Thus the reported t-statistic for $\hat{\beta}$ is:

$$t(\hat{\beta})^2 = (\hat{\beta} - \beta_0)^2 \cdot \left[\hat{\gamma}^2 \cdot \frac{N}{\hat{\sigma}^2} \cdot \frac{m_{xx} \cdot m_{zz} - m_{xz}^2}{m_{xx} + 2\hat{\beta} \cdot m_{xz} + \hat{\beta}^2 \cdot m_{zz}}\right] \tag{14.27}$$

where the null hypothesis is $\beta = \beta_0$. We confine our attention to the case $\beta_0 = 0$, noting that a nonzero value of β_0 simply corresponds to a transformed model. In the standardized regressors case the t-statistic for $\hat{\beta}$ is given by:

$$t(\hat{\beta})^2 = \frac{\hat{\lambda}^2}{\hat{\sigma}^2} \cdot N \cdot (1-\rho^2) \cdot \frac{1}{1+2\hat{\beta}\cdot\rho+\hat{\beta}^2} = t_{\lambda}^2 \cdot \frac{1}{1+2\hat{\beta}\cdot\rho+\hat{\beta}^2}. \quad (14.28)$$

Since the reduced form is a classical linear regression, a test based on $t(\hat{\lambda})$ has correct size and so provides an alternative test of the null hypothesis $\beta = 0$ with correct size. Indeed this is the exact test of Fieller (1954) for a ratio of regression coefficients. As noted by NS, if the two explanatory variables are orthogonal, then in any given sample $t(\hat{\beta})^2 < t(\hat{\lambda})^2$ since the last term in (14.28) must be less than one. In contrast, the effect of strong correlation between x and z, working through the concentration of $\hat{\beta}$ around the value $-\rho$, is to drive $(1+2\hat{\beta}\cdot\rho+\hat{\beta}^2)$ close to zero, making $t(\hat{\beta})$ arbitrarily larger than $t(\hat{\lambda})$. Thus, whether test size is too large or too small depends on the correlation between the regressors, strong correlation of either sign producing an over-sized t-test.

14.6 The ARMA (1,1) case

To obtain the reduced-form test for ϕ, we may rewrite the ARMA(1,1):

$$y_t = \gamma \cdot g(\phi, \vec{\varepsilon}_{t-1}) + \varepsilon_t \quad (14.29)$$

where $g(\phi, \vec{\varepsilon}_{t-1}) = \sum_{i=1}^{\infty} \phi^{i-1}\varepsilon_{t-i}$ and $\vec{\varepsilon}_{t-1} = (\varepsilon_{t-1}, \varepsilon_{t-2}, \ldots)$. Take a linear approximation of $g(.)$ around the null, and the reduced-form test is a t-test for $\lambda = 0$ in the following regression:

$$y_t = \gamma \cdot g(\phi_0, \vec{\varepsilon}_{t-1}) + \lambda \cdot g_{\phi}(\phi_0, \vec{\varepsilon}_{t-1}) + e_t \quad (14.30)$$

where $g_{\phi}(\phi, \vec{\varepsilon}_{t-1}) = \frac{\partial g(\phi, \vec{\varepsilon}_{t-1})}{\partial \phi} = \sum_{i=2}^{\infty}(i-1)\cdot\phi^{i-2}\varepsilon_{t-i}$, and $\lambda = \gamma\cdot(\phi-\phi_0)$.

To make this test feasible, first obtain a consistent estimate for ε through estimation under the restriction of null so as to evaluate the regressors. We generate data with true parameter values $\gamma = 0.1, \phi = 0, \sigma_{\varepsilon} = 1$, and sample size $T = 100$. Estimation is done in EViews™. The rejection frequency of the proposed test is 0.046, at the nominal level 0.05, in contrast to 0.423, that of the standard t-test.

14.7 The ARMA (p, q) case

Consider an ARMA(p, q) model:

$$[1-\phi_p(L)]y_t = [1-\theta_q(L)]\varepsilon_t;\ t = 1, \ldots, T,\ \varepsilon_t \sim \text{i.i.d.}\ N(0, \sigma_{\varepsilon}^2) \quad (14.31)$$

where $\phi_p(L) = \sum_{i=1}^{p} \phi_i L^i, \theta_q(L) = \sum_{i=1}^{q} \theta_i L^i$, and the roots for $1 - \phi(z) = 0$ and $1 - \theta(z) = 0$ are all outside unit circle. A general representation similar to (14.8) may be obtained:

$$y_t = \gamma_1 \cdot [(1 - \theta_m(L))^{-1} \cdot y_{t-1}] + \ldots + \gamma_m \cdot [(1 - \theta_m(L))^{-1} \cdot y_{t-m}] + \varepsilon_t \quad (14.32)$$

where $\gamma_k = \phi_k - \theta_k$, $1 \le k \le m$, $m = \max(p, q)$, and $\phi_k = 0$ for $p < k \le m$ or $\theta_k = 0$ for $q < k \le m$. To test the null $\theta_k = \theta_{k,0}$, $1 \le k \le q$, simply linearize the last term associated with y_{t-m} to obtain the following regression with q augmented terms:

$$\begin{aligned} y_t = {} & \gamma_1 \cdot [(1 - \theta_{m,0}(L))^{-1} \cdot y_{t-1}] + \ldots + \gamma_m \cdot [(1 - \theta_{m,0}(L))^{-1} \cdot y_{t-m}] \\ & + \lambda_1 \cdot [(1 - \theta_{m,0}(L))^{-2} \cdot y_{t-(m+1)}] + \ldots \\ & + \lambda_q \cdot [(1 - \theta_{m,0}(L))^{-2} \cdot y_{t-(m+q)}] + e_t \end{aligned} \quad (14.33)$$

where $\lambda_k = \gamma_k \cdot (\theta_k - \theta_{k,0})$, $1 \le k \le q$. If the null is correct the first m terms on the right hand side of (14.33) are enough to capture the serial correlation. Note to compute the regressors for nonzero $\theta_{k,0}$s, the coefficients $\varphi_{l,j}$s in $\sum_{j=0}^{\infty} \varphi_{l,j} L^j = (1 - \theta_{m,0}(L))^{-l}$, $l = 1, 2$ may be obtained as the $(1, 1)$ element of matrix $(F_l)^j$, where F_l is the $(l \times m)$ by $(l \times m)$ transition matrix $(1 - \theta_{m,0}(L))^l$, $l = 1, 2$ in the state-space representation of the ARMA model.

We experiment with this idea on the ARMA$(2, 2)$ model. With true parameter values $\phi_1 = 0.01, \phi_2 = 0.01, \theta_1 = 0, \theta_2 = 0, \sigma_\varepsilon = 1$, and sample size $T = 100$ we find that the standard t-test for $\hat{\theta}_1$ and $\hat{\theta}_2$ has empirical sizes of 0.571 and 0.698 at a nominal level 0.05. In contrast, the reduced-form test for $\lambda_1 = 0$ and $\lambda_2 = 0$ based on regression (14.33) gives rejection frequencies of 0.049 and 0.049 respectively. Notice here since the null is $\theta_1 = 0$ and $\theta_2 = 0$, our proposed test is equivalent to testing the third and fourth lag in an AR(4) regression.

ACKNOWLEDGMENTS

The authors wish to thank Manfred Deistler, Michael Dueker, Walter Enders, Walter Fisher, Andrew Harvey, Evan Koenig, Junsoo Lee, James Morley, Christian Murray, David Papell, Hashem Pesaran, Benedikt Pötscher, Neil Shephard, Leopold Soegner, Richard Startz, James Stock, Martin Wagner, Arnold Zellner, Eric Zivot, and anonymous referees for helpful comments, but responsibility for all errors is ours. Support from the Ford and Louisa Van Voorhis Endowment at the University of Washington is gratefully acknowledged.

15 Generalized linear spectral models

Tommaso Proietti and Alessandra Luati

15.1 Introduction

Any continuous monotonic transformation of the spectral density function can be expanded as a Fourier series. When the power transformation is considered, the coefficients of the Fourier expansion are interpreted as generalized autocovariances. In a recent paper (Proietti and Luati, 2012), we introduced the generalized autocovariance function (GACV) as the inverse Fourier transform of the power transformation of the spectral density, and illustrated some of its uses.

The GACV and the corresponding generalized autocorrelation function are useful for model identification and discrimination, spectral estimation, and for constructing method of moments estimators of time series parameters.

This paper focuses on two parametric methods, both based on Whittle's likelihood (Whittle, 1961) for the estimation of the generalized spectrum and autocovariance function.

The first estimates directly the generalized autocovariances from the truncated Fourier representation of the power transformation of the spectrum. This method offers two main advantages: the GACV is estimated by iteratively weighted least squares, and the asymptotic covariance matrix of the GACV estimator is available in closed form, enabling e.g., interval estimation of the spectrum and hypothesis testing on the GACVs. On the other hand, the main limitation is that the non-negativity constraints for the spectrum could be violated during the optimization. As a matter of fact, the generalized autocovariances are not variation free, as the GACV is a semidefinite function (see, e.g., Brockwell and Davis, 1991, Ch. 4). This characteristic imposes nonlinear inequality restrictions on the coefficients of a linear spectral model based on the Fourier expansion.

The second method enforces the necessary restrictions by parameterizing the power of the spectrum as the spectrum of an autoregressive-moving average (ARMA) process of finite order, with coefficients in the stationarity and invertibility region. The restrictions are enforced using a transformation of the ARMA parameters due to Barndorff-Nielsen and Schou (1973) and Mohanan

(1984). The power of the spectrum is interpreted as the spectral density of a stochastic process whose Wold representation is obtained by a power transformation of an infinite moving average polynomial, which is approximated by an ARMA rational polynomial. The GACV is derived from the estimated ARMA coefficients.

Two empirical applications illustrate the estimation and the interpretation of the GACV and generalized spectrum. The first deals with the estimation of the inverse autocorrelation function for the series of a chemical process concentration readings popularized by Box and Jenkins (1970), as well with the characterization of the memory of the series provided by the generalized autocorrelations. The second deals with the cyclical properties of an output growth series, and with the identification of spectral peaks.

The cyclical nature of the U.S. gross domestic product (GDP) series has represented a long debated issue. Andrew Harvey has contributed significantly to the topic of extracting cycles from macroeconomic time series, starting with his 1985 paper (Harvey, 1985), where he considered the annual gross national product series, taken from the popular Nelson and Plosser (1982) dataset.

In a highly influential paper, Harvey and Jäger (HJ, 1993) fitted a structural time series model to quarterly U.S. GDP, postulating an orthogonal decomposition into a local linear trend component and a stochastic cycle. The time domain maximum likelihood estimated parameters implied an integrated random walk representation for the trend and the reduced form of the model was ARIMA(2,2,3). The estimated cycle is an important component of U.S. output fluctuations, as its amplitude could reach up to $\pm 5\%$ of trend output. This finding was consistent with the Hodrick-Prescott cycle (Hodrick and Prescott, 1997), of which HJ in a sense provided a model-based validation. Harvey and Trimbur (2003) propose a more general class of model based filters for extracting trend and cycles in macroeconomic time series, showing that the design of low-pass and band-pass filters can be considered as a signal extraction problem in an unobserved components framework.

A different conclusion was reached by Morley, Nelson, and Zivot (MNZ, 2003), using a trend-cycle decomposition with correlated disturbances and an ARIMA(2,1,3) reduced form. They concluded that the deviation cycle represents a minor component of the series, and that its structural real time estimate is identical to the Beveridge-Nelson decomposition (Beveridge and Nelson, 1981) of the unrestricted ARIMA(2,1,3) reduced form model fitted to the series. See, also, Proietti (2006) for further discussion and interpretation of models with correlated trend and cycle disturbances.

Luati and Proietti (2010), using a hyperspherical cycle for GDP growth with ARMA(4,3) reduced form, identified two stochastic stationary cycles, the first characterized by higher amplitude and with a period of two years (about nine quarters), and the second, more important in the 1980s and the 1990s, with a period of about six years (twenty four quarters).

In this paper, we shall fit a class of generalized spectral ARMA models, SpARMA(λ, p, q) models, depending on the power transformation parameter λ and on the AR and MA orders p and q, respectively. The different approaches imply different spectral models for the U.S. GDP growth series and small variations of either the sample period considered or the orders of the SpARMA model lead to very different spectral properties.

The paper is organized as follows. Section 15.2 introduces the generalized spectrum and autocovariance function, along with their interpretation and main properties. Direct estimation of the truncated Fourier approximation of the power of the spectrum is considered in Section 15.3, whereas Section 15.4 deals with the class of generalized spectral models, SpARMA(λ, p, q). The empirical applications are in Section 15.5, where spectrum averaging across different spectral autoregressive models for different λ and p is also discussed. In Section 15.6 we offer our conclusive remarks.

15.2 Generalized spectrum and autocovariances

Let us consider a discrete time stationary stochastic process $\{x_t, t \geq 1\}$, with Wold representation $x_t = \psi(B)\xi_t$, where $\xi_t \sim \text{IID}(0, \sigma^2)$ and $\psi(B) = 1 - \psi_1 B + \psi_2 B^2 + \ldots$, with coefficients satisfying $\sum_{j=0}^{\infty} |\psi_j| < \infty$.

The spectral density function, $f(\omega), \omega \in [-\pi, \pi]$, and the autocovariance function $\gamma(k), k = 0, \pm 1, \pm 2, \cdots$, are defined as the Fourier pair

$$2\pi f(\omega) = \gamma(0) + 2\sum_{j=1}^{\infty} \gamma(j)\cos(\omega j), \qquad \gamma(k) = \int_{-\pi}^{\pi} f(\omega)\cos(\omega k)d\omega.$$

Assuming that the powers of $f(\omega)$ exist for any real power parameter λ, are integrable, and uniformly bounded away from zero, we define the generalized spectrum, $f_\lambda(\omega) = [f(\omega)]^\lambda$ and autocovariance function, $\gamma_\lambda(k)$, as the Fourier pair:

$$[2\pi f(\omega)]^\lambda = \gamma_\lambda(0) + 2\sum_{j=1}^{\infty} \gamma_\lambda(j)\cos(\omega j),$$

$$\gamma_\lambda(k) = \frac{1}{2\pi}\int_{-\pi}^{\pi} [2\pi f(\omega)]^\lambda \cos(\omega k)d\omega. \tag{15.1}$$

The generalized autocovariance (GACV) function is thus the inverse Fourier transform of the power transformation of the spectral density (Proietti and Luati, 2012). For $\lambda = 1$, $\gamma_1(k)$ is the traditional autocovariance function of the process. The case $\lambda = -1$ yields the inverse autocovariance function (Cleveland, 1972), whereas for $\lambda = 0$, $\gamma_0(k) = I(k = 0)$. Dividing by the

GACV at lag zero for λ given yields the generalized autocorrelation function (GACF), $\rho_\lambda(k) = \gamma_\lambda(k)/\gamma_\lambda(0)$.

As $[f(\omega)]^\lambda \geq 0, \forall\lambda$, the GACV is non-negative definite—see Fuller (1996, Section 1.4), and by Brockwell and Davis (1991, Corollary 4.3.2), it can be interpreted as the traditional autocovariance function of the power-transformed process:

$$u_{\lambda t} = \left[\sigma\psi\left(B^{\mathrm{s}(\lambda)}\right)\right]^\lambda \xi_t^*, \tag{15.2}$$

where $\xi_t^* = \sigma^{-1}\xi_t$, s$(\lambda)$ is the sign of λ, and $\left[\sigma\psi\left(B^{\mathrm{s}(\lambda)}\right)\right]$ is an infinite lag polynomial whose coefficients can be derived from Gould (1974).

These generalizations can be useful for model identification. For instance, for integer $\lambda < 0$, the GACV of an autoregressive (AR) process of order p has a cutoff at $k = -\lambda p$. For the MA(1) process $x_t = (1 + \theta B)\xi_t, |\theta| < 1$, if we take $\lambda = 2$, we have

$$\gamma_2(0) = \sigma^4(1 + 4\theta^2 + \theta^4), \gamma_2(1) = \sigma^4\theta(2 + 2\theta^2),$$
$$\gamma_2(2) = \sigma^4\theta^2, \gamma_2(k) = 0, k > 2,$$

and, more generally, for integer $\lambda > 0$, the GACV has a cutoff at lag $k = \lambda$.

The GACV and GACF of the fractional noise process will be derived and illlustrated in Section 15.5 with reference to an empirical illustration.

15.3 Direct Whittle estimation of the generalized autocovariances

Consider the following truncated generalized linear spectral model:

$$\begin{aligned}[2\pi f(\omega)]^\lambda &= \gamma_\lambda(0) + 2\textstyle\sum_{k=1}^K \gamma_\lambda(k)\cos(\omega k) \\ &= z(\omega)'\gamma_\lambda,\end{aligned} \tag{15.3}$$

where $z(\omega) = [1, 2\cos\omega, 2\cos(2\omega), \dots, 2\cos(K\omega)]'$ and $\gamma_\lambda = [\gamma_\lambda(0), \gamma_\lambda(1), \dots, \gamma_\lambda(K)]'$.

Direct estimation of the generalized autocovariances can be carried out by approximate maximum likelihood, where the criterion function is represented by Whittle's likelihood (Whittle, 1961), which is a large sample approximation to the true likelihood when x_t is assumed to be a linear Gaussian process; see Harvey (1989), Section 4.3.1. for a derivation.

Let $\{x_t, t = 1, 2, \dots, n\}$ denote a time series, which is a sample realization from a stationary Gaussian process, and let $\omega_j = \frac{2\pi j}{n}, j = 1, \dots, [n/2]$, be the Fourier frequencies, where $[\cdot]$ denotes the integer part of the argument.

The periodogram (sample spectrum) is defined as

$$I(\omega_j) = \frac{1}{2\pi n}\left|\sum_{t=1}^{n}(x_t - \bar{x})e^{-\imath\omega_j t}\right|^2,$$

where $\bar{x} = n^{-1}\sum_t x_t$.

The following large sample distributional result holds in the short memory case (Brockwell and Davis, 1991, Ch. 10):

$$\frac{I(\omega_j)}{f(\omega_j)} \sim \text{IID}\frac{1}{2}\chi_2^2, 0 < \omega_j < \pi \tag{15.4}$$

whereas $\frac{I(\omega_j)}{f(\omega_j)} \sim \chi_1^2, \omega_j = 0, \pi$, where χ_r^2 denotes a chi-square random variable with r degrees of freedom (equivalently, $I(\omega_j)$ is exponentially distributed with mean $f(\omega_j)$).

The same distributional result would be true for a linear non-Gaussian process; see, e.g., Fuller (1996), Theorem 7.1.2.

For a given transformation parameter λ, the log-likelihood function of γ_λ based on the exponential density for $I(\omega_j)$, for $0 < \omega_j < \pi$ and $j = 1, \ldots, N$ where $N = [(n-1)/2]$, is

$$\ell(\gamma_\lambda) = -\sum_{j=1}^{N}\left(\ln f(\omega_j) + \frac{I(\omega_j)}{f(\omega_j)}\right) \tag{15.5}$$

with

$$f(\omega_j) = \frac{1}{2\pi}\left[z(\omega_j)'\gamma_\lambda\right]^{1/\lambda}.$$

The score vector and the Hessian matrix are

$$s(\gamma_\lambda) = \frac{\partial \ell(\gamma_\lambda)}{\partial \gamma_\lambda} = -\sum_j z^*(\omega_j)u(\omega_j),$$

$$H(\gamma_\lambda) = \frac{\partial^2 \ell(\gamma_\lambda)}{\partial \gamma_\lambda \partial \gamma_\lambda'} = -\sum_j W(\omega_j)z^*(\omega_j)z^*(\omega_j)',$$

where

$$z^*(\omega_j) = \frac{\partial \ln f(\omega_j)}{\partial \gamma_\lambda} = \frac{z(\omega_j)}{\lambda z(\omega_j)'\gamma_\lambda},$$

$$u(\omega_j) = 1 - \frac{I(\omega_j)}{f(\omega_j)},$$

and

$$W(\omega_j) = \frac{I(\omega_j)}{f(\omega_j)} - \lambda u(\omega_j),$$

so that the expected Fisher information is $\mathcal{I}(\gamma_\lambda) = -\mathrm{E}[H(\gamma_\lambda)] = \sum_j z^*(\omega_j)z^*(\omega_j)'$.

Estimation can be carried out by the Newton-Raphson algorithm, i.e., iterating until convergence

$$\tilde{\gamma}_{\lambda,i+1} = \tilde{\gamma}_{\lambda,i} - [H(\tilde{\gamma}_{\lambda,i})]^{-1}s(\tilde{\gamma}_{\lambda,i})$$

or by the method of scoring:

$$\tilde{\gamma}_{\lambda,i+1} = \tilde{\gamma}_{\lambda,i} + [\mathcal{I}(\tilde{\gamma}_{\lambda,i})]^{-1}s(\tilde{\gamma}_{\lambda,i}).$$

In the former case, $\tilde{\gamma}_{\lambda,i+1}$ is obtained equivalently by a weighted least squares regression of $z^*(\omega_j)'\tilde{\gamma}_{\lambda,i} - u(\omega_j)/W(\omega_j)$ on $z^*(\omega_j)$ with weights $W(\omega_j), j = 1,\dots,N$, where both $u(\omega_j)$ and $W(\omega_j)$ are evaluated at $\tilde{\gamma}_{\lambda,i}$.

Whittle estimation of Bloomfield's exponential spectral model (Bloomfield, 1973) by iteratively weighted least squares has been considered by Cameron and Turner (1987). The model is based on the Fourier truncated representation of the logarithm of the spectral density, $\ln[2\pi f(\omega)] = z(\omega)'c_0$, where $c_0 = [c_0(0), c_0(1),\dots,c_0(K)]'$. The coefficients $c_0(k)$ are known as the *cepstral* coefficients and the sequence $c_0(k), k \geq 0$ as the *cepstrum*; see Bogert et al. (1963), and Bloomfield (1973). The interpretation of the cepstral coefficients as pseudo-autocovariances is also discussed in Bogert, Healy, and Tukey (1963) and essentially follows from the Fourier pair $2\pi \ln f(\omega) = c_0(0) + 2\sum_{j=1}^{\infty} c_0(j)\cos(\omega j)$ and $c_0(k) = \frac{1}{2\pi}\int_{-\pi}^{\pi} \ln[2\pi f(\omega)]\cos(\omega k)d\omega$.

In the long memory case, the distributional result (15.4) cannot be invoked to derive the likelihood function, as the large sample distribution of the periodogram (normalized by dividing for the spectral density) is no longer i.d.d. exponential in the vicinity of the zero frequency, see Künsch (1986), Hurvich and Beltrao (1993), and Robinson (1995). However, for a Gaussian series generated by a long memory process $y \sim \mathrm{N}(0, \Gamma_n)$, where the covariance matrix $\Gamma_{\lambda n}$ depends on the parameter vector, an asymptotic approximation to the exact log-likelihood

$$\ell^*(\gamma_\lambda) = -\frac{n}{2}\ln 2\pi - \frac{1}{2}\ln|\Gamma_{\lambda n}| - \frac{1}{2}y'\Gamma_{\lambda n}^{-1}y,$$

is, apart from a constant (Whittle, 1953):

$$-\frac{1}{2\pi}\int_{-\pi}^{\pi}\left[\ln f(\omega) + \frac{I(\omega)}{f(\omega)}\right]d\omega,$$

so that $\ell(\gamma_\lambda)/N$, where $\ell(\gamma_\lambda)$ is given by (15.5), converges in probability to the exact log-likelihood. Hence, the maximizer of $\ell(\gamma_\lambda)$ is asymptotically equivalent to the maximum likelihood estimator also in the long memory case. As pointed out in Beran et al. (2013, pp. 428–9), the Whittle estimator is computed by iteratively reweighted least squares and all the generalized linear models tools are meaningful.

The Whittle likelihood estimator of the GACV coefficients is asymptotically normal and efficient (Dahlhaus, 1989 and 2006).

$$\sqrt{n}(\tilde{\gamma}_\lambda - \gamma_\lambda) \to_d \mathrm{N}(0, V),$$

$$V^{-1} = 2 \lim_{N\to\infty} \sum_{j=1}^{N} z^*(\omega_j) z^*(\omega_j)' = \frac{1}{4\pi\lambda^2} \int_{-\pi}^{\pi} \frac{1}{[2\pi f(\omega)]^{2\lambda}} z(\omega) z(\omega)' \mathrm{d}\omega.$$

For $\lambda = 1$ we recover the result

$$\sqrt{n}(\tilde{\gamma}(k) - \gamma(k)) \sim N\left(0, 4\pi \int_{-\pi}^{\pi} f(\omega)^2 \cos^2(\omega) \mathrm{d}\omega\right).$$

The selection of the order K is the main specification issue raised by the representation (15.3): information criteria like Akaike information criterion (AIC) and Schwar's Bayesian information criterion (BIC) can be used for that purpose. Finally, diagnostic checking can be based on the Pearson's residuals $\frac{I(\omega_j)}{\tilde{f}(\omega_j)} - 1 = -u_j$.

Notice that the power transformation used in this paper does not encompass Bloomfield's exponential (EXP) model, as $\lambda \to 0$. On the contrary, if we considered a linear spectral model for the Box-Cox transformation of the spectral generating function, $\lambda^{-1}\{[2\pi f(\omega)]^\lambda - 1\} = z(\omega)'\gamma_\lambda^*$, the EXP model would arise as the limiting case for $\lambda \to 0$ and the coefficients γ_0^* would be interpreted as the cepstral coefficients c_0.

15.4 Spectral ARMA models

Direct unconstrained estimation of the coefficients $\gamma_\lambda(k)$ is rather problematic, as the generalized autocovariances are subject to constraints, due to the non-negative definite nature of the GACV.

The constraints are enforced for $\lambda \neq 0$ if the power transformed spectral density is assumed to have an ARMA(p, q) representation:

$$[2\pi f(\omega)]^\lambda = \sigma_\lambda^2 \frac{|\theta_\lambda(e^{-\imath\omega})|^2}{|\phi_\lambda(e^{-\imath\omega})|^2}, \tag{15.6}$$

where

$$\theta_\lambda(e^{-\imath\omega}) = 1 + \theta_{\lambda 1} e^{-\imath\omega} + \theta_{\lambda 2} e^{-\imath 2\omega} + \cdots + \theta_{\lambda q} e^{-\imath q\omega}$$

and

$$\phi_\lambda(e^{-\imath\omega}) = 1 - \phi_{\lambda 1} e^{-\imath\omega} - \phi_{\lambda 2} e^{-\imath 2\omega} - \cdots - \phi_{\lambda p} e^{-\imath p\omega}.$$

The spectral model

$$f(\omega) = \frac{1}{2\pi}\left[\sigma_\lambda^2 \frac{|\theta_\lambda(e^{-\imath\omega})|^2}{|\phi_\lambda(e^{-\imath\omega})|^2}\right]^{1/\lambda} \tag{15.7}$$

will be called a SpARMA(λ, p, q) model. It arises from the ARMA(p, q) parametric representation for the power transformed process $u_{\lambda t}$ of Section 15.2:

$$u_{\lambda t} = \sigma_\lambda \frac{\theta_\lambda(B)}{\phi_\lambda(B)}\xi_t^*, \xi_t^* \sim \text{WN}(0, 1),$$

where $\phi_\lambda(B) = 1 - \phi_{\lambda 1}B - \phi_{\lambda 2}B^2 - \cdots - \phi_{\lambda p}B^p$ and $\theta_\lambda(B) = 1 + \theta_{\lambda 1}B + \theta_{\lambda 2}B^2 + \cdots + \theta_{\lambda q}B^p$. It should be noted that the *SpARMA*(λ, p, q) model is observationally equivalent to the *SpARMA*($-\lambda, q, p$) model.

This approach is closely related to the generalized ARMA models proposed by Shitan and Peiris (2011), and Pillai et al. (2012), who derive the time series properties of processes defined as $(1 - \phi B)^{\lambda_1}x_t = (1 + \theta B)^{\lambda_2}\xi_t$, $\xi_t \sim WN(0, \sigma^2)$, with $\lambda_1, \lambda_2 > 0$. Whittle estimation of the transformation parameters, along with the other parameters is considered in Shitan and Peiris (2008).

Cycle models with power transformed AR(2) polynomials $(1 - 2\rho\cos\omega_c B + \rho^2)^\lambda x_t = \xi_t$ with integer $\lambda \geq 1$ have been considered by Harvey and Trimbur (2003), with the purpose of performing model-based band-pass filtering.

Our SpARMA model (15.7) is estimated for a given λ without restriction on the orders; however, we restrict the MA and AR parameters to lie in the stationary and invertible region. The SpARMA models for different triplets λ, p, q, can be compared using an information criterion, such as AIC or BIC.

For a given λ, the model (15.6) features $p + q + 1$ parameters, which can be estimated by maximizing the Whittle likelihood, under the restriction that the spectral density $[2\pi f(\omega)]^\lambda$ is bounded away from zero and infinity at all frequencies. This is achieved by enforcing that $0 < |\phi_\lambda(e^{-\imath\omega})|^2 < \infty$, and $0 < |\theta_\lambda(e^{-\imath\omega})|^2 < \infty$, $\forall\omega$, or, equivalently, that the roots of the lag polynomials $\phi_\lambda(B)$ and $\theta_\lambda(B)$ are outside the unit circle.

Hence, for the purpose of estimation, we reparameterize the AR coefficients in terms of partial autocorrelations (Barndorff-Nielsen and Schou, 1973). This is done by the Durbin-Levinson recursive algorithm: letting $\pi_{\lambda k}, k = 1, \ldots, p$ denote p coefficients, such that $|\pi_{\lambda k}| < 1$, then for $k = 1, \ldots, p$ and $j = 1, \ldots, k-1$, compute

$$\phi_{\lambda j}^{(k)} = \phi_{\lambda j}^{(k-1)} - \pi_{\lambda k}\phi_{\lambda, k-j}^{(k-1)}, \quad \phi_{\lambda k}^{(k)} = \pi_{\lambda k}.$$

The final iteration returns coefficients $\phi_{\lambda p}^{(p)} = \phi_{\lambda j}$ that are in the stationary region.

The same algorithm can be used for constraining the $\theta_{\lambda j}$ coefficients in the invertibility region, as shown by Monahan (1984): given q coefficients $\varpi_{\lambda k}$, $k = 1, \ldots, q$, such that $|\varpi_{\lambda k}| < 1$, the coefficients of the invertible MA polynomial are obtained from the last iteration of the Durbin-Levinson recursion

$$\theta_{\lambda j}^{(k)} = \theta_{\lambda j}^{(k-1)} + \varpi_{\lambda k}\theta_{\lambda,k-j}^{(k-1)}, \quad \theta_{\lambda k}^{(k)} = \varpi_{\lambda k},$$

for $k = 1, \ldots, q$ and $j = 1, \ldots, k-1$.

The coefficients $\pi_{\lambda j}$ and $\varpi_{\lambda j}$, that are constrained in the range (-1,1), are in turn obtained as the Fisher inverse transformations of unconstrained real parameters ϑ_j, $j = 1, \ldots, p+q$, e.g., $\pi_{\lambda j} = \frac{\exp(2\vartheta_j)-1}{\exp(2\vartheta_j)+1}$ for $j = 1, \ldots, p$. Also, we set $\vartheta_0 = \ln(\sigma_\lambda^2)$.

The Whittle likelihood,

$$\ell(\vartheta) = -\sum_{j=1}^{N}\left(\ln f(\omega_j) + \frac{I(\omega_j)}{f(\omega_j)},\right) \tag{15.8}$$

can be maximized with respect to the unconstrained parameter vector ϑ by a quasi-Newton optimization algorithm, using numerical first and second derivatives.

15.5 **Empirical applications**

We illustrate the two estimation methods and the interpretation of the results with two empirical applications. The first deals with a time series popularized by Box and Jenkins (1970), concerning a sequence of $n = 200$ readings of a chemical process concentration, known as *Series A*. We illustrate the estimation of the inverse autocovariance function using the direct GACV estimator of Section 15.3, as well as the use of the generalized autocorrelations with varying λ for investigating the long memory feature of the series.

The second case study deals with the quarterly growth rate of U.S. GDP. The aim is to illustrate that a value of λ greater than one can be used to elicit the spectral peaks corresponding to the business cycle. All the computations have been carried out by the programming software Ox of Doornik (2007).

15.5.1 BOX AND JENKINS SERIES A

Series A is plotted in Figure 15.1, along with the sample autocorrelations. The latter decline very slowly and indeed, as we shall discuss later, it is an open issue whether the series features long memory.

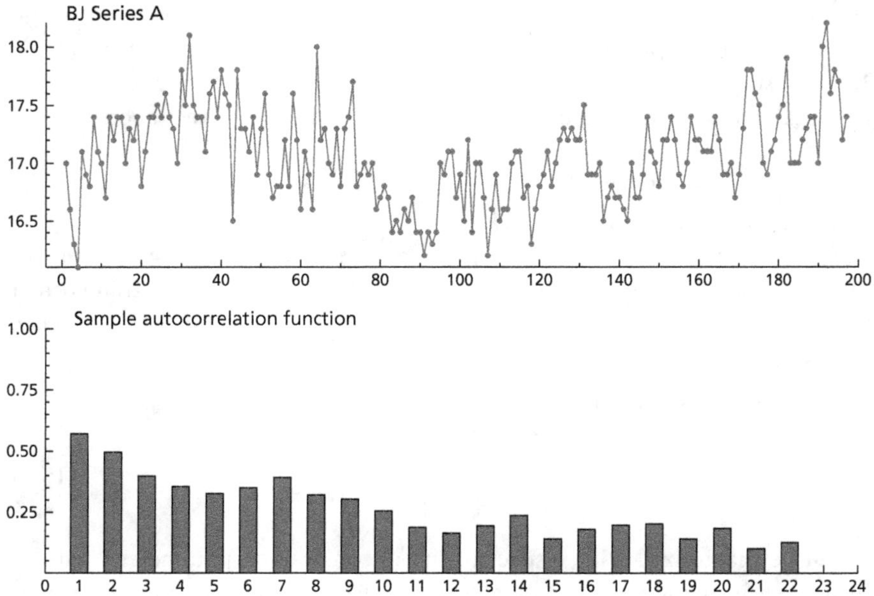

Figure 15.1. BJ Series A: Chemical process concentration readings

Selection of a linear interpolator

The optimal interpolator of x_t, $\mathrm{E}(x_t|x_{t\pm k}, k \geq 1)$ is

$$\hat{x}_t = -\sum_{k=1}^{\infty} \rho_{-1}(k) x_{t\pm k},$$

see Grenander and Rosenblatt (1957), Cleveland (1972), and Battaglia (1983), where $\rho_{-1}(k)$ is the inverse autocorrelation function of x_t, i.e., the GACF obtained for $\lambda = -1$.

The coefficients $\rho_{-1}(k)$ can be estimated using the generalized spectral model (15.3) for $\lambda = -1$. The number of nonzero inverse autocorrelation to consider can be selected by choosing the value K that minimizes the AIC or BIC. Table 15.1 reports the values of the two information criteria for different values of K and suggests taking $K = 1$:

Table 15.1 Values of the two information criteria for values of k

K	1	2	3	4	5	6	7	8
AIC	6.49	6.64	6.67	6.70	6.73	6.76	6.82	6.85
BIC	6.52	6.70	6.75	6.80	6.86	6.92	7.01	7.06

The estimated model parameters are $\hat{\gamma}_{-1}(0) = 12.38$ (standard error 1.39) and $\hat{\gamma}_{-1}(1) = -5.36$ (standard error .77). The corresponding estimate of the inverse autocorrelation function is $\hat{\rho}_{-1}(1) = -0.43$, $\hat{\rho}_{-1}(k) = 0, k > 1$. In conclusion, Whittle likelihood inference on the GACV suggests a linear interpolator based solely on the first inverse autocorrelation: $\tilde{x}_t = 0.43x_{t-1} + 0.43x_{t+1}$. Notice that this is very similar to the optimal linear interpolator for a random walk process.

The interpolation error variance, $\mathrm{E}(\hat{x}_t - x_t)^2$, can be estimated as

$$\text{i.e.v.} = \frac{1}{\hat{\gamma}_{-1}(0)} = 0.078,$$

which is about 49% of the unconditional variance of x_t.

Estimation of a SpARMA model

Box and Jenkins (1970) fitted an ARMA(1,1) model to the levels of the series and an AR(1) to the first differences, by Gaussian maximum likelihood estimation in the time domain. The estimated prediction error variance (p.e.v.) was .097 and .101, respectively.

Table 15.2 presents the main results of the estimation of a SpARMA$(\lambda, 1, 1)$ for different values of λ. The spectral model is

$$f(\omega) = \frac{1}{2\pi}\left[\sigma_\lambda^2 \frac{1+\theta_\lambda^2 + 2\theta_\lambda \cos\omega}{1+\phi_\lambda^2 - 2\phi_\lambda \cos\omega}\right]^{1/\lambda}, \lambda \neq 0.$$

We do not report the values of λ in the negative range, as the model results for the SpARMA$(\lambda, 1, 1)$ are symmetric about 0, except for the AR and MA coefficients, for which $\phi_{-\lambda} = -\theta_\lambda$ and $\theta_{-\lambda} = -\phi_\lambda$.

The maximum of the Whittle likelihood with respect to λ occurs around $\lambda = 1.80$; for larger values the optimization routine fails to converge (as the parameters tend toward the nonstationary region).

Table 15.2 SpARMA$(\lambda, 1, 1)$ model. Estimated AR and MA coefficients, prediction error variance, predictability and interpolability measures

λ	$\hat{\phi}_\lambda$	$\hat{\theta}_\lambda$	$\ell(\hat{\vartheta})$	p.e.v	Pred	Int
0.25	0.73	−0.63	305.24	0.1002	37.77	48.42
0.50	0.78	−0.58	305.36	0.1001	38.22	48.82
0.75	0.82	−0.52	305.80	0.0999	38.98	49.51
1.00	0.87	−0.46	305.24	0.0997	40.10	50.50
1.25	0.91	−0.40	306.11	0.0995	41.72	51.89
1.50	0.95	−0.33	306.46	0.0993	44.21	54.00
1.75	0.98	−0.25	306.81	0.0993	49.59	58.50

The p.e.v. is estimated as $\left(\hat{\sigma}_\lambda^2\right)^{1/\lambda}$. The unconditional variance of x_t implied by the SpARMA model is

$$\hat{V} = \frac{1}{2\pi}\left(\hat{\sigma}_\lambda^2\right)^{1/\lambda}\int_{-\pi}^{\pi}\left(\frac{1+\hat{\theta}_\lambda^2+2\hat{\theta}_\lambda\cos\omega}{1+\hat{\phi}_\lambda^2-2\hat{\phi}_\lambda\cos\omega}\right)^{1/\lambda} d\omega,$$

whereas the implied interpolation error variance is

$$\text{i.e.v.} = 2\pi\left(\hat{\sigma}_\lambda^2\right)^{1/\lambda}\left[\int_{-\pi}^{\pi}\left(\frac{1+\hat{\phi}_\lambda^2-2\hat{\phi}_\lambda\cos\omega}{1+\hat{\theta}_\lambda^2+2\hat{\theta}_\lambda\cos\omega}\right)^{1/\lambda} d\omega\right]^{-1}.$$

The predictability index and the interpolability index are defined respectively as

$$P = 100\left(1-\frac{\text{p.e.v.}}{\hat{V}}\right), I = 100\left(1-\frac{\text{i.e.v.}}{\hat{V}}\right).$$

The second measure has been introduced by Battaglia (1983) and Battaglia and Bhansali (1987), who refer to I as an index of linear determinism, as the maximum is obtained when the series is perfectly interpolable (with zero i.e.v.) from the past and the future.

The estimated spectrum corresponding to the choice $\lambda = 1.75$, which is close to the optimal λ, differs from that obtained by fitting an ordinary ARMA(1,1) to the series around the zero frequency and implies higher persistence, as it can be seen from Figure 15.2. As a matter of fact, $\phi_{1.75}$ gets close to one; also, applying a power parameter greater than one to the AR polynomial implies an autocorrelation pattern that decays very slowly and it is more consistent with the pattern of the sample ACF reported in the bottom panel of Figure 15.1.

The behavior of the estimated generalized autocorrelation as a function of λ, plotted in Figure 15.3, is similar to that of a fractional noise process with long memory parameter around .43; compare with Figure 15.4, where the values of $\rho_\lambda(k)$ as a function of λ are plotted for the case when the memory parameter is $d = .4$. The GACV at lags $k \geq 2$ is zero for $\lambda = -2.5$. This behavior is characteristic of a fractional noise process. Indeed, the GACV and GACF of the fractional noise process, $(1-B)^d x_t = \xi_t$, where $\xi_t \sim \text{WN}(0, \sigma^2)$, with memory parameter $d < .5$, are defined for $\lambda d < .5$ as

$$\gamma_\lambda(k) = \sigma^{2\lambda}\frac{\Gamma(1-2d\lambda)\Gamma(k+d\lambda)}{\Gamma(1-d\lambda)\Gamma(d\lambda)\Gamma(1+k-d\lambda)},$$

$$\rho_\lambda(k) = \frac{\Gamma(1-d\lambda)\Gamma(k+d\lambda)}{\Gamma(1+k-d\lambda)\Gamma(d\lambda)}.$$

It should be noticed that $\rho_\lambda(k) = 0$ for $k \geq 2$ and $\lambda = -d^{-1}$.

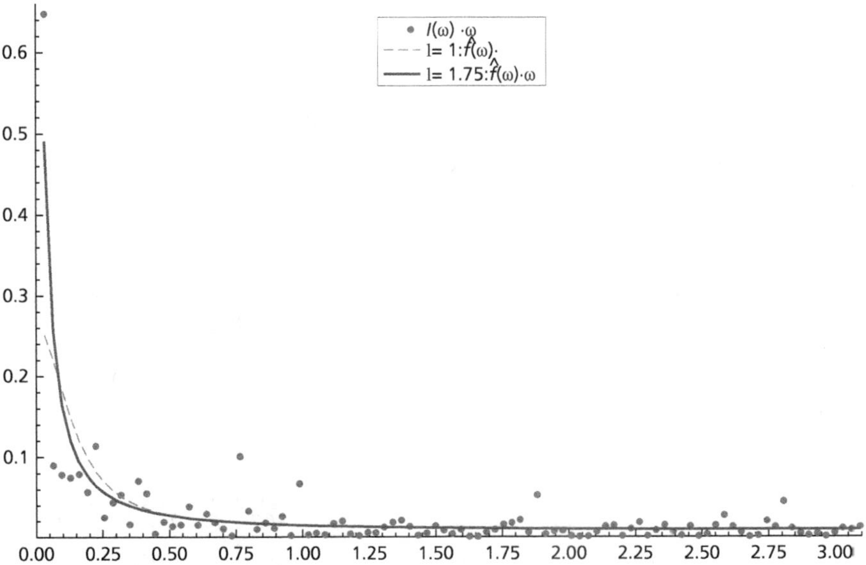

Figure 15.2. BJ Series A: Periodogram $I(\omega)$ and SpARMA$(\lambda, 1, 1)$ estimated spectral densities for the values $\lambda = 1$ and $\lambda = 1.75$

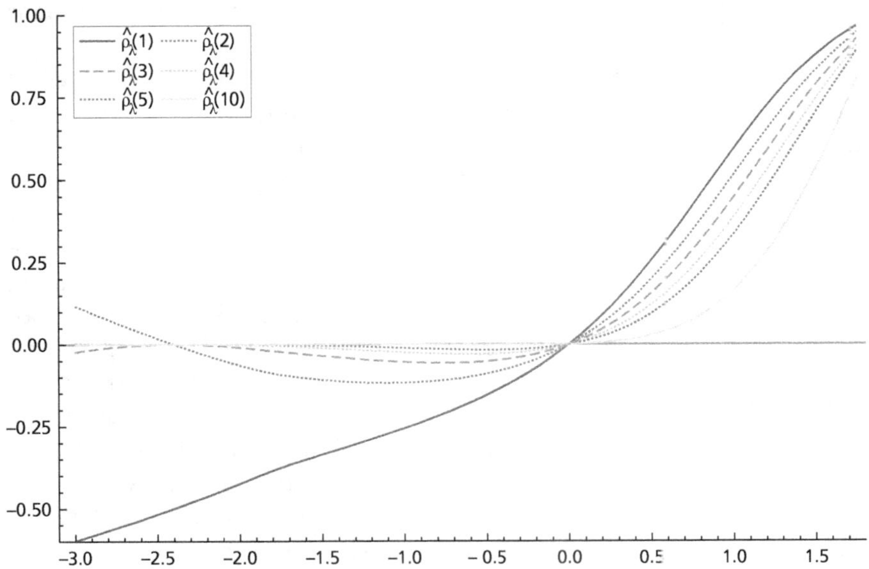

Figure 15.3. BJ Series A: Plot of the estimated generalized autocorrelations, $\hat{\rho}_\lambda$ versus λ

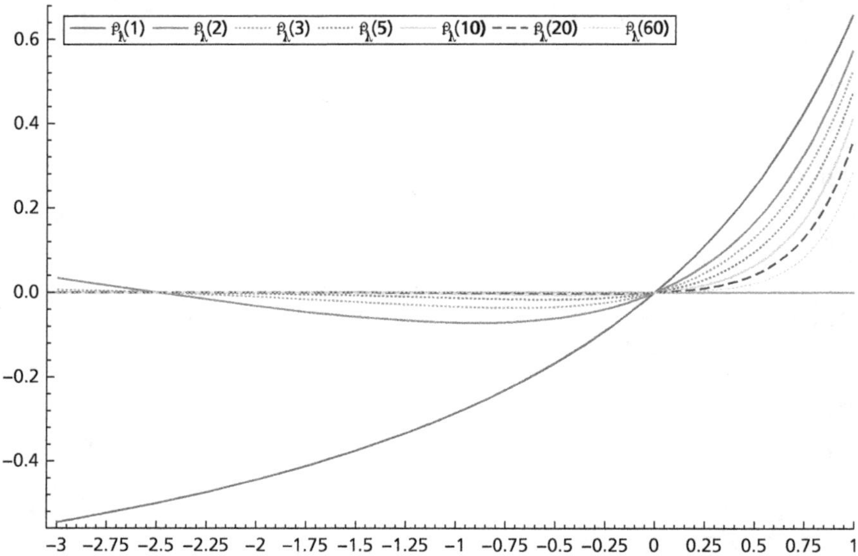

Figure 15.4. Generalized autocorrelation for a fractional noise process with $d = 0.4$

15.5.2 THE CYCLE IN U.S. GDP GROWTH

Our discussion focuses on the most recent available U.S. GDP growth series, plotted in Figure 15.5, along with its sample correlogram, with observations ranging from 1947Q2 to 2012Q1.

In the upper left panel of Figure 15.6 we display the spectrum estimated by fitting a SpARMA(1, 2, 3) model, which corresponds to Whittle estimation of an ARMA(2,3) model for U.S. GDP growth. This is the spectrum implied by the MNZ model, and it features a distinctive peak at the frequency corresponding to a period of two and a half years.

We next consider a generalization of the approach known as autoregressive spectral estimation, according to which the spectrum is estimated by fitting an autoregressive model by Whittle approximate likelihood estimation; see Percival and Walden (1993, Ch. 9) for a discussion of parameteric spectral estimation via autoregressive models. The AR specification is seen as a simple and easily handled specification, but potentially prone to misspecification. The representation that minimizes the AIC is SpARMA(2.5, 3, 0), whereas BIC selects a SpARMA(1.5, 1, 0) representation. The latter cannot account for a cycle in GDP growth; its implied spectrum is plotted in the lower left panel of Figure 15.6. On the contrary, the former implies a spectral peak around the frequency .55, corresponding to a period of about three years.

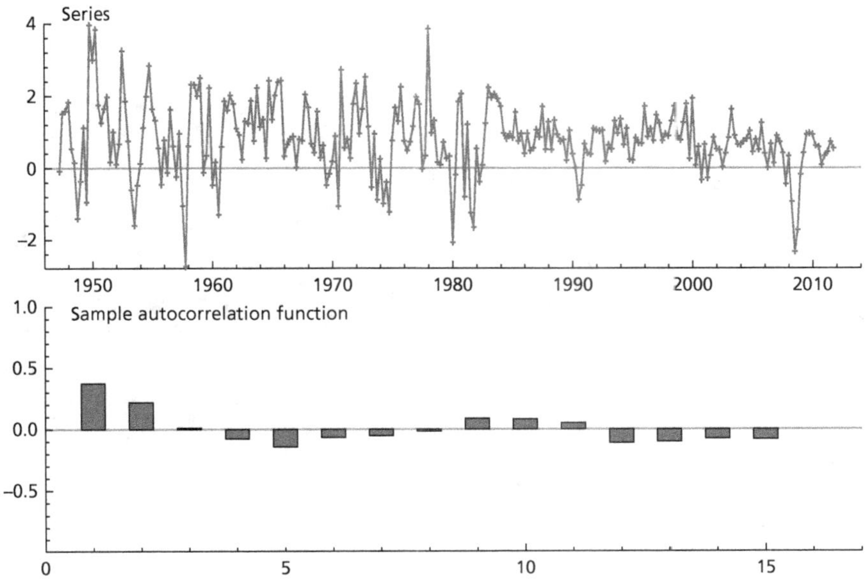

Figure 15.5. U.S. GDP quarterly growth series

Both SpARMA models carry a value of λ greater than one. When a model, such as an AR(3), is misspecified, the possibility of varying the power transformation parameter adds further flexibility in fitting the spectral model. Figure 15.7 plots the estimated spectral density of a SpARMA(λ, 3, 0) as λ varies. A noticeable feature is that for low values of λ, including $\lambda = 1$, the spectrum is monotonically decreasing with the frequency, the maximum occurring at $\omega = 0$, as in the AR(1) case. For $\lambda > 1$ the spectral peak moves away from zero. Hence, the generalized spectrum and the associated GACF may be useful devices for detecting spectral peaks from simple, though potentially misspecified, models.

The variety of spectral fits suggests that spectrum averaging could be an effective strategy. Here, in particular, we focus on the geometric average

$$\bar{f}(\omega) = \exp\left(\sum_j w_j \ln \hat{f}_j(\omega)\right), w_j \propto \exp(-AIC_j),$$

where the individual $\hat{f}_j(\omega)$ arise from fitting all the possible SpARMA($\lambda, p, 0$) models for $p = 1, \ldots 15$ and λ taking values in $(-1, 2.5)$ with step 0.25 ($\lambda = 0$ is excluded). The average is constructed using weights calculated on the basis of the AIC, for the purpose of illustration. The spectral average is reported in the last panel of Figure 15.6.

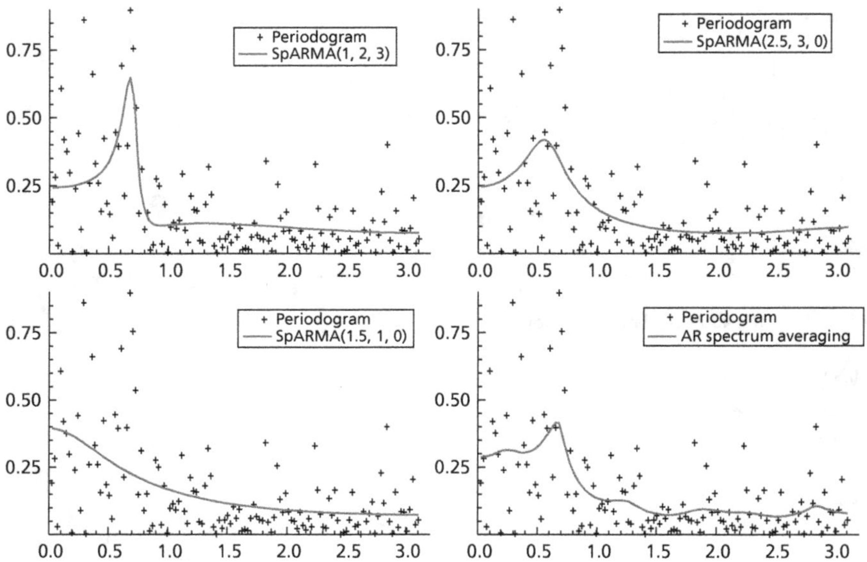

Figure 15.6. U.S. GDP: Estimated spectral density $\hat{f}(\omega)$ estimated using selected SpARMA models, and spectrum averaging

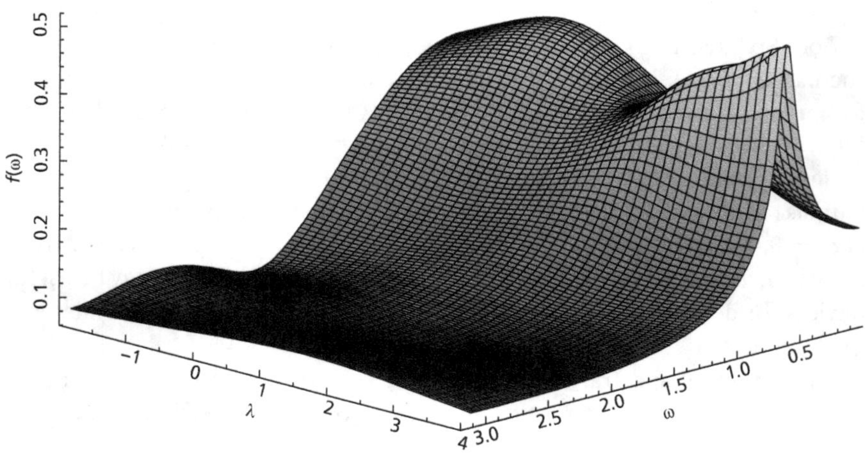

Figure 15.7. U.S. GDP: Estimated spectral density $\hat{f}(\omega)$ estimated by a SpARMA(λ, 3, 0) model, plotted as a function of λ and ω

15.6 **Conclusions**

The generalized autocovariance function and its dual, the power transformation of the spectrum, are useful for characterizing the properties of a stationary stochastic processes. The underlying motivation is that by tuning the power parameter λ, certain properties of the series may be emphasized or muted for a particular purpose.

A power transformation parameter greater than one emphasizes the high density point of the spectrum and will be useful to identify spectral peaks corresponding to a cyclical feature. On the contrary, negative values will highlight the spectral ordinates close to zero. Moreover, the GACV has potential in the long memory case. Further exploration in this promising direction is required.

We have presented two parametric estimators based on Whittle likelihood and have illustrated a few potential uses. An additional important use is in assessing goodness of fit: when applied to the residuals of fitting a time series model our methodology enables tests of spectral flatness for a varying power transformation parameter to be carried out.

BIBLIOGRAPHY

Ackerson, G. A. and K. S. Fu (1970). On state estimation in switching environments. *IEEE Transactions on Automatic Control 15*, 10–7.

Akashi, H. and H. Kumamoto (1977). Random sampling approach to state estimation in switching environments. *Automatica 13*, 429–34.

Altissimo, F., R. Cristadoro, M. Forni, M. Lippi, and G. Veronese (2010). New eurocoin: tracking economic growth in real time. *Review of Economics and Statistics 92*, 1024–34.

Amengual, D., G. Fiorentini, and E. Sentana (2013). Sequential estimators of shape parameters in multivariate dynamic models. *Journal of Econometrics 177*, 233–49.

Amengual, D. and E. Sentana (2010). A comparison of mean-variance efficiency tests. *Journal of Econometrics 154*, 16–34.

Amengual, D. and M. W. Watson (2007). Consistent estimation of the number of dynamic factors in a large *n* and *t* panel. *Journal of Business and Economic Statistics 25*, 91–6.

Anderson, B. D. O. and J. B. Moore (1979). *Optimal Filtering*. New Jersey: Prentice Hall.

Anderson, T. W. and H. Rubin (1949). Estimation of the parameters of a single equation in a complete system of stochastic equations. *Annals of Mathematical Statistics 20*, 46–63.

Anderson, T. W. and H. Rubin (1956). Statistical inference in factor analysis. In J. Neyman (Ed.), *Proceedings of the III Berkeley Symposium on Mathematical Statistics and Probability*, pp. 111–50. Berkeley: University of California.

Andrews, D. W. K. (1993). Tests for parameter instability and structural change with unknown change point. *Econometrica 61*, 821–56.

Andrews, D. W. K. (2001). Testing when a parameter is on the boundary of the maintained hypothesis. *Econometrica 69*, 683–734.

Andrews, D. W. K. (2003). End-of-sample instability tests. *Econometrica 61*, 1661–94.

Andrews, D. W. K. and X. Cheng (2012). Estimation and inference with weak, semi-strong, and strong identification. *Econometrica 80*, 2153–211.

Andrews, D. W. K. and J. Kim (2006). Tests for cointegration breakdown over a short time period. *Journal of Business and Economic Statistics 24*, 379–94.

Andrews, D. W. K., I. Lee, and W. Ploberger (1996). Optimal changepoint tests for normal linear regression. *Journal of Econometrics 70*, 9–38.

Andrews, D. W. K. and W. Ploberger (1994). Optimal tests when a nuisance parameter is present only under the alternative. *Econometrica 62*, 1383–414.

Andrews, I. and A. Mikusheva (2015). Maximum likelihood inference in weakly identified models. *Quantitative Economics* 6, 123–52.

Andrieu, C., A. Doucet, and R. Holenstein (2010). Particle Markov chain Monte Carlo methods (with discussion). *Journal of the Royal Statistical Society, Series B 72*, 1–33.

Ansley, C. F. and R. Kohn (1986). Prediction mean squared error for state space models with estimated parameters. *Biometrika 73*, 467–73.

Ansley, C. F. and P. Newbold (1980). Finite sample properties of estimates for autoregressive moving average processes. *Journal of Econometrics 13*, 159–84.

Applebaum, D. (2004). *Lévy Processes and Stochastic Calculus*. Cambridge: Cambridge University Press.

Arellano, M. and S. Bonhomme (2009). Robust priors in nonlinear panel data models. *Econometrica 77*, 489–536.

Aruoba, S., F. Diebold, M. Kose, and M. Terrones (2011). Globalization, the business cycle, and macroeconomic monitoring (with discussion). In R. Clarida and F. Giavazzi (Eds.), *NBER International Seminar on Macroeconomics 2010*, pp. 245–301. Chicago: University of Chicago Press.

Aruoba, S. B., F. X. Diebold, and C. Scotti (2009). Real-time measurement of business conditions. *Journal of Business & Economic Statistics 27*, 417–27.

Atkeson, A. and L. E. Ohanian (2001). Are Phillips curves useful for forecasting inflation? *Federal Reserve Bank of Minneapolis Quarterly Review 25*, 2–11.

Backus, D., P. Kehoe, and F. Kydland (1995). International business cycles: theory and evidence. In T. F. Cooley (Ed.), *Frontiers of Business Cycle Research*, pp. 331–57. Princeton: Princeton University Press.

Bai, J. (1997). Estimation of a change point in multiple regression models. *The Review of Economics and Statistics 74*, 551–63.

Bai, J. and S. Ng (2004). A panic attack on unit roots and cointegration. *Econometrica 72*(4), 1127–77.

Bai, J. and S. Ng (2008a). Forecasting economic time series using targeted predictors. *Journal of Econometrics 146*, 304–17.

Bai, J. and S. Ng (2008b). Large dimensional factor analysis. *Foundations and Trends in Econometrics 3*, 89–163.

Bai, J. and P. Perron (1998). Estimating and testing linear models with multiple structural changes. *Econometrica 66*, 47–78.

Ball, L. and S. Mazumder (2011). Inflation dynamics and the great recession. *Brookings Papers on Economic Activity 42*, 337–405.

Banbura, M. and G. Runstler (2011). A look into the factor model black box: publication lags and the role of hard and soft data in forecasting GDP. *International Journal of Forecasting 27*, 333–46.

Bansal, R. and A. Yaron (2004). Risks for the long run: a potential resolution of asset pricing puzzles. *Journal of Finance 59*, 1481–509.

Barndorff-Nielsen, O. E. and D. R. Cox (1996). Prediction and asymptotics. *Bernoulli 2*, 319–40.

Barndorff-Nielsen, O. E., P. R. Hansen, A. Lunde, and N. Shephard (2008). Designing realised kernels to measure the ex-post variation of equity prices in the presence of noise. *Econometrica 76*, 1481–536.

Barndorff-Nielsen, O. E. and N. Shephard (2001). Non-Gaussian Ornstein–Uhlenbeck-based models and some of their uses in financial economics (with discussion). *Journal of the Royal Statistical Society, Series B 63*, 167–241.

Barndorff-Nielsen, O. E. and N. Shephard (2002). Econometric analysis of realised volatility and its use in estimating stochastic volatility models. *Journal of the Royal Statistical Society, Series B 64*, 253–80.

Barndorff-Nielsen, O. E. and G. Shou (1973). On the parametrization of autoregressive models by partial autocorrelations. *Journal of Multivariate Analysis 3*, 408–19.

Battaglia, F. (1983). Inverse autocovariances and a measure of linear determinism for a stationary process. *Journal of Time Series Analysis 4*, 79–87.

Battaglia, F. and R. Bhansali (1987). Estimation of the interpolation error variance and an index of linear determinism. *Biometrika 74*, 771–9.

Baxter, M. (1995). International Trade and Business Cycles. In S. Grossman and K. Rogoff (Eds.), *Handbook of International Economics*, pp. 1801–64. Amsterdam: North Holland.

Beaumont, M. (2003). Estimation of population growth or decline in genetically monitored populations. *Genetics 164*, 1139.

Bera, A. and M. J. Yoon (1993). Specification testing with locally misspecified alternatives. *Econometric Theory 9*, 649–58.

Beran, J., Y. Feng, S. Ghosh, and R. Kulik (2013). *Long-Memory Processes Probabilistic Properties and Statistical Methods*. Berlin: Springer.

Beran, R. (1990). Calibrating prediction regions. *Journal of the American Statistical Association 85*, 715–23.

Berger, J. O. (2006). The case of objective Bayesian analysis. *Bayesian Analysis 1*, 385–402.

Berger, J. O. and R. Yang (1994). Noninformative priors and Bayesian testing for the AR(1) model. *Econometric Theory 10*, 461–82.

Beveridge, S. and C. R. Nelson (1981). A new approach to decomposition of economic time series into permanent and transitory components with particular attention to the measurement of the business cycle. *Journal of Monetary Economics 7*, 151–74.

Blanchard, O. and J. Simon (2001). The long and large decline in U.S. output volatility. *Brookings Papers in Economic Activity*, 135–74.

Blitz, D. and J. Huij (2012). Evaluating the performance of global emerging markets equity exchange-traded funds. *Emerging Markets Review 13*, 149–58.

Bloomfield, P. (1973). An exponential model for the spectrum of a scalar time series. *Biometrika 60*, 217–26.

Bogert, B. P., M. J. R. Healy, and J. W. Tukey (1963). The frequency analysis of time series for echoes: cepstrum, pseudo-autocovariance, cross-cepstrum, and saphe cracking. In M. Rosenblatt (Ed.), *Proceedings of the Symposium on Time Series Analysis*, pp. 209–43. New York: John Wiley & Sons.

Boivin, J. and S. Ng (2006). Are more data always better for factor analysis? *Journal of Econometrics 132*, 169–94.

Bollerslev, T. (1986). Generalised autoregressive conditional heteroskedasticity. *Journal of Econometrics 51*, 307–27.

Bollerslev, T. (1987). A conditional heteroskedastic time series model for speculative prices and rates of return. *The Review of Economics and Statistics 69*, 542–7.

Bollerslev, T. and J. M. Wooldridge (1992). Quasi maximum likelihood estimation and inference in dynamic models with time varying covariances. *Econometric Reviews 11*, 143–72.

Bos, C. and N. Shephard (2006). Inference for adaptive time series models: stochastic volatility and conditionally Gaussian state space form. *Econometric Reviews 25*, 219–44.

Box, G. E. P. and G. M. Jenkins (1970). *Time Series Analysis: Forecasting and Control*. San Francisco: Holden-Day.

Box, G. E. P., G. M. Jenkins, and G. C. Reinsel (2008). *Time Series Analysis: Forecasting and Control* (4th edn.). New Jersey: John Wiley & Sons.

Box, G. E. P. and D. A. Pierce (1970). Distribution of the autocorrelations in autoregressive moving average time series models. *Journal of the American Statistical Association 65*, 1509–26.

Breusch, T. S. and A. R. Pagan (1980). The Lagrange multiplier test and its application to model specification in econometrics. *Review of Economic Studies 47*, 239–53.

Brier, G. W. (1950). Verification of forecasts expressed in terms of probabilities. *Monthly Weather Review 78*, 1–3.

Brockwell, P. J. (2001). Lévy-driven CARMA processes. *Annals of the Institute of Statistical Mathematics 53*, pp. 113–24.

Brockwell, P. J. and R. A. Davis (1991). *Time Series: Theory and Methods.* New York: Springer-Verlag.

Brockwell, P. J. and A. Lindner (2009). Existence and uniqueness of stationary Lévy-driven CARMA processes. *Stochastic Processes and their Applications 119*, 2625–44.

Broemeling, L. and M. Land (1984). On forecasting with univariate autoregressive processes: a Bayesian approach. *Communications in Statistics—Theory and Methods 13*, 1305–20.

Brown, P. and P. de Jong (2001). Nonparametric smoothing with the state space model. *Canadian Journal of Statistics 29*, 37–50.

Brown, R. L., J. Durbin, and J. M. Evans (1975). Techniques for testing the constancy of regression relationships over time with comments. *Journal of the Royal Statistical Society, Series B 37*, 149–92.

Buckle, D. J. (1995). Bayesian inference for stable distributions. *Journal of the American Statistical Association 90*, 605–13.

Busetti, F. and A. C. Harvey (2007). Testing for trend. *Econometric Theory 24*, 72–87.

Busetti, F. and A. C. Harvey (2010). Tests of strict stationarity based on quantile indicators. *Journal of Time Series Analysis 31*, 435–50.

Busetti, F. and A. C. Harvey (2011). When is a copula constant? A test for changing relationships. *Journal of Financial Econometrics 9*, 106–31.

Busetti, F. and A. M. R. Taylor (2004). Tests of stationarity against a change in persistence. *Journal of Econometrics 123*, 33–66.

Caggiano, G., G. Kapetarias, and V. Labhard (2011). Are more data always better for factor analysis? *Journal of Forecasting 30*, 736–52.

Caldara, D., J. Fernandez-Villaverde, P. Guerron, J. F. Rudio-Ramirez, and Y. Wen (2012). Computing DSGE models with recursive preferences and stochastic volatility. *Review of Economic Dynamics 15*, 188–206.

Calzolari, G., G. Fiorentini, and E. Sentana (2004). Constrained indirect estimation. *Review of Economic Studies 71*, 945–73.

Camacho, M., G. Perez-Quiros, and P. Poncela (2015). Extracting nonlinear signals from several economic indicators. *Journal of Applied Econometrics 30*, Forthcoming.

Camacho, M. and G. Perez-Quiros (2010). Introducing the euro-sting: Short term indicator of euro area growth. *Journal of Applied Econometrics 25*(4), 663–94.

Cameron, M. A. and T. R. Turner (1987). Fitting models to spectra using regression packages. *Applied Statistics 36*, 47–57.

Canova, F., M. Ciccarellu, and E. Ortega (2007). Similarities and Convergence in G-7 Cycles. *Journal of Monetary Economics 53*, 850–78.

Cappé, O., S. Godsill, and E. Moulines (2007). An overview of existing methods and recent advances in sequential Monte Carlo. *Proceedings of the IEEE 95*(5), 899–924.

Carter, C. K. and R. Kohn (1994). On Gibbs sampling for state space models. *Biometrika 81*, 541–53.

Chamberlain, G. and M. Rothschild (1983). Arbitrage, factor structure and mean-variance analysis in large asset markets. *Econometrica 51*, 1305–24.

Chambers, J., C. Mallows, and B. Stuck (1976). A method for simulating stable random variables. *Journal of American Statistical Association 71*, 340–4.

Chatfield, C. (1993). Calculating interval forecasts (with discussion). *Journal of Business & Economic Statistics 11*, 121–44.

Chatfield, C. (1996). Model uncertainty and forecast accuracy. *Journal of Forecasting 15*, 495–508.

Chen, R. and J. S. Liu (2000). Mixture Kalman filters. *Journal of the Royal Statistical Society, Series B 62*, 493–508.

Chib, S. and I. Jeliazkov (2006). Inference in semiparametric dynamic models for binary longitudinal data. *Journal of the American Statistical Association 101*, 685–700.

Chib, S., F. Nardari, and N. Shephard (2002). Markov chain Monte Carlo methods for stochastic volatility models. *Journal of Econometrics 108*, 281–316.

Chow, G. (1974). Multiperiod predictions from stochastic difference equations by Bayesian methods. In S. E. Fienberg and A. Zellner (Eds.), *Studies in Bayesian Econometrics and Statistics*, Chapter 8, pp. 313–24. Amsterdam: North-Holland.

Chow, G. C. (1960). Tests of equality between sets of coefficients in two linear regressions. *Econometrica 28*, 591–605.

Clark, P. K. (1987). The cyclical component of U.S. economic activity. *The Quarterly Journal of Economics 102*, 797–814.

Clark, T. E. and M. W. McCracken (2001). Tests of equal forecast accuracy and encompassing for nested models. *Journal of Econometrics 105*, 85–110.

Clements, M. P. and D. F. Hendry (1999). *Forecasting Non-stationary Economic Time Series*. Cambridge, MA: The MIT Press.

Clements, M. P. and J. H. Kim (2007). Bootstrap prediction intervals for autoregressive time series. *Computational Statistics & Data Analysis 51*, 3580–94.

Cleveland, W. S. (1972). The inverse autocorrelations of a time series and their applications. *Technometrics 14*, 277–293.

Cochrane, J. H. (1991). Volatility tests and efficient markets: a review essay. *Journal of Monetary Economics 27*, 661–76.

Cogley, T., G. Primiceri, and T. J. Sargent (2010). Inflation-gap persistence in the U.S. *American Economic Journal: Macroeconomics 2*, 43–69.

Cogley, T. and T. J. Sargent (2002). Evolving post-World War II U.S. inflation dynamics. In B. S. Bernanke and K. Rogoff (Eds.), *NBER Macroeconomics Annual*, pp. 331–88. Cambridge, MA: MIT Press.

Cogley, T. and T. J. Sargent (2005). Drifts and volatilities: monetary policies and outcomes in the post World War II U.S. *Review of Economic Dynamics 8*, 262–302.

Cogley, T. and A. M. Sbordone (2008). Trend inflation, indexation, and inflation persistence in the new Keynesian Phillips curve. *American Economic Review 98*, 2101–26.

Connor, G. (1984). A unified beta pricing theory. *Journal of Economic Theory 34*, 13–31.

Connor, G., L. R. Goldberg, and R. A. Korajczik (2010). *Portfolio Risk Analysis*. Princeton: Princeton University Press.

Cox, D. R. and E. J. Snell (1989). *The Analysis of Binary Data* (2nd edn). London: Chapman & Hall.

Creal, D. (2012). A survey of sequential Monte Carlo methods for economics and finance. *Econometric Reviews 31*, 245–96.

Creal, D., S. J. Koopman, and A. Lucas (2013). Generalized autoregressive score models with applications. *Journal of Applied Econometrics 28*, 777–95.

Creal, D., S. J. Koopman, and E. Zivot (2010). Extracting a robust U.S. business cycle using a time-varying multivariate model-based bandpass filter. *Journal of Applied Econometrics 25*, 695–719.

Crowder, M. J. (1976). Maximum likelihood estimation for dependent observations. *Journal of the Royal Statistical Society B 38*, 45–53.

Czado, C. and P. X. K. Song (2008). State space mixed models for longitudinal observations with binary and binomial responses. *Statistical Papers 49*, 691–714.

D'Agostino, A., L. Gambetti, and D. Giannone (2013). Macroeconomic forecasting and structural change. *Journal of Applied Econometrics 28*, 82–101.

Dahlhaus, R. (1989). Efficient parameter estimation for self similar processes. *Annals of Statistics 17*, 1749–66.

Dahlhaus, R. (2006). Correction: Efficient parameter estimation for self similar processes. *The Annals of Statistics 34*, 1045–7.

Datta, G. and R. Mukerjee (2003). Probability matching priors for predicting a dependent variable with application to regression models. *Annals of the Institute of Statistical Mathematics 55*, 1–6.

Datta, G. and R. Mukerjee (2004). *Probability Matching Priors: Higher Order Asymptotics*. New York: Springer.

Davidson, R. and J. G. MacKinnon (1998). Graphical methods for investigating the size and power of tests statistics. *The Manchester School 66*, 1–26.

Davies, R. B. (1977). Hypothesis testing when a nuisance parameter is present only under the alternative. *Biometrika 64*, 247–54.

Davies, R. B. (1987). Hypothesis testing when a nuisance parameter is present only under the alternative. *Biometrika 74*, 33–43.

de Jong, P. (1991a). The diffuse Kalman filter. *Annals of Statistics 19*, 1073–83.

de Jong, P. (1991b). Stable algorithms for the state space model. *Journal of Time Series Analysis 12*, 143–58.

de Jong, P. and N. Shephard (1995). The simulation smoother for time series models. *Biometrika 82*, 339–50.

De Rossi, G., D. Jessop, and J. Sefton (2010). How to pick the right ETF. Technical report, UBS Investment Research, November.

Delcoure, N. and M. Zhong (2007). On the premiums of iShares. *Journal of Empirical Finance 14*, 168–95.

Delle Monache, D. and A. C. Harvey (2011). On the effect of misspecification in models for extracting trends and cycles. Working paper, Economics Department, Cambridge University.

Demos, A. and E. Sentana (1998). Testing for GARCH effects: a one-sided approach. *Journal of Econometrics 86*, 97–127.

Dempster, A., N. Laird, and D. Rubin (1977). Maximum likelihood from incomplete data via the EM algorithm. *Journal of the Royal Statistical Society, Series B 39*, 1–38.

Diebold, F. X. (1993). On the limitations of comparing mean squared forecast errors: comment. *Journal of Forecasting 12*, 641–2.

Diebold, F. X. (2012). Comparing predictive accuracy, twenty years later: a personal perspective on the use and abuse of Diebold-Mariano tests. Working paper, Department of Economics, University of Pennsylvania.

Diebold, F. X., T. A. Gunther, and T. S. Tay (1998). Evaluating density forecasts with applications to financial risk management. *International Economic Review 39*, 863–83.

Diebold, F. X. and J. A. Lopez (1996). Forecast evaluation and combination. In G. S. Maddala and C. R. Rao (Eds.), *Handbook of Statistics*, pp. 241–68. Amsterdam: Elsevier.

Diebold, F. X. and R. S. Mariano (1995). Comparing predictive accuracy. *Journal of Business & Economic Statistics 13*, 253–63.

Diebold, F. X. and K. Yilmaz (2009). Measuring financial asset return and volatility spillovers, with application to global equity markets. *The Economic Journal 119*, 158–71.

Diebold, F. X. and K. Yilmaz (2012). Better to give than to receive: predictive directional measurement of volatility spillovers (with discussion). *International Journal of Forecasting 28*, 57–66.

Diebold, F. X. and K. Yilmaz (2014). On the network topology of variance decompositions: measuring the connectedness of financial firms. *Journal of Econometrics* 182, 119–34.

Doornik, J. A. (2007). *Object-Oriented Matrix Programming Using Ox* (3rd edn). London: Timberlake Consultants Press and Oxford: www.doornik.com.

Dotsey, M., S. Fujita, and T. Stark (2011). Do Phillips curves conditionally help to forecast inflation? Unpublished paper, Federal Reserve Bank of Philadelphia.

Doucet, A., N. de Freitas, and N. J. Gordon (Eds.) (2001). *Sequential Monte Carlo Methods in Practice*. New York: Springer-Verlag.

Doucet, A., S. J. Godsill, and C. Andrieu (2000). On sequential Monte Carlo sampling methods for Bayesian filtering. *Statistics and Computing 10*, 197–208.

Doucet, A. and A. Johansen (2011). A tutorial on particle filtering and smoothing: fifteen years later. In D. Crisan and B. Rozovsky (Eds.), *The Oxford Handbook of Nonlinear Filtering*. Oxford: Oxford University Press.

Doyle, B. and J. Faust (2005). Breaks in the variability and comovement of G-7 economic growth. *Review of Economics and Statistics 87*, 721–40.

Doz, C., D. Giannone, and L. Reichlin (2011). A two-step estimator for large approximate dynamic factor models based on Kalman filtering. *Journal of Econometrics 164*, 188–205.

Doz, C., D. Giannone, and L. Reichlin (2012). A quasi maximum likelihood approach for large approximate dynamic factor models. *Review of Economics and Statistics 94*, 1014–24.

Doz, C. and F. Lenglart (1999). Dynamic factor models: test of the number of factors and estimation with an application to the French industrial business survey. *Annales d'Economie et Statistique 54*, 91–128.

Drinkwater, G. C. (1939). *The Boat Race*. London: Blackie and Son.

Duchesne, P. and S. Lalancette (2003). On testing for multivariate ARCH effects in vector time series models. *Canadian Journal of Statistics 31*, 275–92.

Dueker, M. J. (1997). Markov switching in GARCH processes and mean-reverting stock-market volatility. *Journal of Business & Economic Statistics 15*, 26–34.

Dufour, J. M. (1997). Some impossibility theorems in econometrics with application to structural and dynamic models. *Econometrica 65*, 1365–88.

Dufour, J. M., L. Khalaf, and M. C. Beaulieu (2010). Multivariate residual-based finite-sample tests for serial dependence and ARCH effects with application to asset pricing models. *Journal of Applied Econometrics 25*, 263–85.

Dufour, J. M. and E. Renault (1998). Short-run and long-run causality in time series: theory. *Econometrica 66*, 1099–125.

Dufour, J. M. and A. Taamouti (2010). Short and long run causality measures: theory and inference. *Journal of Econometrics 154*, 42–58.

Dungey, M., V. L. Martin, and A. R. Pagan (2000). A multivariate latent factor decomposition of international bond yield spreads. *Journal of Applied Econometrics 15*, 697–715.

Durbin, J. and S. J. Koopman (1997). Monte Carlo maximum likelihood estimation of non-Gaussian state space model. *Biometrika 84*, 669–84.

Durbin, J. and S. J. Koopman (2000). Time series analysis of non-Gaussian observations based on state space models from both classical and Bayesian perspectives. *Journal of the Royal Statistical Society Series B 62*, 3–56.

Durbin, J. and S. J. Koopman (2001). *Time Series Analysis by State Space Methods*. Oxford: Oxford University Press.

Durbin, J. and S. J. Koopman (2002). A simple and efficient simulation smoother for state space time series analysis. *Biometrika 89*, 603–16.

Durbin, J. and S. J. Koopman (2012). *Time Series Analysis by State Space Methods* (2nd edn). Oxford: Oxford University Press.

Eickmeier, S. (2007). Business cycle transmission from the US to Germany—a structural factor approach. *European Economic Review 51*, 521–51.

Eickmeier, S. (2009). Comovements and heterogeneity in the Euro area analyzed in a non-stationary dynamic factor model. *Journal of Applied Econometrics 24*, 933–59.

Elliott, G. and U. K. Muller (2006). Efficient tests for general persistent time variation in regression coefficients. *Review of Economic Studies 73*, 907–40.

Engle, R. F. (1982). Autoregressive conditional heteroskedasticity with estimates of the variance of United Kingdom inflation. *Econometrica 50*, 987–1007.

Engle, R. F. (1984). Wald, likelihood ratio and Lagrange multiplier tests in econometrics. In S. Griliches and M. D. Intrilligator (Eds.), *Handbook of Econometrics, vol. II*, pp. 775–826. Amsterdam: North Holland.

Engle, R. F. and D. Sarkar (2006). Premiums-discounts and exchange traded funds. *Journal of Derivatives 13*, 27–45.

Engle, R. F. and M. W. Watson (1981). A one-factor multivariate time series model of metropolitan wage rates. *Journal of the American Statistical Association 76*, 774–81.

Fang, K.-T., S. Kotz, and K.-W. Ng (1990). *Symmetric Multivariate and Related Distributions*. London: Chapman and Hall.

Fearnhead, P. and P. Clifford (2003). On-line inference for hidden Markov models via particle filters. *Journal of the Royal Statistical Society, Series B 65*, 887–99.

Feller, W. (1966). *An Introduction to Probability Theory and its Applications* (2nd edn). New York: Wiley.

Fernandes, C. and A. C. Harvey (1990). Modelling binary time series. Unpublished paper, Department of Statistics, London School of Economics.

Fernandez-Villaverde, J., P. Guerron, J. F. Rudio-Ramirez, and M. Uribe (2010). Risk matters: the real effects of volatility shocks. *American Economic Review 101*, 2530–61.

Fieller, E. C. (1932). The distribution of the index in a normal bivariate population. *Biometrika 24*, 428–40.

Fieller, E. C. (1954). Some problems in interval estimation. *Journal of the Royal Statistical Society, Series B 16*, 175–85.

Fiorentini, G. and E. Sentana (2009). Dynamic specification tests for static factor models. Working paper 0912, CEMFI.

Fiorentini, G. and E. Sentana (2010a). New testing approaches for mean-variance predictability. Unpublished paper, CEMFI.

Fiorentini, G. and E. Sentana (2010b). On the efficiency and consistency of likelihood estimation in multivariate conditionally heteroskedastic dynamic regression models. Unpublished paper, CEMFI.

Fiorentini, G. and E. Sentana (2012). Tests for serial dependence in static, non-Gaussian factor models. Working paper 1211, CEMFI.

Fiorentini, G. and E. Sentana (2013). Dynamic specification tests for dynamic factor models. Working paper 1306, CEMFI.

Fiorentini, G., E. Sentana, and G. Calzolari (2003). Maximum likelihood estimation and inference in multivariate conditionally heteroskedastic dynamic regression models with student *t* innovations. *Journal of Business and Economic Statistics 21*, 532–46.

Fiorentini, G., E. Sentana, and N. Shephard (2004). Likelihood-based estimation of latent generalised ARCH structures. *Econometrica 72*, 1481–517.

Firth, D. (1993). Bias reduction of maximum likelihood estimates. *Biometrika 80*, 27–38.

Flury, T. and N. Shephard (2011). Bayesian inference based only on simulated likelihood: particle filter analysis of dynamic economic models. *Econometric Theory 27*, 933–56.

Forni, M., M. Hallin, M. Lippi, and L. Reichlin (2003). Do financial variables help forecasting inflation and real activity in the Euro area? *Journal of Monetary Economics 50*, 1243–55.

Forni, M., M. Hallin, M. Lippi, and L. Reichlin (2005). The generalized dynamic factor model: one sided estimation and forecasting. *Journal of the American Statistical Association 100*, 830–40.

Frale, C., M. Marcellino, G. Mazzi, and T. Proietti (2011). Euromind: A monthly indicator of the euro area economic conditions. *Journal of the Royal Statistical Society, Series A 174*, 439–70.

Franzini, L. and A. C. Harvey (1983). Testing for deterministic trend and seasonal components in time series models. *Biometrika 70*, 673–82.

Fraser, D. A. S., N. Reid, E. Marras, and G. Y. Yi (2010). Default priors for Bayesian and frequentist inference. *Journal of the Royal Statistical Society, Series B 72*, 631–54.

Fuhrer, J. C. and G. Olivei (2010). The role of expectations and output in the inflation process: an empirical assessment. *FRB Boston Public Policy Brief Series*, 10–2.

Fuller, W. A. (1996). *Introduction to Statistical Time Series*. New York: John Wiley & Sons.

Fuller, W. A. and D. P. Hasza (1981). Properties of predictors for autoregressive time series. *Journal of the American Statistical Association 76*, 155–61.

Gallant, A. R. and G. Tauchen (1996). Which moments to match? *Econometric Theory 12*, 657–81.

Gastineau, G. L. (2001). Exchange-traded funds: an introduction. *Journal of Portfolio Management 27*, 88–96.

Gençağa, D., E. E. Kuruoğlu, A. Ertüzün, and S. Yildrim (2008). Estimation of time-varying autoregressive symmetric alpha stable processes by particle filters. *Signal Processing 88*, 2564–72.

Geweke, J. (1977). The dynamic factor analysis of economic time series. In D. J. Aigner and A. S. Goldberger (Eds.), *Latent Variables in Socio-Economic Models*, pp. 365–82. Amsterdam: North-Holland.

Geweke, J. (1989). Bayesian inference in econometric models using Monte Carlo integration. *Econometrica 57*, 1317–39.

Geweke, J. (2005). *Contemporary Bayesian Econometrics and Statistics*. New Jersey: John Wiley & Sons.

Geweke, J. and C. Whiteman (2006). Bayesian forecasting. In C. W. J. Elliott, G. Granger, and A. Timmermann (Eds.), *Handbook of Economic Forecasting, vol. 1*, Chapter 1, pp. 3–80. Amsterdam: Elsevier B.V.

Ghysels, E., A. C. Harvey, and E. Renault (1996). Stochastic volatility. In C. R. Rao and G. S. Maddala (Eds.), *Statistical Methods in Finance*, pp. 119–91. Amsterdam: North-Holland.

Gilks, W., S. Richardson, and D. J. Spiegelhalter (1996). *Markov chain Monte Carlo in practice*. London: Chapman & Hall.

Gnedenko, B. and A. Kolmogorov (1968). *Limit Distributions for Sums of Independent Random Variables* (2nd edn). Cambridge, MA: Addison-Wesley.

Godfrey, L. G. (1988). *Misspecification Tests in Econometrics: The Lagrange Multiplier Principle and Other Approaches*. Cambridge: Cambridge University Press.

Godsill, S. (1999). MCMC and EM-based methods for inference in heavy-tailed processes with alpha-stable innovations. In *Proceedings of the IEEE Signal Processing Workshop on Higher-order Statistics*, pp. 228–32, Caesarea, Israel.

Godsill, S. (2001). On the relationship between MCMC model uncertainty methods. *Journal of Computational and Graphical Statistics 10*, 230–48.

Godsill, S. and E. Kuruoğlu (1999). Bayesian inference for time series with heavy-tailed symmetric alpha-stable noise processes. In Proceedings of ASA-IMS conference on *Applications of Heavy Tailed Distributions in Economics, Engineering and Statistics*, American University, Washington DC, USA.

Godsill, S. and P. Rayner (1998). *Digital Audio Restoration*. Berlin: Springer.

Goetzmann, W., Z. Ivkovic, and K. G. Rowenhorst (2001). Day trading international mutual funds: evidence and policy solutions. *Journal of Financial and Quantitative Analysis 36*, 287–309.

Gómez, V. (2001). The use of Butterworth filters for trend and cycle estimation in economic time series. *Journal of Business and Economic Statistics 19*, 365–73.

Gordon, N. J., D. J. Salmond, and A. F. M. Smith (1993). A novel approach to nonlinear and non-Gaussian Bayesian state estimation. *IEE-Proceedings F 140*, 107–13.

Gordon, R. J. (1998). Foundations of the goldilocks economy: supply shocks and the time-varying NAIRU. *Brookings Papers on Economic Activity 2*, 297–333.

Gould, H. W. (1974). Coefficient identities for powers of Taylor and Dirichlet series. *The American Mathematical Monthly 81*(1), 3–14.

Gourieroux, C. and A. Monfort (1996). *Simulation Based Econometric Methods*. Oxford: Oxford University Press.

Granger, C. W. J. and Y. Jeon (2011). The evolution of the Phillips curve: a modern time series viewpoint. *Economica 78*, 51–66.

Granger, C. W. J. and R. Joyeux (1980). An introduction to long memory time series models and fractional differencing. *Journal of Time Series Analysis 1*, 15–39.

Grenander, U. and M. Rosenblatt (1957). *Statistical Analysis of Stationary Time Series*. New York: John Wiley & Sons.

Grigoletto, M. (1998). Bootstrap prediction intervals for autoregressions: some alternatives. *International Journal of Forecasting 14*, 447–56.

Guilkey, D. K. (1974). Alternative tests for a first-order vector autoregressive error specification. *Journal of Econometrics 4*, 95–104.

Gupta, R. and A. Kabundi (2011). A large factor model for forecasting macroeconomic variables in South Africa. *International Journal of Forecasting 27*, 1076–88.

Hafner, C. M. and J. V. K. Rombouts (2007). Semiparametric multivariate volatility models. *Econometric Theory 23*, 251–80.

Hamilton, J. D. (1986). A standard error for the estimated state vector of a state space model. *Journal of Econometrics 33*, 387–97.

Hannan, E. J. (1982). Testing for autocorrelation and Akaike's criterion. In J. M. Gani and E. J. Hannan (Eds.), *Essays in Statistical Science*, pp. 403–12. Sheffield: Applied Probability Trust.

Hansen, J. (2010). Goddard Institute for Space Studies. http://data.giss.nasa.gov/gistemp/tabledata/GLB.Ts.txt, accessed on August 12, 2015.

Harvey, A. C. (1981). *Time Series Models* (1st edn). New York: Philip Allan.

Harvey, A. C. (1982). A test of misspecification for systems of equations. Discussion paper, LSE Econometrics Programme.

Harvey, A. C. (1985). Trends and cycles in macroeconomic time series. *Journal of Business and Economic Statistics 3*, 216–27.

Harvey, A. C. (1989). *Forecasting, Structural Time Series Models and the Kalman Filter*. Cambridge: Cambridge University Press.

Harvey, A. C. (1990). *The Econometric Analysis of Time Series* (2nd edn). Cambridge, MA: MIT Press.

Harvey, A. C. (1993). *Time Series Models* (2nd edn). New York: Harvester Wheatsheaf.

Harvey, A. C. (2011). Modelling the Phillips curve with unobserved components. *Applied Financial Economics 21*, 7–17.

Harvey, A. C. and T. Chakravarty (2008). Beta-t-(e)-GARCH. Technical report, Cambridge Working Papers in Economics 08340.

Harvey, A. C. and J. Durbin (1986). The effects of seat belt legislation on British road casualties: A case study in structural time series modelling. *Journal of the Royal Statistical Society, Series A 149*, 187–227.

Harvey, A. C. and C. Fernandes (1989). Time series models for count data or qualitative observations. *Journal of Business and Economic Statistics 7*, 407–17.

Harvey, A. C. and A. Jäger (1993). Detrending, stylized facts and the business cycle. *Journal of Applied Econometrics 8*, 231–47.

Harvey, A. C. and S. J. Koopman (1997). Multivariate structural time series models. In C. Heij, H. Schumacher, B. Hanzon, and C. Praagman (Eds.), *Systematic Dynamics in Economic and Financial Models*, pp. 269–98. Chichester: John Wiley & Sons.

Harvey, A. C. and S. J. Koopman (2000). Signal extraction and the formulation of unobserved components models. *Econometrics Journal 3*, 84–107.

Harvey, A. C., E. Ruiz, and E. Sentana (1992). Unobserved component time series models with ARCH disturbances. *Journal of Econometrics 52*, 129–58.

Harvey, A. C., E. Ruiz, and N. Shephard (1994). Multivariate stochastic variance models. *Review of Economic Studies 61*, 247–64.

Harvey, A. C. and M. Streibel (1998). Testing for a slowly changing level with special reference to stochastic volatility. *Journal of Econometrics 87*, 167–89.

Harvey, A. C. and T. Trimbur (2003). General model-based filters for extracting cycles and trends in economic time series. *Review of Economics and Statistics 85*, 244–55.

Hauser, M. A., B. M. Pötscher, and E. Reschenhofer (1999). Measuring persistence in aggregate output: Arma models, fractionally integrated arma models and nonparametric procedures. *Empirical Economics 24*, 243–69.

Hendry, D. F. and F. Pretis (2013). Some fallacies in econometric modelling of climate change. Technical report, Department of Economics, University of Oxford.

Hendry, D. F. (1971). Maximum likelihood estimation of systems of simultaneous regression equations with errors generated by a vector autoregressive process. *International Economic Review 12*, 257–72.

Hendry, D. F. (2010). Climate change: lessons for our future from the distant past. Technical report, Department of Economics, University of Oxford.

Hinkley, D. V. (1969). On the ratio of two correlated normal random variables. *Biometrika 56*, 635–9.

Hodgson, D. J. and K. P. Vorkink (2003). Efficient estimation of conditional asset pricing models. *Journal of Business and Economic Statistics 21*, 269–83.

Hodrick, R. J. (1992). Dividend yields and stock returns: alternative procedures for inference and measurement. *Review of Financial Studies 5*, 357–86.

Hodrick, R. J. and E. C. Prescott (1997). Post-war U.S. business cycles: an empirical investigaton. *Journal of Money, Credit and Banking 29*, 1–16.

Hong, Y. and R. S. Shehadeh (1999). A new test for ARCH effects and its finite sample performance. *Journal of Business and Economic Statistics 17*, 91–108.

Hosking, J. R. M. (1981). Lagrange-multiplier tests of multivariate time series models. *Journal of the Royal Statistical Society B 43*, 219–30.

Hsu, P. H. and C. M. Kuan (2005). Reexamining the profitability of technical analysis with data snooping checks. *Journal of Financial Econometrics 3*, 606–28.

Hurvich, C. M. and K. I. Beltrao (1993). Asymptotics for the low-frequency ordinates of the periodogram of a long-memory time series. *Journal of Time Series Analysis 14*, 455–72.

Inoue, A. (2001). Testing for distributional change in time series. *Econometric Theory 17*, 156–87.

Jazwinski, A. H. (1970). *Stochastic Processes and Filtering Theory*. New York: Academic Press.

Jegadeesh, N. (1989). On testing for slow decaying components in stock prices. Unpublished paper, Anderson Graduate School of Management, University of California at Los Angeles.

Jegadeesh, N. and G. G. Pennacchi (1996). The behavior of interest rates implied by the term structure of eurodollar futures. *Journal of Money, Credit and Banking 28*, 426–46.

Jones, P. (2010). Climate Research Unit at the University of East Anglia. http://www.cru.uea.ac.uk/cru/data/temperature/CRUTEM3-gl.dat, accessed on August 12, 2015.

Jones, P. D. and M. E. Mann (2004). Climate over past millennia. *Reviews of Geophysics 42*, 1–42.

Jungbacker, B. and S. J. Koopman (2007). Monte Carlo estimation for nonlinear non-Gaussian state space models. *Biometrika 94*, 827–39.

Jungbacker, B. and S. J. Koopman (2015). Likelihood-based dynamic factor analysis for measurement and forecasting. *Econometrics Journal 18*, C1–C21.

Jungbacker, B., S. J. Koopman, and M. van der Wel (2011). Maximum likelihood estimation for dynamic factor models with missing data. *Journal of Economic Dynamics & Control 35*, 1358–68.

Kabaila, P. and K. Syuhada (2008). Improved prediction limits for AR(p) and ARCH(p) processes. *Journal of Time Series Analysis 29*, 213–23.

Kano, Y. (1983). Consistency of estimators in factor analysis. *Journal of the Japan Statistical Society 13*, 137–44.

Kim, J. H. (2004). Bootstrap prediction intervals for autoregression using asymptotically mean-unbiased estimators. *International Journal of Forecasting 20*, 85–97.

Kim, S., N. Shephard, and S. Chib (1998). Stochastic volatility: likelihood inference and comparison with ARCH models. *Review of Economic Studies 65*, 361–93.

King, M. A., E. Sentana, and S. B. Wadhwani (1994). Volatility and links between national stock markets. *Econometrica 62*, 901–33.

King, M. L. and G. H. Hillier (1985). Locally best invariant tests of the error covariance matrix in the linear regression model. *Journal of the Royal Statistical Society, Series B 47*, 98–102.

Koenker, R. (1981). A note on studentizing a test for heteroskedasticity. *Journal of Econometrics 17*, 107–12.

Kohn, R., C. F. Ansley, and C.-M. Wong (1992). Nonparametric spline regression with autoregressive moving average errors. *Biometrika 79*, 335–46.

Koop, G., M. H. Pesaran, and S. M. Potter (1996). Impulse response analysis in nonlinear multivariate models. *Journal of Econometrics 74*, 119–47.

Koopman, S. J. and C. Bos (2004). State space models with a common stochastic variance. *Journal of Business and Economic Statistics 22*, 346–57.

Koopman, S. J. and A. C. Harvey (2003a). Computing observation weights for signal extraction and filtering. *Journal of Economic Dynamics and Control 27*, 1317–33.

Koopman, S. J. and A. C. Harvey (2003b). Computing observation weights for signal extraction and filtering. *Journal of Economic Dynamics and Control 27*, 1317–33.

Koopman, S. J., N. Shephard, and D. Creal (2009). Testing the assumptions behind importance sampling. *Journal of Econometrics 149*, 2–11.

Kose, M., C. Otrok, and C. Whiteman (2003). International business cycles: world, region, and country-specific factors. *American Economic Review 93*, 1216–39.

Kose, M., C. Otrok, and C. Whiteman (2008). Understanding the evolution of world business cycles. *Journal of International Economics 75*, 110–30.

Künsch, H. R. (1986). Discrimination between monotonic trends and long-range dependence. *Journal of Applied Probability 23*, 1025–30.

Kuruoğlu, E. (2003). Analytical representation for positive alpha-stable densities. In *The IEEE International Conference on Acoustics, Speech and Signal Processing (ICASSP), vol. 6*, Hong Kong, pp. 729–32.

Lam, C., Q. Yao, and N. Bathia (2011). Estimation of latent factors using high-dimensional time series. *Biometrika 98*(4), 901–18.

Lancaster, P. and L. Rodman (1995). *Algebraic Riccati Equations*. Oxford: Oxford University Press.

Lawley, D. N. and A. E. Maxwell (1971). *Factor Analysis as a Statistical Method* (2nd edn). London: Butterworths.

Lee, S. and S. Na (2005). A nonparametric test for the change of the density function in strong mixing processes. *Statistics and Probability Letters 66*, 25–34.

Lehmann, B. and D. Modest (1988). The empirical foundations of the arbitrage pricing theory. *Journal of Financial Economics 21*, 213–54.

Lemke, T. and S. Godsill (2011). Enhanced Poisson sum representation for alpha-stable processes. In *The IEEE International Conference on Acoustics, Speech and Signal Processing (ICASSP)*, Prague, pp. 4100–3.

Lemke, T. and S. Godsill (2012). Linear Gaussian computations for near-exact Bayesian Monte Carlo inference in skewed alpha-stable time series models. In *The IEEE International Conference on Acoustics, Speech and Signal Processing (ICASSP)*, Prague, pp. 3737–40.

Levin, A. T. and J. M. Piger (2004). Is inflation persistence intrinsic in industrial economics? working paper, series 334, Frankfurt: European Central Bank.

Lindley, D. V. and A. F. M. Smith (1972). Bayes estimates for the linear model. *Journal of the Royal Statistical Society, Series B*, 1–41.

Lindzen, R. (2009). Deconstructing global warming. http://www.globalwarming.org/2009/10/27/deconstructing-global-warming-presentation-by-dr-richard-s-lindzen/, accessed on August 12, 2015.

Liu, J. S. and R. Chen (1998). Sequential Monte Carlo methods for dynamic systems. *Journal of the American Statistical Association 93*, 1032–44.

Liu, S. I. (1994). Multiperiod Bayesian forecasts for AR models. *Annals of the Institute of Statistical Mathematics 46*, 429–52.

Lombardi, M. and S. Godsill (2006). On-line Bayesian estimation of AR signals in symmetric alpha-stable noise. *IEEE Transactions on Signal Processing 54*(2), 775–9.

Luati, A. and T. Proietti (2010). Hyper-spherical and elliptical stochastic cycles. *Journal of Time Series Analysis 31*, 169–81.

Ma, J. (2013). Long-run risks and its implications for the equity premium puzzle: new evidence from a multivariate framework. *Journal of Money, Credit, and Banking 45*, 121–45.

Ma, J., C. R. Nelson, and R. Startz (2007). Spurious inference in the GARCH(1,1) model when it is weakly identified. *Studies in Nonlinear Dynamics & Econometrics 11*, 1–27.

Ma, J. and M. E. Wohar (2014). Expected returns and expected dividend growth: time to rethink an established empirical literature. *Applied Economics 46*, 2462–76.

Magnus, J. R., B. Melenberg, C. Muris, and M. Wild (2013). Statistical climate-change scenarios. *Journal of Environmental Statistics 5*, 1–18.

Magnus, J. R. (1988). *Linear Structures*. New York: Oxford University Press.

Magnus, J. R., B. Melenberg, and C. Muris (2011). Global warming and local dimming: the statistical evidence. *Journal of the American Statistical Association 106*, 452–64.

Magnus, J. R. and H. Neudecker (1988). *Matrix Differential Calculus with Applications in Statistics and Econometrics*. Chichester: John Wiley & Sons.

Mann, M. E. (2004). On smoothing potentially non-stationary climate time series. *Geophysical Research Letters 31*, 710–3.

Mann, M. E. (2008). Smoothing of climate time series revisited. *Geophysical Research Letters 35*(16), L16708.

Mardia, K. V. (1970). Measures of multivariate skewness and kurtosis with applications. *Biometrika 57*, 519–30.

Maronna, R., D. Martin, and V. Yohai (2006). *Robust Statistics - Theory and Methods*. Chichester: John Wiley & Sons.

Masarotto, G. (1990). Bootstrap prediction intervals for autoregressions. *International Journal of Forecasting 6*, 229–39.

McConnell, M. and G. Perez-Quiros (2000). Output fluctuations in the United States: what has changed since the early 1980s? *American Economic Review 90*, 1464–76.

Mencía, J. and E. Sentana (2009). Multivariate location-scale mixtures of normals and mean-variance-skewness portfolio allocation. *Journal of Econometrics 153*, 105–21.

Mencía, J. and E. Sentana (2012). Distributional tests in multivariate dynamic models with normal and student t innovations. *Review of Economics and Statistics 94*, 133–52.

Mesters, G., S. J. Koopman, and M. Ooms (2016). Monte Carlo maximum likelihood estimation for generalized long-memory time series models, *Econometric Reviews*, forthcoming.

Moise, C. E. and J. R. Russell (2012). The joint pricing of volatility and liquidity. Working paper, Booth School of Business, University of Chicago.

Monahan, J. F. (1984). A note enforcing stationarity in autoregressive-moving average models *Biometrika 71*, 403–4.

Morley, J. C., C. R. Nelson, and E. Zivot (2003). Why are Beveridge-Nelson and unobserved components decompositions of GDP so different? *Review of Economics and Statistics 85*, 235–43.

Muth, J. F. (1960). Optimal properties of exponentially weighted forecasts. *Journal of the American Statistical Association 55*, 299–306.

Nelson, C. R. (1988). Spurious trend and cycle in the state space decomposition of a time series with a unit root. *Journal of Economic Dynamics & Control 12*, 475–88.

Nelson, C. R. and C. I. Plosser (1982). Trends and random walks in macroeconomic time series. *Journal of Monetary Economics 10*, 139–62.

Nelson, C. R. and R. Startz (2007). The zero-information-limit condition and spurious inference in weakly identified models. *Journal of Econometrics 138*, 47–62.

Newey, W. K. (1985). Maximum likelihood specification testing and conditional moment tests. *Econometrica 53*, 1047–70.

Newey, W. K. and D. L. McFadden (1994). Large sample estimation and hypothesis testing. In R. F. Engle and D. L. McFadden (Eds.), *Handbook of Econometrics, vol. IV*, pp. 2111–245. Amsterdam: Elsevier.

Newey, W. K. and K. D. West (1987). A simple positive semi-definite, heteroskedasticity and autocorrelation consistent covariance matrix. *Econometrica 55*, 703–8.

Nijman, T. and E. Sentana (1996). Marginalization and contemporaneous aggregation of multivariate GARCH processes. *Journal of Econometrics 71*, 71–87.

Nolan, J. (2005). Modeling financial data with stable distributions. In S. Rachev (Ed.), *Handbook of Heavy Tailed Distributions in Finance*. Amsterdam: Elsevier-North Holland.

Nyblom, J. (1989). Testing the constancy of parameters over time. *Journal of the American Statistical Association 84*, 223–30.

Nyblom, J. and A. C. Harvey (2000). Tests of common trends. *Econometric Theory 16*, 176–99.

Nyblom, J. and T. Makelainen (1983). Comparisons of tests for the presence of random walk coefficients in a simple linear model. *Journal of the American Statistical Association 78*, 856–64.

Omori, Y., S. Chib, N. Shephard, and J. Nakajima (2007). Stochastic volatility with leverage: fast and efficient likelihood inference. *Journal of Econometrics 140*, 425–49.

Onatski, A. (2012). Asymptotics of the principal components estimator of large factor models with weakly influential factors. *Journal of Econometrics 168*, 244–58.

Palma, W. (2007). *Long Memory Time Series*. New York: John Wiley & Sons.

Pascual, L., J. Romo, and E. Ruiz (2004). Bootstrap predictive inference for ARIMA processes. *Journal of Time Series Analysis 25*, 449–65.

Peña, D. and P. Poncela (2004). Forecasting with nonstationary dynamic factor models. *Journal of Econometrics 119*, 291–321.

Peña, D. and P. Poncela (2006). Nonstationary dynamic factor analysis. *Journal of Statistical Planning and Inference 136*, 1237–57.

Percival, D. B. and A. T. Walden (1993). *Spectral Analysis for Physical Applications*. Cambridge: Cambridge University Press.

Pesaran, M. H. and Y. Shin (1998). Generalized impulse response analysis in linear multivariate models. *Economics Letters 58*, 17–29.

Pfeffermann, D. and R. Tiller (2005). Bootstrap approximation to prediction MSE for state-space models with estimated parameters. *Journal of Time Series Analysis 26*, 893–916.

Phillips, P. C. B. (1979). The sampling distribution of forecasts from a first-order autoregression. *Journal of Econometrics 9*, 241–61.

Phillips, P. C. B. (1991). To criticize the critics: an objective Bayesian analysis of stochastic trends. *Journal of Applied Econometrics 6*, 333–64.

Picard, D. (1985). Testing and estimating change-points in time series. *Advances in Applied Probability 17*, 841–67.

Pillai, T. R., M. Shitan, and M. S. Peiris (2012). Some properties of the generalized autoregressive moving average GARCH(1,1) model. *Communications in Statistics Theory and Methods 41*, 699–716.

Piskunov, N. (1969). *Differential and Integral Calculus*. Moscow: Mir Publisher.

Pitt, M. K. and N. Shephard (1999). Filtering via simulation: auxiliary particle filter. *Journal of the American Statistical Association 94*, 590–99.

Pitt, M. K., R. Silva, P. Giordani, and R. Kohn (2012). On some properties of Markov chain Monte Carlo simulation methods based on the particle filter. *Journal of Econometrics 171*, 134–51.

Ploberger, W., W. Kramer, and K. Kontrus (1989). A new test for structural stability in the linear regression model. *Journal of Econometrics 40*, 307–18.

Poncela, P. and E. Ruiz (2012). *More is not always better: back to the Kalman filter in dynamic factor models*. Working paper 12–23, Universidad Carlos III de Madrid.

Pope, P. F. and P. K. Yadav (1994). Discovering errors in tracking error. *Journal of Portfolio Management 20*, 27–32.

Poterba, J. M. and J. B. Shoven (2002). Exchange-traded funds: a new investment option for taxable investors. *American Economic Review 92*, 422–7.

Prado, R. and M. West (2010). *Time Series—Modeling, Computation, and Inference*. Boca Raton: Chapman & Hall/CRC.

Press, W. H., S. A. Teudolsky, W. T. Vetterling, and B. P. Flannery (2007). *Numerical Recipies: The Art of Scientific Computing* (3rd edn). New York: Cambridge University Press.

Proietti, T. (2006). Trend-cycle decompositions with correlated components. *Econometric Reviews 25*, 61–84.

Proietti, T. (2011). Estimation of common factors under cross-sectional and temporal aggregation constraints. *International Statistical Review 79*, 455–576.

Proietti, T. and A. Luati (2012). The generalised autocovariance function. Unpublished paper, Department of Economics and Finance, Universita Degli Studi Di Roma Tor Vergata.

Proietti, T. and F. Moauro (2006). Dynamic factor analysis with non-linear temporal aggregation constraints. *Journal of the Royal Statistical Society Series C 55*, 281–300.

Quandt, R. E. (1960). Tests of the hypothesis that a linear regression system obeys two separate regimes. *Journal of the American Statistical Association 55*, 324–30.

Quenneville, B. and A. C. Singh (2000a). Bayesian prediction mean squared error for state space models with estimated parameters. *Journal of Time Series Analysis 21*, 219–36.

Quenneville, B. and A. C. Singh (2000b). Bayesian prediction mean squared error for state space models with estimated parameters. *Journal of Time Series Analysis 21*, 219–36.

Quiou, Z. and N. Ravishanker (1998). Bayesian inference for time series with stable innovations. *Journal of Time Series Analysis 19*, 235–50.

R Development Core Team (2012). *R: A Language and Environment for Statistical Computing.* Vienna, Austria: R Foundation for Statistical Computing.

Rao, C. R. (1973). *Linear Statistical Inference and its Applications.* New York: John Wiley & Sons.

Reis, R. and M. W. Watson (2010). Relative good's prices, pure inflation, and the Phillips correlation. *American Economic Journal of Macroeconomics 2*, 128–57.

Richard, J. F. and W. Zhang (2007). Efficient high-dimensional importance sampling. *Journal of Econometrics 141*, 1385–411.

Ripley, B. D. (1987). *Stochastic Simulation.* New York: John Wiley & Sons.

Riskmetrics (1996). *RiskMetrics—Technical Document.* 4th edn, New York: Morgan Guaranty Trust Company.

Robert, C. P. and G. Casella (2005). *Monte Carlo Statistical Methods.* New York: Springer.

Robinson, P. (1995). Log-periodogram regression of time series with long range dependence. *Annals of Statistics 23*, 1048–72.

Rodríguez, A. and E. Ruiz (2009). Bootstrap prediction intervals in state-space models. *Journal of Time Series Analysis 30*, 167–78.

Rodríguez, A. and E. Ruiz (2012). Bootstrap prediction mean squared errors of unobserved states based on the Kalman filter with estimated parameters. *Computational Statistics and Data Analysis 56*, 62–74.

Rojas, A. J. (2011). On the discrete-time algebraic Riccati equation and its solution in closed-form. Technical report, preprints, 18th IFAC World Congress, Milan.

Rosenblatt, M. (1952). Remarks on a multivariate transformation. *Annals of Mathematical Statistics 23*, 470–2.

Rosiński, J. (2001). Series representations of Lévy processes from the perspective of point processes. In O. E. Bandorff-Neilson, T. Mikosch, and S. I. Resnick (Eds.) *Lévy Processes: Theory and Applications*, pp. 401–15, Boston: Birkhäuser.

Rosiński, J. (2007). Tempering stable processes. *Stochastic Processes and Their Applications 117*, 677–707.

Ross, S. A. (1976). The arbitrage theory of capital asset pricing. *Journal of Economic Theory 13*, 341–60.

Ross, S. A. (1978). Mutual fund separation in financial theory—the separating distributions. *Journal of Economic Theory 17*, 254–86.

Rubin, D. B. and D. T. Thayer (1982). EM algorithms for ML factor analysis. *Psychometrika 57*, 69–76.

Samorodnitsky, G. and M. Taqqu (1994). *Stable Non-Gaussian Random Processes: Stochastic Models with Infinite Variance*. London: Chapman & Hall.

Sargent, T. J. and C. A. Sims (1977). Business cycle modeling without pretending to have too much a priory economic theory. In C. A. Sims (Ed.), *New Methods in Business Cycle Research*. Minneapolis: Federal Reserve Bank of Minneapolis.

Sbordone, A., A. Tambalotti, and K. R. Walsh (2010). Policy analysis using DSGE models: an introduction. *FRB of New York Policy Review*, 23–43.

Schoutens, W. (2003). *Lévy Processes in Finance: Pricing Financial Derivatives*. New York: John Wiley & Sons.

Sentana, E. (2000). The likelihood function of conditionally heteroskedastic factor models. *Annales d'Economie et de Statistique 58*, 1–19.

Sentana, E. (2004). Factor representing portfolios in large asset markets. *Journal of Econometrics 119*, 257–89.

Sentana, E., G. Calzolari, and G. Fiorentini (2008). Indirect estimation of large conditionally heteroskedastic factor models, with an application to the Dow 30 stocks. *Journal of Econometrics 146*, 10–25.

Sentana, E. and G. Fiorentini (2001). Identification, estimation and testing of conditionally heteroskedastic factor models. *Journal of Econometrics 102*, 143–64.

Sentana, E. and M. Shah (1993). An index of co-movements in financial time series. Working paper 9415, CEMFI.

Shephard, N. (1994). Partial non-Gaussian state space. *Biometrika 81*, 115–31.

Shephard, N. (2005a). Stochastic volatility. In N. Shephard (Ed.), *Stochastic Volatility: Selected Readings*. Oxford: Oxford University Press.

Shephard, N. (Ed.) (2005b). *Stochastic Volatility: Selected Readings*. Oxford: Oxford University Press.

Shephard, N. and M. K. Pitt (1997). Likelihood analysis of non-Gaussian measurement time series. *Biometrika 84*, 653–67.

Shin, S. and G. Soydemir (2010). Exchange-traded funds, persistence in tracking errors and information dissemination. *Journal of Multinational Financial Management 20*, 214–34.

Shitan, M. and M. S. Peiris (2008). Generalized autoregressive (GAR) model: A comparison of maximum likelihood and whittle estimation procedures using a simulation study. *Communications in Statistics—Simulation and Computation 37*, 560–70.

Shitan, M. and M. S. Peiris (2011). Time series properties of the class of generalized first-order autoregressive processes with moving average errors. *Communications in Statistics—Theory and Methods 40*, 2259–75.

Shively, T. S. (1988). An exact test for a stochastic coefficient in a time series regression model. *Journal of Time Series Analysis 9*, 81–8.

Silverman, B. W. (1986). *Density Estimation for Statistical and Data Analysis*. London: Chapman & Hall.

Sims, C. (1980). Macroeconomics and reality. *Econometrica 48*, 1–48.

Sims, C. A. (1991). Comment by C. A. Sims on 'To criticize the critics', by P. C. B. Phillips. *Journal of Applied Econometrics 6*, 423–34.

Sims, C. A. and H. Uhlig (1991). Understanding unit rooters: a helicopter tour. *Econometrica 59*, 1591–9.

Snyder, R. D., J. K. Ord, and A. B. Koehler (2001). Prediction intervals for ARIMA models. *Journal of Business & Economic Statistics 19*, 217–25.

Soon, W. W. H., D. R. Legates, and S. L. Baliunas (2004). Estimation and representation of long-term (>40 year) trends of northern-hemisphere-gridded surface temperature: a note of caution. *Geophysical Research Letters 31*, L03209.

Staiger, D., J. H. Stock, and M. W. Watson (1997a). How precise are estimates of the natural rate of unemployment? In C. Romer and D. Romer (Eds.), *Reducing Inflation: Motivation and Strategy*, pp. 195–242. Chicago: University of Chicago Press for the NBER.

Staiger, D., J. H. Stock, and M. W. Watson (1997b). The NAIRU, unemployment and monetary policy. *The Journal of Economic Perspectives 11*, 33–49.

Stock, J. H. (1994). Unit roots, structural breaks and trends. In R. F. Engle and D. L. McFadden (Eds.), *Handbook of Econometrics, vol. 4*, pp. 2740–841. Amsterdam: Elsevier Science.

Stock, J. H. and M. W. Watson (1991). A probability model of the coincident economic indicators. In K. Lahiri and G. H. Moore (Eds.), *Leading Economic Indicators: New Approaches and Forecasting Records*. Cambridge: Cambridge University Press.

Stock, J. H. and M. W. Watson (2002a). Forecasting using principal components from a large number of predictors. *Journal of the American Statistical Association 97*, 1167–79.

Stock, J. H. and M. W. Watson (2002b). Macroeconomic forecasting using diffusion indexes. *Journal of Business and Economics Statistics 20*, 147–62.

Stock, J. H. and M. W. Watson (2003). Forecasting output and inflation: the role of asset prices. *Journal of Economic Literature 41*, 788–829.

Stock, J. H. and M. W. Watson (2005). Understanding Changes In International Business Cycle Dynamics. *Journal of the European Economic Association 3*, 968–1006.

Stock, J. H. and M. W. Watson (2007). Why has U.S. inflation become harder to forecast? *Journal of Money, Credit, and Banking 39*, 3–34.

Stock, J. H. and M. W. Watson (2009). Phillips curve inflation forecasts. In J. Fuhrer, Y. Kodrzycki, J. Little, and G. Olivei (Eds.), *Understanding Inflation and the Implications for Monetary Policy*. Cambridge, MA: MIT Press.

Stock, J. H. and M. W. Watson (2010). Modeling inflation after the crisis. In *Macroeconomic Policy: Post-Crisis and Risks Ahead*, proceedings of the Federal Reserve Bank of Kansas City 2010 Jackson Hole Symposium.

Stock, J. H. and M. W. Watson (2011). Dynamic factor models. In M. P. Clements and D. F. Hendry (Eds.), *The Oxford Handbook of Economic Forecasting*, pp. 173–205. Oxford: Oxford University Press.

Tauchen, G. (1985). Diagnostic testing and evaluation of maximum likelihood models. *Journal of Econometrics 30*, 415–43.

Taylor, J. W. (2004). Smooth transition exponential smoothing. *Journal of Forecasting 23*, 385–94.

Thompson, P. A. and R. B. Miller (1986). Sampling the future: a Bayesian approach to forecasting to univariate time series models. *Journal of Business & Economic Statistics 4*, 427–36.

Trigg, D. W. and A. G. Leach (1967). Exponential smoothing with an adaptive response rate. *Operational Research Quarterly 18*, 53–9.

Troughton, P. and S. Godsill (1998). A reversible jump sampler for autoregressive time series. *The IEEE International Conference on Acoustics, Speech and Signal Processing (ICASSP), vol. 4*, Seattle, U.S., pp. 2257–60.

Tsionas, E. (1999). Monte Carlo inference in econometric models with symmetric stable disturbances. *Journal of Econometrics 88*, 365–401.

UAH (April 2010). University of Alabama National Space Science and Technology Center. http://vortex.nsstc.uah.edu/data/msu/t2lt/tltglhmam_5.2.

van Binsbergen, J. H. and R. S. J. Koijen (2010). Predictive regressions: a present-value approach. *The Journal of Finance 65*, 1439–71.

Vardharaj, R., F. J. Fabozzi, and F. J. Jones (2004). Determinants of tracking error for equity portfolios. *Journal of Investing 13*, 37–47.

Vidoni, P. (2004). Improved prediction intervals for stochastic process models. *Journal of Time Series Analysis 25*, 137–54.

Vidoni, P. (2009). A simple procedure for computing improved prediction intervals for autoregressive models. *Journal of Time Series Analysis 30*, 577–90.

Watson, M. W. (2003). Macroeconomic forecasting using many predictors. In M. Dewatripont, L. Hansen, and S. Turnovsky (Eds.), *Advances in Economics and Econometrics: Theory and Applications*, Eighth World Congress of the Econometric Society, vol. III, pp. 87–115. Cambridge: Cambridge University Press.

Wecker, W. E. and C. F. Ansley (1983). The signal extraction approach to nonlinear regression and spline smoothing. *Journal of the American Statistical Association 78*, 81–9.

West, K. (1996). Asymptotic inference about predictive ability. *Econometrica 64*, 1067–84.

West, M., P. J. Harrison, and H. S. Migon (1985). Dynamic generalised models and Bayesian forecasting (with discussion). *Journal of the American Statistical Association 80*, 73–97.

White, H. (2000). A reality check for data snooping. *Econometrica 68*, 1097–126.

Whittle, P. (1953). Estimation and information in stationary time series. *Arkiv för Matematik 2*, 423–34.

Whittle, P. (1961). Gaussian estimation in stationary time series. *Bulletin of the International Statistical Institute 39*, 105–29.

Working, H. (1960). Note on the correlation of first differences of a random chain. *Econometrica 28*, 916–18.

Zellner, A. (1971). *An Introduction to Bayesian Inference in Econometrics*. New York: John Wiley & Sons.

INDEX

The manufacturer's authorised representative in the EU for product safety is Oxford University Press España S.A. of el Parque Empresarial San Fernando de Henares, Avenida de Castilla, 2 – 28830 Madrid (www.oup.es/en or product.safety@oup.com). OUP España S.A. also acts as importer into Spain of products made by the manufacturer.

www.ingramcontent.com/pod-product-compliance
Lightning Source LLC
LaVergne TN
LVHW011906200926
848484LV00002B/548
9780199683666